ELEMENTS OF
CHORDATE
ANATOMY

ELEMENTS OF CHORDATE ANATOMY

Fourth Edition

CHARLES K. WEICHERT

*Late Professor of Zoology and
Dean, McMicken College of Arts and Sciences
University of Cincinnati*

WILLIAM PRESCH

*Department of Biology
California State University,
Fullerton*

**McGRAW-HILL
BOOK COMPANY**

*New York   St. Louis   San Francisco   Auckland   Düsseldorf   Johannesburg
Kuala Lumpur   London   Mexico   Montreal   New Delhi   Panama
Paris   São Paulo   Singapore   Sydney   Tokyo   Toronto*

ELEMENTS OF CHORDATE ANATOMY

2 3 4 5 6 7 8 9   K P K P   7 9 8 7

This book was set in Garamond by Progressive Typographers. The editors were William J. Willey and Richard S. Laufer; the designer was Joseph Gillians; the production supervisor was Thomas J. LoPinto. New drawings were done by Vantage Art, Inc.
The Maple Press Company was printer and binder.

Library of Congress Cataloging in Publication Data

Weichert, Charles Kipp, date
   Elements of chordate anatomy.

   (McGraw-Hill series in organismic biology)
   A condensed version of C. K. Weichert's Anatomy of the chordates.
      1. Chordata—Anatomy. 2. Anatomy, Comparative.
I. Presch, William, joint author. II. Title.
[DNLM: 1. Anatomy, Comparative. 2. Chordata—
Anatomy and histology. QS124 W416a]
QL805.W4   1975   596'.04   74-17151
ISBN 0-07-069008-1

Contents

Preface

This text is a condensed version of a larger and more comprehensive book, Weichert's "Anatomy of the Chordates." This fourth edition of the "Elements of Chordate Anatomy" has been revised in accordance with the larger "Anatomy of the Chordates" and new and updated information has been added.

The rapid advances made in all areas of science in recent years mean that new interpretations are placed on older concepts. Such tools as protein comparison, the role of ecology in interpreting anatomical structures, and the new interest in functional anatomy have clarified our knowledge of structures and interrelationships. New concepts and ideas have been introduced in this volume based on these and interrelated approaches.

However, since this work represents a condensation of information available, a full elaboration and treatment of many ideas has not been possible. To help overcome this, a bibliography at the end of each chapter is provided for fuller and more detailed explanations. A glossary has been added for the convenience of the student. Additional figures have been included to better describe the concept or process under consideration.

Many students of chordate anatomy are biology majors or expect to enter professional schools of medicine, dentistry, or veterinary science. For these fields a knowledge of basic vertebrate structure is desirable or even essential for a clearer understanding of other phases of biology. Since most schools offer separate courses in human anatomy, no effort has been made to include detailed descriptions and functional discussions of human anatomy. Reference to human structures is made with the view of fitting man into the vertebrate organization and plan. A companion volume, Weichert's "Representative Chordate Anatomy," is available for laboratory dissections of such vertebrates as the lamprey, spiny dogfish, the mud puppy *Necturus*, and the cat.

The classification scheme of the chordates presented here represents the arrangement of the various groups as seen by modern systematists. It is by no means complete. A short discussion of the origin of chordates is provided, to lay a foundation for the group and to provide a starting point for the comparison of structures. Emphasis is placed for the most part on modern species, but fossil forms are included when their inclusion helps to clarify the discussion. This material is meant to parallel certain aspects of vertebrate paleontology rather than to include it.

A general review of the phylum Chordata is preceded by a discussion of vertebrate evolution and followed by a brief chapter on development. Each organ system is discussed, in the order integumentary, skeletal, muscle systems, respiratory, digestive, excretory, reproductive, circulatory, integrating (nervous system, endocrine organs, and sense organs); but this separation is for convenience, rather than based on functional or anatomical grounds. In most cases, the subject matter is presented so that material in one chapter does not presuppose a knowledge of previous chapters.

A summary chapter embracing in succinct form the characteristics and advances of members of the phylum concludes the text.

I wish to acknowledge the help of my wife, Judy, for her assistance in the preparation of the manuscript.

Great effort has been made to present the subject matter clearly and accurately. I will appreciate it if readers will call to my attention any errors or inconsistencies which may have escaped my notice.

William Presch

ELEMENTS OF
CHORDATE
ANATOMY

1 INTRODUCTION

This book is an introduction to the structure of the vertebrate animals, a large and complex group of organisms comprising fishes, amphibians, reptiles, birds, and mammals. It is the group of animals to which man belongs. To understand man, a knowledge of his evolutionary relationships to other living things is essential. Instead of discussing only knowledge which relates to man, we have included biological information about living, complex, active, and fossil vertebrates that have successfully adapted to the wide range of environmental conditions on the earth. No area of the biological sciences is more fascinating than vertebrate evolution viewed in the light of developmental patterns and functional and structural patterns of organ systems. An understanding of the function and organization is fundamental to the understanding of the organism and its adaptations to changing environments.

METHOD OF APPROACH TO COMPARATIVE ANATOMY When similarities and differences in structural organization are studied and compared, general principles emerge from which deductive conclusions can be drawn. This is the method of the comparative anatomist, who seeks to explain the variations in structure found in the bodies of animals with a view to tracing their biological relationships.

Although the method of comparative anatomy has been used in studying the relationships of almost all forms in the animal kingdom, the scope of the subject is too vast to be considered in a single volume. We confine our attention here to a comparative study of vertebrates and a few of their closest relatives, all of which are included in the phylum Chordata.

In the laboratory the student dissects representative chordate animals and gains an intimate knowledge of the detailed structure of each. This knowledge does not constitute a science unless the facts are properly correlated. The comparative anatomist recognizes that the vertebrates and some of their close relatives are built upon the same fundamental plan and that there is often a close correspondence, even in detail. Within the general vertebrate plan there are many variations, which are for the most part adaptive in nature. This means that they are modifications which meet particular needs in relation to the environment in which the animal lives. The object of a course in comparative anatomy of vertebrates, therefore, is to acquaint the student with the plan of vertebrate structure.

HOMOLOGY The term *homology* refers to the correspondence in type of structure between parts or organs of different animals. These may have diverged considerably from the same or corresponding part or organ of some remote ancestral form, so that superficially they bear little resemblance to one another. For example, the skeleton of the forelimb of a cat is composed of a single bone in the upper leg, two in the lower leg, and a

number of intermediate bones situated between the lower leg and toes. The wing of a bird, functionally different, at first sight appears in no way like the forelimb of a cat. The skeletons, however, reveal basic similarities in bone structure and arrangement. The arm of man, the flipper of the whale, the hoofed leg of the horse, and the wing of the bat are all built upon the same fundamental skeletal plan and are therefore homologous (Fig. 1.1). In some cases differences in structure are so great that their homologies might not be recognized. Observation of a series of intermediate stages between the two extremes can often indicate their basic similarities. The study of homology of parts and the evolution of these homologous structures is one of the most interesting features of comparative anat-

FIG. 1.1 Skeleton of forelimb, showing modifications of structure. (*A, B, and C, after LeConte, "Evolution," Appleton-Century-Crofts, Inc., by permission.*)

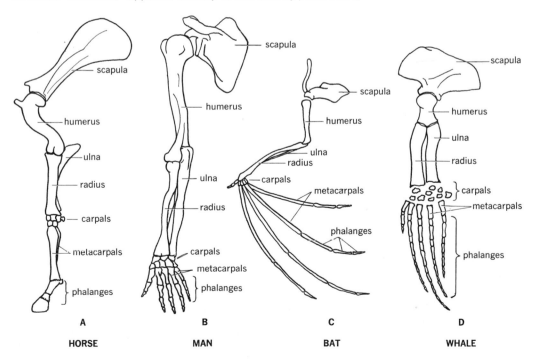

A

HORSE

B

MAN

C

BAT

D

WHALE

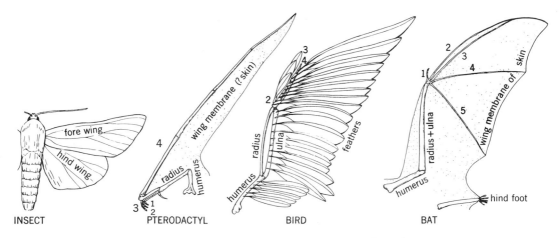

FIG. 1.2 Analogy between wings of insects (no internal skeleton) and of vertebrates (with skeleton), which have a like function but different origins. Homology in the wing bones of vertebrates, all derived from the common pattern of the forelimb in land vertebrates but variously modified. Pterodactyl (extinct reptile) with very long fourth finger; bird with first and fifth lacking, third and fourth partly fused; bat with second to fifth fingers long. (*From Storer et al., "General Zoology," McGraw-Hill Book Company, by permission.*)

omy. Observations of the details of embryonic development have in many instances rendered great service to the comparative anatomist in revealing such homologies. An embryological approach to comparative studies is used in this volume wherever feasible.

ANALOGY The term *analogy* refers to corresponding function of a structure in similar or different organs or organ parts. In general, the development of the structure is very different, however, and may develop from different organ systems (Fig. 1.2). The term implies that the structural development is not homologous. The fin of a fish and the flipper of a whale are analogous, as both may be used for locomotive purpose, but the two are structurally very different.

ONTOGENY AND PHYLOGENY Early embryos of all vertebrates show remarkable similarities in structural development.

During the development of higher forms, certain anatomical features, apparently of little significance, make their appearance. In most cases these become modified, or they may degenerate and disappear. Many speculations about the significance of these structures have been made. For example, at a certain stage of development in all vertebrates a series of pouches pushes out from the pharynx to establish connection with the outer covering of the body. In such lower forms as fishes, perforations occur at the points of contact, becoming gill slits, which persist throughout life. In the embryos of higher vertebrates, however, the pharyngeal pouches usually fail to open to the outside. They persist for a time, but then become modified in various ways, retaining little or none of their original character. Why should pharyngeal pouches appear at all in higher forms? Another illustration is the appearance in the region of the pharynx in all vertebrates during embryonic development of certain blood vessels called *aortic arches.* The structural plan of the aortic arches in the *embryo* of a bird, for example, resembles the condition found in the gill region of *adult* fishes. It is *almost identical* with that found in *embryos* of such diverse forms as fishes, snakes, frogs, and man. In the bird, as well as in other higher forms, the aortic arches soon undergo considerable modification, ultimately forming an arrangement of vessels in the posterior part of the neck region or in the anterior part of the thoracic region, which is characteristic of each vertebrate group.

Several ideas have been advanced to explain why such structures as pharyngeal pouches and aortic arches should make their appearance in the embryos of higher vertebrates. Perhaps the best-known of these is the *recapitulation theory,* or *law of biogenesis.* As originally formulated, it stated that during development an individual passes through certain ancestral stages through which the entire race passed in its evolution and that the embryos of higher forms in certain ways resemble the adults of lower forms. In other words, *ontogeny* (individual life history) recapitulates *phylogeny* (species history). Although

FIG. 1.3 Three comparable stages in the development of embryos of several vertebrates. Such membranous structures as yolk sac, amnion, chorion, and allantois, which appear in some embryos but not in others, have been omitted in order to emphasize similarities and to minimize differences. (*After Haeckel, modified from Daugherty, "Principles of Economic Zoology," W. B. Saunders Company, by permission.*)

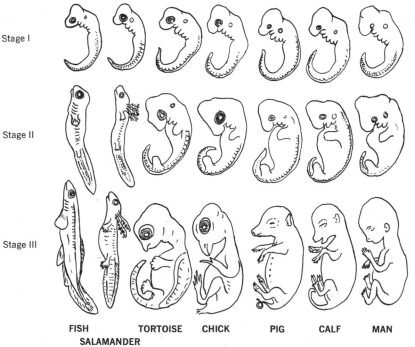

Stage I

Stage II

Stage III

FISH   TORTOISE   CHICK   PIG   CALF   MAN
SALAMANDER

this concept has furnished a good working hypothesis for embryologists in their attempts to work out homologies, biologists today realize that ontogeny does *not* recapitulate phylogeny; at no time does a mammalian embryo, for example, closely resemble an adult fish, amphibian, or any other lower vertebrate. The embryos of all vertebrates, however, show many remarkable similarities. Those of higher vertebrates undergo fundamental changes which are like those of lower forms and which may occur in the same sequence. From such basically similar early embryonic stages, which are believed to be characteristic of ancestral forms, each group of vertebrates diverges in one direction or another, so that when fully developed they differ widely in structure and habits.

It cannot be assumed that during development an animal goes through *every* stage its ancestors did; indeed this is not the case. Some stages are omitted or are passed through so rapidly that they can scarcely be discerned. New characteristics by which they differ have arisen as a result of genetic variation and natural selection, so that ancestral conditions do not always appear in a typical form. Other structures of significance in the adaptation of the embryo to its environment are obviously unrelated to ancestral conditions and therefore do not resemble embryonic stages in other vertebrate groups. In addition, the principle fails in many cases to explain the appearance of new, or *cenogenetic,* structures in certain forms and the cessation of other embryonic conditions before full development has occurred (neotenic development). The theory does, however, account for the fact that the early embryonic stages of all vertebrates are remarkably similar (Fig. 1.3) up to the point where each group diverges in its own characteristic manner. Despite its shortcomings, the concept has been useful to biologists in establishing homologies, in making clear certain phylogenetic relationships, and in helping to give a picture of the probable structure and appearance of ancestral forms and stages in vertebrate evolutionary history.

BIBLIOGRAPHY

Carter, G. S.: "Structure and Habit in Vertebrate Evolution," University of Washington Press, Seattle, 1967.

Darwin, Charles: "On the Origin of Species by Means of Natural Selection, or the Preservation of Favoured Races in the Struggle for Life," London, 1859.

Florey, E.: "General and Comparative Animal Physiology," Saunders, Philadelphia, 1966.

Goodrich, E. S.: "Studies on the Structure and Development of Vertebrates," Dover, New York, 1958.

Gordon, M.: "Animal Physiology: Principles and Adaptations," 2d ed., Macmillan, New York, 1972.

Hoar, W. S.: "General and Comparative Physiology," Prentice-Hall, Englewood Cliffs, N.J., 1966.

Hyman, L. H.: "Comparative Vertebrate Anatomy," University of Chicago Press, Chicago, 1942.

Jollie, Malcolm: "Chordate Morphology," Reinhold, New York, 1968.

Marshall, A. J. (ed.): "Biology and Comparative Physiology of Birds," vol. 2, Academic, New York, 1961.

Mayr, E.: "Principles of Systematic Zoology," McGraw-Hill, New York, 1964.

Oparin, A. I.: The Origin of Life on Earth, *Proc. First Int. Symp. Moscow, 1957,* London, 1959.

Parker, T. J., and W. A. Haswell: "A Text-Book of Zoology," 6th ed., rev. by C. Foster Cooper, Macmillan, London, 1940.

Prosser, C. L.: "Comparative Animal Physiology," 3d ed., Saunders, Philadelphia, 1974.

Romer, A. S.: "Vertebrate Paleontology," 3d ed., University of Chicago Press, Chicago, 1966.

————: "Notes and Comment on Vertebrate Paleontology," University of Chicago Press, Chicago, 1968.

————: "The Vertebrate Body," 4th ed., Saunders, Philadelphia, 1970.

Saunders, J. W., Jr.: "Animal Morphogenesis," Macmillan, New York, 1966.

Simpson, G. G.: "The Meaning of Evolution," Yale University Press, New Haven, Conn., 1949.

————: "The Major Features of Evolution," Columbia University Press, New York, 1953.

Smith, H. M.: "Evolution of Chordate Structure," Holt, New York, 1960.

Thompson, D'Arcy W.: "On Growth and Form," abridged ed. by J. T. Bonner, Cambridge University Press, Cambridge, 1961.

Weichert, C. K.: "Anatomy of the Chordates," McGraw-Hill, New York, 1972.

2 EVOLUTION OF CHORDATES

PHYLUM CHORDATA

The phylum Chordata is usually subdivided into four main groups, known as *subphyla*. The first three of these include a few relatively simple animals which lack a cranium and brain. The term *Acrania* is used by some authors in referring to them collectively. The animals included in this category are believed to show similarities to the ancestors of the chordates and hence are frequently designated as the protochordates. These are the subphyla Hemichordata (acorn worms) (placed in its own phylum, the Hemichordata, by some workers), Urochordata (tunicates), and the Cephalochordata (amphioxus). The fourth subphylum, Vertebrata (Cranata), is a large group embracing chordates having a brain, endoskeleton, and other characteristics discussed below. This book is primarily concerned with members of the subphylum Vertebrata. Members of the phylum Chordata are commonly referred to as chordates. Three characteristics, of prime diagnostic importance, are possessed by all chordates:

1 A notochord present sometime during life. All vertebrates possess an *endoskeleton*, which provides protection to delicate vital organs, gives support to the body, and is jointed, making movement possible. A primitive endoskeletal structure, the *notochord,* is present in all chordates during early embryonic life. It is a pliant, rodlike structure composed of a peculiar type of connective tissue. The notochord is located along the middorsal line, where it forms the axis of support for the body. In some animals it persists as such throughout life, but in most chordates it serves as a foundation around which the vertebral column is built. A salamander larva or a frog tadpole, for example, at first possesses a notochord (Fig. 2.1), but it is generally and gradually replaced, first by a cartilaginous vertebral column and then by a bony one in frogs or directly replaced by bone in salamanders. By the time that a larva has metamorphosed into an adult, the notochord has, for the most part, disappeared.

2 A hollow, dorsal nerve tube present sometime during life. The central nervous system of chordates, usually made up of brain and spinal cord, is located in a dorsal position. It lies just above the notochord (Fig. 2.1) and is a tubular structure, having a hollow canal from one end to the other. The dorsal, hollow nerve tube persists throughout life in almost all chordates, but in a few lower forms it degenerates before maturity is attained.

3 Gill slits, or traces of them, connecting to the pharynx present sometime during life. Most aquatic chordates respire by means of gills made up of vascular lamellae or filaments lining the borders of

FIG. 2.1 Cross section of salamander larva, showing notochord lying beneath spinal cord.

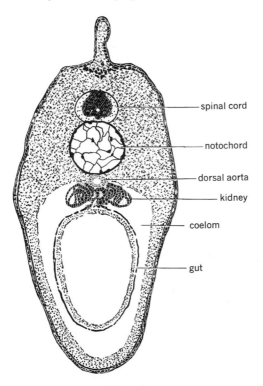

- spinal cord
- notochord
- dorsal aorta
- kidney
- coelom
- gut

gill slits which connect to the pharynx and open directly or indirectly to the outside (Fig. 2.2). Even terrestrial chordates, which never breathe by means of gills, nevertheless have traces of gill slits present as transient structures during early embryonic development. No vascular lamellae line these temporary structures, nor do they open to the outside, but the fact that they are present in all chordates is of primary importance in denoting close relationship.

The three characteristics listed above are possessed by most members of the phylum Chordata. In addition, chordates have certain features common to members of some other phyla as well. They are bilaterally symmetrical, they are metameric, they possess a true body cavity, or coelom, lined with mesoderm (one of the primitive embryonic germ layers), and they show cephalization, i.e., concentration of nervous tissue and specialized sense organs in or toward the head. Moreover, the direction of blood flow is such that blood is pumped anteriorly from the ventrally located heart and then forced to the dorsal side. It then courses posteriorly and returns to the heart by means of veins, the larger ones lying ventral to the main arteries.

The Hemichordata represent a problem. Though they do possess gill slits through the wall of the pharynx, the adult lacks a notochord and a hollow dorsal nerve cord. Such structures have not been definitely shown to exist in either the adult or larval form. This has led some workers to place the Hemichordates in a separate phylum but related to the Chordata.

SUBPHYLUM I. HEMICHORDATA

The hemichordates are usually considered to have the simplest organization of chordate life, although there is some dispute about their

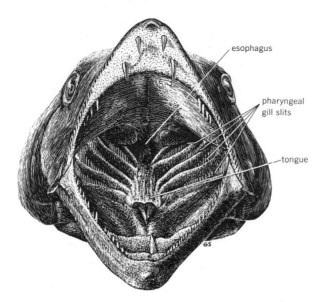

FIG. 2.2 The open mouth of a fish, the barracuda, showing the typical relation of gill slits to the pharynx.

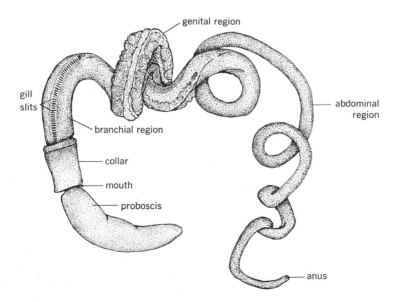

FIG. 2.3 Lateral view of *Dolichoglossus kowalevskii*.

classification, as mentioned above. They are of particular interest, for they give a clue to the link which bridges the gap between the chordates and some of the other animal phyla. A free-swimming larval stage, referred to as the *tornaria,* occurs during development of certain members of this subphylum. The tornaria is very similar in appearance and structure to the larvae of some echinoderms.

A typical example (Fig. 2.3) is *Dolichoglossus kowalevskii,* a fragile, burrowing, wormlike animal found in the Atlantic Coast region. Three body regions are present: (1) a proboscis; (2) a collar; (3) an elongated trunk. The trunk, in turn, is composed of an anterior branchial region bearing a row of transverse gill slits on either side, a genital region of irregular outline, and a posterior abdominal region. The only obvious chordate character possessed by this animal is its pharyngeal gill slits. A peculiar forward extension of the gut into the proboscis has in the

past been homologized with the notochord of higher forms. Most zoologists today doubt whether such a homology exists. *Dolichoglossus* possesses both dorsal and ventral nerve strands. The dorsal strand, which is the larger, is tubular only in the collar region. Nevertheless, it may be homologous with the dorsal, hollow nerve tube of higher forms.

SUBPHYLUM II. UROCHORDATA

The urochordates, or tunicates, include a peculiar group of widely distributed marine animals commonly known as sea squirts, or ascidians. They are of no economic importance. Many of them become sessile after a short free-swimming larval period and attach themselves to a wharf pile, rock, or similar object. *Molgula manhattensis* (Fig. 2.4) is a fairly typical example of the subphylum. In its adult state *Molgula* bears little resemblance to a typical chordate. It lacks a notochord, and its

FIG. 2.4   Internal organs of *Molgula manhattensis: A,* as viewed from the left side and, *B,* as viewed from the right side.

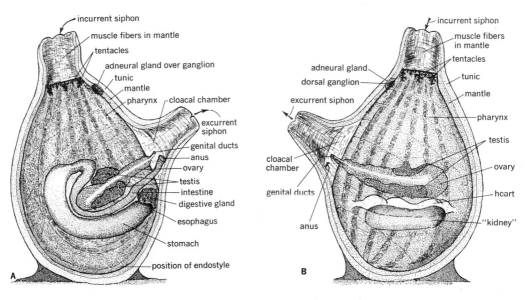

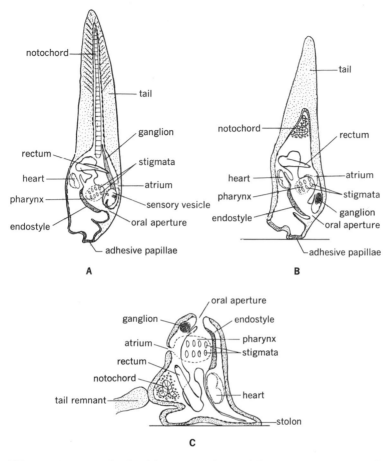

FIG. 2.5   Metamorphosis of free-swimming ascidian larva to the sessile form: *A,* free-swimming larva; *B,* larva soon after attachment; *C,* later attached stage but before disappearance of notochord. (*After Seeliger from Korschelt and Heider.*)

nervous system is reduced to a small ganglionic mass. Only the pharynx with numerous perforations, or gill slits, in its walls and the presence of an endostyle give evidence of its chordate relationships. An adult specimen is about an inch long, round or oval in shape, with two openings, or *siphons,* at its free end. A stream of water enters the larger *incurrent siphon* and leaves through the smaller *excurrent siphon.* The animal is invested by a rather

tough capsule, or *tunic.* This is composed of a substance called *tunicin,* secreted by the cells of the *mantle* beneath. The greater part of the cavity enclosed by the mantle is occupied by the pharynx. The nerve ganglion lies embedded in the mantle in the region between the siphons. This is the dorsal side of the animal. Almost completely surrounding the ganglion is an *adneural gland,* which opens by means of a duct into the pharynx. Some zoologists have homologized the adneural gland with the pituitary gland of vertebrates, but opinions differ concerning this. A ciliated groove, the *endostyle,* extends along the mid-ventral line of the pharynx to the esophagus. Food particles that enter the pharynx with the incurrent stream of water are gathered by the endostyle and swept down by a cilia-produced current to the esophagus. Some biologists believe that the endostyle may be the forerunner of the thyroid gland of higher forms. Certain other structures of *Molgula* are indicated in Fig. 2.4.

Sea squirts were formerly classified as mollusks because of their peculiar structure in the adult stage. It was not until 1866 that the proper classification was determined when Kowalevski, a Russian biologist, found that the ascidian *larva* possesses typical chordate characteristics. During its larval existence a sea squirt somewhat resembles a tadpole in appearance. It has a notochord, which is confined to the tail, a hollow, dorsal nerve cord, and a pharynx perforated with gill slits. After a free-swimming period it finally attaches itself to some object and becomes sessile. The animal then undergoes a metamorphosis during which the notochord disappears as the tail is absorbed and the nerve cord is reduced to a simple ganglion (Fig. 2.5). The only one of the three essential chordate characters which persists is the pharynx with its numerous gill slits communicating indirectly with the exterior.

Sessile urochordates may be either solitary or colonial. Not all species are sessile, however, for some are free-swimming, transparent, pelagic forms throughout life.

SUBPHYLUM III. CEPHALOCHORDATA

There are but two genera and about 20 species of animals included in the subphylum Cephalochordata. The common name amphioxus is applied to members of the more abundant genus, *Branchiostoma.* Amphioxus is of importance to the biologist because of its generally primitive structure. Furthermore, its early development illustrates, with almost diagrammatic simplicity, processes which occur during the embryonic stages of higher chordates.

The animal (Figs. 2.6 and 2.7) is a small, marine form about 2 in. long. It is rather sedentary in habit and generally lies partly buried in the sand of the ocean floor with its anterior end protruding. The body of amphioxus is pointed at either end and during life is semitransparent. At the anterior end an integumentary fold, the *oral hood,* surrounds a cavity, or *vestibule,* which leads to the mouth. The oral hood is encircled by 22 papillalike projections, the *buccal cirri.*

A median *dorsal fin* extends almost the entire length of the body and is continuous with the *caudal fin* posteriorly. On the ventral side the caudal fin continues anteriorly as far as an opening, the *atriopore,* and then divides, ex-

FIG. 2.6 Lateral view of amphioxus.

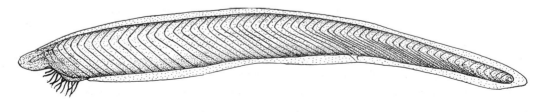

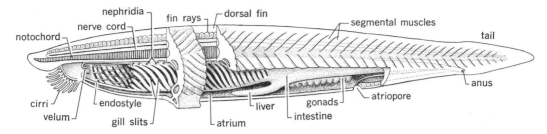

FIG. 2.7 Subphylum Cephalochordata. The lancelet, or amphioxus, *Branchiostoma.* Adult partly dissected from left side. Natural size about 2 in. long. (*From Storer et al., "General Zoology," McGraw-Hill Book Company, by permission.*)

tending farther forward as a pair of *metapleural folds* (Fig. 2.8). These folds were at one time considered to represent the beginning of paired appendages of vertebrates, a theory which is no longer tenable (page 159). The notochord extends practically the entire length of the body, projecting even farther forward than the anterior end of the dorsal, hollow nerve cord. The latter lies

FIG. 2.8 Cross section through posterior pharyngeal region of amphioxus (anterior view).

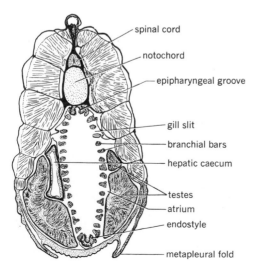

directly above the notochord. A slight enlargement, the *cerebral vesicle,* is formed at the anterior end. Metamerism is clearly indicated by the arrangement of the <-shaped muscle segments, or *myotomes,* of which there are between 50 and 65, depending upon the species. They make up the greater part of the body wall. Those on one side are arranged alternately with those of the other.

The digestive system of amphioxus is simple in structure. The vestibule leads to the mouth opening located in a membranous *velum* and encircled by 12 short *velar tentacles.* The mouth leads to a large pharynx bearing numerous gill slits. A system of ciliated and glandular grooves filled with a mucoid secretion from the endostyle, directs food particles to the intestine. A pouch, the *hepatic cecum,* projects forward from the ventral side of the intestine to the right side of the pharynx. The intestine terminates at the anus, located on the left side at some distance back of the atriopore. Each of the numerous gill slits is secondarily divided by a *tongue bar.* The gill slits of amphioxus do not open directly from the pharynx to the outside in the adult. Water passes through the gill slits into an *atrium,* or *peribranchial chamber,* which surrounds the greater part of the digestive tract except on the dorsal side. This chamber should not be confused with the coelom. The atrium communicates with the outside through the atriopore.

A pulsating tube, homologous with the hearts of higher chordates, lies ventral to the pharynx. It pumps blood anteriorly and then dorsally through the gill region, where it is oxygenated.

The anatomy of amphioxus indicates in every way that the animal is truly a chordate. Further details concerning its structure are discussed under appropriate headings in the following pages.

PROTOCHORDATE PHYLOGENY

The relationship of the protochordates to the vertebrates seems well documented on embryological and structural grounds of larvae and adult. The placement of the Hemichordata within the phylum Chordata or in its own phylum reflects the lack of agreement by workers on the precise relationship of this group. However, all workers agree that the Hemichordata are important in understanding vertebrate origins.

Clearly the three protochordate groups are today represented by highly specialized forms which tend to obscure their relationships with other animal groups. To which group of invertebrate animal must we turn our attention to find an ancestor from which the protochordates may have arisen?

At one time nearly every major phylum was considered: the Annelida, primarily due to the segmentation of the body, bilateral symmetry, and a mass of brainlike cells at the anterior end of the body; the Arachnida, in addition to the above mentioned, were considered, due to the large number of progressive and successful groups in this phylum. However, several major points of organization exhibited by these groups remove them from serious consideration: such problems as the presence of a ventral nerve cord, total organ segmentation in each body segment, and the pattern of embryonic formation of mesoderm and coelomic spaces; a lack of a notochord or internal gills; and in the arachnids numerous pairs of jointed legs. There is little if any reason today to believe the vertebrates descended from either of these two major groups (Fig. 2.9).

A third major invertebrate group, the echinoderms, presents the most striking evidence of a relationship to chordate ancestry.

FIG. 2.9 Basic differences in relative positions of the nervous system, digestive tract, and heart, between (*A*) a nonchordate (insect) and (*B*) a chordate (salamander); diagrammatic. (*From Storer et al., "General Zoology," McGraw-Hill Book Company, by permission.*)

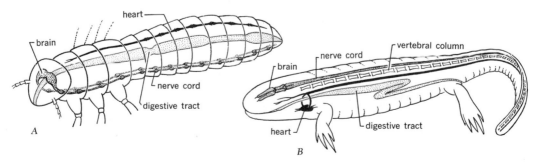

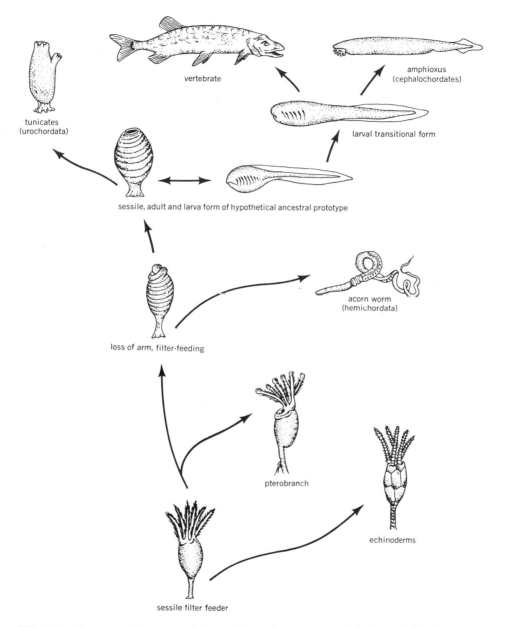

FIG. 2.10 Diagrammatic representation of the major steps in vertebrate evolution from a primitive echinodermlike ancestor. (*Adapted from Romer, "The Vertebrate Body," W. B. Saunders Company, by permission.*)

The formation of mesoderm is complex in vertebrates, but Amphioxus exhibits mesodermal formation like that of echinoderms. Lower chordates and echinoderms share three body segments, and the form of certain hemichordate larva is of the same type as that of echinoderms. In addition biochemistry provides evidence for a relationship, that is, similarity of blood serum between echinoderms and chordates, muscle chemistry, and the presence of phosphocreatine and phosphoargenine in both echinoderms and chordates as energy suppliers in muscle activity.

Phosphocreatine and phosphoargenine are not present together in other invertebrate groups. These are but a few of the characters which point to the echinoderms as relatives of the chordates.

How then are these two major groups related? Certainly the chordates were not derived from modern echinoderms with their armor plates, radial adult symmetry, and specialized sense organs. Such forms as starfish and sea urchins are far off the main line. Most echinoderms are free-living forms capable of some locomotion. It would be hard indeed to derive free-swimming, bilaterally symmetrical vertebrates from modern echinoderms. However, the fossil record of echinoderms shows many sessile forms, represented today by sea lilies and crinoids which are fixed to the bottom by stalks, spreading featherlike arms, gathering food, and carrying it to the mouth.

This life-style is duplicated by a group of hemichordates, the pterobranchs, which lack all diagnostic characters of chordates except the pharyngeal gill slits, but which are important to vertebrate ancestry. The most simple of the protochordates, the pterobranchs, are sessile nonmotile forms with only active short-lived larval forms. We must look for the origin of the vertebrates not among free-living active groups but among small sessile filter-feeding forms. The outline of characters presented above for each of the protochordate groups allows the formation of several hypotheses. The most generally accepted suggestion currently (Fig. 2.10) is that the protochordates represent specialized offshoot stages along the main line of evolution of the vertebrates from a shared ancestor with echinoderms. The earliest sessile filter-feeding forms, represented today by hemichordates, gave rise to a mobile (at least in the larval condition) group, represented today by the urochordates, and to a group which abandoned the sessile adult condition through *neotenic development,* and represents the ancestor from which two groups, the cephalochordates, and the earliest vertebrates arose.

That is by no means the only hypothesis that has been put forth. Much discussion revolves around the place of the cephalochordates and their relationship to the urochordates. Until more evidence is gathered no definite arrangement can be accepted.

BIBLIOGRAPHY

Barrington, E. J. W.: "The Biology of Hemichordata and Protochordata," Freeman, San Francisco, 1965.

Clark, R. B.: "Dynamics in Metazoan Evolution," Clarendon Press, Oxford, 1964.

Halstead, L. B.: "The Pattern of Vertebrate Evolution," Freeman, San Francisco, 1968.

Stebbins, G. L.: "Processes of Organic Evolution," Prentice-Hall, Englewood Cliffs, N.J., 1966.

Young, J. Z.: "The Life of Vertebrates," Oxford Press, New York, 1961.

3 AGE OF VERTEBRATES

Before studying the various features of vertebrate anatomy one should have a working knowledge of the animals included in the subphylum Vertebrata. Zoologists interested in taxonomy have grouped these animals in an orderly fashion according to their evolutionary, natural, and anatomical affinities. This arrangement serves as a basis for classifying all members of the subphylum. It is important that the student of comparative anatomy have some knowledge of the classification and evolution of the vertebrates; otherwise a study of their anatomy from a comparative viewpoint will have little meaning. Since in a brief book like this it is not feasible to discuss all the animals concerned, emphasis has been placed on living forms of which some knowledge is almost essential to an understanding of the principles underlying the subject of comparative anatomy.

The earliest vertebrates appear in the fossil record during the Ordovician period, some 500 million years ago. From this beginning the major vertebrate groups trace their origin. The age and domination of the various ver-

tebrate classes and the present accepted vertebrate phylogeny are presented in Fig. 3.1. Vertebrate evolution was not a singular event but corresponded with other major evolving groups among the plants and insects. A framework of major biotic evolution through geologic time is presented in Table 1.

CLASSIFICATION

The principle used in classifying animals involves the division, or separation, of larger groups into smaller groups according, in the main, to certain distinctive traits which they possess in common. Phyla are thus divided into classes, classes into orders, orders into families, families into genera, and genera into species. The characteristics which separate the lower divisions are of lesser taxonomic importance than those used as a basis for separating the larger, or higher, divisions. Often intermediate groupings may be used, particularly when the number of animals included in a certain category is large. Such groups are re-

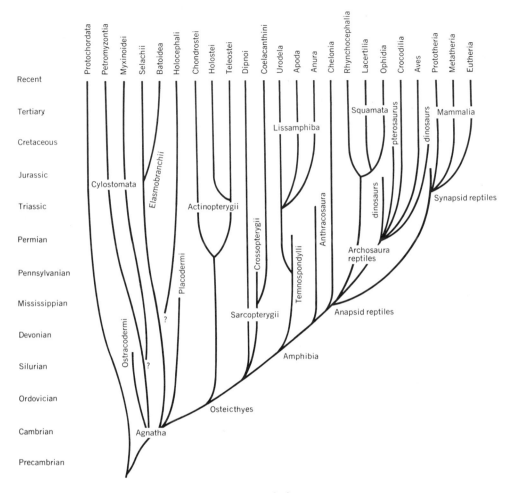

FIG. 3.1 A phylogenetic scheme of vertebrate evolution.

ferred to as subphyla, superclasses, subclasses, suborders, etc.

The scientific name of an animal is made up of its generic and specific names, the genus always being written with a capital letter and the species with a small letter. Thus the common leopard frog is *Rana pipiens.* This system of using the generic and specific names was first devised by Linnaeus and is known as *binomial nomenclature.*

Authorities differ somewhat in the weight they give to certain categories in the general scheme of classification. These are usually of a minor nature but are, nevertheless, confusing to the student. An attempt has been made, therefore, in the present volume to conform in general to the schemes used by certain biologists and paleontologists who are preeminent in their respective fields and who have made a real effort to bring about some uniformity in an otherwise perplexing area of science.

TABLE 1 Major Features of Biotic Evolution through Geologic Time

Era and (duration)	Period	Time from present to beginning of periods, millions of years	Epoch	Major biotic and geological events
Cenozoic, 65 million years	Quaternary	2+	Recent Pleistocene	Most modern species evolve, decline of large mammals
	Tertiary	65	Pliocene Miocene Oligocene Eocene Paleocene	Rise of modern orders, families, genera of mammals, birds, lizards, snakes; evolution of grasses, with reduction of tropics and subtropical flora and fauna; great mountain building; continents in present position
Mesozoic, 165 million years	Cretaceous	130		Angiosperm domination; extinction of dinosaurs by end of period; modern orders of mammals appear; primitive birds become extinct; origin of snakes; continents formed and moving toward present positions
	Jurassic	180		Reptilian domination (on land, sea, and air); birds appear; primitive mammals and angiosperms present; origin of lizards; beginning of continental formation and movement
	Triassic	230		Dominance of mammal-like reptiles; appearance of early mammals; conifers dominate; turtles appear; first dinosaurs; two major land masses, Laurasia and Gondwanaland
Paleozoic, 340 million years	Permian	280		Reptile radiation; appearance of mammal-like reptiles; modern insect orders appear; breakup of Pangea
	Pennsylvanian (Carboniferous age)	310		Adaptive radiation of amphibians; origin of reptiles; coal swamps, ferns, and seed-plant forms dominate
	Mississippian (Carboniferous age)	340		Shark radiation; primitive amphibians; winged insects, gymnosperms appear
	Devonian	400		Age of primitive fishes (Osteichthyes); mountain building; radiation of land plants; first trees; early amphibians appear
	Silurian	450		Jawed fishes (Placodermi, Chondrichthyes) appear; land plants established; arthropods dominate land; land flat with slow uplifting
	Ordovician	500		Appearance of jawless fishes (Ostracoderm); terrestrial plants beginning; invertebrates dominate; greatest marine inundation known
	Cambrian	570		All major phyla present; invertebrates dominate, especially marine forms; blue-green and green algae dominate, lands low; one ma or land mass, Pangea

SUBPHYLUM VERTEBRATA (CRANIATA)

All chordate animals except the protochordates are included among the vertebrates. Five characteristics are possessed by all vertebrates in addition to those found in chordates in general:

1 The anterior end of the dorsal, hollow nerve cord is enlarged to form a brain.

2 A protective and supporting endoskeleton is present.

3 The eyes are paired.

4 Red blood cells (hemoglobin) are present.

5 The notochord does not extend forward under the brain.

6 A ventrally placed heart.

7 A sympathetic nervous system is present.

8 A hepatic portal system is present.

With the appearance of the brain, a skeletal structure called the *neurocranium* or simply the *cranium* develops and serves to protect that delicate organ. The cranium forms part of the skull. Openings, or *foramina,* are present, serving for the passage of cranial nerves, blood vessels, and the spinal cord. The latter is also protected by a series of metamerically arranged skeletal elements called *vertebrae* associated with the notochord in their development. Openings for the passage of spinal nerves are present between adjacent vertebrae or exit from intravertebral foramina.

Vertebrates are usually divided into two large groups, or superclasses. The first, the superclass Pisces, includes the lower forms commonly known as fish. The second superclass, Tetrapoda, embraces the remaining vertebrates, which are basically four-footed animals, although in some the limbs have been lost or modified secondarily.

Superclass I. Pisces

Fishes are entirely aquatic and in most cases respire by means of gills lining pharyngeal gill slits. Included in this large superclass are some jawless forms, exemplified by living lampreys and hagfishes as well as ostracoderms. We shall consider the jawless fishes to be members of the lowest class of fishes, the Agnatha.

Class I. Agnatha

In both living and fossil agnathans, jaws are absent. Living species and most fossil forms also lack paired appendages, but some ostracoderms had paired spiny structures, lateral body lobes, or flaplike flippers posterior to the head region (Fig. 3.2), which have been interpreted as possible forerunners of paired pectoral appendages. The primitive ostracoderms are the earliest vertebrates on record. Some possessed a single, dorsal nostril between the eyes; others lacked it. Unlike those of living agnathans, the bodies of ostracoderms were covered with a heavy armor of bony, dermal plates. The presence of bone in the skin therefore is considered to be a primitive character. Modern forms with cartilaginous skeletons and lacking dermal bone are believed to be descendants of bony ancestors.

ORDER I. CYCLOSTOMATA   Living agnathans, the lampreys and hagfishes, are included in the order Cyclostomata. Although primitive in many respects, they are considered to be modified and derived offshoots of

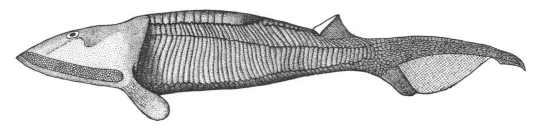

FIG. 3.2 Lateral view of an ostracoderm, *Hemicyclaspis.*

the ancestral vertebrate stock. In some respects they are so highly specialized that it is difficult to homologize them with members of the remaining vertebrate classes.

The cyclostomes have rounded, eel-like bodies with compressed tails. Median fins are present, supported by cartilaginous fin rays. In the absence of jaws there is a round, suctorial mouth. The skin is soft and devoid of scales.

SUBORDER I. PETROMYZONTIA
Lampreys should not be confused with eels, which are true fishes belonging to a different class. Both freshwater and marine forms exist. Some of the freshwater forms are non-parasitic, but most lampreys are parasitic, the adults preying upon fishes. The marine lamprey, *Petromyzon marinus* (Fig. 3.3), a parasitic form, is frequently used for dissection in courses in comparative anatomy. The animal, which attains a length of about 3 ft, lives in the ocean but ascends rivers to spawn. The sexes are separate. The larval form, known

as the *ammocoetes,* lives in burrows in the sand for 3 years or more before undergoing metamorphosis. Young lampreys migrate downstream and out to sea, where they remain 3 or 4 years. At that time they become sexually mature and are ready to ascend rivers to the spawning ground.

Seven pairs of gill slits are present, each opening separately to the outside. A suctorial funnel at the anterior end is ventral in position and beset with horny teeth. When feeding, the lamprey fastens its funnel to the side of a fish, usually just in back of the pectoral fin. Horny teeth on the tongue are used in rasping, and the animal feeds on blood and mucus secured in this way.

The ammocoetes larva, which in itself is harmless, is of great importance to the comparative anatomist, for it is a very primitive and generalized vertebrate. Many complicated structures of higher vertebrates can be traced back to its simplified structural organization.

The marine lamprey has invaded the upper Great Lakes, where it has become firmly es-

FIG. 3.3 Lateral view of lamprey, *Petromyzon marinus.*

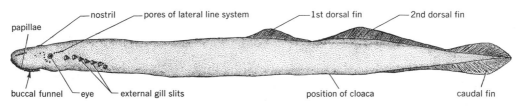

tablished. The fishing industry of the entire region has been virtually destroyed, presumably by the ravages of the lamprey, although it is possible that other factors may be at least partially responsible. It has been found that 3-trifluormethyl-4-nitrophenol (TFM), added to the water of lamprey-spawning streams in proper concentration, will effectively destroy the ammocoetes larvae without harming young fishes. TFM was first used in 1958 on streams flowing into Lake Superior. By 1962 the first signs of success became evident. As adult animals die, they are not being replaced. It is hoped that by using this method to control lampreys the fish population of the Great Lakes may eventually be restored.

SUBORDER II. MYXINOIDEA *Myxine glutinosa,* the hagfish (Fig. 3.4), is the most commonly known species included in the order. It is a marine animal, somewhat like a lamprey in appearance. The mouth is more terminal than that of the lamprey and has four pairs of tentacles about its margin. Only a single pair of external gill openings is present, located some distance from the head. The animal feeds by rasping a hole in the side of a large fish and then crawling inside the body cavity, where it eats the viscera, leaving only the skin and skeleton.

In members of the genus *Bdellostoma* there are 6 to 14 pairs of external gill openings, depending upon the species.

Members of the remaining groups of vertebrates, frequently referred to collectively as the *gnathostomes,* have paired pectoral and pelvic appendages, true upper and lower jaws, paired nostrils, and a well-developed endoskeleton. The cranium is better developed than in the agnathans.

Thus the other classes of fishes have paired appendages. These are fins, of which there are usually two pairs. Median fins are also present. Dermal scales of various types are to be found in the skin with few exceptions.

Class II. Placodermi

Placoderms are known only through fossil remains. During Devonian times, approximately 400 million years ago, ostracoderms gradually became extinct, and primitive placoderm fishes replaced them. The placoderms developed primitive jaws and paired fins but retained, to varying degrees, the bony armor of ostracoderms. The armor consisted of large bony plates, arranged in different patterns, which are believed to have given rise to the several types of dermal scales found in more modern fishes. Placoderms were not successful

FIG. 3.4 The hagfish, *Myxine glutinosa.*

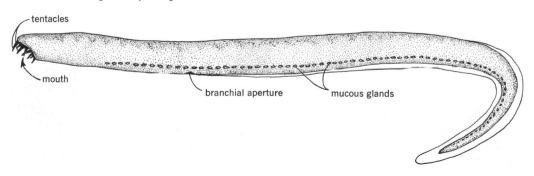

from a long-range point of view. They gradually disappeared as two new types of fishes evolved—the cartilaginous (Chondrichthyes) and the bony fishes (Osteichthyes).

Class III. Chondrichthyes

The fishes belonging to the class Chondrichthyes include sharks, dogfishes, skates, rays, and chimaeras. Most are marine forms with cartilaginous skeletons, placoid scales, and ventral, subterminal mouths. All species have tails of the heterocercal type, in which the dorsal flange is larger than the ventral (Fig. 6.50). The cartilaginous fishes of today are descendants of those which lived in Devonian times and which evolved from placoderms or placodermlike ancestors.

The skeleton of members of this class of fishes is primitive in several respects but not in the fact that it lacks bone and is entirely cartilaginous. The lack of bone in the skeleton, formerly considered to be primitive, is now believed to represent a specialized condition. The bone in the basal plates of placoid scales in the skin of most cartilaginous fishes

is all that remains of the ancient bony armor found among the placoderms.

Members of the class Chondrichthyes lack swim bladders, thus differing markedly from the bony fishes.

Subclass I. Elasmobranchii

The elasmobranch fishes, as members of this order are generally called, include those cartilaginous fishes in which the gill slits open separately to the outside. There are five to seven pairs of gill slits in addition to a modified pair, the spiracles, opening on top of the head, just posterior to the eyes.

ORDER I. SELACHII Sharks and dogfish living today, as well as certain extinct forms, belong to the order *Selachii*. They differ from skates and rays in that their pectoral fins are distinctly marked off from cylindrical bodies and their gill slits are laterally placed (Fig. 3.5). Since many structures in elasmobranchs are primitive or generalized in character, the shark and dogfish are ideal animals for laboratory study as an introduction to the anatomy of higher vertebrates.

FIG. 3.5 Great hammerhead shark. (*Courtesy of the Chicago Natural History Museum.*)

Dogfish, which are among the smaller species, are plentiful. Among the better-known sharks are the hammerhead, white shark, sand shark, and whale shark, the latter being the largest living fish.

ORDER II. BATOIDEA Skates and rays are included in the order Batoidea. These fishes may be regarded as modified sharks which have become flattened in a dorsoventral direction. Head and trunk are widened

FIG. 3.6 Adaptive radiations among elasmobranchs.

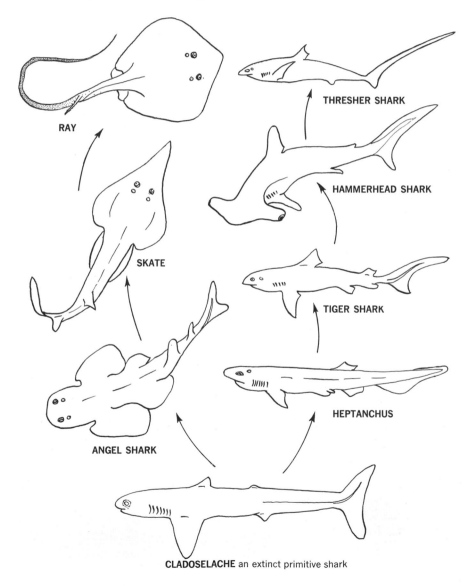

RAY

THRESHER SHARK

HAMMERHEAD SHARK

SKATE

TIGER SHARK

HEPTANCHUS

ANGEL SHARK

CLADOSELACHE an extinct primitive shark

considerably, and the pectoral fins are not sharply marked off from the body. There is a distinct demarcation between body and tail, however. The gill slits are located ventrally. Skates are egg-laying forms, whereas rays give birth to living young. Among Batoidea are included such forms as the gigantic devilfish, or Manta ray; the sting ray, or stingaree; the *Torpedo,* or electric ray; and the sawfish. The body of the last is not flattened to the same degree as in most batoideans, but the gill slits are ventral in position, nevertheless.

The varied body forms of skates and rays represent adaptive radiations from primitive, sharklike ancestors (Fig 3.6).

Subclass II. Holocephali

A small group of fishes, called chimaeras (Fig. 3.7), is included in the subclass Holocephali. Most species inhabit the deep sea. They are of special interest to the comparative anatomist because they occupy a position intermediate to elasmobranchs and higher fishes. Among primitive features are a persistent notochord and poorly developed vertebrae. Advance is indicated by the presence of a flap, or operculum, which covers the gill chamber on each side. Scales are practically absent in the adult stage.

CLASS IV. OSTEICHTHYES This class includes the bony fishes with skeletons bony to some degree. Several types of dermal scales are to be found within the group. The mouth is usually terminal, and an operculum covers each gill chamber. These fishes are divided into two subclasses, the distinctions between them being based primarily on fin structure. Members of the first subclass, Actinopterygii, are commonly known as ray-fins; those of the second, or Sarcopterygii, as the lobe-fins. The latter were formerly referred to as the Choanichthyes because certain members of the subclass possessed internal nares, or choanae. Since some living sarcopterygians do not possess choanae (a fairly recent discovery), the term Choanichthyes scarcely seems appropriate and has generally been discarded.

Subclass I. Actinopterygii

In members of the subclass Actinopterygii all the fins, paired and unpaired, are sup-

FIG. 3.7 Lateral view of *Chimaera monstrosa. (Drawn by A. Page.)*

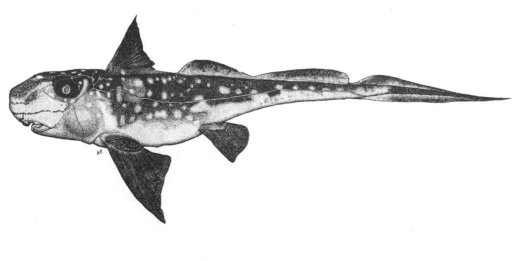

ported by skeletal elements, the dermal fin rays.

SUPERORDER I. CHONDROSTEI

Only a few representatives of this ancient superorder are living today. Two genera, *Polypterus* (Fig. 3.8) and *Calamoichthys,* are found only in Africa. It has been asked whether these fishes should more properly be included in the subclass *Sarcopterygii* since their pectoral fins have a fleshy basal portion. However, closer examination shows this suggested change to be incorrect. The structure, however, is somewhat different from that of the other lobe-finned fishes. The caudal fin, unlike the heterocercal structure of the Chondrichthyes, is symmetrical but not typically diphycercal (Fig. 6.50). These fishes also were known as the "fringe-finned ganoids," because of the fringe like appearance of the eight or more dorsal fin elements and the presence of ganoid scales (page 84).

Other members of the superorder Chondrostei include the spoonbill, or paddlefish, and the sturgeons. They, too, are believed to be survivors of primitive ray-finned fishes which have lost the greater part of their bony skeletons. The skeleton is almost completely cartilaginous. Their tails are of the heterocercal type.

SUPERORDER II. HOLOSTEI

Only two genera of living fishes belong to the superorder Holostei. Both are freshwater forms and survivors of ancient groups of fishes which formerly inhabited the ocean. They include *Amia,* the mudfish, freshwater dogfish, or bowfin, and *Lepisosteus,* the garpike.

SUPERORDER III. TELEOSTEI

All the remaining and most familiar ray-finned fishes are included in the superorder Teleostei. Ninety-five percent of all the fishes in the world come under this category. About 30,000 species have been identified. The posterior, or occipital, region of the skull is bony in all. The tail is usually of the homocercal type (Fig. 6.50), although in some species it is diphycercal. It is never heterocercal.

Division of the subclass Actinopterygii into three superorders, as described above, is not agreed upon by all authorities, since certain characters are not limited entirely to one group or another. Nevertheless, the scheme presented here forms a good working basis for the student of contemporary zoologists and paleontologists.

Subclass II. Sarcopterygii

The lobe-finned fishes, of which this subclass is composed, are known chiefly from ancient fossil remains. Within the subclass Sarcopterygii are to be found the first vertebrates in which nasal passages connect the mouth cavity with the outside. This characteristic does not appear in all, however,

FIG. 3.8 *Polypterus bichir. (Courtesy of the American Museum of Natural History.)*

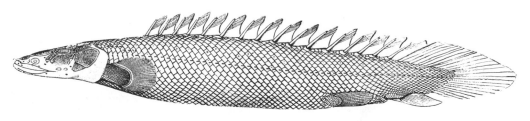

FIG. 3.9 *Latimeria chalumnae,* a surviving coelacanth. *(Drawn by Roswitha Schugt.)*

Although it is possible that some of the extinct members of the subclass used their nasal passages in connection with lung respiration, there is no clear-cut evidence that species living today actually employ them in such a manner. Nevertheless, the presence of choanae is of great importance from an evolutionary point of view, for it suggests that amphibians, the first air-breathing terrestrial vertebrates, may have evolved from similar fishes. The subclass is divided into two orders, which include the typical lobe-fins and the true lungfishes, respectively.

ORDER I. CROSSOPTERYGII Crossopterygians are the typical lobe-finned fishes which make up a small but ancient group known mainly from fossil forms. They appeared on earth at the beginning of the Devonian period. The fins of crossopterygians differ from those of all other fishes in that each is borne on a fleshy, lobelike, scaly stalk extending from the body. Pectoral and pelvic fins have articulations resembling those of tetrapod limbs. These fishes apparently took two different lines of descent and are accordingly separated into two groups or suborders, the rhipidistians and the coelacanths.

SUBORDER I. RHIPIDISTIA The rhipidistian crossopterygians became extinct near the end of the Devonian period. There is considerable evidence that it was through members of this suborder that evolutionary progress toward tetrapods was made. The skeletal elements in the paired fins of the fossil *Eusthenopteron,* one of the better-known rhipidistians, have been rather clearly homologized with those in the limbs of tetrapods (page 162). Fossils of primitive amphibians have been found together with those of rhipidistians in the same geological deposits.

SUBORDER II. COELACANTHINI The coelacanth crossopterygians were a more stable and conservative group which have shown but little change for millions of years. Until 1938 it was believed that coelacanths had become extinct about 70 million years ago. In that year a strange fish was caught off the east coast of South Africa at a depth of 40 fathoms. To the surprise of zoologists and paleontologists this animal proved to be a coelacanth. It was named *Latimeria chalumnae* (Fig. 3.9). The specimen was poorly preserved, and little of its anatomy could be ascertained. In more recent years several addi-

FIG. 3.10 The Australian lungfish, *Epiceratodus*. (*Courtesy of the Chicago Natural History Museum.*)

tional specimens have been obtained. It is not yet quite clear whether these represent more than one species. At any rate, there is no doubt that *Latimeria* is a survivor of one of the oldest stock represented among living vertebrates. In all the time that has elapsed since coelacanths were plentiful on the earth, *Latimeria* has remained relatively unchanged. In addition to its rather complex paired pectoral and pelvic fins it has two unpaired dorsal fins, a joint in the skull, and a single, ventral, anal fin. All except the first dorsal fin are typical lobe-fins. The tail is diphycercal. Internal nares are lacking in *Latimeria*.

ORDER II. DIPNOI Three different genera of lungfishes are included in the order Dipnoi. They include the Australian lungfish, Epiceratodus (Fig. 3.10), the South American lungfish, *Lepidosiren,* and the African lungfish, *Protopterus* (Fig. 3.11). Internal nares, or choanae, are present in members of this order. *Epiceratodus* is the most primitive of the living lungfishes. Most members of the class Osteichthyes possess an internal saclike structure, the air bladder. This arises as a single or paired diverticulum of the digestive tract in the region of the pharynx. It may retain its connection with the pharynx (physo-

FIG. 3.11 The African lungfish, *Protopterus*. (*Courtesy of the American Museum of Natural History.*)

stomous condition) or lose it and become a blind sac (physoclistous condition). In the true lungfishes the air bladder is of the physostomous type. It is well developed, highly vascularized, and used as a lung in respiration. Such air bladders are more efficient respiratory organs than the lungs of many amphibians. Nevertheless, because of certain other characteristics, notably the structure of the fins, it is believed that rhipidistian crossopterygians, rather than ancient dipnoans, were ancestors of amphibians.

Superclass II. Tetrapoda

Tetrapods are those members of the subphylum Vertebrata having paired appendages in the form of limbs rather than fins. In some, the limbs have been lost or modified. The basic plan from which all tetrapod limbs have evolved is the five-toed, or pentadactyl, appendage. Among other characteristics which distinguish tetrapods from fishes are a cornified, or horny, outer layer of skin; nasal passages which communicate with the mouth cavity and which transport air; lungs used in respiration; and a bony skeleton. There has also been a reduction in number of skull bones.

The first tetrapods are believed to have evolved from rhipidistian crossopterygian fishes, which they rather closely resembled. The fossil remains of primitive tetrapods (*Ichthyostegus*) have been found in the eastern part of Greenland in deposits which date back to the end of the Devonian period. These specimens possess characteristics which place them in a category intermediate between late crossopterygians and early amphibians. They were far different from any tetrapods living today.

The superclass Tetrapoda is divided into four classes made up of amphibians, reptiles, birds, and mammals, respectively.

Class I. Amphibia

The living representatives of the class Amphibia subclass Lissamphibia include salamanders, newts, frogs, toads, and some less-familiar, burrowing, legless forms, the caecilians. Frogs and toads, often thought of as primitive tetrapods, are actually far removed from the original amphibian stock and are highly specialized animals.

Some skeletal structures in certain amphibians show a higher degree of specialization than the corresponding parts of lizards, which belong to the class Reptilia. Not all amphibians, therefore, are at a lower level in the evolutionary scale than all reptiles.

The class Amphibia is composed of tetrapods in which the transition from aquatic to terrestrial life is clearly indicated. Amphibians are the first vertebrates to live on land, although they lay their eggs in water or in moist situations. Aquatic larvae with integumentary external gills develop from the eggs, though direct development, without larval stage, is exhibited in both tropical frogs and salamanders. In one group of amphibians, the family Typhlonectidae, no eggs are laid, the entire developmental process occurring within the uterus. After a varying period of time, depending upon the species, metamorphosis usually occurs, following which the animal may spend the greater part of its life on land, although generally in a moist environment. However, some forms (*Scaphiopus, Phyllomedusa sauvagii*) have adapted to a dry environment. Upon metamorphosis, among other changes, the gills are usually lost and lungs develop, supplementing the vascular skin as organs of respiration. Some salamanders never develop lungs even though they lose their gills. A few salamanders, like the mud puppy *Necturus,* never undergo complete metamorphosis. They develop lungs but retain their gills throughout life.

The chief differences between amphibians and the truly terrestrial vertebrates lie in their aquatic reproductive habits, the lack of certain embryonic membranes, and the rather poorly developed corneal layer of the epidermis of the skin. The ends of the digits lack claws. The failure of amphibians to rise to a more dominant position in the world lies chiefly in their mode of development. They are, so to speak, "chained" to the water. Although certain forms have developed structures which fit them fairly well for life on land, nevertheless their conquest of the land has been only partially successful.

The most primitive amphibians, known from fossil remains, are the Labyrinthodonts, the name being based upon the complex folding of the enamel layer of the teeth. These animals are sometimes called Stegocephalians because of the solid roofing of the skull. In certain features they resembled rhipidistian crossopterygian fishes. Many labyrinthodonts had a ventral armor of overlapping bony plates. In a few forms dorsal bony plates were also present. It seems that from such an ancestral stock there may have evolved two main groups, the Temnospondyli and the Anthraco-

sauria, the members of which were distinguished chiefly by differences in the detailed structure of their vertebral columns. One of the groups within the temnospondyl line is currently thought to have given rise to the Lissamphibia (frogs, salamanders, and caecilians). The anthracosaurian line gave rise to primitive reptiles and hence to higher forms of vertebrate life. In evolutionary terms, those amphibians living today are relatively unimportant forms which represent dead ends insofar as evolutionary progress is concerned. The three amphibian types are divided into three orders.

ORDER I. ANURA Frogs and toads, which come under this category, exhibit head and trunk fused, and no neck region (Fig. 3.12). Two pairs of well-developed limbs are present, the hind pair being particularly adapted for leaping. The feet may be webbed and adapted for swimming or resemble long fingers with suction pads for climbing. Frogs and toads are the first vertebrates to have vocal cords for sound production. The anuran larva, or tadpole, does not resemble the parent. Head and body are fused into a single, egg-shaped mass, and the long tail is equipped with a median fin. Horny jaws are used in lieu of teeth in feeding. Metamorphosis is clearly defined. At this time not only are the gills lost as lungs develop, but legs appear and the tail is resorbed. Anurans as a group are better fitted for terrestrial existence than are other amphibians. Certain species inhabit desert regions, where they spend the dry season in a state of estivation. No superficial characters distinguish frogs from toads. Frogs usually have a smoother and moister body than toads, with their warty skin.

The order Anura, together with two other groups of extinct ancestral forms, are often grouped in a separate superorder, Salientia.

FIG. 3.12 Female green frog, *Rana clamitans*. (*Courtesy of the American Museum of Natural History.*)

ORDER II. URODELA (CAUDATA)

The urodele, or caudate, amphibians include salamanders (Fig. 3.13) and newts, the latter being small, semiaquatic forms. Urodeles are found, for the most part, in temperate and subtropical climates in the Northern Hemisphere but do reach the tropics in the New World. The elongated body consists of head, trunk, and a well-developed tail, the tail being retained throughout life. Two pairs of limbs are present in most species. Larvae resemble adults except for the presence of gills and, like adults, possess teeth in both upper and lower jaws. Salamanders are common in the United States, being found under rotten logs in wooded areas and in moist situations. The mud puppy, or water dog, *Necturus,* is a urodele amphibian.

ORDER III. APODA (GYMNOPHIONA)

Members of the order Apoda living today are pantropical in distribution. They include the caecilians (Fig. 3.14), which are burrowing forms, with wormlike bodies, lacking limbs. The tail is very short, and the anus is almost terminal. Unlike other amphibians, some caecilians have dermal scales. Adults lack gills and gill slits. The very small eyes are buried beneath the skin or under the

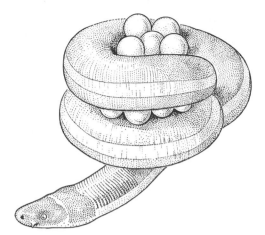

FIG. 3.14 A tropical caecilian or limbless amphibian (*Ichthyophis glutinosus,* order Gymnophiona) with eggs. (*From Storer et al., "General Zoology," McGraw-Hill Book Company, by permission.*)

skull bones. Large caecilians may attain a length of 26 in. or more. Very little is known about this group. Reproductive morphology and some sensory adaptations have been described, but life history information is very limited. Because of the presence of an intromittent organ in males, internal fertilization is assumed. In some caecilians, eggs are laid, which upon hatching spend variable amounts of time as free-swimming larvae. In other caecilians, the eggs are retained within the female, metamorphosis occurring before birth.

Amniota and Anamniota

The three remaining classes of tetrapods, reptiles, birds, and mammals, are referred to collectively as the *Amniota.* The lower classes together form a group called the *Anamniota.* The essential difference between the two lies in the presence of certain membranes as-

FIG. 3.13 Female marbled salamander, *Ambystoma opacum,* guarding eggs in nest.

sociated with the embryos of terrestrial forms during development. These membranes are the *amnion, chorion,* and *allantois.* The name *Amniota* is, of course, taken from the first of these. The amnion is a fluid-filled sac which comes to surround the embryo completely, soon after development has begun. It first appears in reptiles and is of definite advantage to the embryo in preventing desiccation and otherwise offering protection during development. The remaining membranes are concerned with respiration and other processes essential to the developing amniote embryo.

Class II. Reptilia

Reptiles are represented today by five orders of living forms which include turtles and tortoises, alligators and crocodiles, lizards and snakes, and a lizardlike animal, the tuatara, *Sphenodon punctatum,* of New Zealand. Reptiles are believed to have made their first appearance on earth in the coal-swamp faunas existing about 310 million years ago. Today most reptiles live in tropical and subtropical regions. They are small and insignificant compared with the tremendous monsters of prehistoric times. During the Age of Reptiles, which began about 280 million years ago, reptiles dominated the earth. The extinct forms, known only by their fossil remains, have been grouped into numerous orders. Some lived on land, others took to the air, and still others were adapted to an aquatic habitat.

The most ancient reptiles were the cotylosaurs, sometimes called the *stem reptiles.* Several groups of reptiles seem to have evolved from the cotylosaurs. Some were important for a time and then lost ground and disappeared; others have persisted to the present day with little change; still others evolved further and gave rise not only to most of our

modern reptiles but to birds and mammals as well.

Among early forms were such large aquatic reptiles as the short-necked ichthyosaurs and long-necked plesiosaurs. Modern turtles were derived from an early side shoot, stemming from the cotylosaurs, which has undergone little change. Another group tracing its origin to an early branch of the cotylosaurs was that which gave rise to mammals. Among the earliest of these mammal-like reptiles were the therapsids, the skulls and teeth of which more nearly resembled those of mammals than those of other reptiles. Still another stem from the early base is believed to have branched in two directions. One branch gave rise to Sphenodon and to the lizards and snakes of today. The other was made up of a great group of reptiles, the archosaurs, sometimes called the ruling reptiles. The varied types of dinosaurs and the flying pterosaurs belong to the subclass Archosauria. The only reptilian survivors of archosaurian origin are crocodiles and alligators. Even they have become considerably modified from the ancestral stock. Birds evolved from another branch of archosaurs, which, like pterosaurs (Fig. 3.15), developed the ability to fly but used wings of an entirely different construction and feathers for this purpose.

At the close of the Age of Reptiles, great changes in climate and geology occurred, with the result that both animal and plant life were much affected. Several thousand years elapsed before the great dinosaurs, their smaller relatives, and the pterosaurs disappeared and mammals, heretofore an insignificant group, began to flourish.

Although it is believed that reptiles and amphibians evolved from a common ancestral stock, nevertheless the gap between the two groups is large. This is most strikingly shown

FIG. 3.15 Restoration of pterosaur by E. M. Fulda. (*Courtesy of the American Museum of Natural History.*)

by the manner of development. Eggs of reptiles are always laid on land, and a resistant shell of leathery or limy consistency covers the large-yolked eggs. Of even greater significance is the presence of the embryonic membranes, previously mentioned.

As in lower forms, the body temperature of reptiles is dependent on an external heat source. Temperature in both amphibians and reptiles is regulated through behavioral and physiologic responses, which allows the animal to maintain a preferred body temperature. In this respect they are said to be *poikilothermous*. The skeletons of reptiles are more completely bony than those of amphibians; their bodies are covered with dry, horny scales, and the skin is practically devoid of glands. Except in snakes and in certain lizards

which lack appendages, the limbs are characteristically pentadactyl. Gill pouches are present only in the embryo, and respiration is entirely by means of lungs except in certain aquatic turtles in which cloacal respiration (page 238) is employed as well. The young resemble their parents, and metamorphosis does not occur.

ORDER I. CHELONIA Chelonians are not closely related to other modern reptiles, since they represent a very old group which has persisted with little change for some 200 million years. Terrestrial forms are usually called tortoises, whereas turtles are generally semiaquatic. Terrapins are edible freshwater species.

The bodies of chelonians are relatively short and wide. The typically pentadactyl limbs have been modified in certain marine forms and are called flippers. Teeth are lacking, but the jaws are covered with sharp, horny beaks. The cloacal opening is in the form of an elongated slit. Most noteworthy is the presence of a shell which encases the body. The rounded dorsal portion is the *carapace;* the flatter ventral region is the *plastron.* The shell has an underlying layer of bone over which horny scales are arranged in a more or less similar manner in most species.

Variations in size are extreme among turtles, some large marine forms attaining a weight of 1,500 to 2,000 lb.

ORDER II. RHYNCHOCEPHALIA The tuatara, *Sphenodon punctatum* (Fig. 3.16), of New Zealand and about 20 surrounding islands, is the only living member of the order. *Sphenodon* has been called a living fossil because of its primitive and generalized structure. It is one of the oldest types of reptiles known, even when extinct forms are taken into consideration. Adults attain a length of

FIG. 3.16 *Sphenodon punctatum. (Courtesy of the American Museum of Natural History.)*

about 21 in. A significant feature is the presence of a well-developed median, parapineal, or parietal, third eye in the middle of its forehead. Some lizards have a similar median eye (see page 394), but it is never so well developed as that of the tuatara. The cloacal opening is a transverse slit.

Despite the lizardlike appearance of *Sphenodon,* certain features of its skull indicate a lack of close relationships to modern lizards.

ORDER III. SQUAMATA The order Squamata includes lizards and snakes, which are placed in separate suborders. These reptiles apparently came from the same ancestral stock that gave rise to *Sphenodon* but have flourished and evolved further instead of remaining at a stationary level. In evolutionary terms they represent the most recent of all reptiles.

The skin of lizards and snakes is covered with horny epidermal scales. Sometimes bony dermal plates lie under these scales. The cloacal opening is always in the form of a transverse slit.

SUBORDER I. SAURIA (LACERTILIA) In lizards there are usually two pairs of pentadactyl limbs. Some lizards, known as *glass snakes,* as well as certain burrowing forms lack legs, so that this feature cannot be used to characterize lizards as a group. The eardrum, or tympanic membrane, is not always at the surface, and external ear pits are visible. Unlike snakes, lizards have movable upper and lower eyelids as well as a nictitating membrane. The two halves of the lower jaw are united, and the animals lack the ability to open their mouths the way snakes do.

Lizards and salamanders are commonly confused by laymen. The shape of the body in these two groups is essentially similar, but the smooth, moist skin of the salamander and its toes without claws are quite in contrast to the dry scaly skin of the lizard, the toes of which bear claws.

One of the characteristics for which certain species of lizards are well known is their ability to change color rapidly so as to blend in with the background of their environment (*metachrosis*). Not all lizards have this ability, however.

SUBORDER II. SERPENTES (OPHIDIA) Snakes have lost their legs in the evolution of the group, but some, like pythons and boa constrictors, still retain skeletal vestiges of the pelvic girdle and limbs.

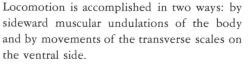

Locomotion is accomplished in two ways: by sideward muscular undulations of the body and by movements of the transverse scales on the ventral side.

There are no external ear openings or tympanic membranes in this group of reptiles. The eyelids are immovably fused. The eye is covered by a transparent scale which gives the animal a glassy stare. The eye beneath the scale is capable of considerable movement, however. The loose ligamentous attachment of the jawbones to each other and to the cranium enables the mouth of the snake to stretch to a remarkable degree.

ORDER IV. CROCODILIA (LORICATA)

Crocodiles and alligators are the only survivors of the terrestrial archosaurs. In many of the archosaurs the hind legs were elongated, and the animals were adapted for bipedal locomotion. Crocodilians are modified descendants of a group of bipedal archosaurs which have assumed an amphibious existence and are quite unlike other reptiles living on earth today.

With the exception of certain giant marine turtles, crocodiles and alligators are the largest living reptiles. Caymans and gavials are also included in the order. The feature that distinguishes crocodiles from alligators is not size or the shape of the snout, as is commonly believed. In an alligator the fourth tooth on each side of the lower jaw fits into a pit in the upper jaw when the mouth is closed. In most crocodiles the fourth tooth of the lower jaw fits into a notch on the *outer* side of the upper jaw and is exposed when the mouth is closed.

Members of this order have compressed tails and two pairs of short legs. There are five toes on the forefeet and four on the hind feet. The toes are webbed. The tympanic membrane is exposed but protected by a fold of skin. Eyes, nostrils, and ears are in a straight line on top of the head. This enables the animal to use its major sense organs when only a small part of the body is exposed above water. The anal opening, like that of turtles, is a longitudinal slit. The skin is thick, with bony plates underlying the horny scales on the back and, in caymans, on the ventral side as well. The snout of the American crocodile is much narrower than that of the American alligator.

Class III. Aves

Birds are the only animals which possess feathers. We treat them here as a separate class, but in many ways they are no more removed from the early reptilian stock than other reptilian orders. Feathers are modifications of the reptilian type of epidermal scale. Together with the scales on the feet and legs of birds, they indicate a close relationship between birds and reptiles. This relationship is borne out in many other features. We have already said that birds did not evolve from pterosaurs, which utilized an entirely different principle in flying. They arose from another type of archosaur which had bipedal locomotion and in which the scales covering the body had become modified into feathers. Feathers on the forelimbs, then, rather than membranous wings, were used to resist the air in flight (Fig. 6.63).

Birds may be considered to be the most highly specialized of the vertebrates. As a group they have become specialized to aerial life, although there are a number of exceptions. Adaptive features include the light, hollow bones; loss of right ovary and oviduct in females of most species; the exceptionally well-developed eyes; the highly specialized lung and air-sac system; the modification of the forelimbs to form wings; and the presence of feathers, which, although light in weight, offer an effective resistance to air when the bird is in flight. All these adaptive specializations are related to flight. Birds have a high and constant body temperature and are said to be *homoiothermous*. There is less deviation in structure within the entire class than can be found within a single order in some of the other classes.

Although the forelimbs of most birds have been modified into wings, they are not always adapted for flight. The hind limbs of different forms show a great deal of variation. The feet usually have four toes which terminate in claws.

Birds are divided into two subclasses, one of which contains but one or possibly two species known only by fossil remains. The other subclass includes all other birds, both extinct and modern.

Subclass I. Archaeornithes

The best-known species of this subclass is *Archaeopteryx* (*Archaeornis*) *lithographica* (Fig.

3.17). It is known from two specimens and the impression of a single feather found in the lithographic limestone in a quarry at Solenhofen in Bavaria, Germany. *Archaeopteryx* was about the size of a crow. The long, jointed tail bore feathers along each side. Three clawed digits on the wings undoubtedly aided the bird in climbing about in the trees. Its hind feet had four toes, as is characteristic of modern birds. Teeth set in sockets were present in both jaws. Riblike dermal bones, called *gastralia,* found in numerous reptiles, but not encountered in other birds, were present in *Archaeopteryx*. In many respects, then, this fossil bird showed distinct reptilian characteristics. The presence of feathers, however, makes it imperative to classify it with birds, for in this feature it differs from any known reptile. *Archaeopteryx* is an almost ideal connecting link, for it definitely bridges the gap separating two great classes of tetrapods, the Reptilia and the Aves.

FIG. 3.17 Fossil remains of *Archaeopteryx*. (*Courtesy of the American Museum of Natural History.*)

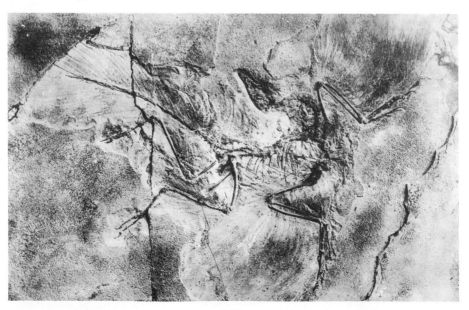

Subclass II. Neornithes

In members of the Neornithes the wing bones have been reduced in number and, with rare exceptions, there are no free, clawed digits on the wings. A few primitive fossil birds, some extinct flightless species, and all modern forms are included in the subclass.

SUPERORDER I. ODONTOGNATHAE The fossil Neornithes include *Hesperornis regalis* (Fig. 3.18), a flightless swimming bird, and *Ichthyornis victor,* with wings well adapted for flight. Both birds had teeth.

SUPERORDER II. PALEOGNATHAE (RATITAE) To this superorder belong the flightless, toothless, running birds. A few, such as the giant moas and elephant birds, have become extinct within the memory of man. The ostrich, rhea, emu, cassowary, and kiwi (Fig. 3.19) are living examples of these birds. Their wings are rudimentary or too small and weak to be used in flight. The tinamous of Central and South America are exceptions, capable of limited flight.

SUPERORDER III. NEOGNATHAE All modern birds except the *Paleognathae* belong to the superorder Neognathae. They are aerial flying birds, except for the penguins, in which the forelimbs have been modified into paddlelike swimming organs (Fig. 3.20). No teeth are present.

FIG. 3.19 The kiwi, *Apteryx australis,* with egg. (*Courtesy of the Chicago Natural History Museum.*)

Numerous adaptations possessed by birds are of importance to the comparative anatomist. Those which fit them for aerial life as well as for life on land or water, the extraordinary migrations of many species, their methods of courtship and nest building, have all had careful consideration. The feet and bills of birds show many adaptive variations.

Class IV. Mammalia

Mammals are homoiothermous tetrapods which have hair and mammary glands. All mammals have some hair, and only mammals have hair. Mammary glands, found only in mammals, are specialized skin glands which secrete milk used to nourish the young.

Biologists believe that mammals evolved from the mammal-like cynodont reptiles, previously mentioned. It is possible, however, that several groups of mammal-like reptiles may have contributed to the pedigree of early mammals. Many such groups are known through fossil remains of skulls, jaws, teeth,

FIG. 3.18 *Hesperornis regalis.* (*Courtesy of the Chicago Natural History Museum.*)

FIG. 3.20 Jackass penguin. (*Courtesy of the Chicago Natural History Museum.*)

etc. Only forms living today are included in the following account.

It is probable that the intricate mechanism which ensures a constant body temperature independent of environmental changes has had much to do with the present wide distribution of mammals. Birds, the only other homoiothermous animals, are also widely distributed. They have specialized along lines which enable them to fly. Mammals have specialized in another direction, i.e., development of the cerebral hemispheres of the brain.

In size, mammals vary from a small shrew, the body of which is less than 2 in. long, to the blue whale, which may reach a length of 103 ft or more.

Mammals have undergone a tremendous

adaptive radiation to a diversity of habitats. Some, like moles, live almost entirely underground; whales are at home only in the water; bats rival birds in their ability to fly; most monkeys are arboreal; but the majority of mammals walk or run on solid ground. Many mammals, sometimes entire orders, have become extinct.

A few mammals (three genera) lay eggs; the rest give birth to living young. Accordingly, the class Mammalia has been divided into two subclasses, of which the first contains only the egg-laying forms.

Subclass I. Prototheria

The eggs of prototherians are incubated outside the body, but the young are nourished by milk from mammary glands. No nipples are present, the ducts of the glands opening directly onto the surface of the skin. The three living genera are included in a single order.

ORDER I. MONOTREMATA To this order belong the duckbill platypus, *Ornithorhynchus anatinus,* and two genera of spiny anteaters, or *echidnas*: *Tachyglossus* of Australia, Tasmania, and New Guinea; and *Zaglossus* of New Guinea and Papua.

The platypus of Australia (Fig. 3.21) is semiaquatic in habit, its beaverlike body, tail, and webbed feet adapting it to life in the water. Its ducklike bill is soft, flexible, and moist and abundantly supplied with sense organs. It bears only a superficial resemblance to the bills of certain birds.

In the echidnas (Fig. 3.22) the skin is covered with spines and coarse hair. A short tail and external ears are present. The long, protrusible tongue is used for feeding on ants and termites.

Monotremes do not have a highly devel-

FIG. 3.21 Duckbill platypus, *Ornithorhynchus anatinus*. (*Courtesy of the American Museum of Natural History.*)

oped mechanism for controlling body temperature, which is, therefore, rather inconstant. This group of mammals apparently has no close relatives. Several features indicate that they have retained certain reptilian characters. They are basically primitive forms, even though they possess certain specialized features, and should be regarded as occupying an intermediate position between the mammal-like reptiles and higher mammals.

Subclass II. Theria

Therians bring forth their young alive and have mammary glands provided with nipples. The subclass is divided into two infraclasses.

Fig. 3.22 Spiny anteater, *Tachyglossus*. (*Courtesy of the Chicago Natural History Museum.*)

Infraclass I. Metatheria

In the Metatheria the young are usually born in an extremely immature condition and undergo further development in a marsupial pouch on the ventral side of the mother. The term *mammary fetus* is sometimes applied to the pouch young. The marsupial pouch is reduced or wanting in certain metatherians.

ORDER I. MARSUPIALIA This strange group of mammals is today almost entirely confined to Australia and the neighboring islands. Exceptions include the opossums of North, Central, and South America, and the small opossum rat, *Caenolestes,* of Ecuador and Peru. The latter lacks a marsupial pouch. The Australian group embraces such forms as the thylacine, or marsupial wolf; the wombat, a small bearlike animal; the bandicoot, which resembles a rabbit; the marsupial mole; the flying squirrel; and a host of additional forms which show likenesses to placental mammals such as rodents, carnivores, and insectivores but only in superficial ways.

The common Virginia opossum, *Didelphis marsupialis,* of eastern and southern United States, is the most familiar form. The young at birth (Fig. 3.23) are more like the embryos of placental mammals. They make their way to

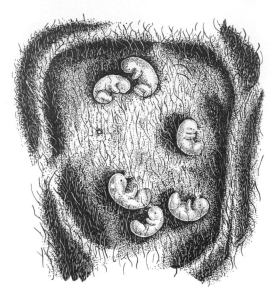

FIG. 3.23 Open marsupial pouch of Virginia opossum, *Didelphis marsupialis,* showing young at about the third day after birth.

the marsupial pouch without aid from the mother. Here they undergo the rest of their development. After having grown still further, they leave the pouch, at first temporarily and then permanently. The Virginia opossum is sometimes referred to as a "living fossil," having undergone little change in 120 million years. Probably because of its relatively small size and generalized structure it has been able to survive in competition with more highly developed and specialized placental mammals.

In both sexes of most marsupials a pair of long, flat *epipubic bones* extends from the lower anterior part of the pelvic girdle. They are embedded in the muscles of the lower part of the abdominal wall, which they help support. Similar bones are found in monotremes. The fact that certain primitive reptiles also possessed a skeletal support of the ventral abdominal wall offers a possible clue to the ancestral relations of metatherians.

The peculiar geographical distribution of marsupials is difficult to explain and has aroused many interesting speculations. Fossil remains of numerous marsupials have been found in South America, indicating that they were once a thriving group on that continent. It is known from the fossil record that primitive marsupials, of little relative importance, were present at least in Europe and North America. Placental mammals, however, were predominant. In Australia and South America the mammalian populations differed from those of other continents. The Australian forms were probably all marsupials, whatever their origin may have been. In South America, herbivorous placental mammals and carnivorous marsupials seem to have coexisted for some time. Later on, when South America and North America became connected, there was an invasion of competitively superior carnivorous placental mammals from the north, and the marsupials in South America gradually became extinct except for a few surviving forms. Those in Australia, on the other hand, were not in competition with placentals, flourished, and slowly evolved into the many types now indigenous to that continent.

Many biologists today are of the opinion that both marsupial and placental mammals originally sprang from the same basic mammalian stock, which underwent an early dichotomy. Neither type was more primitive or more advanced than the other, and both were equally progressive in their evolution and ability to adapt themselves efficiently to changing environmental conditions. The extinction of South American marsupials in competition with placental carnivores does not imply that marsupial mammals were necessarily inferior to placentals or more primitive, but rather that the types of placental mammals which invaded South America from North America had already become competitively superior.

Infraclass II. Eutheria (Placentalia)

Eutherians are often referred to as placental mammals, since their developing young are nourished by means of a placenta attached to the lining of the uterus of the mother. An exchange of gases, nutritional substances, and excretory products between mother and young takes place at the placenta. Marsupial pouches and epipubic bones are never found in eutherians. To this infraclass belong most of the familiar mammals. Although as many as 28 separate orders, including fossil forms, have been recognized, we shall list only 16, which include the more familiar forms living today.

ORDER I. INSECTIVORA To this rather primitive order belong the moles, shrews, and hedgehogs. They are rather small mammals, usually with elongated snouts.

Moles (Fig. 3.24) live underground in burrows. Their eyes are extremely small and hidden in the fur. The heavily clawed, paddle-shaped forelimbs are used in digging; the hind limbs are small and weak.

Shrews include the smallest of all mammals and are frequently confused with mice. They are voracious and pugnacious.

In America the name hedgehog is often used to refer to the porcupine, which belongs

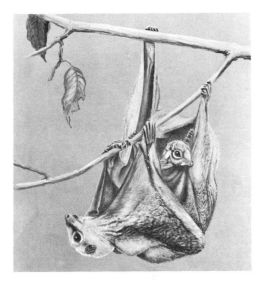

Fig. 3.25 Flying lemur, *Galeopithecus*. (*From a painting of a live animal by W. A. Weber. Courtesy of the Chicago Natural History Museum.*)

to the order Rodentia. The European hedgehog, *Erinaceus,* is a true insectivore, the body of which is covered with short spines intermingled with hair.

ORDER II. DERMOPTERA Only one genus, consisting of two species, is included in the order. *Galeopithecus,* the flying lemur, or colugo, of the Malay region (Fig. 3.25), is representative. It is larger than any of the insectivores and approximately the size of a cat. A well-developed fold of skin, the *patagium,* extends along either side of the body from neck to tail, enclosing the limbs. The feet are webbed. The presence of the patagium enables the animal to soar for some distance.

ORDER III. CHIROPTERA The bats are highly specialized mammals which show unmistakable relations to insectivores. Bats are the only mammals that can actually fly. The wing of the bat (Fig. 6.63) is constructed entirely differently from the wing of a bird.

FIG. 3.24 Common mole, *Scalopus aquaticus.* (*Courtesy of the American Museum of Natural History.*)

The fingers are greatly elongated and connected by a web which offers resistance to the air. The web is attached to the body, hind limbs, and tail, if a tail is present.

Whereas in the prehistoric flying reptiles (pterosaurs) the fourth finger of the hand was greatly elongated (Fig. 6.63), in bats the third, fourth, and fifth fingers are very long and the first two are short. The first finger is free, but the rest are fastened to the web. The sternum is carinate (keeled) like that of the flying birds and serves for the attachment of the strong pectoral muscles used in flight.

ORDER IV. PRIMATES It may seem strange that the order of mammals to which man belongs should be placed among the lowest orders. Although the nervous system, especially the cerebral portion of the brain, is best developed of all in primates, nevertheless in other respects they show many generalized characteristics. The development of the human brain far exceeds that of other primates, but aside from this, man is clearly related, even in minor details, to the rest of the animals making up the order.

In primates the limbs are usually long and the pentadactyl hands and feet are relatively large. In many forms both thumb and big toe are opposable, although in man this applies only to the thumb. The limbs, in general, are adapted for arboreal existence. The eyes are generally directed forward. The order is divided into four suborders.

SUBORDER I. LEMUROIDEA The lemurs (Fig. 3.26) of today are confined to Madagascar, Ethiopia, and parts of Asia. Fossil remains indicate, however, that they were once widely distributed. Lemurs are arboreal animals of moderate size and crepuscular or nocturnal in habit. The head is foxlike, with a pointed muzzle; the tail is long but not prehensile. The second digit on each hind foot bears a sharp claw, whereas the other digits

FIG. 3.26 Sportive lemur. (*Courtesy of the Chicago Natural History Museum.*)

on both forefeet and hind feet are provided with flattened nails.

SUBORDER II. TARSIOIDEA Tarsiers (Fig. 3.27) are small, lemurlike creatures with long ears, large and protruding eyes, and elongated heels. The second and third digits of the hind feet bear claws; those of the forefeet and the remainder on the hind feet have nails. These animals occur in the Philippines and in the islands between India and Australia. They are arboreal and nocturnal.

SUBORDER III. CATARRHINI Old World monkeys, apes, and man, which are included in this order, bear digits with flattened or slightly rounded nails and have a

FIG. 3.27 Tarsier, *Tarsius spectrum*. (*Courtesy of the American Museum of Natural History.*)

narrow septum between the downward-directed nostrils. They are arboreal or terrestrial and diurnal in habit. Within the group there is a tendency to walk upright, but only man has actually attained the upright posture. The increased specialization of the nervous system, with its epitome in man, is the most prominent feature of the group.

The term *ape* may be used to designate any monkey, but it is generally applied to the larger Old World forms. The anthropoid apes are those which most closely resemble man.

The well-developed brain, the generalized condition of parts of the skeleton and other body structures, together with the fact that he is homoiothermous and can adapt himself to almost any climatic condition, all account for the success of *Homo sapiens* (man) in attaining his present leading position in the animal world.

Biologists do not believe that man evolved from the apes but rather that man and the apes sprang from a common ancestral stock, each having diverged in a different direction. Many fossils have been found which are considered as the ancestral form. Discovery of the remains of manlike apes in South Africa within recent years indicates that in this region paleontologists may ultimately find all the links in the human pedigree.

SUBORDER IV. PLATYRRHINII New World monkeys are placed here and exhibit a wide distance between outwardly directed nostrils and a prehensile tail in many forms.

ORDER V. EDENTATA The lack of teeth or the presence of only poorly developed molar teeth characterizes the edentates. Some forms that are toothless as adults have teeth when young, but they are lost as the animals grow older. Among the edentates are the South American anteaters (Fig. 3.28), armadillos, and sloths. Only one species, the nine-banded armadillo, is a native to the United States. It is found in the southwestern part of the country but is spreading eastward. Armadillos are the only living mammals which possess secondary dermal skeletal structures, the body being covered with a bony case of armor. There are large, bony plates on the head, shoulders, and hindquarters. The shoulder and hindquarter plates are connected by a series of bony rings, the number of which is characteristic of each species.

ORDER VI. PHOLIDOTA Only a single genus, *Manis,* the panolin, or scaly anteater (Fig. 3.29), is included in the order. Its

FIG. 3.28 Great South American anteater. (*Courtesy of the Chicago Natural History Museum.*)

native habitat is in eastern Asia and Africa. The animal is covered with large, horny, overlapping scales, among which a few hairs are interspersed. Teeth are lacking.

ORDER VII. RODENTIA The gnawing teeth of rodents are their most outstanding feature. These are two chisel-shaped incisor teeth in the front of both upper and lower jaws. They continue to grow throughout life. Canine teeth are lacking, but

FIG. 3.29 Asiatic pangolin, or scaly anteater. (*Courtesy of the Chicago Natural History Museum.*)

grinding, posteriorly located premolar and molar teeth are present.

Rodents are by far the most numerous of all mammals and are distributed all over the world. Among the more familiar forms are rats, mice, squirrels, beavers, chipmunks, porcupines, and guinea pigs. The largest rodent of all is the capybara, *Hydrochoerus,* a South American form.

ORDER VIII. LAGOMORPHA Rabbits, hares, and pikas are included among the lagomorphs. Like rodents they lack canine teeth, and their incisors grow continually. They differ from true rodents, however, in other respects. They have four incisor teeth, rather than two, in the upper jaw. The second pair lies behind the first and is smaller. Their tails are short and stubby.

ORDER IX. CARNIVORA Carnivores are flesh eaters, for the most part, although some are omnivorous. All have three pairs of small incisor teeth in both upper and lower jaws and large, well-developed canine teeth. Their limbs are typically pentadactyl. There are two suborders, the distinctive features of which have to do with the condition of the toes. In the suborder Fissipedia the toes are separated, whereas in the Pinnipedia they are webbed, forming flippers.

The fissiped carnivores include bears, doglike mammals, cats, raccoons, weasels, skunks, and others. Seals, sea lions, and walruses are placed in the suborder Pinnipedia. They are well adapted for aquatic life but move about on land with difficulty. Most are fish eaters. There has been a recent tendency among taxonomists to place the fissiped and pinniped carnivores in separate orders.

ORDER X. CETACEA Whales, porpoises, and dolphins are members of the order Cetacea. Although porpoises and dolphins are relatively small, in this group are found the largest animals inhabiting the earth. Cetaceans have undergone profound modifications in adapting themselves to marine conditions. Nevertheless, they are true mammals. They are homoiothermous, breathe air by means of lungs, bring forth their young alive, and nurse them with mammary glands. Only traces of a hairy covering remain, shown by a few scattered hairs in the muzzle region of the snout. The webbed pectoral appendages

are flippers. Pelvic appendages are lacking except for rudiments of the pelvic girdle, embedded in the flesh. The tail is flattened horizontally into lobes referred to as the *flukes*. Since whales are homoiothermous and have a thick, insulating layer of fat, or blubber, beneath the skin, they are at home even in Arctic and Antarctic seas.

Whales are separated into two large groups: the toothed whales and the whalebone whales. The toothed whales are primarily fish eaters. They are usually not of exceptional size and include the dolphin, porpoise, and grampus. The sperm whale (Fig. 3.30) and bottle-nosed whale are among the larger species. The whalebone whales (Fig. 3.31) have no teeth and are characterized by the great sheets of baleen, or whalebone, which hang from the roof of the mouth. These plates are used in straining from the water the great numbers of microscopic plants and animals which form the chief food supply of these huge mammals. The giant whales are members of this group. They include the gray, blue, sulfur-bottomed, humpbacked, and right whales.

The so-called "spout" of the whale is not a column of water, nor is it merely a condensa-

FIG. 3.30 Sperm whale. (*Courtesy of the Chicago Natural History Museum.*)

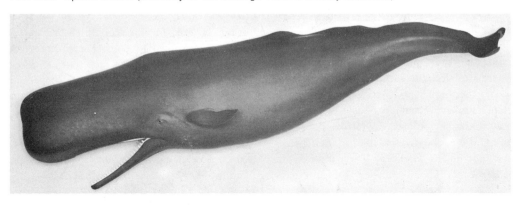

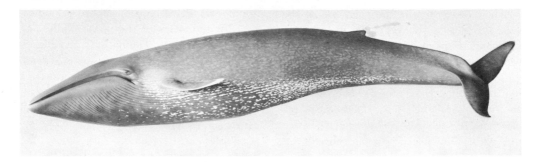

FIG. 3.31 Blue whale. (*Courtesy of the Chicago Natural History Museum.*)

tion of moisture from the warm, exhaled air coming from the lungs. It is composed largely of a mixture of mucus, gas, and emulsified oil that gathers in the lungs. This is expelled in the form of foam through the nostrils located on top of the head.

ORDER XI. TUBULIDENTATA The aardvark, *Orycteropus,* is the only living representative of this order. It is a rather large, African, termite-eating, burrowing mammal, with a thick-set body, large, pointed ears, and a long snout. There are a few poorly developed permanent teeth. Incisors and canines are lacking. The aardvark is not related to other anteaters previously mentioned.

ORDER XII. PROBOSCIDEA Elephants are characterized by their large size, thick skin, and scanty hair coat. The nose and upper lip have been extended to an extreme degree to form a long, prehensile proboscis, called the trunk, with nostrils at its free end. There are two large species of elephant living today; the Indian and African forms. Indian elephants have small ears; tusks in the male only; a single, pointed, fingerlike tip at the end of the proboscis; five toes on the forefeet and four on the hind feet. These animals are rather easily tamed. African elephants have large, fanlike ears; tusks in both sexes but larger in males; two opposable fingerlike tips at the end of the proboscis; four toes on the forefeet and three on the hind feet. They are savage and not easily tamed. The pygmy elephants of the Congo are small animals of the African type. Extinct members of the order include the mastodon and the mammoth. Mastodons were older and more primitive. The proboscideans of today seem to be on their way to extinction. Their ancestors were at one time present in great numbers and in great variety on all the continents of the earth except Australia. Paleontologists have gathered a great deal of information on the evolutionary history of the proboscideans which has proved to be rather complex.

ORDER XIII. HYRACOIDEA Only one genus, *Hyrax (Procavia),* belongs to this order. These small herbivorous creatures, commonly known as conies, inhabit certain regions of Arabia, Syria, and Africa. They somewhat resemble guinea pigs in shape and size. Four digits are present on the forefeet and three on the hind feet. All digits are provided with hooflike nails, with the exception of the second digit on the hind foot, which bears a claw. The upper incisor teeth, like those of rodents, grow continually.

ORDER XIV. SIRENIA The manatees (Fig. 3.32) and dugongs, commonly called sea cows, are the only representatives of the order. They are herbivorous, and whalelike in external form, but are actually more closely related to members of the orders Hyracoidea and Proboscidea. Their bones are heavy and dense, a feature of importance in connection with their bottom-feeding habits. Manatees are about 9 or 10 ft long. They inhabit the rivers along the Atlantic coasts of South America and Africa. Some go as far north as the Everglades of Florida. Dugongs are oriental and Australian forms. An extinct species, Steller's sea cow, reached a length of 24 ft.

ORDER XV. PERISSODACTYLA This order of odd-toed, hoofed mammals includes horses, donkeys, zebras, tapirs, and rhinoceroses. All are herbivorous. They walk on their nails (hoofs) and usually only on the middle finger or toe. The functional axis of

Fig. 3.33 Rhinoceros. (*Courtesy of the Chicago Natural History Museum.*)

the leg passes through the middle toe. The toes of the tapir have not been reduced to the extent of those of the horse family, there being four toes (one of them small) on the forefeet and three on the hind feet. The nose and upper lip have been drawn out into a relatively short proboscis.

There are one-horned and two-horned species (Fig. 3.33) of rhinoceroses. The horns

FIG. 3.32 Florida manatees. (*Courtesy of the Chicago Natural History Museum.*)

are located on the median line of the snout. Here, also, there are four toes on the forefeet (one of them small) and three on the hind feet.

ORDER XVI. ARTIODACTYLA

These are the even-toed hoofed mammals. All members of the order walk on the nails (hoofs) of the third and fourth toes. The other digits are greatly reduced or absent. The functional axis of the leg passes between the third and fourth toes. Members of the orders Perissodactyla and Artiodactyla are often called *ungulates,* because of their unguligrade foot posture in which only the hoof is in contact with the ground.

In this order there is a division into two groups: the cud chewers, or *ruminants,* and the *nonruminants.* The first include cattle, sheep, goats, camels, llamas, antelopes, deer, and giraffes. Among the nonruminants are pigs, hippopotamuses, and peccaries. Cud chewers first swallow their food, which is later regurgitated into the mouth for thorough mastication. Many members of the order bear horns or antlers projecting from the frontal bones of the skull.

SUMMARY OF CLASSIFICATION

PHYLUM CHORDATA
    Subphylum I. Hemichordata. *Dolichoglossus*
    Subphylum II. Urochordata. *Molgula*
    Subphylum III. Cephalochordata. Amphioxus
    Subphlyum IV. Vertebrata (Craniata)
      Superclass I. Pisces
        Class I. Agnatha
            Order I. Cyclostomata
                Suborder I. Petromyzontia. Lampreys
                Suborder II. Myxinoidea. Hagfishes
        Class II. Placodermi. Fossil placoderms
        Class III. Chondrichthyes
          Subclass I. Elasmobranchii
            Order I. Selachii. Sharks; dogfishes
            Order II. Batoidea. Skates; rays
          Subclass II. Holocephali. Chimaeras
        Class IV. Osteichthyes
          Subclass I. Actinopterygii. Ray-finned fishes
            Superorder I. Chondrostei. *Polypterus; Calamoichthys; Polyodon;* sturgeons
            Superorder II. Holostei. *Amia; Lepisosteus*
            Superorder III. Teleostei. Teleost fishes
          Subclass II. Sarcopterygii. Lobe-finned fishes
            Order I. Crossopterygii
                Suborder I. Rhipidistia. *Eusthenopteron*

Suborder II. Coelacanthini. *Latimeria*

Order II. Dipnoi. Lungfishes: *Epiceratodus; Protopterus; Lepidosiren*

Superclass II. Tetrapoda

Class I. Amphibia

Order I. Anura. Frogs and toads

Order II. Urodela (Caudata). Salamanders; newts

Order III. Apoda (Gymnophiona). Caecilians

Class II. Reptilia

Order I. Chelonia. Turtles; tortoises

Order II. Rhynchocephalia. *Sphenodon*

Order III. Squamata

Suborder I. Sauria (Lacertilia). Lizards

Suborder II. Serpentes (Ophidia). Snakes

Order IV. Crocodilia (Loricata). Alligators; crocodiles

Class III. Aves

Subclass I. Archaeornithes. Fossil *Archaeopteryx*

Subclass II. Neornithes

Superorder I. Odontognathae. Fossils: *Hesperornis; Ichthyornis*

Superorder II. Paleognathae (Ratitae). Ostrich; kiwi

Superorder III. Neognathae. Most familiar birds

Class IV. Mammalia

Subclass I. Prototheria

Order I. Monotremata. Platypus; echidnas

Subclass II. Theria

Infraclass I. Metatheria

Order I. Marsupialia. Opossum; kangaroo

Infraclass II. Eutheria (Placentalia)

Order I. Insectivora. Moles; shrews

Order II. Dermoptera. Flying lemur

Order III. Chiroptera. Bats

Order IV. Primates

Suborder I. Lemuroidea. Lemurs

Suborder II. Tarsioidea. Tarsiers

Suborder III. Catarrhini. Old World monkeys, ape, and man

Suborder IV. Platyrrhini. New World monkeys

Order V. Edentata. Sloths; South American anteaters

Order VI. Pholidota. Scaly anteater

Order VII. Rodentia. Rats; mice; squirrels

Order VIII. Lagomorpha. Rabbits; hares; pikas

Order IX. Carnivora. Dogs; cats; seals; walrus

Order X. Cetacea. Whales; porpoise

Order XI. Tubulidentata. Aardvark

Order XII. Proboscidea. Elephants

Order XIII. Hyracoidea. Cony
Order XIV. Sirenia. Manatee; dugong
Order XV. Perissodactyla. Horse; zebra
Order XVI. Artiodactyla. Cattle; deer; sheep

BIBLIOGRAPHY

Alexander, R. McN.: "Functional Design in Fishes," Hutchinson University Press, London, 1967.

Bellaris, A. A.: "Biology of Reptiles," New York University Press, New York, 1970.

Blair, W. F.: "Vertebrates of the United States," McGraw-Hill, New York, 1957.

Brodal, A., and R. Fange: "The Biology of Myxine," Universitetsforlaget, Oslo, 1963.

Lagler, K. F., J. E. Bardach, and R. R. Miller: "Ichthyology," Wiley, New York, 1962.

Noble, G. K.: "The Biology of the Amphibia," Dover, New York, 1954.

Ørvig, T. (ed.): "Current Problems of Lower Vertebrate Phylogeny," Wiley-Interscience, New York, 1968.

Romer, A. S.: "The Vertebrate Story," Saunders, Philadelphia, 1959.

Simpson, G. G.: "Principles of Animal Taxonomy," Columbia University Press, New York, 1961.

Stahl, B.: "Vertebrate History: Problems in Evolution," McGraw-Hill, New York, 1974.

Welt, J. C.: "The Life of Birds," Saunders, Philadelphia, 1962.

Young, J. Z.: "The Life of Mammals," Oxford Press, New York, 1957.

4 EARLY EMBRYONIC DEVELOPMENT AND HISTOGENESIS

In both plants and animals there are two general types of reproduction, asexual and sexual. The asexual, or agamic, method of reproduction in animals is confined to members of some of the lower phyla. Budding and fission are examples of this method.

Sexual reproduction is the rule in the phylum Chordata. Parthenogenesis, the development of an egg without fertilization, is considered to be a form of sexual reproduction since a sexual element, the egg, is involved.

EARLY EMBRYOLOGY

In the usual type of sexual reproduction the union of egg and spermatozoon results in a fertilized egg, or *zygote*. Fertilization of the egg brings about changes which, under proper conditions, result in the development of an embryo. The embryo is built up by a series of cell divisions in which the resulting cellular units do not dissociate but remain attached to each other and become differentiated or specialized later to form the various tissues and organs of the adult. Although some stages in embryonic development differ in the several chordate groups, nevertheless the significant phases are basically similar in all. Such variations as occur are primarily related to the differences in size of the egg cells. The size of the egg depends, for the most part, upon the quantity of yolk present.

YOLK CONTENT OF EGGS Eggs may be classified according to the amount of yolk which they contain. *Alecithal* eggs are those without yolk or with a very small amount. Many mammals and numerous invertebrates have eggs of the alecithal type. *Meiolecithal* eggs, as those of amphioxus, contain a small amount of yolk. *Mesolecithal* ova, typical of amphibians (Fig. 11.12), have a medium amount of yolk, quite in contrast to the large, yolk-laden, *polylecithal* eggs of elasmobranch fishes, reptiles, and birds.

DISTRIBUTION OF YOLK Eggs are also classified according to the manner in which the yolk is distributed within the ovum. *Isolecithal,* or *homolecithal,* eggs are those in

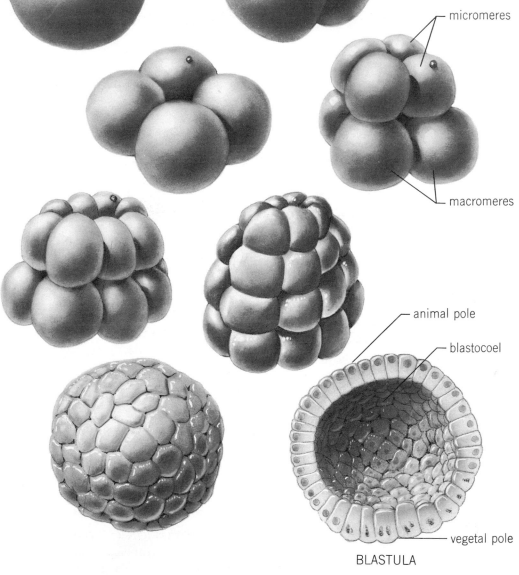

polar body

micromeres

macromeres

animal pole

blastocoel

vegetal pole

BLASTULA

FIG. 4.1 Cleavage stages of amphioxus egg. Lower right, cross section of a blastula. (*Drawn by G. Schwenk.*)

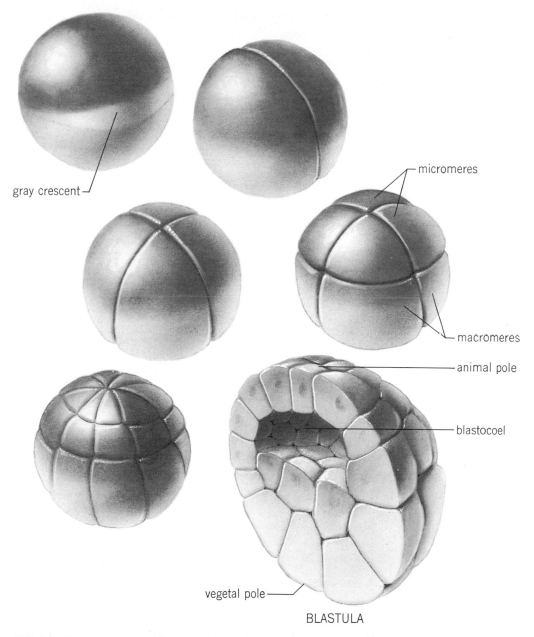

gray crescent

micromeres

macromeres

animal pole

blastocoel

vegetal pole

BLASTULA

FIG. 4.2 Cleavage stages of frog egg. Lower right, section of an early blastula. (*Drawn by G. Schwenk.*)

which the yolk is evenly distributed throughout the cytoplasm. Microscopic mammalian eggs are of this type. *Telolecithal* eggs, on the other hand, are those in which the yolk is more concentrated on one side than another. This arrangement occurs in the eggs of fishes, amphibians, reptiles, and birds.

POLARITY   Even in the one-cell stage, and during subsequent development, it is possible, in most forms, to recognize a definite axis in the embryo. The extremes of the axis consist of unlike *poles,* called the *animal* and *vegetal poles,* respectively. The existence of such an axis is termed *polarity.* It is first expressed by the arrangement of egg substances (yolk, nucleus, etc.) and later by differential rates of cleavage. In most eggs there is a tendency for the nucleus to be located nearer the animal pole and for the yolk to concentrate at the vegetal pole.

CLEAVAGE   The term *cleavage* is applied to the early mitotic divisions of the zygote, resulting in a stage known as a *blastula.* During cleavage the original cell first divides into two smaller cells. These in turn divide into smaller and smaller units, with no perceptible increase in the total mass. When growth and differentiation of parts begins, the period of cleavage is over. The rate of cleavage depends to a great extent on the amount of yolk present in the egg. Cleavage stages of amphioxus and frog eggs are shown in Figs. 4.1 and 4.2.

TYPES OF CLEAVAGE   The type of cleavage exhibited by an ovum is also governed largely by the amount of yolk present. Alecithal, meiolecithal, and mesolecithal ova show *complete,* or *holoblastic, cleavage.* The first cleavage plane completely bisects the zygote, passing through the animal and vegetal poles. The second cleavage plane also passes through these poles but at right angles to the

first. There are two variations of holoblastic cleavage, *equal* and *unequal.* These terms refer to the result of the third cleavage. The third cleavage plane is latitudinal in relation to the poles. After the third cleavage the ovum is in the eight-cell stage. If the eight cells are of equal size, the cleavage is said to be of the *holoblastic equal* type. In meiolecithal and mesolecithal ova the third cleavage plane is nearer the animal pole than the vegetal pole and there is a difference in the size of the resulting blastomeres. The four cells near the animal pole are smaller than the four yolk-laden cells near the vegetal pole. They are referred to as *micromeres* and *macromeres,* respectively. This type of cleavage is termed *holoblastic unequal cleavage.*

In polylecithal ova complete cleavage does not occur. Instead, the nucleus and the clear cytoplasm undergo a series of incomplete divisions which result in the formation of a small disclike area of cleaving cells at the animal pole. Cleavage of this type is referred to as *incomplete,* or *meroblastic, discoidal cleavage.* It is typical of the eggs of fishes, reptiles, and birds (Fig. 4.3).

BLASTULA   As cleavage progresses and the original ovum becomes subdivided, the cells usually become arranged so as to form a sphere with a single layer of cells surrounding a central cavity (Figs. 4.1 and 4.2), the *blastocoel.* At this stage the developing embryo is referred to as a *blastula.* The blastula of polylecithal ova (Fig. 4.3) differs in that it is not spherical but consists of a disc of cells lying on the yolk. Polarity is still evident in the blastula stage.

GASTRULA: EPIBLAST AND HYPOBLAST   The blastula is succeeded by the *gastrula;* the process that brings about the change is called *gastrulation.* In amphioxus, as a result of gastrulation, a double-walled,

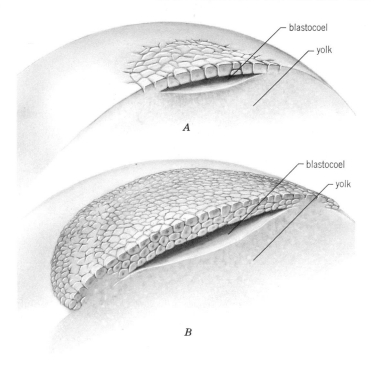

— blastocoel

— yolk

A

— blastocoel

— yolk

B

FIG. 4.3 Meroblastic discoidal cleavage of the polylecithal bird ovum: *A,* early blastula stage; *B,* later blastula stage. (*Drawn by G. Schwenk.*)

cuplike structure is formed (Fig. 4.4). The larger, yolk-laden cells from the original vegetal pole form the *hypoblast,* or inner layer of the cup, and the smaller, animal-pole cells make up the *epiblast,* or outer wall. The terms *endoderm* and *ectoderm* are frequently used for hypoblast and epiblast, respectively. Since, however, in later development the hypoblast of amphioxus gives rise to endoderm, mesoderm, and notochord, it is preferable to apply the terms ectoderm and endoderm at a later stage when all three germ layers as well as the notochord have become established. The original blastocoel is obliterated during gastrulation, and a new cavity, the *gastrocoel,* is formed. This is lined entirely with hypoblast and opens to the outside through the *blastopore.*

The endoderm derived from the hypoblast is destined to form the lining of the digestive tract and its derivatives. In mesolecithal ova the larger amount of yolk in the vegetal-pole cells is responsible for the fact that in the gastrula stage a plug of yolk-filled cells usually protrudes through the blastopore (Figs. 4.5 and 4.7). Moreover, in such gastrulae the cells of the hypoblast lining the gastrocoel are not of equal size. In mesolecithal amphibian gastrulae at this stage of development the hypoblast lining the gastrocoel are not of equal size. In mesolecithal amphibian gastrulae at this stage of development the hypoblast forming the roof of the gastrocoel is often referred to as *chordamesoderm,* since it is destined to give rise to the notochord and some prechordal

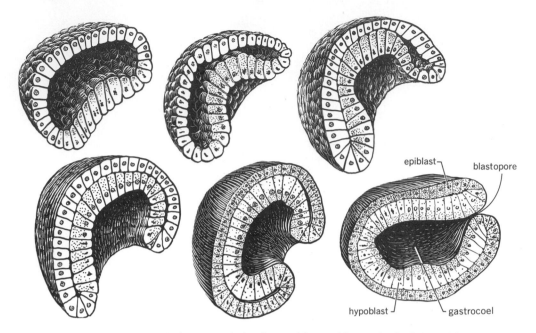

FIG. 4.4 Successive stages during gastrulation in amphioxus. (*Drawn by G. Schwenk.*)

mesoderm in the midline, with somitic and lateral mesoderm on either side. The latter proliferates and spreads laterally and ventrally. The cells of the hypoblast making up the floor and side walls of the gastrocoel may now be considered to be endodermal. As the mesodermal cells along the marginal borders of the endoderm proliferate and grow upward, ultimately to meet dorsally. The gastrocoel thus becomes completely lined with endoderm (Fig. 4.7).

The way that blastula and gastrula form in polylecithal ova differs considerably from the above description, the differences being related to the enormous amount of yolk present. It is not necessary for our purposes here to discuss the details of these processes about which there is still some contention. Suffice it to say that the gastrula stage consists of a two-layered disc lying on the yolk with which it is in contact only at the periphery.

The upper layer comprises the epiblast; the lower, the hypoblast. The small space between epiblast and hypoblast represents the blastocoel; that between hypoblast and yolk is the future gastrocoel. Soon after the egg is laid and incubation has begun, the area of the blastoderm expands. Concomitantly it may be observed that a thickened area appears in one quadrant. It is destined to become the caudal end of the embryo. The thickened area, where epiblast and hypoblast are in contact, soon begins to show an anterior-posterior elongation, extending along the midline of the developing structure to a point somewhat anterior to the center. This longitudinal structure is called the *primitive streak*. It establishes the anterior-posterior axis of the embryo. The entire blastoderm has now assumed an elliptical shape. Epiblast and hypoblast are in contact at the primitive streak and at the peripheral margins of the blastoderm.

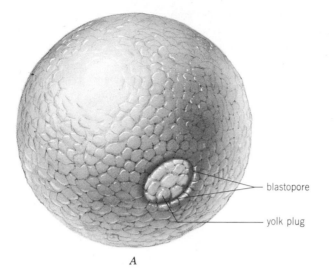

A

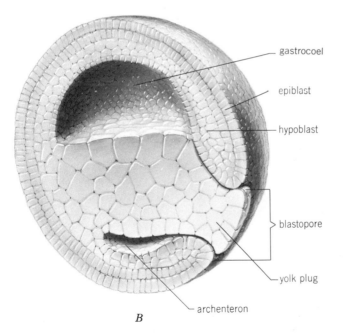

B

FIG. 4.5 Gastrula stage in development of frog: *A*, external view showing yolk plug fill-ing the blastopore; *B,* median sagittal section of same, showing internal arrangement of cells making up epiblast and hypoblast. (*Drawn by G. Schwenk.*)

CLOSURE OF THE BLASTOPORE In the usual type of gastrula an elongation in the direction of the longitudinal axis passing through the blastopore next takes place. It is due chiefly to rapid cell divisions, particularly on the dorsal side or area forming the margins of the blastopore where epiblast and hypoblast are in continuity. As development progresses the blastopore becomes smaller and smaller. In most species the edges finally coalesce. In the embryos of certain invertebrates the blastopore remains open to become the anus, but in chordates it closes and an anal aperture is established at a later stage.

MESODERM After gastrulation is complete or is nearing completion, a third germ layer, designated as mesoderm, appears and comes to lie between ectoderm and endoderm. In amphioxus the mesoderm first appears as a series of pouches which pinch off from the dorsolateral angles of the hypoblast lining the gastrocoel (Fig. 4.6). In higher forms, except in amphibians as described above, it usually originates by a proliferation of cells in the region of the blastopore. The mesodermal cells push between epiblast and hypoblast and fill the crevice which represents the remnant of the original blastocoel. Usually the mesoderm is at first a solid mass, but it soon splits into two layers with a space, the *coelom,* lying between. In amphioxus the cavities within the original mesodermal pouches become the coelom. With the establishment of the mesoderm and notochord (page 59), the terms ectoderm and endoderm are appropriately used. The outer, or *somatic, layer* of mesoderm is closely applied

FIG. 4.6 Method of mesoderm formation in amphioxus.

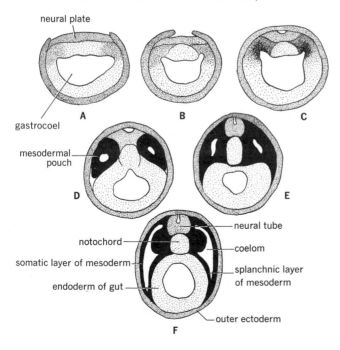

to the ectoderm, the two together being destined to form the definitive body wall. The inner, or *splanchnic, layer* of mesoderm is closely applied to the endoderm, the two together giving rise to the digestive tract and its derivatives. The coelom, which is thus lined entirely with mesoderm, represents the future peritoneal, pericardial, and pleural cavities of higher forms, as the case may be.

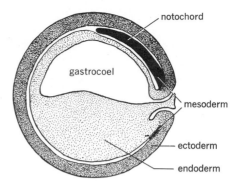

FIG. 4.7 Diagram of median sagittal section of frog gastrula in late yolk-plug or neural-plate stage, showing germ-layer relationships.

DIFFERENTIATION　The cells making up the germ layers of an embryo are said to be undifferentiated; i.e., they do not possess distinctive or individual characteristics. Further development of an embryo entails, among other things, a differentiation, or specialization, of various groups of cells to form the several types of tissues and organs which make up the body of the individual.

NOTOCHORD　At about the time that the mesoderm is differentiating into somatic and splanchnic layers, the notochord begins to form. In amphibians, as previously noted, this event takes place somewhat earlier. In amphioxus it arises as a middorsal thickening of the wall of the gastrocoel (Fig. 4.6). In many chordates, however, it originates from cellular proliferation in the region of the dorsal lip of the blastopore. It forms an elongated, rodlike structure which comes to lie in the middorsal region of the body, between ectoderm and endoderm, and separates the dorsal mesoderm of the two sides (Fig. 4.7). In most chordates the vertebral column later forms about the notochord, which is gradually reduced or replaced.

NEURAL TUBE　In the details of its development the nervous system of amphioxus differs somewhat from that of higher chordates and need not be considered here. In vertebrates, as the above-mentioned

changes are going on, the flattened layer of ectoderm along the middorsal side of the gastrula becomes thickened. It then becomes known as the *neural plate.* Proliferative changes in the cells of this region result in the formation of a longitudinal depression, the *neural groove,* along the middorsal line, flanked on either side by an elevated *neural tube* (Fig. 7.4). During its formation the neural tube gradually sinks down to a deeper position in the embryo. The neural tube is the forerunner of the brain and spinal cord. The outer ectodermal fold on each side meets its partner along the midline above the neural tube, forming a continuous layer from which the neural tube soon becomes completely separated. Thus the neural tube lies under the outer surface of the body.

Small dorsolateral masses of ectodermal cells grow out on each side in the crevices formed between neural tube and outer ectoderm (Figs. 4.8 and 7.4). They are termed *neural crests* and are of significance in the development of numerous nerve ganglia and in the formation of certain pigment cells, known as *chromatophores,* found in many lower

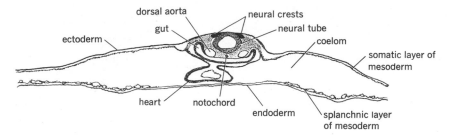

FIG. 4.8 Cross section of 33-hour chick embryo showing neural tube and neural crests in relation to other embryonic structures.

vertebrates. Neural crests, at least in amphibians and birds, also appear to contribute to the mesenchyme from which a number of cartilages in the branchial region are derived.

FURTHER DIFFERENTIATION OF MESODERM Mesoderm exists in the body either in the form of sheets of cells lining cavities of one kind or another, or as masses of loose, undifferentiated branching cells called *mesenchyme.* The sheets of cells lining the body cavities are known as *mesothelia;* those lining the cavities of blood vessels, lymphatic vessels, and the heart are *endothelia.* Both epiblast and hypoblast may originally contribute to mesenchyme formation. Mesenchymal cells may migrate from their place of origin and fill in spaces between other structures. The intercellular substance between mesenchymal cells permits diffusion of nutrients, gases, wastes, and water over a considerable distance at a time when blood vessels are developing and tissues are not, and cannot, be supplied directly with blood.

During the time the above changes have been taking place in vertebrate embryos the mesoderm not only has pushed in between ectoderm and endoderm but has grown dorsally on either side of the notochord and neural tube. Three different regions or levels of mesoderm may then be recognized: (1) an upper, or dorsal, *epimere;* (2) an intermediate

mesomere; and (3) a lower *hypomere* (Fig. 4.9). The epimeric mesoderm becomes marked off into a metameric, longitudinal series of blocklike masses, the *mesodermal somites* which form in succession, beginning at the anterior end of embryo (Fig. 4.10). The somites later become separated from the remainder of the mesoderm. Although the mesoderm of the mesomere shows evidence of metamerism in some lower forms, this is not evident in embryos of

FIG. 4.9 Cross section of hypothetical vertebrate embryo, illustrating the three different regions, or levels, of mesoderm.

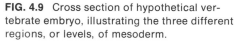
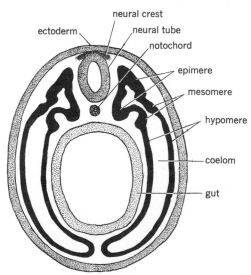

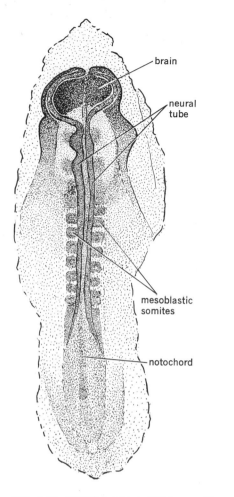

brain

neural tube

mesoblastic somites

notochord

FIG. 4.10 Rabbit embryo, 190 hours after copulation, showing series of paired mesoblastic somites in process of forming.

higher vertebrates, except, perhaps, toward the anterior end of the body. The hypomeric mesoderm does not become segmented.

DERIVATIVES OF THE PRIMARY GERM LAYERS

All structures of the body trace their origin to one or more of the three primary germ layers. Further development is the result of operation of a number of processes which are effective in shaping the body and in forming the various tissues and organs. Multiplication of cells goes on at a rapid rate, and the newly formed cells increase in size. Some clumps of cells form localized thickenings, or swellings; others undergo a thinning, or spreading out. Certain masses of cells may evaginate (push outward) or invaginate (push inward) from a surface. Some groups of cells migrate to different parts of the body and become rearranged. Diverse masses of cells may form temporarily and then disappear. Various cell aggregates differentiate to form tissues of different types.

Growth of the embryo and differentiation of organs from germ layers begin before the germ layers are completely formed. The entire process of development is gradual, and although the various phases occur in a logical, step-by-step order, they are not in themselves distinct but one merges imperceptibly into the next. Experimental work has shown that the determination of cells, germ layers, tissues, and organs is not invariably fixed, but is directed, in the main, by relationships to other cells and by the influence of organizers released by surrounding cells. The different germ layers and the structures arising from them are, nevertheless, considered to be homologous in all vertebrates. The germ-layer concept is of importance chiefly because it furnishes a convenient method of classifying organs according to their embryonic derivation and of tracing their homologies.

ECTODERM The entire outer covering of the body of the developing vertebrate embryo consists of a layer of ectodermal cells. During development when the somites, which have formed from the epimeric region of mesoderm, are undergoing differentiation, the outer portion of each somite becomes mesenchymatous. These mesenchymal cells proliferate and migrate in such a manner as to underlie the outer ectodermal covering of the body. The two layers together will form the future skin, or integument. The ectoderm gives rise to the *epidermis* of the skin. Many structures associated with the skin are derived from the epidermis and are therefore of ectodermal origin. They include various kinds of skin glands, epidermal scales, hair, feathers, nails, claws, hoofs, and horns. Certain specialized smooth-muscle fibers of ectodermal origin, referred to as the *myoepithelial cells,* are associated with certain glands, e.g., sweat, mammary, salivary, and lacrimal. They aid in emptying the glands of their secretory products. The origin of these cells from ectoderm is unlike that of most smooth muscle, which, in general, is of mesodermal origin.

The gastrocoel, which is destined to form the digestive tract, is lined with endoderm. At first it opens to the outside through the blastopore, which soon closes, leaving the gastrocoel as a blind tube without mouth or anus. At the anterior end of the embryo an ectodermal invagination occurs, pushing inward to meet the endoderm of the gastrocoel. The ectodermal invagination is referred to as the *stomodaeum,* or primitive mouth. A two-layered membrane, the *oral plate,* at first separates the two cavities. This persists for a short time and then ruptures, so that the stomodaeum becomes continuous with the remainder of the digestive tract. A similar ectodermal invagination forms at the posterior end of the embryo. This is the *proctodaeum.* An *anal plate* at first separates gastrocoel and proctodaeum. With the perforation of the anal plate a complete tube is established which terminates anteriorly with the mouth and posteriorly with the anus. Derivatives of the ectoderm lining the stomodaeum include the lining of the lips and mouth, enamel of the teeth, glands of the oral cavity, covering of the tongue, and the anterior and

intermediate lobes of the pituitary gland. The proctodaeal ectoderm gives rise to the lining of the anal canal or, in most forms, to a portion of the cloacal lining. It is very difficult to ascertain the exact point of transition between endoderm and proctodaeal ectoderm. For this reason there is some question as to the precise germ layer from which some cloacal derivatives arise. Certain anal and cloacal glands are believed to be derived from the ectoderm of the proctodaeum.

The origin of the neural tube and the neural crests has already been reviewed. The neural tube is the primordium of the brain and spinal cord, which make up the *central nervous system.* Early evidence of metamerism in the neural tube is indicated by the temporary appearance of segments known as *neuromeres.* The anterior end of the neural tube enlarges rapidly to form the brain, the remainder becoming the spinal cord. The brain has three primary divisions, referred to, respectively, as *forebrain (prosencephalon), midbrain (mesencephalon),* and *hindbrain (rhombencephalon).* From the posterior part of the forebrain a pair of lateral evaginations, the *optic vesicles,* grows out to come in contact with the superficial ectoderm of the head. The optic vesicles soon become flattened and then indented in such a manner that an incompletely double-walled *optic cup* is formed (Fig. 15.2) This gives rise to the *retina* and to part of the *iris* of the eye. The lens of the eye develops from the outer, or superficial, layer of ectoderm opposite the optic cup. With the exception of the conjunctiva and a portion of the cornea, both of which are derived from superficial ectoderm, and the muscles of the iris which come from the outer layer of the optic cup, the other structures of the eye are of mesodermal origin.

The sensory parts of the remaining sense organs also come from ectoderm. The organs for the sense of smell first make their appearance as a pair of thickenings, the *olfactory placodes,* situated on either side of the head. These thickenings invaginate to become *olfactory pits.* In those vertebrates which breathe by means of lungs, and in a few others, the olfactory pits establish connection with the stomodaeum. A membranous *olfactory plate* at first separates the olfactory pit from the stomodaeum, but this soon ruptures, and the two cavities become continuous. Some of the cells lining the olfactory pit become sensory and send out fibers which grow back to the forebrain, with which they make connection. These fibers form the *olfactory nerve.*

On each side of the head opposite the hindbrain, a thickened mass of superficial ectodermal cells, the *auditory placode,* is destined to form the inner ear. An invagination, the *auditory pit,* appears, which soon pinches off from the outer ectodermal layer. It is then called the *auditory vesicle* (Fig. 15.3). The auditory vesicle undergoes rather complex changes, ultimately forming the semicircular ducts and other portions of the inner ear, which in their aggregate make up the *membranous labyrinth.* Groups of cells in various portions of the membranous labyrinth develop sensory functions. In lower aquatic vertebrates similar placodes on the head and along the sides of the body give rise to the sensory elements of the *lateral-line system.*

The organs for the sense of taste arise, for the most part from stomodaeal ectoderm, but some are of endodermal origin. They consist of clusters of taste cells and supporting nonsensory cells (Fig. 15.21). Fibers of the seventh and ninth cranial nerves supply the gustatory receptors.

The various *cranial* and *spinal nerves* which form connections with the brain and spinal cord, as the case may be, make up the *peripheral nervous system.* The fibers of the peripheral nerves, depending upon the nerve in question, arise from neural-crest cells, cells from the ventrolateral parts of the neural tube, or from thickened placodes of superficial ectoderm. The *autonomic portions* of the peripheral nervous system, regulating those activities of the body under involuntary control, are derived, for the most part, from neural-crest cells, but in some forms the ventrolateral portions of the neural tube contribute certain components. The peripheral nervous system is entirely ectodermal in origin, since the neural crests, neural tube, and placodes, from which it arises, are all ectodermal structures.

Some neural-crest cells migrate to a position near the kidneys, where they become differentiated as suprarenal tissue or, in mammals, as the medulla, or central portion, of the adrenal glands. These tissues are endocrine in nature and secrete substances of great physiological importance into the bloodstream.

Associated with tissues of the central nervous system and giving them support and protection are some other components known as *neuroglia.* Several types of neuroglia cells, all of ectodermal origin, are recognized. They are derived originally from the neural tube.

Two or three other structures of ectodermal origin should also be mentioned here. A ventral evagination of the floor of the brain in this region is the *infundibulum.* It grows down to meet an evagination from the roof of the stomodaeum called *Rathke's pocket.* A part of the infundibulum together with Rathke's pocket give rise to the pituitary gland. One or two dorsal evaginations grow out from the posterior part of the forebrain. In some forms the more anterior of the two, the *parietal body,* becomes a median, eyelike organ. The posterior evagination differentiates into a structure called the *pineal body.* The former is lacking in most higher vertebrates.

ENDODERM The gastrocoel, or primitive digestive tube, is lined entirely with endoderm. Its communication with the stomodaeum anteriorly and with the proctodaeum posteriorly has already been discussed. All endodermal structures of the body are derived from the gastrocoel. This structure elongates as the embryo grows. With progressive development it gradually differentiates into various parts, which, beginning from the anterior end, include the pharynx, esophagus, stomach, intestine, and the greater part of the cloaca when such a structure is present. The inner portion of the splanchnic, or visceral, layer of mesoderm surrounding the archenteron becomes differentiated into mesenchyme. From this come the connective tissues and smooth-muscle coats which, together with the endoderm, make up the wall of the digestive tract and of the various structures derived from it.

Two important digestive glands, liver and pancreas, arise as evaginations, or diverticula, of the digestive tract. The endodermal cells push out into masses of mesenchyme, where they branch profusely. These organs are thus actually made up of tissues coming from two germ layers, the epithelial cells alone being of endodermal origin. The ducts of the liver and pancreas usually open into the digestive tract at the points where the original evaginations occurred.

The pharyngeal region is of particular interest, since many important organs are derived from it. Several pairs (usually four or five in higher vertebrates) of lateral pharyngeal pouches push out from the walls of the pharynx through the mesoderm until they come in contact with the outer, superficial, ectodermal covering of the body with which they fuse. In cyclostomes, fishes, and amphibians the pouches break through to the outside and thus form a series of gill slits. In cyclostomes and fishes, but not in amphibians, gill lamellae, richly supplied with blood vessels, arise from the walls of the pharyngeal pouches. Larval amphibians develop external gills of a different type. Although in higher vertebrates (amniotes) the pouches form in the characteristic manner, they rarely break through to the outside. Their existence is temporary, except for the first pouch, which persists in modified form as the middle ear and Eustachian tube. From the others are budded off groups of cells which go to form portions of such structures as the palatine tonsils, thymus, and parathyroid glands and the small, irregular ultimobranchial bodies frequently found in the neck region. The appearance of pharyngeal pouches in higher forms of vertebrates is indicative of their ancestry from gill-breathing, aquatic predecessors.

The thyroid gland originates as a small, midventral evagination of the pharynx. In some groups certain pharyngeal pouches contribute to the formation of the thyroid gland.

Another midventral outgrowth of the pharynx in air-breathing vertebrates develops into larynx, trachea, and lungs. It soon divides into two branches, each of which may divide many times to form the lobes of the lungs.

The yolk sac is an additional endodermal structure which is attached to the midventral part of the gastrocoel somewhat near its middle portion. In embryos in which abundant yolk is present it grows out to surround the yolk. The yolk sac is a typically embryonic structure. It becomes smaller and smaller as the yolk which it contains is utilized in nourishing the developing embryo. Eventually it disappears, except in unusual and abnormal cases.

Still another endodermal structure of great significance in the development of amniotes is an embryonic, membranous sac, the <u>allantois,</u> which consists of a more or less extensive outpocketing of the posterior end of the gastrocoel. The allantois is lost at birth. A portion of the allantois may be utilized in the development of the <u>urinary bladder</u>, although the greater part of this organ is derived from cloacal endoderm, as is the urethra, which carries urine from the bladder to the outside.

MESODERM The origin and differentiation of mesoderm into <u>epimere, mesomere,</u> and <u>hypomere</u> has already been discussed. The segmentally arranged somites, derived from the epimeric mesoderm, become mesenchymatous in their dorsolateral and midventral portions. The cells of the dorsolateral mesenchyme (*dermatome*) proliferate and migrate so as <u>to lie under the layer of ectoderm</u> <u>covering the body</u>. They give rise to the dermis of the skin. The mesenchyme from the midventral portion of the somite (*sclerotome*) grows in and <u>fills the spaces about the no-</u> <u>tochord and neural tube</u>. In most forms it is destined to form the <u>various elements</u> of the <u>vertebral column</u>. The remaining portion of the somite is called the *myotome,* or muscle segment. Adjacent myotomes are separated from each other by partitions of <u>connective</u> <u>tissue,</u> the *myocommata.* During development each myotome grows down in the body wall between the superficial ectoderm and the somatic layer of mesoderm to meet its partner from the other side at the midventral line of the embryo (Fig. 7.5). With some exceptions the myotomes give rise to the greater part of the <u>voluntary musculature.</u> The precise origin of the voluntary muscles of the limbs in most vertebrates has been debated. Whether or not they are derived from myotomes, they are, at any rate, of mesodermal origin.

The mesomere, which, except in lower forms, is for the most part unsegmented, is concerned with the development of the urogenital organs and their ducts. The terminal parts of these ducts may, however, be lined with epithelia of ectodermal or, in some cases, of endodermal origin.

The mesothelial splanchnic layer of the hypomere surrounding the gastrocoel, or primitive gut, becomes mesenchymatous on the side adjacent to the endoderm. From this mesenchyme arise the involuntary muscles and connective tissues of the gut, and other structures which ultimately come to <u>surround</u> the <u>endodermal lining of the digestive tract</u> and its derivatives. The heart itself is derived from splanchnic mesoderm. The remainder of the splanchnic layer, together with the somatic layer, forms the mesothelium lining the coelom and contributes to the <u>pericardium,</u> <u>pleura, or peritoneum,</u> as the case may be. <u>Mesenteries and omenta</u> are also derived from the splanchnic mesoderm.

The parts of the skeleton other than the vertebral column, whether made up of <u>carti-</u> <u>lage, bone, or other connective tissues,</u> are all derived from mesenchyme in different parts of the body and of various origins. Mesenchyme also gives rise to <u>blood and lymphatic</u> <u>vessels, blood corpuscles, lymph glands, and</u> other <u>blood-forming tissues</u>. The heart, however, comes from the splanchnic mesoderm of the hypomere.

Other structures derived from mesenchyme include <u>various parts</u> of the eye, dentine of the teeth, and in mammals, the cortices of the adrenal glands.

Numerous mesenchymal cells remain undifferentiated even in adult life. They form a reserve from which connective tissues of various kinds may be formed and which may be called upon for repair of injured structures.

SUMMARY

The sexual method of reproduction is the rule in members of the phylum Chordata. A fertilized egg or zygote undergoes a series of cell divisions resulting in the development of an embryo. The manner of development is influenced greatly by the amount and distribution of yolk within the egg cell. An embryo during early development passes through blastula and gastrula stages, during which two layers, epiblast and hypoblast, are formed. A third layer soon appears which comes to lie between the other two. Three germ layers are thus established: ectoderm, mesoderm, and endoderm. All structures in the body trace their origin to one or more of these three primary germ layers. The following outline summarizes the derivation of the various structures of the body from the three primary layers.

I. **ECTODERM**

    A. Skin. Epidermis, skin glands, hair, feathers, nails, claws, hoofs, horns, epidermal scales, covering of external gills

    B. Lining of mouth. Enamel of teeth, glands of the mouth, covering of tongue and lips, anterior and intermediate lobes of the pituitary gland

    C. Nervous system. Brain and spinal cord, cranial and spinal nerves, autonomic portion of peripheral nervous system, sensory parts of all sense organs, medulla of adrenal gland, infundibulum, and posterior lobe of the pituitary gland

    D. Miscellaneous. Lens of eye, intrinsic eye muscles, neuroglia, pineal and parapineal bodies, lining of anal canal and derivatives, lining of a portion of cloaca, myoepithelial cells of certain glands (sweat, mammary, salivary, lacrimal)

II. **ENDODERM**

    A. Alimentary canal. Pharynx, esophagus, stomach, intestine, liver, pancreas, lining of most of cloaca

    B. Pharyngeal derivatives. Larynx, trachea, lungs, gills of the internal type, middle ear, Eustachian tube, tonsils, thyroid, parathyroids, thymus, ultimobranchial bodies

    C. Miscellaneous. Allantois, urinary bladder, urethra, yolk sac

III. **MESODERM**

    A. Muscles. Smooth, striated, cardiac

    B. Skeleton. Cartilage, bone, other connective tissues

    C. Excretory organs. Kidneys and their ducts

    D. Reproductive organs. Gonads, ducts, accessory structures

    E. Circulatory system. Heart, blood vessels, blood, spleen, lymphatics, blood-forming tissues

F. Miscellaneous. Dentine of teeth, dermis of skin, cortex of adrenal glands, lining of body cavities, mesenteries and omenta, portions of the eye

The cells making up the germ layers do not possess distinctive differences and are said to be undifferentiated. Further development entails, among other things, a differentiation or specialization of groups of cells which make up the body.

BIBLIOGRAPHY

Arey, L. B.: "Developmental Anatomy," 7th ed., Saunders, Philadelphia, 1965.

Balinsky, B. I.: "An Introduction to Embryology," 3d ed., Saunders, Philadelphia, 1970.

Berrill, N. J.: "Developmental Biology," McGraw-Hill, New York, 1971.

Billingham, R. E.: Tissue Transplantation: Scope and Prospect, *Science,* 153:266–270 (1966).

Burns, R. K., Jr.: Urogenital System in B. H. Willier, P. A. Weiss, and V. Hamburger (eds.), "Analysis of Development," Saunders, Philadelphia, 1955.

Ebert, J. E.: "Interacting Systems in Development," Holt, New York, 1965.

Etkin, W., and Lawrence I. Gilbert (eds.): "Metamorphosis: A Problem in Developmental Biology," Appleton-Century-Crofts, New York, 1968.

Nelsen, Olin E.: "Comparative Embryology of the Vertebrates," McGraw-Hill, New York, 1953.

Patten, B. M.: "Foundations of Embryology," 3d ed. McGraw-Hill, New York, 1974.

Waddington, C. H.: "Principles of Development and Differentiation," Macmillan, New York, 1966.

5 INTEGUMENTARY SYSTEM

The integument, or outer covering of the body, is commonly referred to as the skin. Together with its derivatives it makes up the integumentary system. It is continuous with the mucous membrane lining the mouth, eyelids, nostrils, and the openings of the rectum and urogenital organs. The skin functions primarily to cover and protect the tissues lying beneath it. The integument forms the interface between the organism and the external environment. It is the part that the predator sees first and which offers the first line of defense. It is abundantly supplied with sensory nerve endings, which are affected by environmental stimuli and play an important role in communication. The part played by the integument in the general metabolism of the body, temperature regulation, water loss, and defense is of vital importance. The character of the skin and its derivatives shows much variation in different regions of the body, in different individuals, in the same individual as age advances, and in different groups of vertebrates. The type of environment, whether aquatic or terrestrial, is of importance in connection with these variations. Nevertheless, basic similarities exist in the integuments of all vertebrates. The relationships of function and structure are very well shown in a study of the integument.

INTEGUMENT PROPER

In such invertebrates as arthropods, annelids, mollusks, and some others, the integument consists of a single layer of cells, the *epidermis,* together with an outer noncellular *cuticle* (Fig. 5.1) secreted by the cells. The cuticle may be very thin, as in annelids, or a heavy layer composed of chitin, calcareous material, or other substances. In arthropods the rigid cuticle makes up the *exoskeleton.* The term *ecdysis* refers to the periodic shedding of this outer layer. The integument of vertebrates (Fig. 5.2) consists of an outer layer, the *epidermis,* composed of cells derived from ectoderm, and an underlying layer of mesodermal origin known as the *dermis.* In tetrapods, ecdysis consists of shedding, or

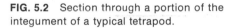

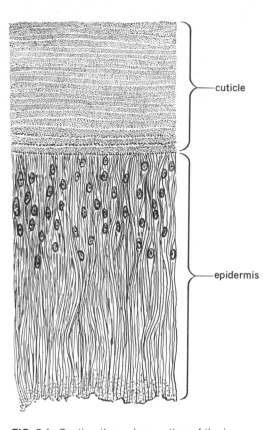

FIG. 5.1 Section through a portion of the integument of a crayfish, a typical invertebrate.

sloughing, of the outermost layer of the epidermis. In some forms this is shed as a whole, but in others it is given off in small fragments of various sizes. Under the dermis lies a loose layer of connective tissue, the *subcutaneous tissue.* In such places as the palms of the hands and soles of the feet the fibers of the dermis are rather tightly interwoven with those of the subcutaneous layer so that the skin in these regions is more firmly attached.

SKIN OF MAN In order to understand more readily the modifications of the integument found in various vertebrates, it is convenient first to discuss that of the human being, since it is most familiar.

The integument proper, under normal conditions, is germproof and serves as an effective barrier against disease organisms which might otherwise gain access to the body. It is, to a great extent, impervious to water, helps to regulate body temperature, and upon exposure to ultraviolet light can manufacture vitamin D.

Epidermis The outer epithelium, or epidermis, is made up entirely of cells which are arranged in more or less distinct layers. It is closely applied to the dermis beneath. In the deepest layer, called the *stratum germinativum* or *Malpighian layer,* the cells are columnar in shape and arranged perpendicular to the dermis. These cells frequently undergo mitosis. As new cells are formed, they gradually

FIG. 5.2 Section through a portion of the integument of a typical tetrapod.

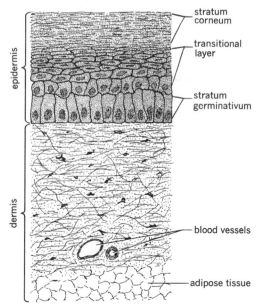

stratum corneum

transitional layer

stratum germinativum

epidermis

dermis

blood vessels

adipose tissue

approach the surface, becoming flattened as they do so. The region where this occurs is called the *transitional layer.* On the outer surface the cells are scaly and dead and have lost their nuclei. The outer layer of flattened cells is the *stratum corneum,* or *horny layer,* of the skin. Its chief constituent is *keratin,* a very hard, tough, insoluble protein.

The cells of the stratum germinativum lie on the basement membrane which is in direct contact with the dermis, with its rich supply of blood vessels, from which they derive nourishment. Since new cells are constantly being formed, the outer flattened cells would ultimately make up a very thick layer if they were not continually being sloughed, or worn off, from the surface and replaced by new cells. The rate of proliferation of the cells is approximately the same as that at which the outer cells are being desquamated, so that the thickness of the epidermis is relatively constant. An increased amount of friction in a given area of epidermis seems to stimulate the cells to divide more rapidly and results in the formation of *calluses.* Such thickened areas give greater protection to the delicate tissues beneath.

In many tetrapods the corneal layer is shed at intervals as a continuous sheet, but in man, small fragments are given off and shedding is ordinarily not noticeable. Shedding of the corneal layer seems to be at least partially controlled by the secretion of the thyroid gland.

The epidermis in certain parts of the body in man shows modifications from the condition described above. For example, in the thick skin on the palms and soles (Fig. 5.3) the transition from the columnar cells of the stratum germinativum to the flattened corneal cells is not so abrupt. Three fairly distinct regions represent the so-called transitional layer. The lowest, or *stratum spinosum,* also called the prickle-cell layer, is in contact with

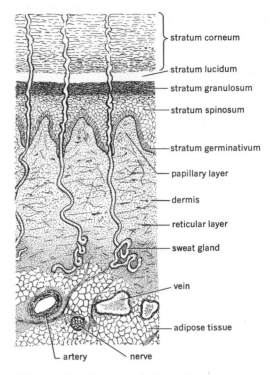

stratum corneum

stratum lucidum

stratum granulosum

stratum spinosum

stratum germinativum

papillary layer

dermis

reticular layer

sweat gland

vein

adipose tissue

artery   nerve

FIG. 5.3 Section through thick integument on the palm of a human being.

the stratum germinativum. Peripheral to this is a granular layer, the *stratum granulosum.* The third layer, or *stratum lucidum,* lies just above the stratum granulosum in these regions and forms the remainder of the transitional layer. The stratum lucidum is translucent and resists ordinary stains in sections prepared for histological study. These tough layers serve as a protection in regions most likely to come in contact with external objects.

Dermis The dermis is best developed in mammals. It is not composed entirely of cells but consists mostly of connective-tissue fibers extending in all directions and forming a fairly elastic covering. Cells are scattered among the fibers. In addition, the dermis contains nerves, smooth-muscle fibers, blood vessels, and certain glands. In preparing leather the epidermis is first removed by maceration and the connective-tissue fibers are thickened and toughened by the action of tannin, alum, chromium salts, or other so-called tanning agents.

PATTERNS On the surface of the human skin are many small grooves and ridges which intersect so as to bound small triangular and quadrangular areas. On the palms and soles the grooves and ridges generally run parallel to one another. The side of the dermis in contact with the epidermis is thrown into rows of *papillae,* which are most prominent on the palms and soles. The papillae are quite definitely arranged so that a double row lies beneath one of the external ridges (Fig. 5.4). In other regions the papillae show variations in size and arrangement. The ridges on the palms and soles are so arranged as to provide the greatest possible friction with a surface. The finely detailed patterns formed in this manner, with their endless variations, are useful in recording fingerprints and footprints. The patterns do not change from birth to death.

PIGMENT Pigment in vertebrates occurs in certain cells in the form of small granules of *melanin.* The human integument does not contain the kinds of chromatophores (pigment-bearing cells) that are found in cyclostomes, fishes, amphibians, and reptiles, which are ectodermal in origin (being derived from neural crests), manufacture their own pigment, and respond to stimulation by changing the distribution of the pigment granules within the cells (Fig. 5.5). Instead, the color of the skin depends upon the presence of certain cells called *melanocytes** in the lower layers of the epidermis. These also are derived from neural crests, manufacture their own pigment, and may have branched extensions, but pigment distribution within the cells is *not* variable. Other deep epidermal cells may receive pigment which diffuses from the melanocytes which lie among them. In true albinos there is a total lack of pigment in all parts of the body. Certain cells in the dermis, often unfortunately referred to as dermal chromatophores, would seem to be pigment cells but are actually phagocytes, engulfing pigment granules rather than manufacturing them. They are mesenchymal in origin.

FIG. 5.4 The relation of dermal papillae to epidermis in friction area of skin.

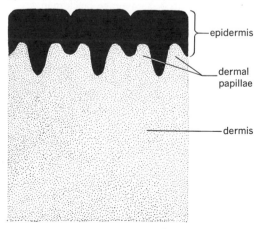

epidermis

dermal papillae

dermis

* The terminology used for cells containing the pigment *melanin* is often confusing. The nomenclature used here has been officially adopted by the National Research Council's Committee on Pathology [T. B. Fitzpatrick and A. B. Lerner, *Science,* **117**:640(1953)]. *Melanoblasts* are immature cells which give rise to *melanocytes; melanophores* are mature melanin-forming cells; *melanophores* are chromatophores in which distribution of the pigment melanin varies under different conditions; *melanophages* are actually macrophages of mesenchymal origin, located in the dermis, which engulf melanin pigment granules.

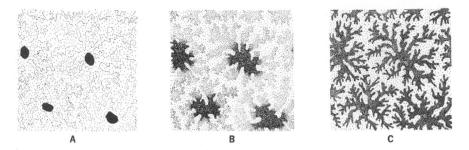

FIG. 5.5 Three stages in concentration and dispersion of pigment in chromatophores in the skin of certain vertebrates: *A,* pigment concentrated in center of cells; *B,* intermediate condition; *C,* pigment dispersed throughout cytoplasm of chromatophores.

The chromatophores of lower forms contain pigments of various colors. *Melanophores* contain black or brown pigment; *erythrophores,* red pigment; *xanthophores,* yellow pigment. Red, yellow, and orange types are sometimes referred to collectively, as *lipophores.* Chromatophores have many irregular branching processes. When pigment granules are dispersed throughout the cell, it displays the greatest amount of color (Fig. 5.5C). When the granules become concentrated about the nucleus, only a small spot of color is visible. Various colors may be produced by combinations and blending of chromatophores bearing different kinds of pigment granules and by various degrees of dispersion of the granules. The background color of the stratum germinativum is also of importance. The ability to change color is known as *metachrosis.* Two methods of control have been discovered, endocrine and nervous. Secretions from the medullary components of the adrenal gland and from the intermediate lobe of the pituitary gland are endocrine factors of importance. Neurohumors given off by nerve endings are involved in nervous control.

It might be supposed that all the cells of the epidermis would contain pigment since they are derived from the stratum germinativum. This, however, is not the case, for most of the pigment is lost as the cells approach the surface.

Wide variations in skin color exist among men and even among individuals of the same population. The color of the skin actually depends upon three varying factors: its basic color is yellowish; the blood vessels in the dermis give it a reddish hue; the pigment melanin in different concentrations causes various shades of brown. The total amount of pigment is relatively small. It has been estimated that there is about 1 g of pigment in the entire skin of an average black man.

The degree of pigmentation seems to be associated with protection from the sun's rays. Mild exposure to ultraviolet light stimulates pigment formation. In lighter-colored peoples this is commonly known as "tanning." A dense pigmentation gives its possessor a certain advantage in enduring ultraviolet radiation. Tanning is at best a slow process. In sunburn a severe, rapid, and harmful action occurs in response to excessive exposure to ultraviolet light.

In man such regions as the axillae, nipples, external genitalia, groin, and circumanal area all show a greater degree of pigmentation than other parts of the body. A pigmented area appears in the skin over the sacral region of young Mongolian babies. No satisfactory explanation has been offered to throw any light on the function of pigment in these regions.

COMPARATIVE ANATOMY OF THE INTEGUMENT

Amphioxus

The skin of amphioxus represents chordate integument at its simplest form. The epidermis consists of a single layer of columnar epithelial cells without glands which secrete a thin, noncellular cuticle which in adults is perforated by minute pores. The dermis is thin and composed of soft connective tissue. Pigment is lacking.

Cyclostomes

The epidermis is made up of several layers of cells, but there is no dead stratum corneum. The outer cells are living, have nuclei, and are active enough to secrete a thin cuticle. The dermis is even thinner than the epidermis and firmly united, at metameric intervals, to connective-tissue partitions, the *myocommata,* between successive muscle segments. Chromatophores are located in the dermis.

Fishes

The integument of fishes shows little change over that of cyclostomes except that dermal scales or bony plates are usually embedded in the dermis. Unlike the loose-fitting skin of amphibians and birds, the skin of fishes fits smoothly and tightly over the underlying muscle. The epidermis is thin and contains glandular elements. The dermis is thicker and divided into two parts: a superficial, fibrous connective-tissue layer, and a deep, loose connective-tissue layer.

The color of fishes is due primarily to chromatophores in the dermis, which sometimes wander into the epidermis. Crystals of guanin located in cells called *iridocytes* or *guanophores* are responsible for the iridescence often observed.

Amphibians

The epidermis of amphibians (Fig. 5.6) is composed of several layers of cells and is the first to have a dead stratum corneum. The epidermis is six to eight cells in thickness and is divisible into three layers: stratum corneum, stratum germinativum, and a basal portion in contact with the basement membrane. Molting (desquamation) occurs periodically and consists of removing a unicellular sheet of stratum corneum. A dead corneal layer is an adaptation to terrestrial life, protecting the body and preventing excessive loss of moisture. The dermis is relatively thin in amphibi-

FIG. 5.6 Section through integument of frog *Rana pipiens.* Note upper loose layer (stratum spongiosum) and lower compact layer (stratum compactum) of dermis.

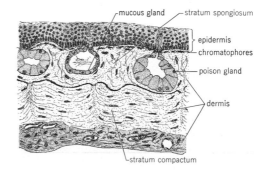

ans. It is composed of two layers: an outer, less compact layer, the *stratum spongiosum,* and an inner, more compact layer, the *stratum compactum.* A great number of gland types are recognized. Mucous glands are numerous, their secretion helping to maintain a moist skin. A second group of glands, the poison glands, secrete a number of substances that are distasteful and/or poisonous in some tropical frogs. The amphibian skin is an important organ of respiration. Blood vessels, lymph spaces, glands, and nerves are abundant in the stratum spongiosum. Chromatophores of complex structures lie, for the most part, between epidermis and dermis.

Reptiles

The epidermis of reptiles is a complex structure characterized by its extremely well-developed corneal layer. The scales of reptiles are derived from the horny stratum corneum. Between the scales of snakes, a thinner, more flexible area of epidermis permits movement.

The dermis consists of two layers: a superficial, and a deep layer. Glands are greatly reduced in number, those remaining having special functions related to reproductive behavior and protection. In some groups of lizards ossified elements, termed *osteoderms,* are present in the dermis. Leather of high commercial value can be prepared from the skins of certain reptiles.

Birds

The epidermis of birds is thin and delicate except in such exposed regions as the legs and feet, where corneal scales are present. The skin is adapted for free movement over the underlying muscle and is almost entirely free of glands. An exception is the uropygial gland located at the base of the tail on the dorsal surface. This gland secretes an oily substance used by the bird to preen its feathers.

Feathers, of course, cover and protect the bodies of birds. They are modified corneal scales derived from the epidermis of reptiles. The dermis is also thin, but muscle fibers are particularly abundant, being used to raise and lower the feathers. A rich supply of sensory organs, mainly tactile, are also present.

Pigment in the skin of birds is generally confined to the feathers and leg scales. Chromatophores of the type found in lower forms are not present.

Mammals

The human skin, previously discussed, is representative of mammalian integument, though variations in mammals do exist. A major distinction of mammalian dermis is that it is much thicker than the overlying epidermis.

STRUCTURES DERIVED FROM THE INTEGUMENT

GLANDS

One of the most important functions of epithelial tissues is the part they play in the metabolism of the body. Absorption of external substances and liberation of others may take place through the skin. Modifications of the epithelium in the form of glands aid in carrying on its secretory function. The glands of the skin are not essentially different from those found in other parts of the body, but the stratum germinativum alone is concerned with their formation. All are *exocrine,* or externally secreting, glands which pour their secretions onto an epithelial surface.

Structure

According to structure, integumentary glands are of two general types, *unicellular* or *multicellular,* according as they consist of isolated cellular units or numbers of similar cells joined together to form the glandular element.

UNICELLULAR GLANDS   The simplest skin glands are unicellular glands which are scattered, as modified single cells, among other epithelial cells covering the body in amphioxus, cyclostomes, fishes, and amphibian larvae. They are known, for the most part, as *mucous cells.* They secrete a substance called *mucin,* which is protein. Mucin, together with water, forms a slimy, viscid material called *mucus.* In some cases the elimination of mucus continues gradually. In others, the mass of mucin is liberated and the cell collapses, only to fill up gradually again with secretion, after which the process is repeated. Mucous cells may pass through a number of such phases of secretory activity but finally die and are shed. In the meantime new cells have been formed. The slippery mucus lubricates the surface of the body, thus lessening the friction with surrounding water. It also enables the animal to escape more readily from enemies. The various types of unicellular glands include *goblet cells, granular-gland cells, beaker cells,* and *thread cells,* all of which show distinctive traits. The secretion of unicellular glands in the skin on the snout of amphibian larvae digests the egg capsule and frees the embryo.

MULTICELLULAR GLANDS   Multicellular integumentary glands are formed by ingrowths of the stratum germinativum into the dermis. In some cases the ingrowth may be hollow from the beginning, but in others it is a solid structure in which a lumen appears later. As development progresses, the mass of cells may give off side shoots, resulting in the formation of compound glands. The branches, in most cases, therefore, connect to a single duct which opens onto the outer surface. The glandular portions push deeper down into the dermis, where their actively secreting cells are nourished by blood vessels. Multicellular glands are classified according to shape.

Tubular Glands   The tubular gland is practically uniform in diameter, without any bulblike expansion at its end.

Simple tubular glands   These are short, blind tubes lying partly in the dermis but extending to the surface (Fig. 5.7*A*). Certain specialized glandular areas, such as the thumb pads of male anurans and the mental glands of some male plethodontid salamanders, appear to be masses of simple tubular glands, as are the ceruminous, or wax-producing, glands of the external-ear passage of mammals.

Simple coiled tubular glands   Sweat glands, present only in the skin of mammals, are examples of this type. Each gland consists of a long narrow tube, the distal end of which is coiled into a small ball that lies in the dermis (Fig. 5.7*B*). The openings of sweat glands are often referred to as the *pores* of the skin. The glandular epithelium itself is composed of a single layer of cuboidal or columnar cells.

Simple branched tubular glands   The duct of the simple branched tubular gland divides at its distal end into two or more branches (Fig. 5.7*C*). The terminal portions may or may not be coiled. Some of the large sweat glands in the axillae, or armpits, may be of the latter type.

Compound tubular glands   These consist of a varying number of tubules, the excretory ducts of which unite to form tubules of a higher order. These tubules in turn combine to form others of a still higher order, etc. (Fig. 5.7*D*). The unit structure of the compound

FIG. 5.7 Diagrams representing various types of integumentary glands, all of epidermal origin: *A,* simple tubular gland; *B,* simple coiled tubular gland; *C,* simple branched tubular gland; *D,* compound tubular gland; *E,* simple saccular gland; *F* and *G,* two types of simple branched saccular glands; *H,* compound saccular gland.

tubular gland is like a simple tubular gland. Mammary glands of monotremes are compound tubular glands.

Saccular Glands The saccular, acinous, or alveolar gland has a spherical expansion at the end of a tubular duct. The actual secretory cells are in the expanded portion which lies in the dermis. The duct serves merely as a means through which the secretion is discharged onto the epithelial surface. There are three general types of saccular gland.

Simple saccular glands If only one expanded bulb, or acinus, is at the end of a duct, the gland is of the simple saccular type (Fig. 5.7*E*). Numerous simple saccular mucous and poison glands are found in the skin of amphibians.

Simple branched saccular glands If several acini are arranged along a single excretory duct, as in the *tarsal,* or *Meibomian, glands* of the eyelid, or if a single acinus appears to be divided by partitions into several smaller acini (Fig. 5.7*F* and *G*), as in the sebaceous, or oil, glands of the skin, the gland is said to be of the simple branched saccular type.

Compound saccular glands These consist of several portions called *lobules*. The smallest unit of a lobule corresponds to a simple saccular gland. Several of these unit structures enter a common duct. This unites with similar ducts of other lobules. All finally lead to a main duct, which opens onto the skin surface (Fig. 5.7*H*). The mammary glands of metatherians and eutherians are examples of such glands.

METHOD OF SECRETION

Glands are sometimes classified according to the manner in which they secrete.

MEROCRINE GLANDS The glandular cells are not injured or destroyed during secretion. Unicellular integumentary glands are merocrine glands, as are sweat glands of the eccrine type (see page 80).

HOLOCRINE GLANDS There is an accumulation of secretion within the cell bodies. The cells die and are discharged with their contained secretion. New cells are constantly being produced, so that the process is continuous. Sebaceous glands of the skin are of this type. The term *necrobiotic* is frequently applied to holocrine glands.

APOCRINE GLANDS The secretion gathers at the outer ends of the glandular cells. The accumulated secretion is pinched off with a bit of the cytoplasm. Most of the cytoplasm, as well as the nucleus, remains unchanged, and after a time the process is repeated. Mammary glands and certain sweat glands are of this type. It seems that secretion droplets of apocrine glands, during their discharge, are surrounded by a membrane which is in continuity with the cell membrane.

TYPE OF SECRETION

Glands are sometimes classified according to the type of secretion they produce:

Mucous glands secrete mucus. Unicellular glands of amphioxus, cyclostomes, and fishes, and certain simple saccular glands in the integuments of fishes and amphibians are examples.

Serous glands, such as sweat glands, form a thin, watery secretion.

Mixed glands secrete a mixture of mucus and serous fluids. It is doubtful whether any skin glands are of this type. The submaxillary salivary glands are nonintegumentary examples.

Fat or oil glands are exemplified by the sebaceous glands of mammals and the tarsal glands of the eyelids, which are considered to be modified sebaceous glands.

COMPARATIVE ANATOMY OF INTEGUMENTARY GLANDS

AMPHIOXUS Only unicellular mucous glands of the merocrine type are found in amphioxus. Most are goblet cells scattered among the other epithelial cells covering the body.

CYCLOSTOMES Goblet cells, granular gland cells, beaker cells, and thread cells, all of which are unicellular merocrine glands, are the only integumentary glands found in cyclostomes. Pockets of thread cells are found in the hagfish. A spirally coiled thread of mucus is shot out of these cells upon proper stimulation.

FISHES Both unicellular and simple saccular mucous glands are present in the skin of fishes. The "cocoon," or capsule, formed about the body in the lungfishes, *Lepidosiren* and *Protopterus,* during estivation is composed of the secretion of mucous glands.

A few elasmobranch and teleost fishes have multicellular poison glands of epidermal origin which are used in a protective manner. The common catfish has a poison gland opening at the base of a spine on each pectoral fin. Other fishes may have similar glands associated with spines on the operculum.

Some deep-sea elasmobranchs and teleosts living in almost total darkness have luminous phosphorescent organs, called *photophores,* in the skin (Fig. 5.8). Usually they are arranged in longitudinal lines near the ventral surface of the body but form patterns on the head. Photophores are modified integumentary glands, often of complicated structure. Although many hypotheses have been put forth, the function of photophores remains in doubt. The light emitted by these luminous organs may help certain deep-sea fishes to attract their prey; photophores may be of survival value in that nearsighted predators looking upward from the waters below may not see a sharply silhouetted figure emitting bright points of light. Instead, the light points

may appear to be blended together and the object seen is less likely to be regarded as a possible source of food; the lights may be used to attract mates.

Still another type of integumentary gland found in fishes is the *pterygopodial gland* associated with the clasping organs of certain skates and rays. They are multicellular mucous glands; their function is not clear.

AMPHIBIANS The skin of amphibians is important in respiration and must be kept moist. This is accomplished by means of the secretion of large numbers of simple saccular mucous glands, the bulblike enlargements of which lie in the vascular stratum spongiosum of the dermis.

Large simple saccular poison glands are also present in the skin of amphibians. They are generally larger than the mucous glands and are more distinctly localized, being more abundant on the dorsal side of the body and hind legs. The so-called "warts" of toads, and the *parotoid glands,* which lie near the tympanic membrane, are actually masses of poison glands. Stories of getting warts from handling toads are fictitious. Toads usually have more poison glands than frogs. Some specialized integumentary glands in amphibians have a tubular structure. They include those on the feet of certain tree-dwelling frogs and toads. Suctorial discs on the toes aid in climbing. Other examples include the glandular thumb pads of male frogs and toads and the mental gland of the male in some plethodontid salamanders. Unicellular glands are present on the snouts of tadpoles and urodele larvae. Their secretion has digestive properties, thus aiding in freeing the larva from the egg capsule during early development. Large *glands of Leydig* are unicellular glands of uncertain function in the epidermis of some larval urodeles.

FIG. 5.8 Section through luminous photophore of the fish *Porichthys. (Redrawn from Kingsley, after Greene.)*

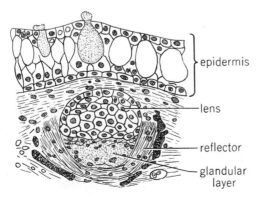

epidermis

lens

reflector

glandular layer

REPTILES Reptiles, in contrast to amphibians, have rough and scaly skins primarily adapted for life on land. Correlated with this is an almost total lack of integumentary glands. Any unnecessary evaporation of water from the body is thus prevented. The integumentary glands present in reptiles are, for the most part, scent glands associated with sexual activity.

Crocodilians have two pairs of skin glands which secrete musk. One pair is located on the throat on the inner halves of the lower jaw. A second pair lies just within the cloacal orifice. They are possessed by both sexes and are undoubtedly hedonic in function. A row of glands of unknown function also runs along either side of the back between the first and second rows of scales.

The skin of lizards is practically devoid of glands. *Femoral glands* are found in males on the undersurfaces of the thighs. Each opens through a short duct which passes through a conical projection of the skin. During the breeding season they give off an abundant yellowish secretion, the chemical makeup of which is not clearly understood. The biological function appears to be involved with behavioral activities.

Some snakes (*Thamnophis*) have glands in the cloacal region which secrete a milky fluid with a nauseating odor used as a defensive tactic.

In certain turtles musk glands are present beneath the lower jaw and along the line of junction of plastron and carapace.

BIRDS Many characteristics of birds resemble those of reptiles. The skin of birds, like that of reptiles, is practically devoid of glands. Usually only a *uropygial gland* is present. This is a simple branched saccular oil gland located on the dorsal side of the body at the base of the *uropygium,* or tail rudiment. A

FIG. 5.9 Section through the uropygial gland of bird. (*After Schuhmacher.*)

septum divides the gland into two halves (Fig. 5.9). The bird, by squeezing out a quantity of oil on its bill, oils its feathers as it preens them, making them quite impervious to water. The uropygial gland is best developed in aquatic birds. Not all birds possess a uropygial gland. It is notably absent in the Paleognathae, parrots, and some varieties of pigeons. The only other skin glands found in birds are certain modified oil glands in the region of the external ear opening. They are found only in a few gallinaceous birds such as the American turkey.

MAMMALS The skin of mammals is particularly abundant in glands, and a considerable variety is found. There are two essential types, from which all are probably derived, sebaceous glands and sweat glands.

Sebaceous Glands Sebaceous, or oil, glands are distributed over the greater part of the surface of the skin, being notably absent from the palms and soles. With few exceptions the duct of the gland opens into a hair follicle. The oily secretion serves to keep the hair and skin smooth and soft and also imparts to the animal an individual scent, or odor. Sebaceous glands are present without being connected with hair follicles on the corners of the mouth and lips, internal surface of the prepuce, labia minora, mammary papillae, and occasionally on the glans penis. The tarsal glands of the eyelids also have no connection with hairs. Their oily secretion forms a film over the layer of lacrimal fluid, which is thus held evenly over the surface of the eyeball. The film also prevents tears from overflowing onto the cheeks under normal conditions. Small *glands of Zeis* on the eyelids are sebaceous glands which open into the hair follicles of the eyelashes. Sebaceous glands are lacking in cetaceans.

Sweat Glands Sweat glands, or sudoriparous glands, are found only in mammals. They are either simple coiled tubular or simple branched tubular glands. Some mammals, including the spiny anteater, moles, sirenians, cetaceans, the scaly anteater, and certain edentates, lack sweat glands. In humans they are abundantly distributed over the surface of the body except on the lips, glans penis, eardrum, and nail bed. Sweat glands have two important functions: evaporation of sweat from the skin surface as a cooling device important in thermal regulation, and removal of salt, urea, and other waste products contained in the sweat gland secretion. Several quarts of sweat may be given off by an average human in a 24-hour period. According to their manner of secreting, sweat glands are of two general types, *eccrine* and *apocrine*.

Eccrine sweat glands These are merocrine glands (page 77) in which the glan-

dular cells are not injured or destroyed during secretion. They are simple, coiled, tubular structures and are restricted, in general, to catarrhine primates. In other mammals a few may be present on those surfaces of the feet which make contact with the environment. Among primates the chimpanzee and gorilla are the only species other than man in which eccrine sweat glands exceed the apocrine type in number. Lemurs and platyrrhine primates lack eccrine sweat glands. It has been estimated that in an average human being there are about $2\frac{1}{2}$ million sweat glands in the skin. They are most numerous in regions devoid of hair or where the hairy coat is scant. On the palms and soles their ducts open onto the surface of the skin along the tops of the small ridges rather than through the furrows between ridges (Fig. 5.10). The activity of sweat glands varies considerably. Those in various areas respond differently to thermal, sensory, and psychic stimuli.

Apocrine sweat glands These are to be

FIG. 5.10 Droplets of sweat accumulating at the openings of sweat glands on friction ridges of the palm of the hand (highly magnified).

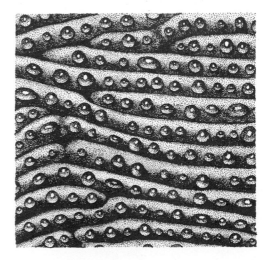

found in most mammals. In man they are, for the most part, confined to certain areas such as the axillae, pubic region, circumanal area, vicinity of the nipples, inner surface of the prepuce, and on the labia minora. Most apocrine sweat glands are simple coiled tubular glands, but those in the axillae may be branched. In man, apocrine sweat glands usually open into hair follicles above the opening of the duct of the sebaceous gland. In lower primates they open directly onto the surface. These glands do not secrete continuously, there being an inactive period of approximately 24 hours between active periods. The secretion of apocrine sweat glands is relatively thick and contains fat droplets as well as pigment granules. This explains why sweat in the axillary region may appear stained. The sweat of the hippopotamus has a reddish color. That of the South African antelope is of a bluish cast. It has been reported that the gorgeous color of the fur of the male red kangaroo, *Macropus rufus,* is caused by red pigment granules in the sweat which dries on the hair.

The horse, bear, and hippopotamus are supplied with an abundance of apocrine sweat glands. Many textbooks and articles state that sweat glands are lacking in the hairy skin of the dog. On the contrary, large numbers of apocrine sweat glands are present. It has been shown that they do not play an important part in regulating body temperature but serve chiefly in protecting the skin from an excessive rise in temperature. In many mammals sweat glands are confined to certain regions. In the platypus they are in the snout region; in the deer they are arranged around the base of the tail. In mice, rats, and cats they are present on the undersides of the paws. The moisture on the muzzles of sheep, pigs, dogs, and others is due to the presence of sweat glands. *Ceruminous,* or *wax-producing, glands*

in the external ear passages are modified apocrine sweat glands. Their secretion together with that of sebaceous glands is known as *cerumen* or *ear wax.* The secretory portions branch. The ducts either open directly into the external-ear passage or join those of large sebaceous glands which open into hair follicles. The secretion protects the tympanic membrane and also prevents the entry of insects.

Glands of Moll, on the eyelids, and *circumanal glands* are large, modified sweat glands.

Scent Glands Many mammals possess scent glands. They are either modified sweat glands or sebaceous glands, but the nature of their secretion is known only in a few cases. They may be useful in attracting members of the same species or of the opposite sex. In some they serve as lures, attracting prospective food items; in others they function as a defense against enemies; in still others, they are used by males in marking out a territory.

The location of scent glands varies greatly. In the deer family they are located on the head in the region of the eyes. Many carnivores have scent glands near the anal opening. Skunks and weasels have saclike scent glands which open into the rectum just inside the anus. They can be everted and their foul-smelling secretion expressed. Scent glands may be located at the openings of the reproductive organs, as in many rodents; on the face, as in some bats; between the hoofs, as in pigs; over the temporal bone, in the elephant; and in still other mammals on the arms, legs, and other parts of the body.

Mammary Glands Milk-producing glands, present in all mammals and only in mammals, are actually modified sweat glands. Their ducts open directly or indirectly onto the surface of the skin. The apocrine method of secretion is typical. Mammary glands are active only at certain times: immediately before the young are born and generally as long as active sucking continues. The development and functioning of the mammary glands are under hormonal control.

In general, the actively secreting mammary gland is made up of many small masses called *lobules*. Each lobule consists of large numbers of *alveoli* composed of secretory cells. The small ducts leading from the alveoli converge to form a larger duct. This unites with similar ducts from other lobules, and the common duct or ducts thus formed lead to the outside. In inactive glands, alveoli and lobules are reduced or wanting, the mammary tissue consisting mainly of branching ducts. Fatty, or adipose, tissue may surround the ducts and alveoli, contributing much to the size of the *mammae,* or *breasts.*

In monotremes the mammary glands are of the compound tubular type. No nipples are present, the glands opening onto a depressed area of the skin. The young animals grasp tufts of hair and obtain their nourishment by lapping, licking, or sucking. Mammary glands of therians lead to nipples or teats which are grasped by the mouths of the young when sucking.

The distinction between a nipple and a teat is not always apparent (Fig. 5.11). A nipple is a raised area on the breast through which the mammary ducts or duct open directly to the outside. In some mammals a single duct leads to the surface, as in certain rodents, marsupials, and insectivores. In others, several ducts may open on the nipple, as in some carnivores and in man, in which as many as 20 separate ducts may be present. In the false nipple, or teat, present in horses, cattle, and others, the skin of the mammary area grows outward to form a large projection. The mammary ducts open into a "cistern" at the base of

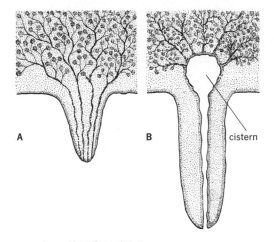

FIG. 5.11 Diagrammatic section: *A,* through nipple; *B,* through teat.

the teat, and the milk is then carried by the secondary duct to the surface.

Many laymen believe that the udder of a cow is a simple sac which fills with milk and must be drained periodically by milking. Nothing is further from the truth. The udder is composed of mammary and other tissues through which the ducts ramify to converge at the bases of the teats (Fig. 5.12). If a milch cow which has been giving several quarts of milk per day were slaughtered, it is doubtful that more than a pint of milk could be obtained from her udder. It is the secretory activity of the cells in the alveoli which forms milk from substances brought to them by the bloodstream.

Although nipples, teats, and mammary tissue may be present in both sexes, functional mammary glands occur normally only in lactating females. It is not uncommon, however, for milk to be present at birth and at puberty in the mammary glands of human beings of both sexes. That which is present at birth is sometimes called "witches' milk."

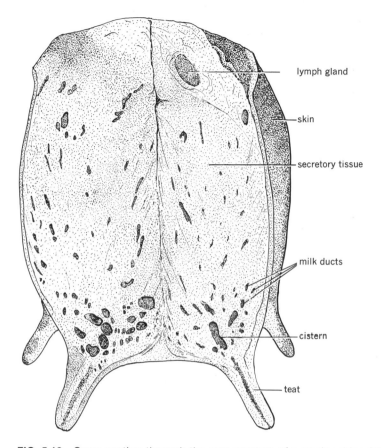

lymph gland

skin

secretory tissue

milk ducts

cistern

teat

FIG. 5.12 Cross section through the rear quarters of a cow's udder. (*From C. W. Turner, Univ. Miss. Agr. Exp. Stn. Res. Bull. 211, by permission.*)

Nipples and teats are usually paired, their number being roughly proportional to the number of young delivered at birth. They appear along the milk ridge, running along the ventral surface of the abdomen on either side. They vary from 1 pair in man to 11 pair in certain insectivores. Occasionally, in a particular species, extra nipples or teats may be present, a condition which harks back to primitive ancestral structures and abnormal embryonic development.

The distribution of mammary glands varies in different groups of mammals and is related to the habit of the mother when nursing her young. Animals such as dogs, cats, and pigs, which lie on their sides when the young are feeding, have mammary glands located in two ventrolateral rows extending from axilla to groin. Horses and cows, which stand up when nursing, have teats in a protected location between the hind legs. In elephants a single pair is located between the front legs, whereas in whales they are in the inguinal region. Primates, with their arboreal habits, have pectoral mammary glands, as do sirenians.

SCALES

In many vertebrates the body is covered with scales which give it protection. Scales, in general, are of two types: epidermal and dermal. The structure and mode of development of these two types are very different.

Epidermal Scales

These are cornified derivatives of the stratum germinativum, found primarily in terrestrial tetrapods. Few examples are to be found in amphibians, but in reptiles, birds, and certain mammals they are very well developed. Epidermal scales, with few exceptions, are usually shed and replaced from time to time. Large epidermal scales, such as those on the shell of turtles and the head of snakes, are called *scutes*.

Dermal Scales

Dermal scales are located in the dermis of the skin and are of mesenchymal origin. They are found for the most part in fishes, where they take the form of small bony plates, or calcareous plates, which fit closely together or overlap. Both epidermal and dermal scales are present in certain reptiles and a few other forms. It is important to distinguish between the two. Dermal scales, as well as the flat membrane bones in the skulls of fishes and higher vertebrates, are apparently remnants of the bony dermal armor of ostracoderms and placoderm fishes.

Comparative Anatomy of Scales

AMPHIOXUS No scales of any kind are present in the skin of amphioxus.

CYCLOSTOMES Although the bodies of ancient ostracoderms were covered with bony dermal plates, or scales, integumentary scales are absent in modern cyclostomes. The epidermal teeth found in the buccal funnel and on the tongue, however, are modified epidermal scales.

FISHES Epidermal scales are lacking in fishes, but dermal scales, making up part of what is called the *dermal skeleton,* are abundant and of several types. Not all fish have scales. They are lacking in the common catfish, in the electric ray, and some others. In chimaeras they occur only in localized regions. The scales of fishes are colorless. Numerous chromatophores, which give the fish its color, are located in the outer part of the dermis both above and below the scales.

Two types of fish scales have been identified among fossils of primitive bony fishes. One type, the *cosmoid scale,* not found in any form living today, was constructed on the plan of ostracoderm or placoderm armor plate, but the scales were in the form of small, separate elements. Cosmoid scales were present in primitive members of the subclass Sarcopterygii. The other type, the *ganoid scale,* found in primitive ray-finned fishes, existed in two forms, the *paleoniscoid ganoid scale,* found today in *Polypterus,* and the *lepisosteoid ganoid scale,* which persists today in the garpike. The two differ somewhat in structure, but both are covered with a hard, shiny, translucent material of mesodermal origin called *ganoin.*

Dermal fish scales of four types, ganoid, placoid, ctenoid, and cycloid, are recognized in species living today. All are fundamentally similar in origin, but their form and structure vary considerably.

Ganoid Scales Ganoid scales of both types mentioned above fit together like tiles on a floor and are arranged in diagonal rows (Figs. 3.8 and 5.13). *Amia* has modified ganoid scales on the head. Sturgeons have similar

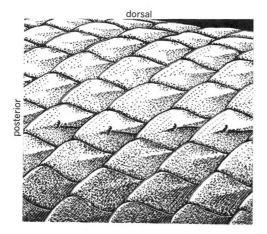

FIG. 5.13 Ganoid scales of the garpike *Lepisosteus.* Note perforations for openings of the lateral-line canal. (*Drawn by G. Schwenk.*)

scales in the tail region. In both these fishes the layer of ganoin is lacking.

Placoid Scales With a few exceptions, placoid scales are found only in the elasmobranch fishes. Each scale consists of a basal bony plate, embedded in the dermis, from which a spine projects outward through the epidermis and points posteriorly (Fig. 5.14). The spine is composed of *dentine* covered with a hard layer of *vitrodentine,* both of mesodermal origin. A pulp cavity lies within the spine, opening through the basal plate. Many authorities have stated in the past that the placoid scale is covered by a layer of enamel. Although the scale may be formed under the organizing influence of an enamel organ (page 252), no actual enamel is formed and the spine is composed of dentine alone. The general similarity in structure of placoid scales to teeth of higher forms should be apparent. Both are considered to be remnants of the bony armor of such primitive vertebrates as ostracoderms and certain placoderms.

The placoid scales of dogfishes and sharks are numerous and set closely together. The teeth on the rostrum, or saw, of the sawfish (Fig. 5.15) are actually extremely large spines of modified placoid scales. In chimaeras placoid scales may develop during embryonic life, but they soon disappear except in a few scattered areas.

Ctenoid Scales The ctenoid scale (Fig. 5.16) is a common type found in most teleost fishes. They are thin, translucent plates composed of an underlying layer of fibrous material covered by a layer which somewhat resembles bone. They might be compared with ganoid scales from which the layer of ganoin has disappeared and the underlying layer of bone modified. Each scale is embedded in a small pocket in the dermis. The scales are obliquely arranged so that the posterior end

FIG. 5.14 Enlarged photograph of skin of shark *Squalus acanthias,* showing numerous placoid scales.

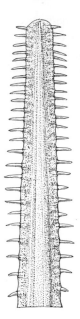

FIG. 5.15 Saw of sawfish *Pristis pectinatus.* The teeth are modified placoid scales.

of one scale overlaps the anterior edge of the scale behind it. The basal end of the ctenoid scale is scalloped; its free edge bears numerous comblike projections. Lines of growth are present, which if examined under a microscope will give a good indication of the age of the fish. The number of lines of growth varies according to the species in question

FIG. 5.16 Ctenoid scale.

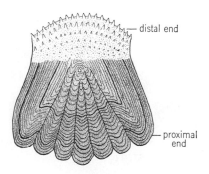

distal end

proximal end

and the part of the scale examined. When several lines of growth are widely separated, it indicates that that portion of the scale was formed in the warm season, when food was plentiful and rapid growth took place. Several lines of growth close together denote the cold season. Thus an area on the scale which covers a series of wide lines and a series of close lines would correspond to a year in the life of the fish.

Cycloid Scales Cycloid scales are roughly circular in outline. They too are located in pockets in the dermis and have lines of growth similar to those of ctenoid scales (Fig. 5.17).

In certain fishes, e.g., flounders, both ctenoid and cycloid scales may be present, those on the underside being cycloid and those on the upper side being ctenoid. Intermediate types of scales are present in a variety of fishes. The scales covering the lateral line are frequently perforated, permitting the passage of small connectives of the lateral-line canal to the outside (Fig. 5.13).

AMPHIBIANS The skin of modern amphibians lacks scales except in a few toads and in some of the burrowing, limbless caecilians. The skin is usually smooth and moist.

Epidermal Scales The spade-foot toads have a highly cornified area of epidermis on the inner side of each hind foot, which is used in digging. In the African clawed toads, *Xenopus*, *Hymenochirus*, and *Pseudohymenochirus*, the dark cornified epidermis at the ends of the first three digits on the hind feet has a clawlike appearance. The salamander *Hynobius* also has similar structures at the ends of the digits.

Dermal Scales In a few toads bony plates are embedded in the skin of the head or back. True dermal scales are found in the

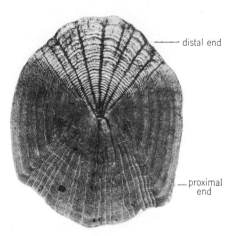

FIG. 5.17 Cycloid scale of white sucker from a fish that has lived through eight winters. (*From a photograph by Dr. W. A. Spoor.*)

integument of certain caecilians. They lie between ringlike folds of skin, these areas alternating with others of a glandular nature. The scales resemble those of fishes in that they are embedded in pockets, but several (four to six), rather than one, may be confined to a pocket.

REPTILES Both dermal and epidermal scales exist in reptiles, the two types frequently being associated with each other.

Epidermal Scales Highly developed epidermal scales are characteristic of members of the class Reptilia. There are two general types, those present in snakes and lizards and those found in turtles and crocodilians. In the former type each scale projects backward and overlaps the scales behind (Fig. 5.18). During development, raised and depressed areas appear in the integument (Fig. 5.19). Each thickened area, which consists of a covering of epidermis together with the dermis beneath, is known as a papilla. The flattened papillae grow upward and posteriorly and become lopsided, since the cells

on the upper and outer surface multiply faster than those beneath. The result is the formation of a complex layering of cell types containing varying amounts of alpha and beta keratin. The scales are continuous with each other at their bases. Snakes and lizards periodically undergo ecdysis. Before this occurs, a new outer layer of cells is formed beneath the old. The old dead outer layer is shed as a whole; in snakes, it is turned inside out. The newly exposed layer is at first soft, but hardens upon exposure to the air. The frequency of shedding depends upon such factors as amount of food eaten, general health, and the activity of the thyroid gland and anterior lobe of the pituitary gland.

The scales on the ventral side of the body in snakes are transversely arranged and are of importance in locomotion (page 35). Those on the dorsal surface of the head are considerably modified, forming a close-fitting layer over the bones. They form a continuous sheet with the other scales and are shed along with them during ecdysis.

Special modifications of epidermal scales of lizards and snakes include the horn cover-

FIG. 5.18 Overlapping epidermal scales of snake (diagrammatic).

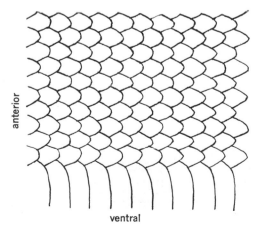
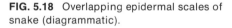

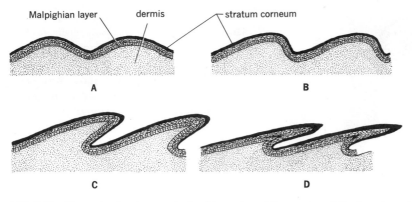

Malpighian layer   dermis   stratum corneum

A   B

C   D

FIG. 5.19   Stages in development of epidermal reptilian scales of the type found in snakes and lizards (diagrammatic).

ing of the horned lizard (page 103), digital lamellae in gekkos for climbing, and the rattle of the rattlesnake. A rattle is made up of a series of old, dried scales, loosely attached to one another in sequence. At the time of ecdysis the scale at the very tip of the tail is not shed with the other scales because of a peculiar bump which causes it to adhere to the new formed scale beneath (Fig. 5.20).

In most turtles epidermal scales cover the plastron and carapace, but the pattern of the scales does not conform to that of the bony plates beneath. Each scale develops separately. Periodically the stratum germinativum under each scale grows peripherally, and thus the area of the scale is increased. A new cornified layer then is formed which pushes the older scales away from the shell. This results in a piling up of the older scales, which then appear from above as a series of irregular concentric rings (Fig. 5.21). In such forms as the painted turtle the old scales peel off, and only the smooth, most recently formed scale is present at any one time. Pigment in the turtle integument is located in the stratum germinativum. The beautiful tortoise shell of commerce owes its color to the various combinations of pigment which have diffused into the

corneal layer. Soft-shelled turtles possess a rather soft, leathery "shell" lacking epidermal scales as well as underlying bony dermal plates.

Epidermal scales of crocodilians (Fig. 5.22) cover the entire body. On the lateral and ventral sides, as well as on the tail, each scale bears a pitlike depression in which a small sensory capsule is located. Crocodilian scales do not undergo periodic ecdyses. Instead, there is a wearing away and a gradual replacement. Patches of a few scales are occasionally sloughed.

Dermal Scales   Perhaps the best example of highly developed dermal scales exists in turtles. The bony plates form a rigid

FIG. 5.20   Rattle of rattlesnake: *A,* median sagittal section of end of tail; *B,* a single horny element of the rattle. (*After Czermak.*)

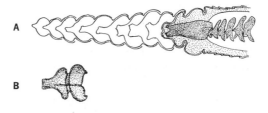

A

B

FIG. 5.21 Arrangement of scales on carapace of wood turtle.

dermal skeleton which becomes intimately fused with the endoskeleton. The part of the carapace beneath the epidermal scales is composed of bony costal plates fused to the neural arches of vertebrae in the midline and laterally to the ribs beneath. The plastron is usually made up of nine dermal plates covered with epidermal scales.

In some lizard groups an ossified bony plate develops in the dermis directly below the epidermal scale. These *osteoderms* provide a certain amount of protection; they are most abundant on the dorsal surface of the skull, where they may fuse together.

In crocodilians the dermis is thick and soft except on the dorsal side and on the throat, where bony plates lie beneath the epidermal scales. In caymans thin, bony scutes also occur under the ventral scales. The dermal plates are small and not fused. Small dermal scales are also present in some snakes and lizards. Crocodilians possess dermal "ribs" called *gastralia*, located in the ventral abdominal region. *Sphenodon* also has gastralia. They are not homologous with true ribs.

BIRDS With the exception of the membrane bones of the skull, dermal skeletal elements are usually lacking in birds. Epidermal scales and their derivatives are exceptionally well developed.

Epidermal Scales Epidermal scales of the reptilian type are confined to the lower parts of the legs and feet and the base of the beak. They generally overlap and are formed in much the same manner as those of lizards and snakes.

A bony projection of the tarsometatarsus in the males of certain species of birds, known as the *spur*, is covered with a horny, scalelike

FIG. 5.22 Arrangement of scales on neck and shoulder region of alligator.

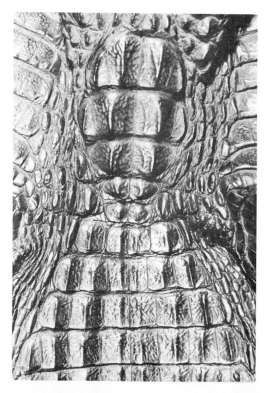

epidermal sheath which may be very sharp and pointed. The spur is used in fighting. Spurs may also occur on the wings (metacarpus) in certain forms.

Webs on the feet of such aquatic birds as geese, ducks, and swans are modified regions of the integument which are characteristically scaled.

Feathers Feathers, which are actually modified reptilian scales formed from the beta keratin layer, are found only in birds. Their relation to scales is very well shown in such varieties of chickens as cochins, brahmas, and langshans, where both feathers and scales appear on the legs and toes. The elaborate covert feathers making up the "tail" of the peacock represent the climax of modification of the stratum corneum.

In general there are three main types of feathers, *filoplumes,* or hair feathers; *plumulae,* or down feathers; and *plumae,* or contour feathers.

Filoplumes Hair feathers appear superficially like hairs but have an entirely different structure and origin. They may usually be observed on a chicken after it has been plucked, and are commonly singed, or burned off. The structure is simple since it consists only of a long, slender shaft which may bear a few barbs at its distal end (Fig. 5.23). The shaft is embedded in the skin and surrounded by the *feather follicle* at its base. Hair feathers are usually scattered over the surface of the body, but in such birds as flycatchers they are concentrated about the mouth, where they aid in catching insects. In peacocks they are of unusual length.

Plumulae Down feathers (Fig. 5.24) are more complex than hair feathers. Each is composed of a basal, short, hollow *quill,* embedded in the integument, and numerous *barbs* which arise from the free end of the quill. The barbs bear tiny *barbules* along their

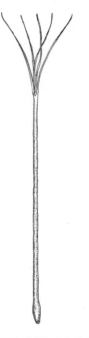

FIG. 5.23 A hair feather.

edges. A transient nestling down is the first feather covering of a young bird. This is shed when the contour feathers emerge later. Down feathers of adult birds make up what is called "powder down." These lie under the larger contour feathers and usually form a warm insulating layer which not only aids in temperature control but assists in warming the eggs during incubation. The down of the eider duck is used for stuffing pillows and has high commercial value.

Plumae Except in penguins, ostriches, and toucans, plumae, or contour feathers, arise from certain areas of the skin called feather tracts or *pterylae* (Fig. 5.25). Large areas between pterylae are known as *apteria.* Hair feathers and down feathers may occur in the apteria, however. Contour feathers give the body its outline, or contour. If they are removed by plucking, the appearance of the

FIG. 5.24 A down feather.

body is decidedly altered. Special contour feathers on the wings are *remiges,* or flight feathers. Those on the tail are *rectrices,* or tail feathers.

A typical contour feather (Fig. 5.26) consists of a long *shaft* and a broad, flat portion called the *vane.* The shaft is made up of two parts, a hollow *quill,* or *calamus,* embedded in the skin, and a solid *rachis,* which bears the vane. At the lower end of the quill is a small opening, the *inferior umbilicus.* At the junction of rachis and quill is another opening, the *superior umbilicus.* On the underside of the rachis and extending from the superior umbilicus to the tip is an *umbilical groove.* The vane is composed of a number of *barbs* which arise from the rachis. Each barb in turn bears small *barbules* on both proximal and distal sides. The lower part of each distal barbule bears tiny *hooklets* which fasten onto the proximal barbules of the next adjacent barb (Fig. 5.27). Hooklets are lacking from the proximal bar-

bules. Thus the barbs are hooked together via their barbules, and the vane offers a flat, wide, resilient, and unbroken surface except at the base, where the distal barbules of a variable number of barbs lack hooklets. If barbs become separated, in preening the bird can draw them together with its beak or bill. In some of the smaller contour feathers of the wings a considerable portion of the lower part of the vane may consist of loose barbs since hooklets are lacking from their barbules.

In many birds, at the junction of rachis and quill is located what appears to be another feather, called the *aftershaft* or *hyporachis.* It also bears barbs and barbules, but hooklets are generally lacking. The aftershaft is usually smaller than the main shaft and may consist only of a downy tuft. In such birds as the emu and cassowary the aftershaft with its accessory

FIG. 5.25 Distribution of feather tracts in the pigeon. (*After Nitzsch.*)

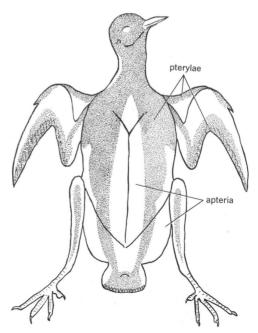

pterylae

apteria

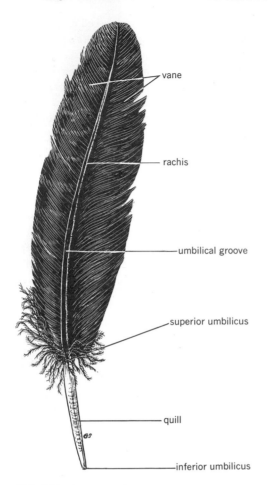

FIG. 5.26 A contour feather from the wing of a chicken. (*Drawn by G. Schwenk.*)

with the lizard and snake type of epidermal scale, but the exact evolution of feathers is unclear.

The first indication of development of a typical down feather is seen in the appearance of a projection, the *dermal papilla,* with its covering of epidermis. Instead of being flattened, as in the reptile, the papilla is elongated and round. The two are essentially the same in structure, however. An *annular groove,* which is the beginning of the feather follicle, appears around the base of the papilla. The blood vessels of the dermal pulp supply nourishment to the developing structure. The outer, thin, cornified layer of the epidermis forms a sheath known as the *periderm.* This is ultimately sloughed. First, however, the epidermis beneath the periderm forms a series of longitudinal folds, or ridges, arranged on the surface of the dermal pulp into which they extend (Fig. 5.29). Actually,

FIG. 5.27 Arrangement by means of which barbs of a feather are held together. (*Drawn by G. Schwenk.*)

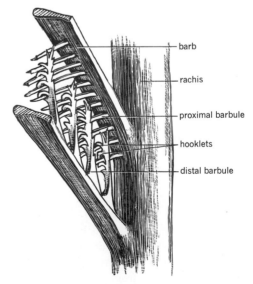

parts may be as long as the main shaft, and the feather appears to be double (Fig. 5.28).

Feathers are shed periodically in a process known as *molting.* During molting feathers are not all shed at once. If such were the case, the bird would be greatly handicapped by being temporarily unable to fly. New feathers are developing as the old ones are shed.

Development A study of the development of a feather emphasizes its homologies

FIG. 5.28 Feather from a cassowary. The main shaft and aftershaft are of similar dimensions.

these ridges arise from a collar of cells, the stratum germinativum, at the base of the papilla. The entire structure is now called the *feather germ*. The epidermal ridges are destined to form the barbs of the feather. The feather germ with its peridermal sheath grows rapidly and soon projects from the feather follicle above the surface of the skin as a *pin-*

feather. The dermal pulp and stratum germinativum gradually retract. Soon the periderm at the apex splits, and the tips of the cornified epidermal ridges dry and crack. The distal ends of the barbs may then be observed projecting from the apex of the sheath. As the feather continues to grow, the proximal parts of the ridges become cornified and separate from one another. The periderm is shed in small flakes, and the barbs spread out. The bases of the barbs remain attached to the quill, which does not split. Small pithy partitions inside the quill represent remnants of the dermal pulp. A small papilla remains at the base of the feather. This will give rise to another feather later on.

Feathers are thought to have originated from the beta keratin layer, one of the complex layers of cells in the epidermis of lizard scales. As in the formation of hair, the epidermal layers or germinal regions of the feather sink down into the dermis, forming a follicle, thus providing a base and giving strength to the structure (Fig. 5.30). Development then proceeds as described above.

The development of a contour feather is basically similar to that of the down feather but is more complex (Fig. 5.31), involving a differential growth on one side of the collar of stratum germinativum at the base of the feather germ. The barbs arise from the sides of the rachis rather than directly from the quill.

MAMMALS Except for the membrane bones of the skull, most mammals lack dermal skeletal structures. In armadillos, bony plates lie under the epidermal scales, which they reinforce. Certain whales have bony plates on the back and dorsal fin. As in birds, there has been a tendency toward elaboration of epidermal structures and a loss or diminution of dermal derivatives.

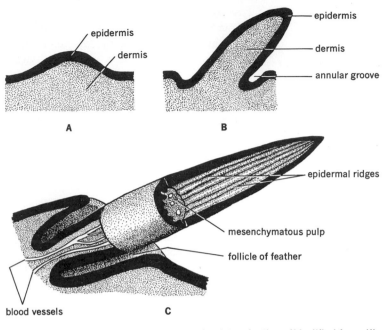

FIG. 5.29 Stages in the development of a down feather. (*Modified from Kingsley, "Comparative Anatomy of Vertebrates," McGraw-Hill Book Company, by permission.*)

Epidermal Scales With the exception of armadillos and scaly anteaters epidermal scales of mammals are generally confined to tails and paws and are usually associated with hairs. The relationship of scales, feathers, and hairs is of interest, but it is clear that feathers are more closely related to scales than hairs are.

In the scaly anteaters the body is covered with large, overlapping, horny epidermal scales except on the ventral side (Fig. 3.29). They are typically reptilian in structure, but ecdysis occurs singly. The large epidermal scales of armadillos have fused to form plates over the head, shoulders, and hindquarters and the ringlike bands which lie in the midbody region except along the midventral line. Instead of true ecdysis there is a gradual

wearing away from the outer surface. Hairs are interspersed among the scales.

On the tails of such rodents as mice, rats, and beavers, imbricated scales are present. They also are reptilian in structure but are not highly cornified, nor does ecdysis occur. Of interest is the arrangement of hairs in relation to scales. The hairs project from beneath the scales in a definite arrangement. On the tail of a rat, for example, three hairs project from beneath each scale (Fig. 5.32). The astonishing fact is that the hairs in other parts of the body have a similar arrangement. The skin of the pig is devoid of scales, but the grouping of the hairs, or bristles, gives evidence of origin from scaled ancestors. Coarse hairs are usually arranged in clusters of three and interspersed with finer hairs. Examination of a pair of pig-

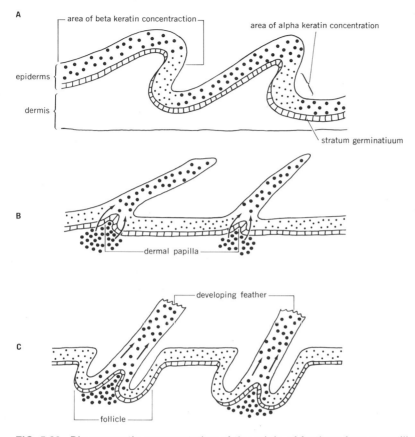

A

area of beta keratin concentraction

area of alpha keratin concentration

epiderms

dermis

stratum germinatiuum

B

dermal papilla

developing feather

C

follicle

FIG. 5.30 Diagrammatic representation of the origin of feathers from a reptilian ancestor: *A*, sagittal section through the skin of modern crocodilian showing relationship of alpha and beta keratin; *B*, hypothetical ancestral condition; *C*, sagittal section through skin of protoavian ancestor.

skin gloves will clearly indicate this relationship. In other mammals, including man, similar arrangements are found, hairs being grouped in twos, threes, fours, and fives. Often this arrangement, although obvious during development, is obscured in the adult.

In most mammals the undersurfaces of the hands and feet are scaled or bear evidence of the former presence of scales. The friction ridges, which form patterns used in making fingerprints, represent scale rudiments. In certain mammals elevated pads on the undersurfaces of hands and feet bear friction ridges known as *tori*. On the foot of a rat, for example (Fig. 5.33), there are 11 tori. In man there are no conspicuous elevations, but in the regions where one might expect them there are groupings of friction ridges which indicate a similar relationship.

Hair True hair is found in all mammals and is composed primarily of alpha keratin.

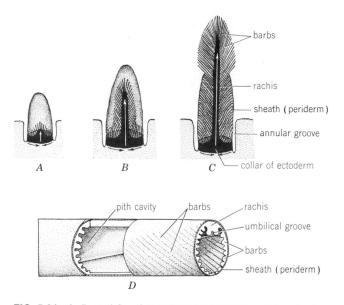

barbs

rachis

sheath (periderm)

annular groove

collar of ectoderm

A   *B*   *C*

pith cavity   barbs   rachis

umbilical groove

barbs

sheath (periderm)

D

FIG. 5.31 *A, B, and C,* schematic representation of developing contour feather from a collar of ectodermal cells which shows differential rates of growth. In *C,* the periderm at the distal end has ruptured. The barbs of the two sides have separated along the mid-ventral line and have flattened out. *D,* a stereogram of part of a developing contour feather. (*A, B, and C, based on figures by Lillie and Juhn, Physiol. Zoöl.,* **11:***442 (1938) by permission of the University of Chicago Press. D, after Kingsley, "Comparative Anatomy of Vertebrates," McGraw-Hill Book Company, by permission.*)

Only mammals have hair, but in some the hairy coat has practically disappeared and only traces remain. In the adults of some of the larger whales, for example, only a few coarse, stubby hairs are present in the snout region.

During fetal development in all mammals, the body at one stage is covered with a coating of fine hair, the *lanugo.* This is absent only on the undersurfaces of the hand and feet. The lanugo hair is transient and is usually shed some time before birth. In the human fetus it is fully developed by the seventh month, but most of it is lost by the time of birth or shortly thereafter. A new growth of hair takes place over most of the rest of the body, forming a fine, downy coat referred to as the *vellus.*

Structure, development, and growth of hair Hair is entirely epidermal in origin, developing from the alpha layer of keratin in reptilian scales. During development a small thickening of the epidermis, which is to become the *hair follicle* from which the hair will arise, pushes down into the dermis and finally becomes cupped at the lower end (Fig. 5.34). The dermis which extends into the cupped depression constitutes the *dermal papilla.* The blood vessels in the papilla bring nourishment to the epidermal cells of the stratum germinativum. Hair follicles usually are formed singly. In man an additional follicle customarily develops on either side of the original so that there is a tendency to form groups of three. Two thickenings appear along the side of a

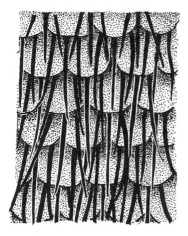

FIG. 5.32 Arrangement of hairs in relation to scales on the tail of the rat.

developing follicle. The proximal one is destined to form a sebaceous gland; the distal one will provide the point of attachment for one of the *arrector pili* muscles, to be described later. In the meantime, as a result of keratinization, the follicle undergoes a change so as to form a central *shaft* surrounded by a space between it and the follicular wall. The duct of the developing sebaceous gland opens into this space. Keratinization of the hair shaft becomes complete about half the distance from the surface. Within the follicle (Fig. 5.35) the hair shaft is surrounded by two layers of cells which do not extend beyond the limits of the follicle. *Huxley's layer* is the inner layer of this sheath; *Henle's layer* is the outer one. The hair shaft is entirely cellular and is composed of a central core, the *medulla*; a middle *cortex*; and an outer covering of scaly cells, the *cuticle*. Fine, downy, vellous hairs lack a medulla. At the base of the follicle the hair is expanded slightly to form a bulb-like enlargement, the "root" of the hair. All growth takes place at the root, where the cells of the stratum germinativum are proliferating.

Beyond this point the cells gradually die, the shaft of the hair thus being composed of dead, cornified cells.

Hairs which grow to a considerable length before they loosen and are shed are said to be *angora*. The head hair of man is of this type. *Definitive* hairs, on the other hand, grow to a certain length and then stop. They are then shed and quickly replaced. The definitive hair follicles have precisely controlled periods of growth and rest. Eyelashes, eyebrows, and body hairs are of the definitive type. Angora hairs may persist for several years, but definitive hairs usually last only a few months.

Hairs do not emerge vertically from the skin but project at an acute angle. Associated with each hair is a small involuntary dermal muscle, the *arrector pili* (Fig. 5.36). It extends from the basal part of the hair to the upper part of the dermis and lies on the side toward which the hair slopes. Contraction of the arrector pili muscles causes the hairs to stand

FIG. 5.33 Arrangement of tori on hind foot of rat.

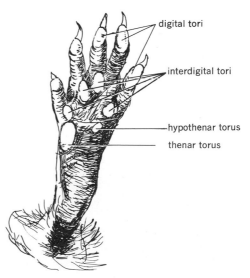

digital tori

interdigital tori

hypothenar torus

thenar torus

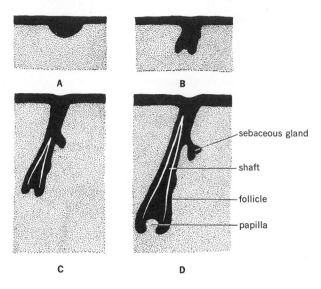

A   B

sebaceous gland

shaft

follicle

papilla

C   D

FIG. 5.34 Semidiagrammatic representation of stages in the development of a hair.

erect. In man this brings about a condition called *cutis anserina* or gooseflesh (duck bumps). In cats and dogs the raising of the hair, particularly along the back of the neck, is often observed when the animal is in danger. This is believed to be a protective adaptation which gives a false contour, or outline, to the body, making the animal appear larger and more "fierce" to an attacker. Arrector pili muscles are not associated with eyelashes or with nasal hairs.

Variations in the shape of hairs in cross section determine the degree of straightness or curliness. Straight hair is round; kinky hair is flat. Oval, or elliptical, hairs show different degrees of curliness or waviness.

Hairs in different parts of the body show some variation in the direction in which they

FIG. 5.35 Diagram showing structure of a hair and its follicle. (*After Kingsley, "Comparative Anatomy of Vertebrates," McGraw-Hill Book Company, by permission.*)

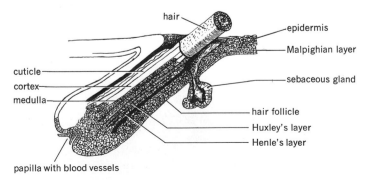

hair

epidermis

Malpighian layer

cuticle

cortex

medulla

sebaceous gland

hair follicle

Huxley's layer

Henle's layer

papilla with blood vessels

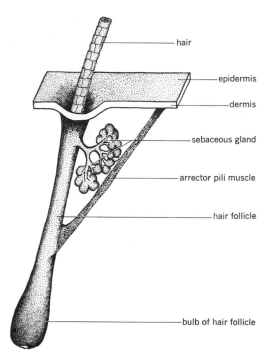

hair

epidermis

dermis

sebaceous gland

arrector pili muscle

hair follicle

bulb of hair follicle

FIG. 5.36 A hair follicle, showing its relationship to the skin, a hair, a sebaceous gland, and an arrector pili muscle.

grow or slant. Hair currents, or patterns, are often evident. When hair currents diverge, *parts* occur. *Crests,* or *tufts,* appear when such currents converge. In some regions *whorls,* or *vortices,* indicate a spiral arrangement. The familiar cowlick of humans is an example of a whorl. Although in most mammals the hair parts along the middorsal line and slopes toward the ventral side, sloths show the opposite condition. Here, associated with the animal's habit of hanging upside down in the trees, parting occurs on the midventral line and the hair slopes dorsally. This makes it easier to shed water in the tropical rain forests where sloths live.

Color of hair The hair in various mammals shows tremendous range in color.

Four factors are responsible for the color and luster of hair: (1) the color of the pigment in the intercellular spaces in the cortex; (2) the amount of pigment; (3) the character of the hair surface, whether smooth or rough; (4) the amount of air in the intercellular spaces of the medulla. Browns seem to predominate in most forms, but in man the color seems to have black and red pigments as its foundation. Absence of one or both, or different combinations or dilutions of the two, apparently bring about the many variations encountered.

White hair is due not to the presence of white pigment but to a lack of all pigment, light being reflected in all directions from air spaces, particularly those in the medulla. This embodies the same principle as when a transparent piece of ice is crushed into many smaller particles, giving the mass a whitish appearance. Gray hair may be the result of reduction of pigment and reflection from an increased number of air spaces. Usually, however, the term is used when white hairs are intermingled with pigmented hairs.

Although pigment deposition generally does not cease until the latter part of life, this is not true of all mammals. The varying hare, arctic fox, stoat, and others are remarkable in that as winter approaches, the old brown pelage is replaced by a white coat which provides protective or aggressive coloration, as the case may be. In these mammals, pigment deposition is seasonal.

In man the deposit of pigment may be interrupted or stopped after sickness or extreme nervous shock. It is of course impossible for hair to turn white "overnight" since the pigment already in the hair cannot be retracted.

Shedding of hair In mammals which show decided differences between summer and winter coats, the hair is shed seasonally.

In many mammals, however, there is a constant shedding and renewal. If a hair is pulled out, the place where it "gives" is usually just above the stratum germinativum at the base of the follicle. A new hair then usually grows in its place. If there is complete destruction of the stratum germinativum and papilla, regeneration will not occur.

In man, unlike many mammals in which there is wholesale shedding of hair, each follicle has its own independent cycle of shedding and renewal. As a growth period terminates, mitotic activity in the stratum germinativum gradually ceases and the dermal papilla atrophies. The base of the hair loosens, and the hair is shed. The follicle then remains quiescent for a variable length of time, and then the cells begin to divide actively again. Baldness in man is due to a hereditary factor together with the presence in the circulatory system of the male hormone testosterone.

Density of the hairy coat The density of the hair on the body shows much variation in different mammals. Those living in cold climates have the heaviest coats of all. Tropical forms are often sparsely covered. A permanent abode in the water is associated with an almost total absence of hair in cetaceans and sirenians. Even among races of men, marked variations exist. Hair in man is of little importance in keeping the body warm. Hair follicles are significant, however, in that they provide points of origin for regeneration of epidermal tissue in the repair of epidermis destroyed by burns or abrasions of one kind or another.

The quality of a fur, or pelt, depends for the most part on the degree of development of the underfur. This consists of large numbers of short, densely arranged hairs interspersed among the longer and coarser ones. Animals living in cold climates have the heaviest underfurs. This special coat of hair prevents heat loss by providing an insulating layer of air between the skin and the outer hair surface. As the cold season approaches, the underfur begins to grow; when it is fully developed, the fur is said to be *prime*.

Special types of hair The type of hair in different parts of the same animal may vary. Manes like those of the horse and lion, crests and tufts in certain regions, long tail hairs, eyelashes and eyebrows, dust-arresting hairs in the nose are all hairs of special types which are, in the main, adaptive in function. They serve as an aid in protection from enemies, in defense from insects, and in guarding delicate membranes from foreign particles. In man the head hair clearly serves the function of protection from sun and rain. Public and axillary hair aid in decreasing friction between the limbs or between limbs and body during locomotion.

Other special hairs include *vibrissae* (whiskers, or feelers) found on the snouts of most mammals. They are best developed in nocturnal mammals. These hairs are unusually sensitive, their follicles having an abundant nerve and blood supply at the base, the latter being in the form of cavernous sinuses closely resembling those of erectile tissue (page 334). Other hair follicles, even though possessing a good nerve supply, are less sensitive than vibrissae since no cavernous sinuses are associated with them. Increase in pressure of blood in the sinuses is transmitted to the hair shaft, stimulating the sensitive nerves supplying the vibrissae.

The spines, or quills, of the porcupine have barblike scales at the tips and are loosely attached at the base. They cannot be projected but are easily pulled out when the barbed ends become embedded in the flesh of an enemy. Practically all the main types of hair are to be found in the porcupine.

Differences in quality and distribution of

hair in different regions are often marked in the sexes and furnish some of the outstanding secondary sex characteristics that distinguish male from female. The beards of men and the hairy covering of the chest, arms, and legs are in contrast to the relatively smooth condition of the skin in women. The heavy mane of the lion clearly distinguishes him from the lioness. Even in rats, where secondary sex characteristics are not very clear cut, the coarser texture of the fur of the male makes it easy to distinguish the sexes by anyone accustomed to handling them.

BEAKS AND BILLS

In turtles and tortoises and in all modern birds, teeth are lacking. Each jawbone is covered with a modified epidermal scale which forms the beak or bill. In turtles and tortoises the beak is hard and sharp.

Among birds, great variation is found in the shape of the beak (Fig. 5.37) correlated with its use in procuring food. Seedeating birds usually have short, rather blunt beaks; insect eaters possess long and narrow beaks which are not so strong as those of seedeaters. The long, strong, hooked beaks of birds of prey are well fitted for their methods of obtaining food.

Applied to birds, the words *beak* and *bill* are used interchangeably. The term *bill* is more commonly used by ornithologists.

The bill of the duckbill platypus is soft and pliant. It is not covered with a modified epidermal scale and should not be confused with the type found in birds.

FIG. 5.37 Types of beaks in birds, showing various adaptations to different kinds of food: *A*, red-tailed hawk (flesh); *B*, robin (worms, insects, berries); *C*, cardinal (seeds); *D*, flycatcher (insects); *E*, red-breasted merganser (fish); *F*, wood duck (aquatic vegetation, fish); *G*, American bittern (fish, crustaceans); *H*, warbler (insects); *I*, woodpecker (grubs); *J*, crossbill (seeds in fir- and pinecones).

CLAWS, NAILS, AND HOOFS

The hard structures at the distal ends of the digits are derived from the horny layer of the integument. They grow parallel to the surface of the skin and wear away at the tip. Claws, nails, and hoofs are built upon the same plan and are homologous. The stratum lucidum of the epidermis is best developed at the base of these structures. True claws are present only in reptiles, birds, and mammals. They have also been described in certain fossil amphibians. Horny epidermal caps at the ends of the digits of the larval stage of the salamander *Onychodactylus* and the clawlike tips on the first three digits of the hind feet of the African clawed toads, *Xenopus, Hymenochirus,* and *Pseudohymenochirus,* may possibly foreshadow the appearance of claws in amniotes. The salamander *Hynobius* also possesses such structures in the adult. These structures are not true claws, however. Mammals are sometimes grouped into those which bear claws, nails, or both, and those possessing hoofs. The hoofed forms are usually referred to as *ungulates.*

CLAWS A claw is composed of a dorsal plate, called the *unguis,* and a ventral plate, the *subunguis,* or *solehorn.* The unguis is the better developed of the two. The claw covers the terminal phalanx of the digit, being thus reinforced. In the typical reptilian claw (Fig. 5.38) the unguis is curved both longitudinally and transversely, enclosing the subunguis between its lower edges. The claw forms a sort of cap over the end of the digit. The outer layers of reptilian claws are shed and renewed periodically.

Claws of birds are typically reptilian in structure, but many variations occur in association with the bird's mode of life. Shedding of the outer layers is unusual, growth and wearing away taking place at a fairly constant

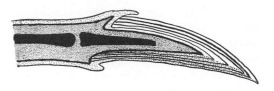

FIG. 5.38 Structure of a typical reptilian claw. (*After Bütschli, from Walter and Sayles, "Biology of the Vertebrates," copyright 1949 by The Macmillan Company and used with their permission.*)

rate. Although claws are generally present only on the feet of birds, the young hoatzin, *Opisthocomus cristatus* (Fig. 5.39), of Guyana and the valley of the Amazon, bears claws on the first two digits of the wings. In this respect the young hoatzin resembles the fossil *Archaeopteryx,* which has three clawed digits on each wing. The claws disappear when the hoatzin gets older.

In mammalian claws the subunguis is reduced and is continuous with the torus, or pad, at the end of the digit. Members of the cat family have retractile claws which, when not in use, are withdrawn into a sheath and are thus protected. In lemurs and tarsiers nails are present on certain digits of the hind feet, and claws, on others (page 42).

NAILS. The unguis of the nail is broad and flattened, the subunguis being reduced to a small remnant which lies under the tip of the nail (Fig. 5.40).

FIG. 5.39 Wing of young hoatzin, showing claws on first two digits. (*After Pycraft.*)

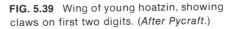

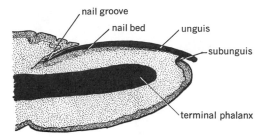

FIG. 5.40 Longitudinal section through end of human finger, showing arrangement of nail.

HOOFS. The unguis in the hoofs of ungulates curves all the way around the end of the digit and encloses the subunguis within it (Fig. 5.41). The torus lies just behind the hoof and is called the *frog*. Since the unguis is of harder consistency than the subunguis, it wears away more slowly, and a rather sharp edge is thus maintained. The surefootedness of many ungulates depends upon this factor. Workhorses are customarily shod to prevent the unguis from wearing away too rapidly. Metal shoes are nailed to this portion of the hoof. Since the hoof grows continuously, the shoes must be removed occasionally. The unguis is then pared down and the shoe nailed on again.

HORNS AND ANTLERS

Except for the hornlike structures in a few lizards, horns are found only in mammals. Certain dinosaurs possessed horns on the head consisting of bony projections, each of which was probably covered with a horny epidermal cap. The so-called horned "toads" of the southwestern United States are lizards with numbers of such structures on their heads. Each bony projection of the skull is covered with a horny epidermal scale, forming a rather sharp spine. These are of high protective value. Among mammals, horns are found only in certain members of the order Artiodactyla and in the rhinoceros of the order Perissodactyla. Four types are recognized, the keratin fiber horn, the hollow horn, the pronghorn, and the antler.

FIG. 5.41 Structure of hoof: *A,* median sagittal section of distal end of leg of horse; *B,* ground surface of hoof. (*After Ellenberger in Leisering's Atlas, based on Sisson, "The Anatomy of the Domesticated Animals," W. B. Saunders Company, by permission.*)

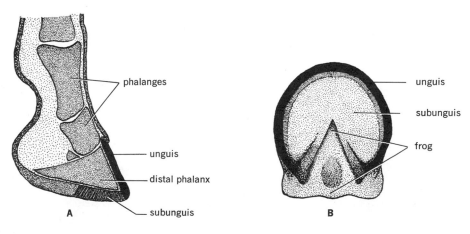

KERATIN FIBER HORNS. This type of horn is found only in the rhinoceros. It is a conical structure composed of a mass of hardened, keratinized cells growing from the epidermal covering of long, dermal papillae. A fiber, somewhat resembling a very thick hair, grows from each papilla. Cells in the spaces between the papillae serve as a cement substance, binding the whole together. The fibers are not true hairs, since their bases are not embedded in follicles extending into the dermis. These weapons are median in position on the head. A roughened, bony excrescence supports the horn. Only one horn is present in the Indian species, but the African form has two, the larger one being the more anterior (Fig. 3.33).

HOLLOW HORNS. The hollow horn is the type found in cattle, sheep, goats, and certain other mammals. In some species they occur only in males. Each horn consists of a

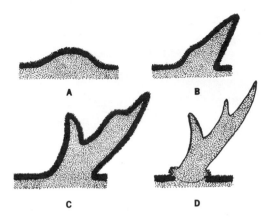

FIG. 5.43 Stages in the development of an antler. *D* shows the final stage after the covering skin has been sloughed.

projection of the frontal bone of the cranium covered by a cornified layer of epidermis. A cavity continuous with the frontal sinus extends into the bony projection. The horny layer is not shed. The powder horns of early settlers of America were prepared by loosening the horny covering from the bone beneath.

PRONGHORNS. A unique type of horn, the pronghorn, is found only in the antelope *Antilocapra americana,* of the western United States. It consists of a projection of the frontal bone covered with a horny epidermal sheath (Fig. 5.42). The sheath usually bears one prong, although as many as three have been observed. The horny covering is shed annually, and a new one forms from the epidermis which persists over the bony projection.

ANTLERS. Males of the deer tribe possess branched antlers which project from the frontal bones. Only in the reindeer, giraffe, and caribou do both sexes have antlers. In its fully developed state an antler is composed of solid bone. It is therefore of

FIG. 5.42 Structure of horn of the pronghorn antelope.

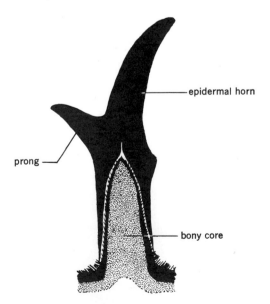

epidermal horn

prong

bony core

mesodermal origin, and properly should not be called a horn.

In the young male an outgrowth of the frontal bone develops on each side (Fig. 5.43). This is covered with soft skin, the blood vessels in the dermis bringing minerals and other materials necessary for growth of the structure. The antler is then said to be "in velvet." It is sensitive and warm to the touch. When it has grown to a certain extent, a wreathlike burr grows out from the base. This cuts off the blood supply to the skin covering the antler. The skin then dries, cracks, and is rubbed off in shreds by the deer. This usually occurs as autumn approaches. At the end of the first season the antler appears as a single spike. In the spring certain degenerative changes occur between the frontal bone and the burr. The bony antlers become loosened and are shed. The skin then closes over the area. Soon a new antler begins to grow, but this time it is branched. The entire process is then repeated. Each year new branches appear, and the structure becomes more complicated. Antler growth is apparently regulated by hormones secreted by both the testes and the anterior lobe of the pituitary gland.

In giraffes the antlers are small and inconspicuous and remain permanently in velvet. They are never shed. Antlers are secondary sex characters in other species, with the exception of the reindeer and caribou. They reach the height of their development just before the mating season when it is customary for the males to fight for domination of the herd.

SUMMARY

1. The vertebrate integument consists of an outer, ectodermal, epithelial layer, called the epidermis, and an underlying mesodermal layer, the dermis, composed largely of connective tissue, blood and lymphatic vessels, nerves, and smooth-muscle fibers.

2. Pigment in the integument occurs in the form of small granules in the cells of the lowest layers of the epidermis, or in special branched cells called chromatophores in which the pigment may become dispersed throughout the cell or concentrated in its center around the nucleus. Chromatophores of this type are usually located between the epidermal and dermal layers. They are not present in birds or mammals.

3. Both epidermis and dermis give rise to various integumentary structures which are, for the most part, protective and adaptive in character.

4. From the epidermis are derived (*a*) skin glands of the following types: mucous, poison, sweat, sebaceous, ceruminous, mammary, and scent (photophores are light-producing modifications of skin glands); (*b*) scales, formed from the hardened, horny, corneal layer. They include epidermal scales of reptiles, birds, and mammals; feathers; hair; claws; nails; hoofs; spurs; true horns; beaks and bills.

5. From the dermis are derived (*a*) bony scales of the ganoid, placoid, ctenoid, and cycloid types; (*b*) bony plates in the skin of certain reptiles and mammals; (*c*) membrane bones of the skull.

6. The dermis, through its blood vessels, supplies nourishment to the developing epidermal structures but is not otherwise involved in their formation. Dermal structures arise within the dermis itself. In certain cases bony dermal plates are closely associated with epidermal scales which lie above them.

7. The evolution of the integument is fundamentally correlated with the transition from aquatic to terrestrial life. The highly developed epidermis of terrestrial forms, with its many modifications fitting the animal for life on land, is indicative of the trend. Dermal derivatives are basically more primitive.

BIBLIOGRAPHY

Cohen, J.: Feathers and Patterns, *Adv. in Morphogen.,* 1:1–38 (1966).

Gross, J.: The Skin, *Sci. Am.,* 204(5):120–126 (1961).

Marples, M. J.: Life of the Human Skin, *Sci. Am.,* 220(1):108–115 (1969).

Montagna, W.: "The Structure and Function of the Skin," Academic, New York, 1962.

Symposium on the Vertebrate Integument, *The American Zoologist,* Vol. 12, n. 1, 1972.

6 SKELETAL SYSTEM

The term *skeleton* refers to the framework of the animal body. In vertebrates it is composed of cartilage or bone, or a combination of the two, and serves for support, attachment of muscles, and protection for certain delicate vital organs which it more or less completely surrounds. Furthermore, it maintains the definite form of the animal and is provided with joints so that movement is possible.

The word *endoskeleton* is used to denote internal skeletal structures. The presence of an endoskeleton is one of the distinguishing characteristics of chordates. Although in most chordates the integument is soft and contains no hard skeletal parts, many members of the phylum have bony elements, derived from the dermis, present in the skin. Dermal scales of integumentary origin, and providing a protective armor, are present in most fishes living today, in a few amphibians, in crocodilians, and in turtles. The term *dermal skeleton* is used in referring to such structures and their derivatives. Sometimes the dermal skeleton is spoken of as the *exoskeleton,* but this term is more properly applied to the skeleton of in-

vertebrates. The presence of bony, dermal plates in the extinct ostracoderms and placoderms indicates that they appeared very early in the evolutionary history of vertebrates and should be considered to be primitive structures. However, in forms having both endoskeletal and dermal skeletal structures, the endoskeleton appears much earlier during embryonic development. Many vertebrates appear to lack a dermal skeleton, but their teeth and the membrane bones of their skulls are dermal derivatives.

CONNECTIVE TISSUES The principal functions of connective tissues are to bind other tissues together and to give support to various structures of the body. In contrast to epithelia, connective tissues are distinguished by the presence, in most cases, of considerable quantities of *intercellular substance,* or *matrix.* The nature of the intercellular material is characteristic of each type. Connective-tissue cells usually make up only an inconspicuous portion of the tissue. It is the intercellular substance, produced by the cells, which constitutes the main bulk of the connective-tissue mass.

Connective tissues are of mesodermal origin and develop from mesenchyme. There are four main types of connective tissue: (1) connective tissue proper; (2) blood and lymph; (3) cartilage; (4) bone. It is not always possible to distinguish sharply between the various types of connective tissues since there are several intergrading forms.

Loose connective tissue (Fig. 6.1) consists of several types showing rather wide differences in the nature of the intercellular substance. It is composed of fibers among which various types of cells are scattered. Mechanical forces undoubtedly are important in influencing the manner and direction in which the fibers are arranged.

An important constituent of connective tissue proper is referred to as the *amorphous ground substance.* This is an intercellular substance composed of microfibrils and mucopolysaccharides, in which cells, fibers, etc., are embedded. It permits diffusion of nutrient substances, water, gases, and wastes over considerable distances. This is of importance in areas where small blood vessels are absent, particularly during developmental stages.

Loosely Organized Connective Tissue Also called *areolar tissue,* this is the most widely distributed type. The subcutaneous tissue between skin and deep structures is the most familiar example. It is composed of loose fibers which ramify in all directions and of scattered cells of various types. Two types of fibers, collagenous and elastic, are usually recognized in areolar tissue.

The cellular elements in areolar tissue are of several types, some of which are apparently *undifferentiated* or *partially differentiated mesenchymal cells.* These may, at various times, give rise to various types of connective-tissue cells, *fibroblasts* or *fibrocytes, macrophages* or *histiocytes, multinucleated giant cells, lymphoid cells, mast cells,* and *fat cells. Chromatophores,* or *pigment cells,* are derived from neural crest. Other cells, present in both epidermis and dermis, may engulf pigment particles instead of manufacturing them, thus behaving like macrophages. This type of connective tissue is of great importance in forming a groundwork for support and provides the materials for reconstruction or repair after injury and for combating infection.

Dense Connective Tissue This is composed essentially of the same materials as loosely organized connective tissue, but the collagenous fibers are thicker and more numerous and are arranged in a particular pattern. They are very closely interwoven to form a dense matting. Cellular elements are relatively few. The bundles of collagenous fibers are woven together with enough elastic fibers to permit a certain degree of stretch. The dermis of the skin, portions of the walls of the digestive tract and blood vessels, and certain parts of the urinary excretory ducts are composed of densely organized connective tissue.

Regularly Arranged Connective Tissue This is sometimes considered to be a variety

FIG. 6.1 Loosely organized connective tissue.

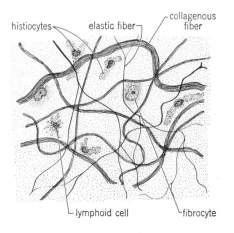

histiocytes — elastic fiber — collagenous fiber

lymphoid cell — fibrocyte

of dense connective tissue. It is made up, for the most part, of densely organized collagenous fibers which show a definite arrangement. The fibers run parallel to each other, and fibrocytes are the only cellular elements present. *Tendons,* by means of which muscles are attached to bones or other structures, are composed of regularly arranged connective tissue. They are very strong and inelastic. *Ligaments,* which connect bones or support internal organs, are similar to tendons, but the collagenous fibers are not so regularly arranged and elastic fibers are also present. The *nuchal ligament* at the nape of the neck of certain grazing mammals has predominantly yellow elastic fibers. It serves to sustain the weight of the head and help to prevent muscle fatigue.

Specialized Connective Tissues Several types are present in addition to those already referred to:

Mucous connective tissue, a type found in embryos.

Elastic tissue, in which parallel yellow elastic fibers predominate, is found in various parts of the body. It can be stretched, and then springs back to its normal position when released, a fact of great importance in the regions where this tissue occurs. Examples are the true vocal cords and the flaval ligaments between adjacent vertebrae which serve to maintain the upright posture of man.

Adipose, or *fatty, tissue* (Fig. 6.2) is really loosely organized connective tissue in which fat cells are particularly abundant. They take the place of other connective-tissue cells, which are therefore far less numerous. Adipose tissue stores fat, which is then available as a source of heat and other forms of energy for the body.

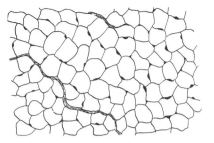

FIG. 6.2 Adipose, or fatty, tissue. In this preparation the fat has been dissolved and removed from the cells so that only a skeletal framework remains.

CARTILAGE Cartilage is a type of connective tissue which forms an important part of the endoskeleton in all vertebrates, although in higher forms it is much reduced in amount, most of it having been *replaced* by bone. In an embryo much of the endoskeleton first appears in cartilaginous form. In elasmobranch fishes the skeleton does not go beyond the cartilage stage. In higher vertebrates the cartilage which is later replaced by bone is spoken of as *temporary cartilage,* whereas that which is retained throughout life is *permanent cartilage.*

It will be recalled that connective tissues are characterized by the presence of intercellular material, or matrix, secreted by the cells. During the development of hyaline cartilage, cartilage cells, or *chondrocytes,* differentiate from mesenchyme. They become rounded and begin to secrete a clear matrix about themselves which forces the cells apart so that they are more or less isolated from one another by the matrix. The cells lie in spaces called *cartilage lacunae* (Fig. 6.3). When a chondrocyte divides, each daughter cell forms matrix so that only one or two cells usually lie in a lacuna. Occasionally three or four cells are observed together, forming what is called an *isogenous group.* With the formation of new matrix, the cartilage grows, this

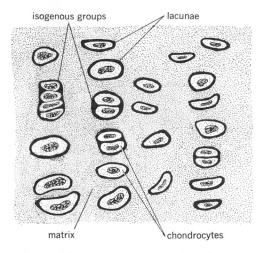

FIG. 6.3 Microscopic structure of hyaline cartilage.

type of growth being known as *interstitial growth*. It is restricted to early stages of development. The surface of the developing cartilage becomes covered by a dense layer of connective tissue, the *perichondrium,* to which muscles, tendons, and ligaments later become attached. Increase in diameter of a cartilage is brought about by proliferation of chondrocytes in the inner, or *chondrogenic,* layer of the

perichondrium. This is termed *growth by apposition.* Cartilage is an avascular tissue. Chondrocytes receive nourishment and dispose of wastes by diffusion through the matrix. Impregnation of the matrix with calcium salts often occurs, rendering it hard and rigid. Bone, which replaces the cartilage, has an entirely different histological structure from that of calcified cartilage. *Hyaline cartilage* is the most abundant type. It is homogeneous, translucent, and of a bluish-green cast. The distal parts of the ribs (*costal cartilages*), the articular surfaces of bones, most cartilages of the larynx, trachea, and bronchi, are examples of hyaline cartilage in man. Other types of cartilage include *elastic cartilage,* in the external ear and epiglottis, and *fibrous cartilage* in the intervertebral discs and places of attachment of tendons and ligaments.

BONE Bone, which forms most of the skeleton in higher vertebrates, is another type of connective tissue. Bone-forming cells, called *osteoblasts,* differentiate from mesenchymal cells. They have numerous branching processes by means of which adjacent cells are at first connected. Intercellular matrix, ap-

FIG. 6.4 Microscopic structure of bone. (*Drawn by G. Schwenk.*)

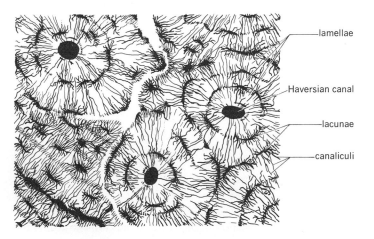

parently secreted by the osteoblasts, is present between the cells. The matrix quickly becomes impregnated with calcium and other inorganic salts which make up about 30 percent of bony tissue. Each bone-forming cell, now known as an *osteocyte,* lies in a space, the *lacuna,* surrounded by bony matrix. Before long the cytoplasmic connecting processes are withdrawn, leaving in their places tiny canals called *canaliculi* (Fig. 6.4). They provide a means by which tissue fluid may gain access to the osteocytes trapped in the lacunae. Contrary to the condition in cartilage, blood vessels and nerves ramify through bony tissue, gaining access through *nutrient foramina.* They then pass through small *Haversian canals.* Arranged concentrically about each Haversian canal are bony *lamellae* composed of matrix. Between adjacent lamellae are the lacunae containing osteocytes. The canaliculi penetrate the lamellae. By this means tissue fluid may reach osteocytes. The canaliculi penetrate the lamellae. By this means tissue fluid may reach osteocytes at some distance from a Haversian canal. A *Haversian system* consists of a Haversian canal together with its surrounding lamellae, lacunae, and canaliculi. Irregularly disposed *interstitial lamellae* fill in spaces between Haversian systems. They represent remnants of old Haversian systems which have been partially removed during reconstructive processes made · necessary during growth of the bone. It is interesting that the Haversian systems of different mammals, although basically similar, show idiosyncrasies in arrangement great enough in magnitude to enable one to distinguish certain species on this basis alone. Bones are tightly surrounded by a dense layer of connective tissue, the *periosteum,* comparable to the perichondrium surrounding cartilage, to which muscles, tendons, and ligaments are attached. Increase in thickness or diameter of bones is the result of activity of bone-forming

cells located on the inner surface of the periosteum.

A fundamental difference between cartilage and bone is that the matrix of bone becomes calcified very soon after it is formed and is rigid. Electron-microscopic studies have also revealed many fine points of difference in the structure of the matrices of calcified cartilage and bone. Growth of bone is by apposition alone, since interstitial growth is precluded by the presence of the rigid matrix.

Bone is formed in two ways: (1) by direct ossification in connective tissue without an intervening cartilage stage, such bone being referred to as *membrane,* or *dermal,* bone; (2) by replacement of preexisting cartilage, in which case it is known as *endochondral, cartilage,* or *replacement, bone.*

Membrane Bone Certain flat bones of the face and of the top of the cranium are typical membrane bones. The frontal and parietal bones are examples. Phylogenetically they represent dermal plates which have moved from their original peripheral location in the integument and have sunken inward to a deeper position, where they have become attached to the true endoskeleton. Such bones are formed by a method known as *intramembranous ossification.* A typical membrane bone consists of two parallel layers of *compact* bone, the *inner* and *outer* laminae, enclosing between them a layer of spongy-appearing *cancellous* bone, the diploe (Fig. 6.5). Periosteum covers the surface of both inner and outer tables. Compact, or *periosteal,* bone and spongy, or *cancellous,* bone are not basically different. They actually represent different arrangements of the same type of tissue. The small spaces between the irregular meshwork of spongy bone are filled with a very vascular tissue known as *bone marrow.*

Endochondral Bone Endochondral bones, unlike membrane bones, go through a

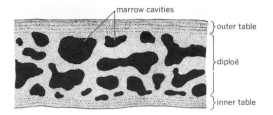

FIG. 6.5 Section through a portion of a membrane bone (parietal bone of man) showing cancellous bone bounded by compact periosteal bone on both inner and outer surfaces.

cartilaginous stage in their development. The cartilage is usually a miniature of the bone which is to replace it. The term *endochondral bone formation* is used to describe this type of bone development, since it takes place within preexisting cartilage.

The main part, or shaft, of a long cartilage bone, such as the tibia, is called the *diaphysis.* The first indication of bone formation consists of the appearance of a ring of bony tissue around the center of the cartilaginous diaphysis (Fig. 6.6). The original perichondrium becomes the periosteum. *Osteoblasts,* originally

FIG. 6.6 Diagrams indicating progressive ossification of a long bone: *A,* collar of compact periosteal bone surrounding cartilage in middle region of diaphysis; *B,* endochondral bone appearing as primary center of ossification; *C,* marrow cavity appearing as a result of dissolution of cancellous bone in center; *D,* appearance of secondary centers of ossification, the epiphyses, at either end of structure, separated from primary center by epiphysial plate; *E,* later stage, similar to *D; F,* closure of epiphyses with disappearance of epiphysial plates; articular cartilage at either end of bone (no further increase in length can take place).

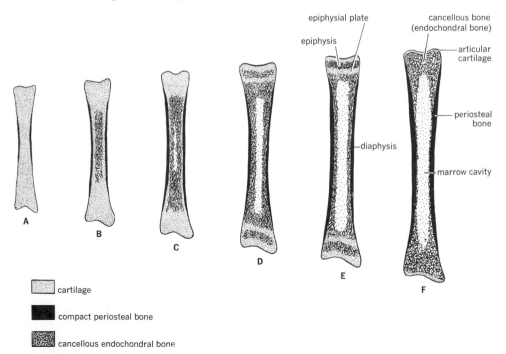

derived from the inner layer of the perichondrium, give rise to compact, periosteal bone. At about the same time that this bony ring is forming, degenerative changes occur within the cartilage in the center of the diaphysis. Capillaries, together with osteoblasts in the form of *periosteal buds,* enter the cartilage and begin to erode it in some unknown manner so as to honeycomb the area with a number of intercommunicating channels which become filled with embryonic bone marrow. The irregular cartilaginous bars, or trabeculae, become calcified and serve as ossification centers about which osteoblasts, derived from the embryonic bone marrow, aggregate. These cells deposit bone around the cartilaginous bars. This forms a bony meshwork of trabeculae, with the result that the center of the diaphysis is composed of a mass of spongy bone. In the meantime osteoblasts beneath the periosteum are depositing compact bone which surrounds the spongy bone. While these processes are going on, the cartilage at the ends of the developing structure continues to grow in length. Increase in diameter is brought about by the continued deposition of periosteal bone. The center of ossification, which begins in the center of the diaphysis, gradually extends toward the ends of the developing bone, which, however, remain cartilaginous and continue to grow until a much later period. In mammals, and to a lesser degree in reptiles, secondary centers of ossification later appear at the ends of the bone. These are known as *epiphyses* (Fig. 6.6D and E). The time of appearance of the epiphysial centers of ossification varies in different bones of the body. Between each epiphysis and the diaphysis there remains a cartilaginous area, the *epiphysial plate.* Until adult stature is attained, increase in length of the bone takes place at the epiphysial plate. New cartilage forms at its terminal portion, while the inner part is being replaced by developing

bone from the diaphysis. Eventually the epiphysial plates are entirely replaced by bone, and the epiphyses are firmly united to the diaphysis. Further growth in length of the bone can then no longer occur. The ends of the bones are covered with articular cartilage throughout life. Some bones, e.g., certain phalanges, have only one epiphysis.

While growth is taking place, constant internal alterations and reconstructions are going on, leading to the formation of a large marrow cavity in the center of the bone. Not only does this space house the bone marrow, but the resulting cylindrical shape of the bone has the mechanical advantage of providing greater strength. Destruction of old bone appears to be the function of certain multinucleated giant cells called *osteoclasts,* which are possibly identical with so-called foreign-body giant cells found in many other parts of the body.

We have thus seen how the center of the diaphysis of a long bone is first composed of hyaline cartilage. This becomes eroded, and the remaining cartilaginous trabeculae become calcified. Spongy bone *replaces* the cartilage and later, through osteoclastic activity, is itself destroyed, so that finally only a marrow cavity remains, surrounded by compact bone of periosteal origin (Fig. 6.7).

BONE MARROW *Hemopoietic tissue,* in which the various kinds of blood cells are formed, occurs in lower vertebrates in a variety of structures. It is in anuran amphibians that the bone marrow is first concerned with blood-cell formation. In these amphibians as well as in reptiles and birds, *all* types of blood cells are formed in the bone marrow. In mammals, however, the hemopoietic function of bone marrow is restricted to the production of *red blood corpuscles (erythrocytes), blood platelets,* and certain *white corpuscles,* the *granular leukocytes.* Other kinds of white cor-

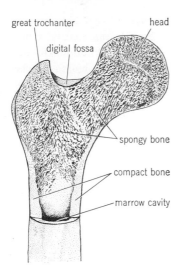

great trochanter

head

digital fossa

spongy bone

compact bone

marrow cavity

FIG. 6.7 Section through proximal end of fully developed human femur, showing relation of spongy bone, compact bone, and marrow cavity.

puscles are formed, for the most part, in lymphoid tissue. In mammalian embryos and in the newborn the only type of bone marrow that is present is *red bone marrow,* or *myeloid tissue,* which has the hemopoietic properties. As an individual grows older, however, the red bone marrow is replaced in many regions by *yellow,* or *fatty, bone marrow,* which is not actively engaged in hemopoiesis but remains capable of blood-cell formation. Estrogens exert an inhibiting action on erythrocyte production. Another substance, *erythropoietin,* probably from the kidneys, stimulates red-cell formation. In adult human beings and other mammals, red bone marrow is confined to the diploe in the membrane bones of the skull, the proximal epiphyses of the humerus and femur, the ribs, sternum, and vertebrae. Other marrow cavities serve primarily as storage places for fat.

JOINTS The term *joint* refers to an articulation between cartilages or bones. Three general types are recognized: (1) *synarthroses,* or immovable joints; (2) *diarthroses,* or freely movable joints; and (3) *amphiarthroses,* or

FIG. 6.8 Method of formation of a diarthrodial joint: *A,* beginning of joint cavity appearing in the mesenchyme between two adjacent cartilages; *B,* section through fully formed joint.

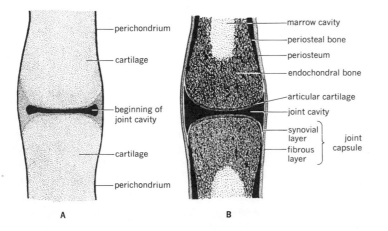

perichondrium

cartilage

beginning of joint cavity

cartilage

perichondrium

marrow cavity

periosteal bone

periosteum

endochondral bone

articular cartilage

joint cavity

synovial layer

fibrous layer

joint capsule

A

B

slightly movable joints. There are several types in each category. Among synarthroses are included such joints as *sutures,* which are the lines of junction of skull bones; *gomphoses,* in which a bony projection fits into a socket, as with teeth in the jawbones; *schindyleses,* in which one bone fits into a slit in another, as in the articulation of the lamina perpendicularis of the ethmoid bone with the vomer. Among diarthroses are *ball-and-socket joints* found in such places as the shoulder and hip; *hinge joints,* in the elbow and knee; *pivotal,* or *rotary, joints,* illustrated by the junction of atlas and axis; and *gliding joints* found in the intercarpal or intertarsal complex. *Amphiarthroses* have no joint cavity and include such structures as the pelvic symphysis between pubic bones.

In the formation of synarthroses the mesenchyme between bones may differentiate into connective tissue proper, cartilage, or bone, depending upon the type of joint and its location.

In the development of a diarthrodial joint (Fig. 6.8) spaces appear in the mesenchyme between adjacent bones, ultimately to form the *synovial,* or *joint, cavity.* This is lined with a *synovial membrane* composed of dense irregular connective tissue. The capsule which surrounds the joint is continuous with the periosteum of the two articulating bones. The synovial cavity contains a sticky, glairy *synovial fluid,* which serves as a lubricant. The ends of the bones are covered with hyaline cartilage. Although the synovial membrane is reflected over the articular cartilages during development, it is lacking in the adult.

DERMAL SKELETON

The dermal skeleton, consisting of bony structures, scales, and plates, originates in the dermis of the skin and is, therefore, of mesen-chymal origin. It should not be confused with *horny* scales and other structures derived from the epidermis. The various types of fish scales, already referred to, represent remnants of the dermal skeleton. They are composed of membrane bone or of substances closely related to bone. In the past it was believed that the placoid scales of elasmobranchs represent a primitive form of dermal derivative. With the realization that bony dermal plates were already present in ancient ostracoderms and placoderms, a different interpretation has been adopted. It is now believed that elasmobranchs, with their cartilaginous skeletons, have lost most of the dermal skeleton and that placoid scales merely represent remnants of the earlier dermal structures.

That teeth are undoubtedly dermal derivatives is indicated in elasmobranchs, in which the transition between placoid scales and teeth occurs at the borders of the mouth. The fact that both placoid scales and teeth in elasmobranchs develop under the influence of enamel organs, even though enamel is not formed over the exposed surfaces of placoid scales or elasmobranch teeth, further suggests a common origin and homology. Tooth structure and formation will be discussed later.

Fin rays supporting the peripheral portions of the fins of fishes are also of dermal origin. However, the cartilaginous or bony elements which support them at the base belong to the endoskeleton.

In the evolution of vertebrates there has been a tendency toward elimination of the dermal skeleton, which is fundamentally a protective device. In terrestrial forms such epidermal derivatives as hair, scales, feathers, etc., serve to protect the body in lieu of dermal structures. The dermal plates of extinct labyrinthodont amphibians and the scales of caecilians are continuing dermal structures. Likewise the bony plates in the skin of crocodilians, turtles, and a few snakes and lizards, together with the ventral dermal "ribs," or *gastralis,* encountered in *Sphenodon,* crocodilians, and *Archaeopteryx,* are of a similar nature.

The homology of the membrane bones of the skull of modern teleosts, amphibians, and amniotes with the dermal armor of the earliest known vertebrates is well established.

Except for the membrane bones of the skull, dermal elements are generally lacking in birds and mammals though dermal elements are found in the pectoral girdle. The antlers of deer are outgrowths of the frontal bones, formed under influence of the integument, and may be considered to belong to the dermal skeleton. Bony plates are present in the back and in the dorsal fins of some whales. Armadillos, however, are the only mammals with a well-developed, secondary derived dermal skeleton. The extinct glyptodons, which were also edentates, possessed a rigid armor. Whether the dermal skeleton of such edentates arose *de novo* or has been derived from reptiles has not been determined.

ENDOSKELETON

The endoskeleton, in which the elements are preformed in cartilage and may remain cartilaginous throughout life, is made up of the visceral and somatic portions. The term *visceral skeleton* refers to the elements which in fishes support the gills and jaws and in higher forms contribute to the skull and ear ossicles. The *somatic skeleton* includes all the remaining endoskeletal structures. The appendicular skeleton is made up of the skeletal parts of the paired pectoral and pelvic appendages, together with their respective girdles, by means of which the appendages are connected directly or indirectly with the *axial skeleton.*

The axial skeleton forms the main axis of the body and is made up of (1) the skull, (2) the vertebral column, (3) the ribs, and (4) the sternum. The last is usually considered to be part of the axial skeleton because of its topographical relationships. Its embryonic origin, however, suggests that it is really a portion of the appendicular skeleton (page 159).

AXIAL SKELETON

Notochord

The primitive axial skeleton consists of the notochord, a stiff but resilient form of connective tissue, which is replaced by the vertebral column and base of the skull in all but a few of the lower vertebrates. It lies beneath the neural tube, extending from the diencephalic region of the brain to the posterior or caudal end of the body. The notochord is usually encased within a sheath of dense fibrous connective tissue. In amphioxus, cyclostomes, and a few fishes it persists throughout life as the main axial support. Even in the lamprey, however, small, paired, segmentally arranged cartilages abut against the dorsal side of the notochord, forming a sort of incomplete arch over the spinal cord. They foreshadow the appearance of neural arches of vertebrae of higher vertebrates.

Spinal Column

The portion of the axial skeleton which protects the spinal cord is composed of a series of segmentally arranged *vertebrae.* In their aggregate they make up the *spinal,* or *vertebral, column,* which extends from the base of the skull to the tip of the tail. The vertebral column gives rigidity to the body and provides a place for the direct or indirect attachment of the appendage girdles and numerous muscles. A *neural arch* on the dorsal side of

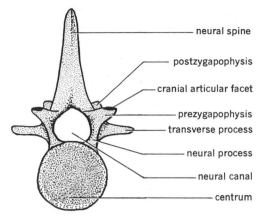

FIG. 6.9 Cranial view of typical mammalian vertebra.

each vertebra may surround and protect the spinal cord. Various projections extend outward from the vertebral column, thus furnishing surfaces for attachment of muscles and ribs. The separate vertebrae are bound together so as to provide rigidity together with a certain degree of flexibility. In some fishes the vertebral column is cartilaginous, but in most forms it is bony, having developed, for the most part, by the endochondral method of ossification.

In lower vertebrates there is much varia-

tion in regard to the elements of which each vertebra is composed, and considerable confusion exists concerning the homologies of their components. In higher forms, such as mammals, the structure is less complex.

A typical mammalian vertebra (Figs. 6.9 and 6.10) is composed of a solid, ventral, cylindrical mass, the *centrum,* dorsal to which is the neural arch surrounding the *neural,* or *vertebral, canal.* The ends of the centrum are smooth and joined to the centra of adjacent vertebrae by means of intervening cartilaginous *intervertebral discs.* The lateral portions of the neural arch consist of two *pedicles,* or *neural processes,* fastened to the dorsal part of the centrum on either side. A flattened bony plate, the *lamina,* extends medially from each pedicle to form the roof of the neural arch. A dorsomedian projection, the *neural spine,* or *spinous process,* is formed by the fusion of the two laminae. Anterior and posterior articular processes, the *prezygapophyses* and *postzygapophyses,* are projections which serve to join the neural arches of adjacent vertebrae. The prezygapophyses of one vertebra articulate with the postzygapophyses of the vertebra immediately anterior to it. Articular facets are present on both pre- and postzygapophyses; those on the prezyga-

FIG. 6.10 Lateral view of three typical vertebrae from young mammal, showing epiphyses (semidiagrammatic).

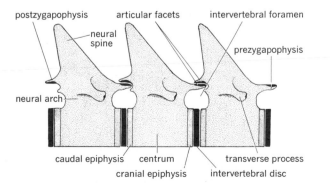

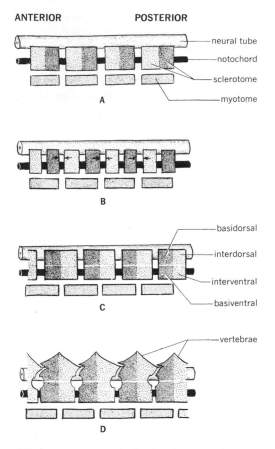

ANTERIOR   POSTERIOR

neural tube
notochord
sclerotome
myotome

A

B

basidorsal
interdorsal
interventral
basiventral

C

vertebrae

D

FIG. 6.11 Diagrams indicating how vertebrae are formed from the metameric sclerotomes of somites in relation to neural tube and notochord. The position of the myotomes is indicated by the rectangular blocks in the lower part of each drawing. *A,* heavy stippling represents caudal compact portion of sclerotome; light stippling represents less dense cranial region. *B,* the two halves of the sclerotome have split vertically. *C,* the dense caudal portion of one sclerotome has united with the less dense cranial portion of the next posterior sclerotome to form the primordium of a vertebra. Four centers of ossification represent arcualia as indicated. *D,* fully formed vertebra.

pophyses are usually on the dorsomedial surface, whereas those of the postzygapophyses are generally on the ventrolateral surface. Various other projections, often referred to as *transverse processes,* extend outward from the lateral sides of the vertebrae. There are several types: (1) *diapophyses,* arising from the base of the neural arch, serve for attachment of the tuberculum, or upper head of a two-headed rib; (2) *parapophyses,* from the centrum, serve for attachment of the lower, or capitular, head of a rib (Fig. 6.46); (3) *basapophyses,* extending ventrolaterally from the centrum, are believed to be remnants of the hemal arch; (4) *pleurapophyses,* projecting laterally, are extensions with which ribs have fused. Another type of projection, the *hypapophysis,* is found in certain forms, extending ventrally from the median part of the centrum. The anterior and posterior ends of the neural arches are indented so as to form *intervertebral notches.* When two vertebrae are in apposition, the notches together form an *intervertebral foramen,* which serves for the passage of a spinal nerve.

The vertebrae, as well as the ribs, are derived from the sclerotome regions of the somites during embryonic development. The mesenchymal cells of the segmentally arranged sclerotomes come to lie in paired masses on either side of the notochord. A single vertebra, however, is derived from parts of two adjacent sclerotomes (Fig. 6.11). Further development consists of a medial extension of mesenchymal cells so that they surround the notochord, forming the primordium of the centrum; some extend dorsally, surrounding the neural tube and forming the primordium of the neural arch; still others migrate in a ventrolateral direction to become *costal processes* concerned with development of the ribs (Fig. 6.12). Mesenchyme between the centra differentiates into the intervertebral discs. In the meantime the notochord gradually disappears, but remnants may persist between vertebrae as the *pulpy nuclei* of the intervertebral discs. The myotome regions of the somites alternate with the vertebrae. The muscle fibers which later develop from a single myotome thus come to pass from one vertebra to the next. Such an arrangement must exist if movement of the vertebral column is to take place.

Centers of chondrification appear in each developing vertebra, and soon a solid cartilaginous structure is formed. Resorption of tissue occurs at the junction of the costal processes and the rest of the vertebra, so that the rib cartilages become separated, at least in the thoracic region. Several centers of ossification appear later, and the cartilage is replaced by bone by the usual method of endochondral ossification. In many vertebrates (e.g., salamander) certain portions may form

by direct ossification of the mesenchyme of the sclerotomes. In mammals, epiphyses form at either end of the centrum much as in the development of long bones.

From the foregoing account it would appear that the formation of the vertebral column is relatively simple. However, comparative studies of the spinal column in different vertebrates point to a rather complex evolutionary history. In certain lower forms the vertebrae originate from a number of elements which may or may not retain their integrity. The presence of these separate components has been more or less obscured in the development of vertebrae of higher forms in which the only evidence of their existence may be indicated by the appearance of separate centers of chondrification and ossification during embryonic development. Paleontological studies indicate that primitively

FIG. 6.12 Cross section through the thoracic region of 16-day rat embryo, showing the appearance of a developing vertebra. (*Drawn by R. Speigle.*)

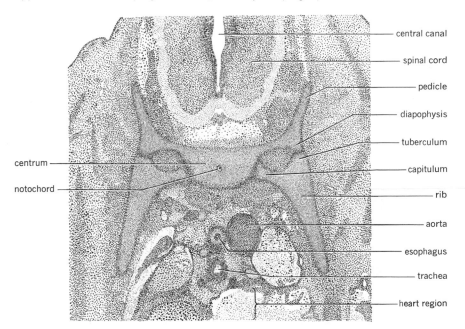

central canal
spinal cord
pedicle
diapophysis
tuberculum
capitulum
rib
aorta
esophagus
trachea
heart region

centrum
notochord

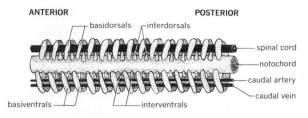

FIG. 6.13 Diagram showing primitive arrangement of cartilaginous arcualia derived from the sclerotomes in relation to spinal cord, notochord, caudal artery, and caudal vein.

each sclerotome on both sides, right and left, gave rise to four separate elements, the *arcualia,* two dorsal and two ventral. The dorsal ones are referred to as *basidorsal* and *interdorsal cartilages,* and the ventral ones as *basiventral* and *interventral cartilages* (Figs. 6.11 and 6.13). Basidorsals and basiventrals become the anterior parts of separate vertebrae. Interdorsals and interventrals, which are the less conspicuous elements, contribute to the posterior parts. Basidorsals fuse with their partners dorsally, forming the neural arches. The basiventrals in the tail region similarly give rise to hemal arches, which enclose the caudal artery and vein. Anterior to the tail region they usually contribute only to the centrum. Interdorsals and interventrals also contribute to

the centrum, although in some cases the interdorsals may extend dorsally and fuse to form an *interneural,* or *intercalary, arch* over the spinal cord (Fig. 6.14). The cartilaginous arcualia may eventually be replaced by bone. In addition to contributions from the arcualia, the centrum is also derived in part from mesenchyme between the arcualia and the notochord. This *perichordal mesenchyme* may undergo direct ossification. In teleosts and amphibians the perichordal component gives rise to the greater part of the centrum and the arcualia contribute but little.

The vertebrae of numerous fossil adult tetrapods possess two bony elements, the exact homologies of which are somewhat questionable. A pair of dorsolateral *pleurocentra* and a

FIG. 6.14 Trunk vertebrae of dogfish *Squalus acanthias: A,* sagittal section through three vertebrae; *B,* end view of vertebra; *C,* lateral view of three intact vertebrae.

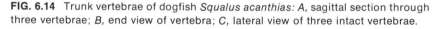
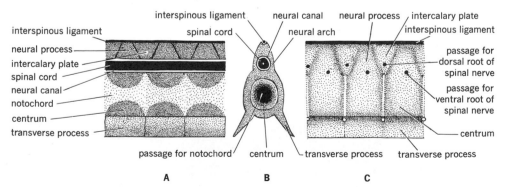

pair of ventral *hypocentra* (Fig. 6.15), present in many forms (Fig. 6.15), seem to have originated in the perichordal mesenchyme but to have fused with the interdorsal and basiventral arcualia, respectively. Many comparative anatomists are of the opinion that the pleurocentra, by further elaboration, and with a corresponding reduction of the hypocentra, have given rise to the centrum as found in reptiles and mammals. On the other hand, the hypocentra have played the more important role in centrum formation in amphibians. Here the pleurocentra have been reduced or have disappeared (Fig. 6.15). Because of insufficient embryological and paleontological evidence,

the exact evolutionary history of these vertebral components continues to be a matter for conjecture.

CYCLOSTOMES The notochord, which is enclosed in a thick *chordal sheath,* persists throughout life in cyclostomes. In lampreys small cartilages, two pairs per segment, are present both in trunk and tail regions. They abut upon the chordal sheath but do not meet above the spinal cord. These cartilages, which probably represent basidorsals and interdorsals, form rudimentary neural arches. Similar cartilages extend ventrally in the tail region to form hemal arches. In hagfishes small neural

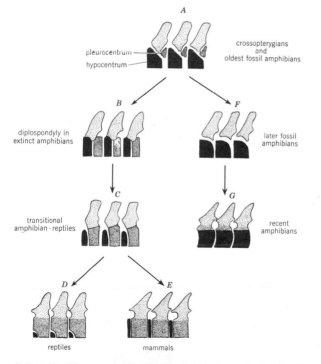

FIG. 6.15 Diagrammatic hypothetical scheme indicating two possible lines of evolution of vertebrae from ancestral crossopterygian fishes and the oldest known fossil amphibians, that on the left leading to the condition found in reptiles and mammals and that on the right leading to the condition found in recent amphibians. Black areas represent the hypocentrum; heavily stippled areas represent the pleurocentrum.

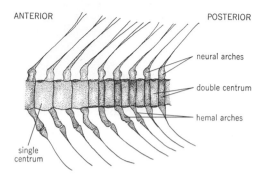

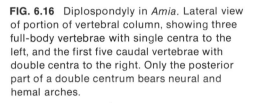

ANTERIOR POSTERIOR

neural arches

double centrum

hemal arches

single
centrum

FIG. 6.16 Diplospondyly in *Amia*. Lateral view of portion of vertebral column, showing three full-body vertebrae with single centra to the left, and the first five caudal vertebrae with double centra to the right. Only the posterior part of a double centrum bears neural and hemal arches.

arch elements seem to be confined to the tail region.

FISHES The spinal column of fishes is composed of two kinds of vertebrae, *trunk* vertebrae, all of which are almost alike, and *caudal* vertebrae, bearing hemal arches and confined to the tail. The position of the anus marks the point of transition between the two types.

Much variation is to be found in the vertebral columns of fishes. The most primitive condition is not much in advance of the lamprey. In sturgeons, for example, the notochord persists, and centra are lacking. Basidorsal cartilages, forming a neural arch, fuse above the spinal cord to form a neural spine; interdorsals lie between the basidorsals. Basiventral cartilages, alternating with interventrals, form incomplete hemal arches in the tail region. In chimaeras and dipnoans the condition is similar, but cartilage has begun to invade the chordal sheath, in which ringlike calcifications may appear. Perichordal cartilage covers the chordal sheath to a minor degree. Interdorsals and interventrals may be lacking.

The cartilaginous spinal column of elasmobranchs is much better developed (Fig. 6.14). Each vertebra bears a biconcave, or *amphicoelous centrum* (Fig. 6.17), through the center of which the constricted notochord runs. The notochord is not constricted between the centra. If it alone were removed, it would resemble a string of beads. The centrum is formed by the invasion of the chordal sheath by cartilage. Basidorsal cartilages form *neural processes,* or *plates,* between which are *intercalary plates,* representing the interdor-

FIG. 6.17 Diagram of sagittal sections of vertebrae showing four types of centra.

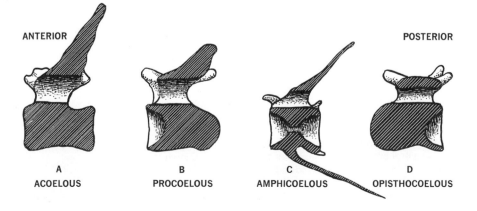

ANTERIOR POSTERIOR

A
ACOELOUS

B
PROCOELOUS

C
AMPHICOELOUS

D
OPISTHOCOELOUS

sals. These are sometimes further subdivided. In the tail region a hemal arch with a *hemal spine* is present. In the trunk region a pair of ventrolateral transverse processes (basapophyses), representing remnants of the hemal arch, extends toward the lateral skeletogenous septum.

In the caudal vertebrae of *Amia,* some selachians, and others, there are two centra per vertebra. This condition is referred to as *diplospondyly.* In *Amia* only the posterior centrum bears neural and hemal arches (Fig. 6.16). There is little agreement among comparative anatomists in regard to the specific elements represented, and the significance of the diplospondylous condition is obscure.

In the garpike the posterior end of each centrum is concave and the anterior end convex, a condition not observed in other fishes. Thus the convex portion of one centrum fits into the concavity of the vertebra anterior to it like a ball-and-socket joint. Such vertebrae are said to be *opisthocoelous* (Fig. 6.17D). The notochord disappears in the adult.

Teleosts, for the most part, have biconcave, or *amphicoelous,* centra (Fig. 6.17C). A remnant of the notochord may persist in a small canal in the center. The concavities are filled with a pulpy material, probably derived from the degenerate notochord. Rather poorly defined pre- and postzygapophyses, arising from the neural arch, first make their appearance in bony fishes. Similar but less-well-developed processes may be found on the hemal arches. Basapophyses, with slender ventral ribs attached, are present on the trunk vertebrae.

AMPHIBIANS With the appearance of limbs in the evolutionary scale, a further differentiation of the spinal column into regions has occurred. This is clearly shown in amphibians, with the exception of caecilians and certain extinct forms. A single *cervical vertebra* articulates with the skull at the anterior face of the vertebra. This is followed by a series of *trunk vertebrae.* A single *sacral vertebra,* connected to the pelvic girdle by the sacral diapophysis, lies between the trunk and caudal vertebrae, which are, of course, found only in tailed forms. All caudal vertebrae except the first bear hemal arches (Fig. 6.18). Amphibian vertebrae bear transverse processes and well-developed zygapophyses. Caecilians and some urodeles have amphicoelous vertebrae, although in most salamanders they are of the opisthocoelous type (Fig. 6.17D). Anurans usually have *procoelous* centra (Fig. 6.17B). Here the anterior face of the centrum is concave and the posterior face convex.

The most striking change among amphibians occurs in anurans, since they lack a tail in the adult stage. The number of vertebrae has been reduced. In modern anurans there are 7 to 10 vertebrae in the column, depending on the group. In common frogs a single cervical vertebra, the *atlas,* articulates with the skull. It has no transverse processes or prezygapophyses. Its anterior face bears two depressions for reception of the occipital condyles of the skull. The next six vertebrae are procoelous trunk vertebrae with no unusual features. The eighth vertebra is amphicoelous. The trans-

FIG. 6.18 Lateral view of caudal vertebra of *Necturus.*

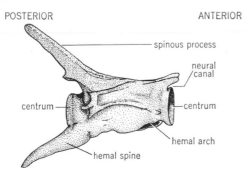

POSTERIOR   ANTERIOR

spinous process

neural canal

centrum

centrum

hemal arch

hemal spine

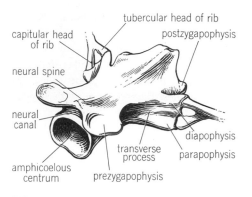

capitular head of rib
tubercular head of rib
neural spine
postzygapophysis
neural canal
diapophysis
transverse process
parapophysis
amphicoelous centrum
prezygapophysis

FIG. 6.19 Trunk vertebra of *Necturus*.

verse processes of the trunk vertebrae are tipped with small cartilaginous ribs (Fig. 6.19). The ninth is the sacral vertebra. Its cranial end is convex, fitting into the posterior concavity of the eighth vertebra. The caudal end of the sacral vertebra bears two convex projections which fit into corresponding concavities of the *urostyle*. The transverse processes (diapophyses) of the ninth vertebra are large and strong and slope posteriorly in an oblique direction (Fig. 6.20). The pelvic girdle articulates with the ends of these sacral diapophyses. The urostyle is a long bone which

FIG. 6.20 *A*, dorsal view of pelvic girdle, last three vertebrae, and urostyle of the bullfrog, *Rana catesbeiana; B*, lateral view of the right innominate bone. (*Drawn by G. Schwenk.*)

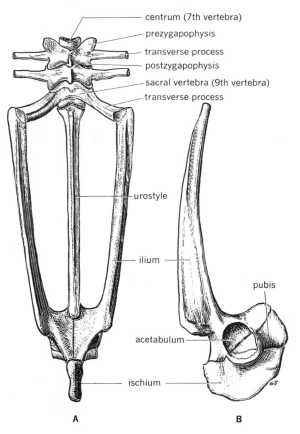

centrum (7th vertebra)
prezygapophysis
transverse process
postzygapophysis
sacral vertebra (9th vertebra)
transverse process
urostyle
ilium
pubis
acetabulum
ischium

A   B

extends from the sacral vertebra to the posterior end of the pelvic girdle. It was formerly believed to be formed by a fusion of several caudal vertebrae. There is little or no embryological evidence for this, since the urostyle actually forms by ossification of a cartilaginous rod which for a time lies beneath the notochord. A pair of small openings near the anterior end is the equivalent of intervertebral foramina. They connect with the neural canal.

REPTILES Associated with the diversity of form and habit of various reptiles, differences of considerable magnitude are to be observed in their vertebral columns. In turtles the number of vertebrae is much less than in other groups. In some large snakes over 400 may be present. Perhaps the most noteworthy advance to be observed in the vertebral column of reptiles is its division in crocodilians and most lizards into *cervical, thoracic, lumbar, sacral,* and *caudal regions.* Cervical, or neck, vertebrae are more freely movable than the others. Small ribs are frequently attached to the cervical vertebrae. Thoracic vertebrae bear arched ribs which articulate ventrally with the sternum anteriorly, followed by free ribs ending in the ventrolateral myocommata. These are succeeded by larger and more freely movable lumbar vertebrae, which do not bear ribs. Next come the sacral vertebrae, to the transverse processes and ribs of which the pelvic girdle is attached. Strength in the sacral region is attained by fusion of *two* vertebrae to form a *sacrum.* Caudal vertebrae follow the sacrum. Small V-shaped *chevron bones* on the ventral side of caudal vertebrae in reptiles and other amniotes probably represent remnants of hemal arches.

Turtles, snakes, and a few limbless lizards do not exhibit the five subdivisions seen in the vertebral columns of crocodilians and other lizards. Turtles lack a lumbar region; snakes and some limbless lizards have only precaudal and caudal regions, the separate vertebrae showing similarity in structure.

All types of centra are encountered among reptiles. *Sphenodon* has amphicoelous vertebrae. A few reptiles have opisthocoelous vertebrae in certain regions. Most, however, have procoelous vertebrae. Turtles may exhibit all vertebral types in the neck vertebrae, the body vertebrae being amphicoelous. In some extinct forms the centra of certain vertebrae had flat ends of the *acoelous,* or *amphiplatyan,* type (Fig. 6.17A). The sacral vertebrae of living crocodilians are acoelous. Two or more types may be found in a single individual.

Among reptiles, for the first time, is found a modification of the first two cervical vertebrae, a condition also found in birds and mammals. The second vertebra, or *axis,* in crocodilians, some snakes, lizards, and chelonians bears a projection, the *odontoid process,* at the anterior end of the centrum. This is actually the centrum of the atlas. It serves as a pivot which permits considerable freedom of movement of the head. The atlas bears an anterior concavity for the reception of the single occipital condyle of the reptilian skull. A separate bone, the *proatlas,* in the form of a neural arch, is found in *Sphenodon,* alligators, and a few other reptiles.

Although two sacral vertebrae are usually present in living crocodilians and lizards, in many prehistoric forms as many as five or six were present. These were firmly fused.

The 10 thoracic vertebrae of turtles lack transverse processes. Their neural arches, as well as those of the sacrum and first caudal vertebra, are rigidly united with the carapace.

BIRDS The most distinctive feature of the spinal column of birds is its rigidity. The cervical region is extremely mobile, but the remainder of the column is capable of little, if

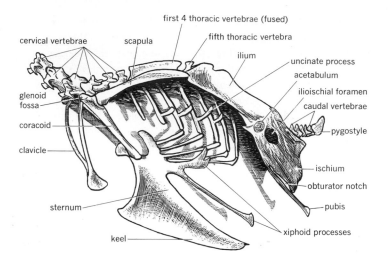

cervical vertebrae
scapula
first 4 thoracic vertebrae (fused)
fifth thoracic vertebra
ilium
uncinate process
acetabulum
ilioischial foramen
caudal vertebrae
glenoid fossa
coracoid
clavicle
pygostyle
ischium
obturator notch
pubis
sternum
xiphoid processes
keel

FIG. 6.21 Lateral aspect of part of chicken skeleton, showing union of vertebrae in thoracic, lumbar, sacral, and caudal regions. The wing and leg bones are not shown. (*Drawn by G. Schwenk.*)

any, movement. The posterior thoracic, lumbar, sacral, and anterior caudal regions are so firmly united that it is difficult to delimit one region from another (Fig. 6.21). Rigidity is of advantage in flight.

Cervical vertebrae are variable in number, ranging from 8 or 9 to 25. The atlas is small, narrow, and ringlike. Its centrum is represented by the odontoid process of the axis. The anterior face of the atlas bears a deep

FIG. 6.22 Heterocoelous cervical vertebra of bird.

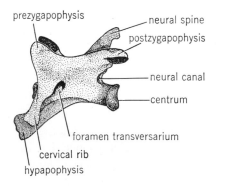

prezygapophysis
neural spine
postzygapophysis
neural canal
centrum
foramen transversarium
cervical rib
hypapophysis

depression for articulation with the single occipital condyle. The transverse processes of the cervical vertebrae are pierced by *foramina transversaria,* which serve for the passage of the vertebral arteries and veins. Only in parrots among birds are epiphyses present at the ends of the centra, a condition which is otherwise confined to mammals and a few reptiles. *Archaeopteryx* had amphicoelous vertebrae, a condition not encountered in other birds except in the caudal region. The ends of the centra in modern birds are usually saddle-shaped, or *heterocoelous* (Fig. 6.22), the anterior face being convex in a dorsoventral direction, but concave from side to side. The posterior face has its curves in the opposite direction, so that proper articulation is provided. Penguins, parrots, and a few other birds possess opisthocoelous vertebrae. Cervical ribs are often well developed in birds, and hypapophyses are common.

Thoracic vertebrae bear ribs which unite ventrally with the sternum. Their rather prominent transverse processes have articular surfaces at their tips. These, together with small projections on the sides of the centra, serve for attachment of ribs. The vertebral end of the rib, therefore, is attached in two places. Not all the thoracic vertebrae are always fused. In the fowl, for example, there are seven thoracic vertebrae. The first four are fused, but the fifth is free. The sixth and seventh are fused to the first lumbar vertebra. Posterior thoracic, lumbar, sacral, and the first few caudal vertebrae are fused into a single bony mass, the *synsacrum.* The number of vertebrae included in the synsacrum varies in different birds, as many as 20 vertebrae sometimes being included. Only two of the vertebrae are originally sacral, except in the ostrich, where there are three.

A few free caudal vertebrae are present in modern birds, although *Archaeopteryx* had 20 tail vertebrae. From 6 to 10 posterior caudal vertebrae are usually fused to form the "plow share" bone, or *pygostyle,* which supports the large tail feathers. A pygostyle is poorly developed or absent in the Paleognathae. The free caudal vertebrae are usually amphicoelous.

MAMMALS Fully developed vertebrae of mammals show little evidence of their ancestral origin from arcualia and other components. During development in mammals many

ancestral stages are glossed over, and this seems to be true of vertebra development. Separate centers of chondrification during development are believed to represent arcualia. Before adult stature is attained, the centra of mammalian vertebrae, with the exception of monotremes and sirenians, bear an epiphysis at either end. In most mammals these finally become fused to the centrum. The mammalian vertebral column is divided into cervical, thoracic, lumbar, sacral, and caudal regions.

Perhaps the outstanding feature of the cervical vertebrae (Fig. 6.23) in mammals is their constancy in number, seven vertebrae being present with few exceptions. Length of neck is determined by the length of individual vertebrae, rather than by increase in number. Three of the four mammals which are exceptions are edentates. They include the two-toed sloth, with six cervical vertebrae; the three-toed sloth, with nine; and *Tamandua,* the ant bear, with eight. The other exception is the manatee, which has six cervical vertebrae. Fusion of some or all of the cervical vertebrae occurs in many whales, armadillos, manatees, jerboas, and the marsupial mole. The atlas has a pair of concavities for articulation with the two occipital condyles. The axis bears an odontoid process. Cervical vertebrae bear small ribs, but these are poorly developed. The transverse processes characteristically arise from two roots, one from the base of the neural arch, the other from the centrum. These transverse processes (pleurapophyses) consist for the most part of ribs which are fused to the vertebrae. As in birds, the foramen transversarium lies between the two points of attachment. This is diagnostic of cervical vertebrae, although the seventh, and sometimes the sixth, may lack these foramina. In most mammals the centra of the cervical vertebrae are acoelous (amphiplatyan), but in the Perissodactyla they are opisthocoelous.

FIG. 6.23 Typical cervical vertebra of cat, posterior view.

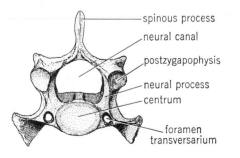

spinous process

neural canal

postzygapophysis

neural process

centrum

foramen transversarium

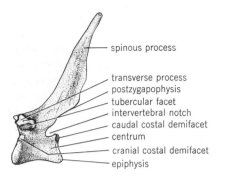

FIG. 6.24 Thoracic vertebra of cat, lateral view.

Thoracic vertebrae (Fig. 6.24) bear ribs which connect directly or indirectly with the sternum. Some of the posterior ribs may not reach the sternum, in which case they are known as *floating ribs.* Special articular facets are present on thoracic vertebrae for articulation with the ribs (Fig. 6.24). These consist of (1) *tubercular facets* on the ventral side of the free ends of the transverse processes (diapophyses) for articulation with the *tubercle* of the rib; (2) *costal demifacets* on the dorsolateral angles of the centrum, for reception of the *head,* or *capitulum,* of the rib. The caudal costal demifacet of one vertebra, together with the cranial costal demifacet of the next, may form a depression which serves as the point of articulation. Thus the capitulum may

FIG. 6.25 Lumbar vertebra of cat, lateral view.

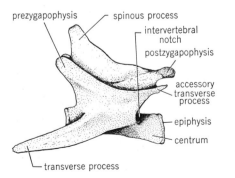

actually articulate with the centra of two thoracic vertebrae, although this is not always true. The number of thoracic vertebrae varies from 9 to 25 in different mammalian species. The centra are amphiplatyan, except in certain ungulates in which they are slightly opisthocoelous.

Lumbar vertebrae (Fig. 6.25) in mammals vary in number from 2 to 24, but generally there are 4 to 7. They are usually large and strong. Their transverse processes (pleurapophyses) are prominent and directed forward. The anterior lumbar vertebrae may bear *mammilary processes* dorsolateral to the prezygapophyses. Accessory processes may be present between the postzygapophyses and transverse processes.

The sacrum usually consists of from three to five vertebrae, firmly united and serving for articulation with the pelvic girdle. Whales lack a sacrum. Anomalous conditions are occasionally found in which more than the usual number of vertebrae form the sacrum. Extra elements are taken over from either the lumbar or caudal regions. The presence of *sacral foramina* on both dorsal and ventral sides provides for the passage of dorsal and ventral rami of sacral spinal nerves. They communicate with the neural canal. The first sacral element is most important in supporting the pelvic girdle and hind limbs. Large *lateral masses* bear cartilage-covered *auricular surfaces* for articulation with the ilium (Fig. 6.26). They are composed of transverse processes and sacral ribs which have become indistinguishably fused.

Caudal vertebrae range in number from 3 or 4, as in man, to around 50 in the scaly anteater. The anterior caudal vertebrae possess neural arches, spines, zygapophyses, etc. These gradually diminish in size toward the end of the tail so that the terminal caudal vertebrae consist of little more than centra. Chevron bones, homologous with those of

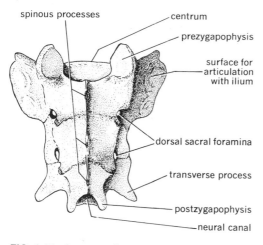

spinous processes
centrum
prezygapophysis
surface for articulation with ilium
dorsal sacral foramina
transverse process
postzygapophysis
neural canal

FIG. 6.26   Sacrum of cat, dorsal view.

reptiles, are usually present. They are particularly prominent in whales and edentates. The few caudal vertebrae in man usually unite to form a single *coccyx.* Sometimes the first one or two coccygeal vertebrae remain free.

Skull

The skeletal framework of the vertebrate head is referred to as the skull. The comparative anatomy of the skull is extremely complex, and homologies are often difficult to ascertain. Confusion stems from the fact that large numbers of separate elements enter into the formation of skulls and arise from such varied sources. Furthermore, certain gaps in the paleontological record, together with the fact that groups living today usually represent specialized or degenerate conditions, make it difficult to construct the true phylogenetic history of skull development.

The part of the skull which surrounds and protects the brain is called the *cranium* (*neurocranium*). Sense capsules associated in development with the olfactory and auditory and visual sense organs become attached to the cranium, as do numerous components of the visceral skeleton. All these in their aggregate make up the skull.

The skull of most vertebrates is derived essentially from three different embryonic components: (1) the *chondrocranium,* composed of cartilage, which contributes chiefly to the base of the skull, includes the sense capsules mentioned above and in most vertebrates is replaced by bone; (2) the *dermatocranium,* made up of membrane bones which roof over the chondrocranium* or are closely applied to its lower surface; (3) the *splanchnocranium,* which is derived from the visceral skeleton and is primitively cartilaginous although some of it becomes invested by, or covered with, membrane bones.

Early Development

Chondrocranium   Soon after the appearance of the brain during embryonic development, the mesenchymal cells which surround it begin to differentiate, forming a membranous investing layer. This is referred to as the *membranous cranium.* The notochord extends forward beneath the brain, terminating near the infundibulum. The membranous layer also surrounds the anterior portion of the notochord. The ventral region of the membranous cranium furnishes the material from which the chondrocranium is derived. The chondrocranium is made up of several components. First there appears a pair of flat, curved cartilages, the *parachordal plates,* which

* The use of these terms by different authors varies somewhat. Some use *skull* and *cranium* interchangeably. They include all parts of the skull derived from cartilage under the heading *chondrocranium* and all membrane bones under the category of *dermatocranium.* The terminology used here is less confusing to the student attempting to understand a rather complex subject. The skeletal elements of the lower jaw are here considered to be parts of the skull.

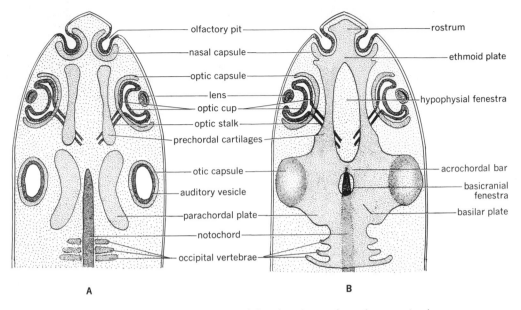

FIG. 6.27 Diagrams illustrating development of the chondrocranium: *A*, separate chondrocranial cartilages surrounding the special sense organs and flanking the notochord and brain; *B*, all the cartilages except the optic capsules have fused to form the early chondrocranium.

flank the notochord on either side (Fig. 6.27). They extend laterally as far as the otic capsules (see below) and posteriorly to the point where the tenth cranial nerve emerges. Caudal to this point are two to four *occipital vertebrae,* with which the parachordal plates later fuse and which are incorporated in the skull. The parachordal plates grow larger, their posterior portions fusing in the midline to form the *basilar plate,* which encloses the tip of the notochord and becomes the floor beneath the midbrain and hindbrain. The anterior ends form a separate connection, being united by a transverse *acrochordal bar.* A small opening, the *basicranial fenestra,* often remains in the center.

Next there appears in front of the parachordals another pair of cartilages, the *prechordal cartilages.* The anterior ends of these fuse with each other, forming a trans-

verse bar, the *ethmoid plate.* This grows anteriorly to become the *rostrum,* which later contributes to the formation of the *internasal septum* between the nasal capsules. Posteriorly the prechordal cartilages unite with the parachordals in such a manner as to leave a prominent opening in the center, the *hypophysial fenestra,* in which the pituitary gland lies and through which the internal carotid arteries pass.

At the same time as the above cartilages are developing, three pairs of capsules appear about the developing olfactory, auditory, and optic sense organs. They are known, respectively, as the *nasal, otic,* and *optic capsules.* With the continued growth of the parachordal and prechordal cartilages a union is effected with the cartilaginous nasal and otic capsules, the entire mass forming the chondrocranium. The optic capsules remain as independent elements, thus permitting un-restricted movement of the eyes. The optic capsules are seldom cartilaginous. In mammals they are of a fibrous nature.

The basal cartilaginous cranium does not exist for long. The side portions soon begin to grow dorsally for a short distance on either side of the brain. In elasmobranchs these dorsally extending parts tend to meet above the brain and fuse, thus enclosing that structure rather completely with a true chondro-cranium. In most vertebrates, however, only the otic capsules and *occipital region* become roofed over as described. The floor and sides of the more anterior region are cartilaginous, but the roof is composed of membrane. This portion of the cranium will later be roofed over by membrane bones of the dermato-cranium. The opening at the posterior end of the occipital portion is the *foramen magnum,* through which the spinal cord emerges. Openings for the passage of the blood vessels represent spaces which persist following fusion of the major cartilaginous components.

Dermatocranium   In addition to the chondrocranium, which appears uniformly

through the vertebrate series, several dermal plates, or membrane bones, contribute to the formation of the skull. These appear first in the head region of bony fishes in the form of large scales. They are similar to bony scales on other parts of the body. The dermal scales gradually sink down into the head, roofing over the more anterior region of the chondro-cranium with which they fuse, thus complet-ing the protective envelope surrounding the brain. Numerous other membrane bones be-come closely applied to the ventral portion of the chondrocranium; others are associated with portions of the visceral skeleton.

Visceral Skeleton   In fishes the series of cartilaginous or bony visceral arches en-circling the pharyngeal portion of the diges-tive tract makes up the visceral skeleton. It serves primarily to support the gills and the floor of the mouth. The arches are arranged one behind the other between the gill slits (Fig. 6.28). They are derived from mesen-chyme. Ectodermal neural crests contribute to this mesenchyme. The first visceral arch is the *mandibular arch.* It becomes divided into dorsal and ventral portions, the *palatoquadrate bar* and *Meckel's cartilage,* which contribute to

FIG. 6.28   Diagram showing relation of visceral arches of an elasmobranch fish to the chondrocranium, vertebral column, and gill slits.

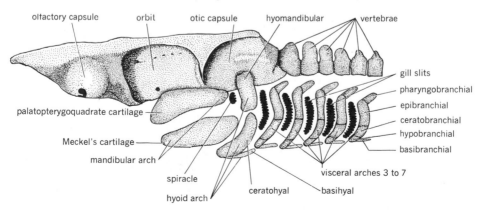

the upper and lower jaws, respectively. The second is the *hyoid arch.* It furnishes support to the anterior visceral arch, if present, but in many fishes its upper division, the *hyomandibular cartilage,* also serves to attach the jaws to the cranium in the otic region by means of a ligamentous connection. This is referred to as the *hyostylic* method of jaw attachment (Fig. 6.30*B* and *F*). The remaining visceral arches vary in number in different species, furnishing support to the rest of the gills. The last visceral arch, however, usually does not bear a gill. The typical number of visceral arches in fishes is seven but may be nine in one shark. Each arch is usually divided into several separate cartilages.

In higher forms there has been a reduction in the number of visceral arches, and the entire visceral skeleton has become greatly modified. The parts of which it is composed are primitively made up of cartilage, but the cartilage, for the most part, tends to disappear, being replaced by cartilage bone or surrounded by membrane bones.

FURTHER DEVELOPMENT In cyclostomes, elasmobranchs, and a few of the higher fishes, no dermal bones (except those associated with scales) are present, and the entire skull remains cartilaginous. In most vertebrates, however, the primitive chondrocranium is replaced, at least to some extent, by bone. So many variations are found that it is difficult to generalize in discussing the origin of the various bones of the skull. Nevertheless, a more or less common pattern of development is found throughout the vertebrate series.

Separate centers of ossification appear in the basal plate of the chondrocranium in what are called, beginning from the posterior end, the *basioccipital, basisphenoid, presphenoid,* and *ethmoid* regions (Fig. 6.29). The ethmoid region lies anterior to the cranial cavity. Other centers appear in the cartilage, which extends a short distance dorsally on either side of the brain. Each center typically gives rise to a separate bony element which may or may not fuse with others. Three bony seg-

FIG. 6.29 Diagram of *A*, ventral, and *B*, dorsal, views of the typical tetrapod chondrocranium, showing the main regions which will later ossify to form cartilage bones. The palatoquadrate bar is also included. (*After Kingsley, "Comparative Anatomy of Vertebrates," McGraw-Hill Book Company, by permission.*)

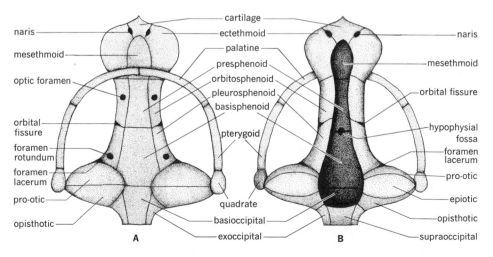

ments are thus formed below and along the sides of the brain; a fourth, or *nasal segment,* develops anterior to the brain. The most posterior is the *occipital segment,* made up of four cartilage bones; a *basioccipital* below, two lateral *exoccipitals,* and a dorsal *supraoccipital.* It will be recalled that only the occipital region of the chondrocranium becomes roofed over by cartilage. Anterior to the occipital segment is the *parietal segment,* which in reptiles and birds is composed of five elements: a *basisphenoid* below, two lateral *pleurosphenoids,* and two dorsal *parietals* (generally fused), the latter being membrane bones which roof over the parietal segment. In mammals the pleurosphenoids are lacking. However, a pair of bones, the *alisphenoids,* representing the epipterygoids (page 135) derived from the palatoquadrate bar, comes to lie on either side of the basisphenoid. Alisphenoids are *not* homologous with pleurosphenoids. The *frontal segment,* lying anterior to this complex, is also composed of five elements, a *presphenoid* below, an *orbitosphenoid* on either side, and two membrane bones, the *frontals,* forming the roof. Several cartilages originate in the region of the nasal capsules and ethmoid plate. This makes up the ethmoid region. An ossification center in the middle of the ethmoid plate gives rise to the *mesethmoid* bone, which contributes to the nasal septum. In some forms this region remains unossified. In fishes the nasal capsules give rise to *lateral ethmoids,* or *ectethmoids. Turbinate* bones are the chief derivatives of the nasal capsules. In mammals the *cribriform plate,* containing numerous *olfactory foramina,* also is derived from this region. Several of these bones may unite to form a single ethmoid bone. Two sets of membrane bones usually develop in this area; a pair of *nasal* bones, roofing over the mesethmoid and a pair of *vomers* below it. Often there is a single median vomer. The region of the skull anterior to the cranial cavity is commonly referred to as the nasal segment. Many variations are encountered in this portion of the skull.

It should be noted that the nasals, frontals, and parietals, being membrane bones, are not comparable to the supraoccipital, which is a cartilage bone. Membrane bones, representing ancestral elements, are sometimes added to the dorsal portion of the occipital. In modern amphibians basioccipital and supraoccipital regions remain ossified. In mammals all four regions fuse to form a single *occipital* bone. The bones of the occipital segment trace their origin to the occipital vertebrae and possibly to the posterior portion of the parachordal plates; the cartilage bones of the parietal segment are derived from the anterior portions of these primitive cartilages, whereas those of the frontal segment arise from the prechordal cartilages. The cartilage bones of the nasal segment are derived from the prechordal cartilages and the nasal capsules.

Certain other bones of the skull trace their origin to the otic capsules, which were parts of the original chondrocranium. Three separate centers of ossification appear in the cartilaginous otic capsules, forming three bones: (1) an anterior *prootic,* (2) a posterior *opisthotic,* and (3) a dorsal *epiotic.* These bones may unite with one another or with neighboring bones. The complex comes to lie on either side between pleurosphenoid and exoccipital. In bony fishes two additional bones, the *sphenotic* and *pterotic,* develop in the otic capsule.

Certain membrane bones, in addition to those already mentioned, are frequently present in the vertebrate skull. These include a *parasphenoid,* lying ventral to the basicranial axis; *prefrontals* and *postfrontals,* situated anterolaterad and posterolaterad, respectively, the frontal bone proper; and *lacrimals,* which help to complete the inner wall of the orbit.

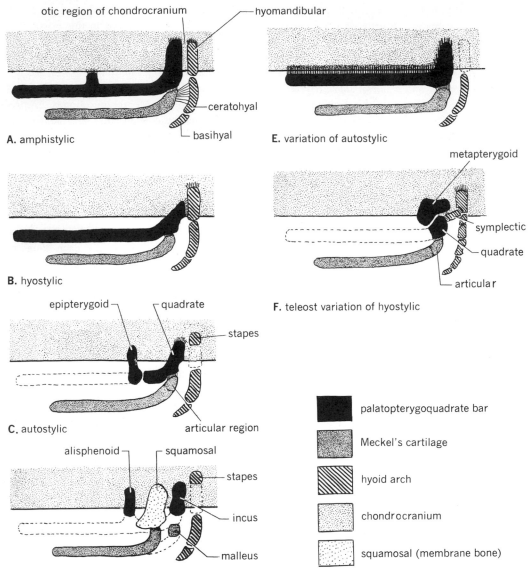

otic region of chondrocranium — hyomandibular

— ceratohyal

— basihyal

A. amphistylic

E. variation of autostylic

B. hyostylic

metapterygoid

symplectic

quadrate

articular

F. teleost variation of hyostylic

epipterygoid — — quadrate

stapes

C. autostylic — articular region

alisphenoid — — squamosal

stapes

incus

malleus

D. autostylic (craniostylic)

	palatopterygoquadrate bar
	Meckel's cartilage
	hyoid arch
	chondrocranium
	squamosal (membrane bone)

FIG. 6.30 Diagrams indicating various methods of jaw suspension: *A,* amphistylic, found in a few primitive elasmobranchs; *B,* hyostylic, found in sharks and sturgeons; *C,* autostylic, found in tetrapods other than mammals; *D,* craniostylic, typical of mammals; *E,* variation of autostylic found in *Polypterus,* holocephalians, and lungfish; *F,* teleost variation of hyostylic, found in *Amia, Lepisosteus,* and teleosts. The membrane bones investing Meckel's cartilages are not indicated.

From the visceral skeleton is also derived a number of separate skeletal elements. The mandibular, or first, visceral arch is highly modified. It is divided on each side into a dorsal *palatoquadrate bar* and a ventral *Meckel's cartilage.* The mouth cavity is bounded by these components. The angle of the mouth is located posteriorly, where the palatoquadrate bar and Meckel's cartilage join at a sharp angle.

In bony fishes, centers of ossification in the posterior portion of the palatoquadrate bar give rise to such cartilage bones as *metapterygoid* and *quadrate.* The anterior part of the bar is reduced and replaced or invested by such membrane bones as the palatine and pterygoid. The lower jaw articulates with the quadrate. In these fishes the quadrate forms a connection with a *symplectic* bone derived from the hyomandibular and attached to it. The hyomandibular, in turn, unites with the otic region of the chondrocranium. This is a variation of the hyostylic method of jaw suspension (Fig. 6.30F).

In tetrapods other than mammals a quadrate bone, with which the lower jaw articulates, is also formed by direct ossification of the posterior end of the palatoquadrate bar. An epipterygoid,* corresponding to the metapterygoid of fishes, develops anterior to the quadrate. The quadrate forms a firm union with the otic region of the chondrocranium.

---

* There is some confusion whether the pterygoid is a cartilage bone or a membrane bone. Misunderstandings are largely a matter of terminology. In this account we shall consider the pterygoid proper to be a membrane bone. Subdivisions of the pterygoid such as the *ectopterygoids* (transpalatines, mesopterygoids) and *endopterygoids* are, when present, also membrane bones. Such bones as the *metapterygoids* of fishes and the *epipterygoids* of certain tetrapods, which develop from ossification centers in the palatoquadrate bar, are clearly cartilage bones.

The hyomandibular cartilage is reduced in tetrapods and is not concerned with jaw suspension (Fig. 6.30C). It becomes the *stapes,* one of the auditory ossicles. This method of suspending the jaws via the quadrate is termed the *autostylic* method, in contrast to the hyostylic method, in which, as already mentioned, suspension is by means of the hyomandibular. Another type of jaw suspension found in a few primitive sharks is called the *amphistylic* method (Fig. 6.30A). Here both the palatoquadrate bar and the hyomandibular cartilage of the hyoid arch are united with the otic region of the chondrocranium.

In some tetrapods, as in *Sphenodon,* lizards, and turtles, the epipterygoid is a separate paired bone. In mammals, however, it comes to lie on either side of the basisphenoid and is commonly referred to as the *alisphenoid.* It will be recalled that pleurosphenoids, which in reptiles and birds develop on either side of the basisphenoid, are lacking in mammals.

In bony vertebrates another arch, the *maxillary arch,* composed entirely of membrane bones, forms outside of and roughly parallel to the palatoquadrate bar (Figs. 6.31 and 6.32). This gives rise to the greater part of the functional upper jaw, which is fused to the rest of the skull in these forms. The most anterior bones to develop in this region are the paired *premaxillaries.* They are followed by the *maxillaries,* and these, in turn, by the *jugals* (*malars, zygomatics*). Next in succession come the *quadratojugals, squamosals,* and *supratemporals.* Only the premaxillaries and maxillaries in this arch bear teeth.

The term *palate* is used to indicate the roof of the mouth and pharynx. In fishes and amphibians this is a flattened area directly beneath the floor of the cranium. In most reptiles and in birds a pair of longitudinal *palatal folds* grows medially for a short distance on either side. These folds do not meet in a median line. In crocodilians and mammals,

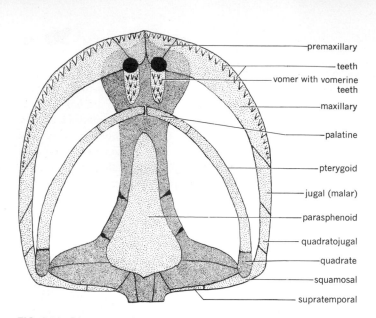

FIG. 6.31 Diagrammatic ventral view of skull of typical tetrapod, showing relationship of maxillary arch to chondrocranium and palatoquadrate bar. Membrane bones are lightly stippled; cartilage bones are heavily stippled. The palatines and pterygoids, as represented here, are membrane bones which have invested the anterior portion of the palatoquadrate bar. (*After Kingsley, "Comparative Anatomy of Vertebrates," McGraw-Hill Book Company, by permission.*)

FIG. 6.32 Diagrammatic lateral view of mammalian skull, showing relationships of the various bones. Membrane bones are lightly stippled; cartilage bones are heavily stippled. The palatines and pterygoids, as represented here, are membrane bones which have invested the greater portion of the palatoquadrate bar. Numbers refer to foramina through which cranial nerves emerge. (*After Flower, from Kingsley, "Comparative Anatomy of Vertebrates," McGraw-Hill Book Company, by permission.*)

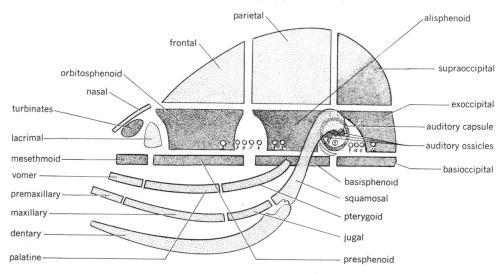

horizontal projections of the premaxillary, maxillary, palatine, and in some cases of the pterygoid bones, extend medially to meet their partners of the opposite side. In this manner a complete *secondary palate* is formed separating the nasal passages above from the mouth cavity below. As a result, the nasal passages communicate with the mouth cavity much farther posteriorly than they would otherwise. The portion of the secondary palate which is bony is called the *hard palate.* In mammals the secondary palate is continued for some distance posteriorly by a *soft palate,* composed mostly of connective tissue without any bony foundation.

The original Meckel's cartilage forming the lower jaw becomes modified in higher vertebrates with bony skeletons. The posterior region is usually replaced by a cartilage bone, the *articular.* This forms an articulation with the quadrate. In anuran amphibians at the point where the two halves of the lower jaw join in front, a small cartilage bone, the *mentomeckelian,* develops on either side. The remainder of Meckel's cartilage serves as a core about which several membrane bones form a sheath (Fig. 6.33). The separate elements which, for the most part, enter this complex are (1) an anterior *dentary* (*dental*), surrounding Meckel's cartilage; (2) a medial *splenial;* (3) a ventral *angular;* (4) a posterolateral *surangular;* (5) a dorsolateral *coronoid;* (6) a prearticular lying below and to the me-

dial side of the articular. Many modifications of the above are found in vertebrates. Only rarely are all these bones present. They exist as separate elements in amphibians and most reptiles, but in birds they begin to unite. In mammals the lower jaw consists of a single bone, the *mandible.* It represents the dentary of lower forms, the other elements having been lost except for the angular, which gives rise to the *tympanic* bone, and the articular, which has become the *malleus,* one of the auditory ossicles. Study of lower jaws of the fossils of some of the mammallike reptiles indicates clearly how the mammalian mandible evolved.

In tetrapods below mammals in the evolutionary scale, the articular bone of the lower jaw articulates with the quadrate of the upper jaw. Both of these are cartilage bones. In mammals the articular and quadrate are modified to form the *malleus* and *incus* bones, respectively. A new articulation of the lower jaw with the *squamosal,* a membrane bone of the maxillary arch, is then established (Figs. 6.30*D* and 6.32). This occurs only in mammals. Despite this difference, the method of jaw suspension in mammals is considered to be autostylic by many authorities. Others believe that a distinction should be made and have applied the terms *craniostylic* and *amphicraniostylic* to the mammalian method.

The hyoid, or second visceral, arch does not become so highly modified as the man-

FIG. 6.33 The position of membrane bones forming about Meckel's cartilage in the development of the lower jaw. (*After Kingsley, "Comparative Anatomy of Vertebrates," McGraw-Hill Book Company, by permission.*)

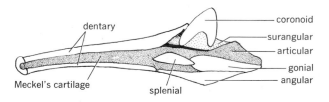

TABLE 2 The Major Bones in the Skull of Vertebrates

DERMOCRANIUM					
Roofing	**Orbital**	**Temporal**	**Maxillary arch**	**Palatal**	**Mandibular arch**
Nasal	Lacrimal	Intertemporal	Premaxilla*	Palatine*	Dentary* (mandible**)
Frontal	Prefrontal	Supratemporal	Septomaxilla	Prevomer* (palatine process of premaxilla**)	Splenial*
Parietal	Postfrontal	Tabular	Maxilla*	Parasphenoid* (vomer**)	Coronoid
Postparietal	Postorbital	Squamosal	Jugal	Pterygoid* (hamulus**)	Angular
			Quadratojugal	Ectopterygoid** (pterygoid**)	Supra-angular
					Prearticular

CHONDROCRANIUM					
Neurocranium proper				**Sense capsules**	
Parachordals	**Trabeculae**	**Ethmoid plate**	**Otic**	**Optic**	**Nasal**
Basioccipital	Basisphenoid	Mesethmoid	Prootic (petrous**)	Sclerotics	Ethmoturbinal (=ectethmoid)
Exoccipital	Presphenoid (=sphenethmoid)		Opisthotic (mastoid**)		Nasoturbinal
Supraoccipital	Orbitosphenoid		Epiotic		Maxilloturbinal
	Pleurosphenoid (=laterosphenoid)		Pterotic		
			Sphenotic		

VISCERAL ARCHES				
I		II	III–VII	
Palatoquadrate	**Meckel's cartilage**			
Quadrate (incus**)	Articular (malleus**)	Hyomandibular = stapes (? = columella)	III thyrohyal	
Epipterygoid (alisphenoid**)	Mentomeckelian	Ceratohyal	IV and V thyroid cartilage	
		Hypohyal	VI epiglottis	
		Basihyal	VII arytenoid	
		Epihyal	VII cricoid	
			VII trachial rings	

* Teeth may be present on this element.
** Element called by this name in mammals.

dibular arch during its evolutionary history. In elasmobranchs it is usually composed of three cartilages, a dorsal *hyomandibular,* a lower *ceratohyal,* and a small, ventral, median *basihyal* which serves to unite the ceratohyals of the two sides. The hyomandibular forms a ligamentous union with the otic region of the skull, serving as a suspensorium of the jaws (hyostylic method). In higher fishes several cartilage bones may be derived from the hyoid arch, although their appearance and number are highly variable. In general they are named as follows in a dorsoventral sequence: *hyomandibular, interhyal, epihyal, ceratohyal,* and *hypohyal.* A small, median *basihyal* unites the hypohyals of the two sides. A *symplectic* bone, derived from the hyomandibular and found only in teleost fishes (Fig. 6.30*F*), extends forward from the ventral end of the hyomandibular near its junction with the interhyal and articulates with the quadrate. In tetrapods the hyomandibular is reduced and gives rise to the *columella,* or to the *stapes* of the middle ear. The ceratohyal is important in elasmobranchs as the chief support of the gill lamellae of the most anterior, or hyoid, hemibranch. In higher forms it gives rise to a portion of the hyoid apparatus which furnishes support to the tongue and larynx.

The operculum, a posteriorly directed flap of tissue which covers the gill chamber in most fishes, begins in the region of the hyoid arch. In bony fishes it is strengthened by a number of thin, opercular membrane bones.

The visceral arches posterior to the hyoid serve an important function in fishes, in which they support the remaining gills. Their ventral ends are united by separate median cartilages or bones, the *basibranchials,* or *copulae.* In higher forms the posterior visceral arches are much reduced, being of importance mainly in contributing elements to the hyoid apparatus and to the larynx.

Our knowledge of the evolution of the skull in the bony fishes is rather incomplete. The comparatively recent discovery of the fossil remains of labyrinthodont amphibians from the Devonian period has furnished important clues. Ancestral crossopterygian fishes, also from the Devonian period, show similarities in skull structure to the labyrinthodonts, indicating that it was through the crossopterygian fishes that evolutionary progress was made. In both groups there was a movable articulation between two main regions of the skull, anterior and posterior. The anterior portion includes the presphenoid and ethmoid regions to which the palatal elements of the upper jaw were fused. The posterior region is composed of derivatives of the parachordal cartilages as well as those from occipital and otic regions. The contribution of the visceral skeleton to skull formation remains obscure. The branchial basket supporting the gill region of cyclostomes (see below) is composed of continuous cartilage and is not broken up into separate elements as in higher forms. Its homologies are doubtful. Biologists can only speculate about the ancestral origin of the visceral skeleton, some of which is derived from ectodermal neural-crest cells. A summary of the origin of cranial bones in vertebrates is presented in Table 2.

Skull in Different Classes of Vertebrates

Cyclostomes   The skull of hagfishes is very primitive, consisting only of a floor of cartilage with side walls and roof of connective tissue. The prechordal cartilages terminate abruptly at the anterior end.

In lampreys the skull is better developed. Dorsolateral extensions of the prechordal cartilages are present, forming side walls and a roof over the brain in this region. The re-

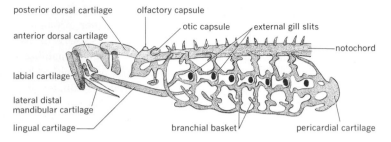

FIG. 6.34 Lateral view of chondrocranium and branchial basket of lamprey *Petromyzon marinus*. (*After Parker, from Schimkewitsch.*)

FIG. 6.35 Dorsal view of skull of elasmobranch *Squalus acanthias*. (*After Senning, outline drawings for "Laboratory Studies in Comparative Anatomy," McGraw-Hill Book Company, by permission.*)

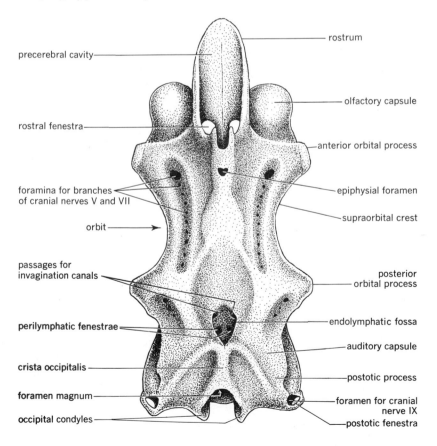

mainder of the brain is covered by fibrous connective tissue. Prechordal and parachordal cartilages as well as otic capsules are homologous with similar structures in higher forms. Other cartilages of doubtful homology are present, supporting the round suctorial mouth, the olfactory organ, and the tongue. *Posterior* and *anterior dorsal cartilages* are located anterior to the chondrocranium. These also appear to be without homologs in higher vertebrates.

The gill region in lampreys is supported by a peculiar *cartilaginous branchial basket.* The cartilage of which it is composed is continuous, not divided into separate parts. Lateral openings for the gill slits are present in addition to other openings above and below. The branchial basket (Fig. 6.34) lies just beneath the skin. It is *not* homologous with the visceral skeleton of other vertebrates. Its posterior end furnishes support for the pericardium. In hagfishes the visceral skeleton is much reduced, but a hyoid arch as well as third and fourth visceral arches are recognizable. They are located far within the head and are interpreted as true visceral arches.

Fishes

Elasmobranchs The skull in elasmobranch fishes is entirely cartilaginous and often calcified. Absence of dermal bones is a secondary development and does not represent a primitive condition. The chondrocranium consists of a single piece of cartilage, the separate elements of which it is formed having fused together. Nasal and otic capsules are indistinguishably united with the braincase. A large, depressed *orbit* on either side serves to receive the eyeball. Openings are present for the passage of blood vessels and cranial nerves (Fig. 6.35). A cuplike depression, the *sella turcica,* in the floor of the chondrocranium, serves to receive the pituitary body. The occipital region of the chondrocranium bears a posterior projection, the *occipital condyle,* on either side. These usually form an immovable articulation with the first vertebra. A peculiar feature of the elasmobranch chondrocranium is the *rostrum,* an anterior troughlike extension derived from the prechordal cartilages. The presence of the rostrum causes the mouth to be subterminal and ventral in position. The rostrum of the sawfish is a very extensive projection, supplied along the sides with numerous large and sharp spines of modified placoid scales.

The visceral skeleton of elasmobranchs (Fig. 6.36) is typically composed of seven cartilaginous visceral arches surrounding the anterior part of the digestive tract and furnishing support for the gills. The first, or mandibular, arch is larger than the rest. From it are derived both upper and lower jaws. On each side the mandibular arch divides into an upper palatoquadrate bar and a ventral Meckel's cartilage. These articulate with each other posteriorly. Both cartilages are suspended by a ligamentous attachment from the hyomandibular cartilage of the hyoid arch, which in turn is attached to the otic region of the chondrocranium. This is the hyostylic method of jaw attachment (Fig. 6.30*B*). Below the hyomandibular cartilage the remainder of the hyoid arch consists of a *ceratohyal* cartilage and a *bisihyal* which unites the ceratohyals of the two sides ventrally. The number of visceral arches posterior to the hyoid varies in elasmobranchs, but usually there are five. Each consists of four or more cartilages on either side.

Other fishes The bony fishes belonging to the class Osteichthyes are, in evolutionary terms, believed to be older than members of the class Chondrichthyes, but this is now being questioned. The presence of a bony skeleton is usually considered to be a primitive feature which has been retained, but evidence against this is now appearing in the lit-

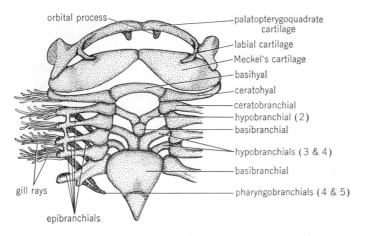

orbital process

palatopterygoquadrate cartilage

labial cartilage

Meckel's cartilage

basihyal

ceratohyal

ceratobranchial

hypobranchial (2)

basibranchial

hypobranchials (3 & 4)

basibranchial

pharyngobranchials (4 & 5)

gill rays

epibranchials

FIG. 6.36 Ventral view of visceral skeleton of elasmobranch *Squalus acanthias.* The hyomandibular cartilage is not shown.

erature. Early in their history the bony fishes had already divided into two branches, the ray-fins, or Actinopterygii, and the lobe-fins, or Sarcopterygii. The latter embraces the order Crossopterygii, some of which were the ancestors of tetrapods.

Among the skulls of fishes are to be found all transitional stages from those which are completely cartilaginous to those which are completely bony.

In holocephalians (*chimaera*) the skull is composed entirely of cartilage. The entire palatoquadrate bar is immovably fused to the cranium. The lower jaw is suspended from the quadrate region, and the hyoid arch plays no part in jaw suspension. This is a variation of the autostylic method of jaw suspension (Fig. 6.30E). The hyoid arch is developed only slightly more than the remaining visceral arches. The ceratohyal component, with its gill rays, supports the *operculum,* which first makes its appearance in this group of fishes.

The skulls of the ray-fins, with the exception of *Polyodon,* are invested with numerous membrane bones, several of which are homol-

ogous with those of crossopterygians and tetrapods. A chief difference is that in the ray-fins the front part of the skull is relatively long, and the posterior region is short. The opposite condition was evident in ancient crossopterygians. Also, in the ray-fins there is a lack of development of a homolog of the squamosal bone of crossopterygians and tetrapods.

The skull of sturgeons is cartilaginous, except for ossifications in the otic and orbitosphenoid regions. Several membrane bones, however, have made their appearance, forming a protective shield over and under the chondrocranium. The prominent rostrum is covered with scales. Sturgeons have the hyostylic method of jaw attachment. *Polypterus* exemplifies an advance toward the more highly developed teleostean skull. A veritable armor of membrane bones covers the chondrocranium, palatoquadrate bar, and Meckel's cartilage. The outer arch of membrane bones is present, and membrane bones cover the operculum. The method of jaw attachment is autostylic (Fig. 6.30E). In Amia and *Lepisosteus* a more advanced stage of teleostean skull development is represented. Both cartilage bones and numerous membrane bones are present. Jaw attachment is of the hyostylic type. A rather wide variation is to be observed among the teleosts. Their skulls have more bones than those of any other vertebrates. Many membrane bones are present, but most characteristic is the presence of several cartilage bones which have formed from ossification centers in the primitive chondrocranium, which still persists to some degree. The palatoquadrate bar in these fishes ossifies, in part, to form such cartilage bones as metapterygoid and quadrate. The posteriorly located quadrate articulates with the *symplectic*. The latter joins the hyomandibular, which in turn is fastened to the cranium. This is a variation of the hyostylic method of jaw suspension and is characteristic of teleosts (Fig. 6.30F). The skull of dipnoans (lungfishes) is not so far advanced as that of teleosts. It is cartilaginous for the most part, but both cartilage bones and a few large membrane bones are found. The palatoquadrate bar, as in holocephalians and *Polypterus,* is fused to the cranium, the method of jaw suspension therefore being autostylic.

In the evolution of the skull from primitive fishes to mammals there has been a gradual reduction in number of the separate bony elements by elimination and fusion. As many as 180 skull bones are present in certain fishes. The human skull consists only of 28, including the auditory ossicles. In some cases cartilage bones and membrane bones fuse together to form a single structure. The autostylic method of jaw attachment or a variation of it is the rule in tetrapods.

Amphibians   Early tetrapods evolved from rhipidistian crossopterygians or some of their close relatives. The ancestors of amphibians were little more than fish possessing primitive "legs" rather than fins. Modern amphibians are far removed from the ancestral stock. The ancient labyrinthodonts gave rise not only to modern amphibians but to primitive reptiles as well.

Although the skulls of early tetrapods retained certain characteristics of their piscine progenitors, those of modern amphibians show considerable deviation (Fig. 6.37) There has been a reduction in the number of bones as well as a general flattening of the skull. The

FIG. 6.37 *A,* dorsal, and *B,* ventral, views of skull of bullfrog, *Rana catesbeiana.* In *A,* the right tympanic annulus has been omitted for clarity; in *B,* the left one has been omitted.

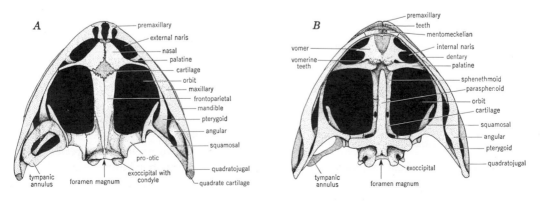

otic capsule bears a ventral opening, the *fenestra ovalis,* into which a cartilaginous or bony plug fits. This is the *stapedial plate* of the *columella,* derived from the hyomandibular cartilage. The columella (stapes) has developed in connection with the evolution of the sense of hearing and with the change from the hyostylic to the autostylic method of jaw suspension in which the hyomandibular bone or cartilage loses its significance as a suspensorium.

The chondrocranium persists to a considerable extent in amphibians, but some of it has been replaced by cartilage bones. Basioccipital and supraoccipital regions are not ossified. The skull is articulated with the atlas vertebra by means of a pair of occipital condyles, projections of the exoccipital bones. Basisphenoid and presphenoid regions also are not ossified. Prootics, and in some cases opisthotics, are ossified and fused to the exoccipitals.

Membrane bones form the greater part of the roof of the skull. They are no longer closely related to the integument and occupy a deeper position in the head than in fishes. A rather large membrane bone, the *parasphenoid,* covers much of the ventral part of the chondrocranium.

The quadrate in amphibians is fused to the otic region of the chondrocranium. Palatine and pterygoid membrane bones, which form about the anterior portion of the palatoquadrate bar, are well developed. The outer arch of membrane bones is represented by premaxillaries and maxillaries. Anurans have a *quadratojugal* in addition.

The lower jaw consists of a core of Meckel's cartilage surrounded by membrane bones. The remaining portion of the visceral skeleton is reduced in comparison with fishes.

Reptiles The "stem" reptiles, or cotylosaurs, which very closely resembled labyrinthodont amphibians, are believed to have given rise to all reptiles, extinct and modern.

The chief differences to be noted in the adult reptilian skull compared with that of amphibians is the greater degree of ossification and the increased density of the bones. Little remains of the embryonic chondrocranium. Only the ethmoid region has retained its primitive cartilaginous character. All four parts of the occipital complex are ossified, but not all necessarily bound the foramen magnum. A single occipital condyle is present, formed from the basioccipital and usually by a contribution from each exoccipital. A cartilaginous or bony *interorbital septum,* forming a partition between the orbits, is present in many reptiles. In others, the anterior part of the cranial cavity lies in this region and the septum is reduced. The prootic region of the auditory capsule is ossified and usually remains separate from epiotics and opisthotics. Parietal and frontal bones are paired in some species but are frequently fused. An *interparietal foramen* is prominent in *Sphenodon* (Fig. 6.38) and many lizards. Membrane bones are more numerous in reptiles than in amphibians.

Birds Birds evolved from the branch of archosaurs, or "ruling" reptiles, which included dinosaurs with bipedal locomotion, rather than from the flying pterosaurs. The scales that covered the body have been modified into feathers, used to resist the air in flight. The skull of birds has not deviated far from the reptilian type (Fig. 6.39). It resembles that of lizards in many ways. The single occipital condyle, instead of being located at the posterior end of the skull, lies somewhat forward along the base, and the skull articulates with the vertebral column at almost a right angle. With the greater development of the brain in birds, the cranial cavity has increased in size, mostly by lateral expan-

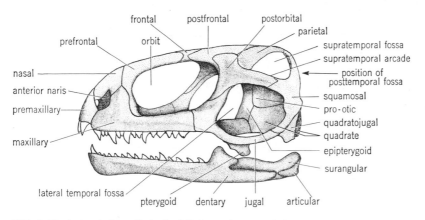

frontal   postfrontal   postorbital
prefrontal   orbit   parietal
supratemporal fossa
supratemporal arcade
position of
posttemporal fossa
nasal
anterior naris
premaxillary
squamosal
pro-otic
quadratojugal
quadrate
maxillary
epipterygoid
surangular
lateral temporal fossa
pterygoid   dentary   jugal   articular

FIG. 6.38 Lateral view of skull of *Sphenodon punctatum*.

sion. The orbit is large, to accommodate the relatively massive eye. The orbits are situated at a level somewhat anterior to the cranium proper. Each is bounded dorsally by a frontal bone, to which the lacrimal is attached, and posteriorly by a pleurosphenoid. A thin interorbital septum is present. It is better developed than in most reptiles. A ring of thin membrane bones, the *scleral ossicles,* embedded in the sclerotic coat of the eye, provides additional protection for the eye.

The skull of birds is unusually light in weight, and the cranial bones are fused together to a high degree. All four components of the occipital complex surround the foramen magnum. Where the supraoccipital joins the parietal, a prominent crest, the lamb- doidal ridge, is present. Epiotics and op- isthotics are fused to the occipital bones. Most conspicuous is the great development of the premaxillary bones, which together with the maxillaries form the upper portion of the beak, or the bill. This is covered by a horny sheath derived from the epidermis. Premax- illaries, maxillaries, and nasals are completely fused. All three bones bound the external nares, which, except in the kiwi, are located at the base of the beak.

Contrary to the condition in crocodilians, there is no complete secondary palate. The inner arch, or palatoquadrate bar, is slender and not connected with its partner across the midline. The quadrate is well developed and moves freely. The outer arch is also slender, consisting of maxillary, jugal, and quadra- tojugal bones. The latter joins the quadrate posteriorly. Thus inner and outer arches are united at the quadrate.

The lower jaw consists of two rather flat- tened halves united anteriorly and articulating with the quadrate posteriorly. Actually, each half is composed of five bones fused together. Only the articular is a cartilage bone. The an- terior dentary makes up the greater part of the lower jaw, which is also covered by a horny sheath in lieu of teeth.

The hyoid apparatus of birds consists of a median portion composed of three bones placed end to end and one or two pairs of cornua, or horns. These represent portions of the second and third visceral arches. The pos- terior cornua are extremely long in wood- peckers.

Mammals The ancestors of mammals di- verged early from the main reptilian line.

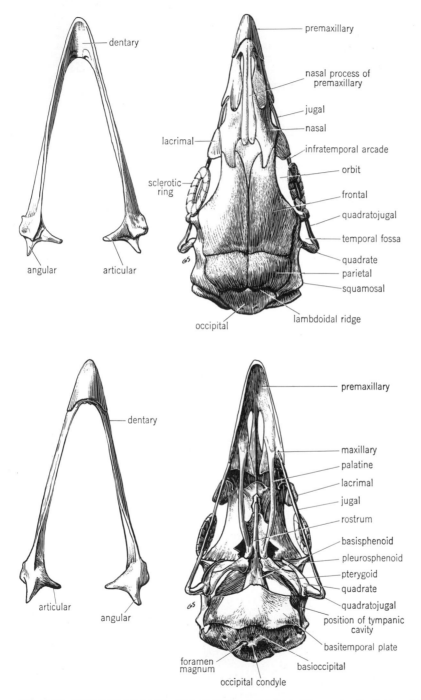

FIG. 6.39 Dorsal and ventral views of skull of chicken. (*Drawn by G. Schwenk.*)

They have little in common with reptiles living today, nor were they closely related to the "ruling" reptiles which at one time dominated the earth. The cynodonts, or mammallike reptiles, are believed to have been in the direct line of ascent of mammals. Even they were rather far removed from the oldest known reptiles, the pelycosaurs, whose skulls differed but little from ancient amphibians. Unlike most reptiles, which have a single occipital condyle, cynodonts possessed two, as do amphibians and mammals.

Although much variation exists in the skulls of different mammals, certain features are shared in common which serve to distinguish them from lower forms.

Most noteworthy is the increased capacity of the cranium, which has expanded in correlation with the much greater size of the mammalian brain. A firm union exists between all the bones of the skull with the exception of the mandible, hyoid, and auditory ossicles. All four components of the occipital complex surround the foramen magnum. Two occipital condyles, derived for the most part from the exoccipitals, articulate with the atlas vertebra. In higher apes and man the occipital condyles are decidedly ventral in position. Articulation of lower jaw is at the *mandibular fossa* of the squamous region of the temporal bone, rather than at the quadrate. This is the *craniostylic* or *amphicraniostylic* method of jaw suspension.

FIG. 6.40 Dorsal view of cat's skull.

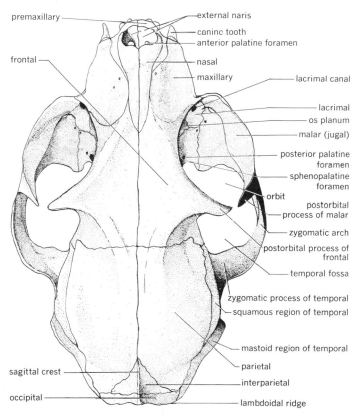

premaxillary
external naris
canine tooth
anterior palatine foramen
frontal
nasal
maxillary
lacrimal canal
lacrimal
os planum
malar (jugal)
posterior palatine foramen
sphenopalatine foramen
orbit
postorbital process of malar
zygomatic arch
postorbital process of frontal
temporal fossa
zygomatic process of temporal
squamous region of temporal
mastoid region of temporal
parietal
sagittal crest
interparietal
occipital
lambdoidal ridge

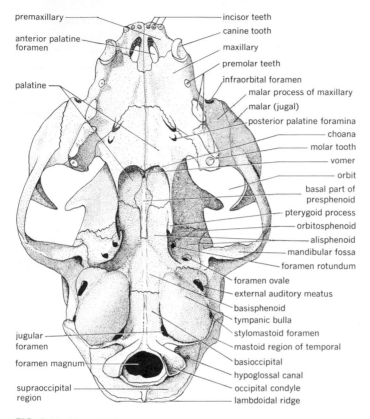

premaxillary

anterior palatine foramen

palatine

jugular foramen

foramen magnum

supraoccipital region

incisor teeth

canine tooth

maxillary

premolar teeth

infraorbital foramen

malar process of maxillary

malar (jugal)

posterior palatine foramina

choana

molar tooth

vomer

orbit

basal part of presphenoid

pterygoid process

orbitosphenoid

alisphenoid

mandibular fossa

foramen rotundum

foramen ovale

external auditory meatus

basisphenoid

tympanic bulla

stylomastoid foramen

mastoid region of temporal

basioccipital

hypoglossal canal

occipital condyle

lambdoidal ridge

FIG. 6.41 Ventral view of cat's skull.

The squamosal usually forms part of the temporal bone. The quadrate is reduced to the incus. A bony tympanic ring, apparently derived from the angular bone of the reptilian lower jaw, also may form part of the temporal bone. It is not homologous with the tympanic annulus of anuran amphibians. Otic bones are fused to form the *petrosal,* which encloses the inner ear. The petrosal may fuse with the squamosal and tympanic and thus becomes the *petrous region* of the temporal bone. The opisthotic is represented by the spongy *mastoid region* of the temporal.

Teeth of the upper jaw are borne only by premaxillary and maxillary bones. The premaxillaries are often fused together. The the-

codont method of tooth attachment is the rule. Except in the toothed whales, which have homodont dentition, the teeth of mammals are heterodont.

The number of bones in the mammalian skull (Figs. 6.40 and 6.41) is less than in most lower forms. There are usually about 35 of them. This is the result of the loss of certain bones and fusion of others. Among the elements that have disappeared in mammals are the prefrontals, postfrontals, and postorbitals. The temporal bone, just referred to, is an example of the result of fusion. Another common area of fusion is in the cartilage bones of the sphenoid complex. The basal portion of the frontal segment frequently unites with the orbitosphenoids to form a presphenoid bone. Similarly, the basisphenoid and alisphenoids may fuse. There are no pleurosphenoids in mammals. Their place is taken by the epipterygoids, cartilage bones derived from the palatoquadrate bar. These are then known as the alisphenoids. In some cases, as in man, all six elements of the sphenoid complex unite to form a single *sphenoid* bone. In most mammals a single occipital bone, containing the foramen magnum, is present in the adult. This also represents a fusion, although the four separate elements of which it is composed may easily be identified during development. Fusion of the membrane bones of the calvarium is common. A membrane bone, the interparietal, may develop between the parietals and supraoccipital. It may fuse later with the supraoccipital or with the parietals. The facial bones of mammals are more firmly united with the cranium than in other vertebrates. This is true both of the original maxillary arch and palatoquadrate bar components. A hard, or bony, secondary palate is typically found in mammals. It is composed of premaxillary, maxillary, and palatine bones. In whales and certain edentates the pterygoids form a portion of the hard palate, but these bones are much reduced in most mammals, so that the choanae (internal nares) are usually bordered by the palatines alone.

The nasal cavities of mammals are large, in correlation with the usually highly developed olfactory sense. Premaxillaries, maxillaries, and nasals surround the greater part of the nasal cavities. The ethmoid complex is well differentiated in most species. The mesethmoid forms a vertical cartilaginous plate which separates right and left nasal passages. The posterior portion becomes the bony *lamina perpendicularis*. The ethmoid also contributes the *cribriform plate,* separating nasal and cranial cavities. This bears numerous small *olfactory foramina* for the passage of fibers of the olfactory nerve. Ethmoturbinates are scroll-like outgrowths of the ectethmoids. Other turbinates are contributed by nasal and maxillary bones. The vomer, which is paired only during early developmental stages, lies ventral to the median nasal septum. It passes anteriorly and downward in an oblique direction from the basal portion of the cranium to join the hard palate. The vomer of mammals possibly is homologous with the parasphenoid of lower forms rather than with their *paired* vomers, the latter often being referred to as *prevomer* bones.

Many interesting variations from the above general description of the mammalian skull are to be observed. Among these is the presence of hollow, bony projections of the frontal bones in many members of the order Artiodactyla. They are covered with horn derived from the epidermis. The antlers of deer, which develop under the organizing influence of the integument in response to stimulation by certain hormones, when fully developed are naked projections of the frontal bones. In some mammals a prominent sagittal crest extends forward from the posterior part of the skull on the middorsal line. A lambdoidal ridge may pass transversely across the skull along the line of junction of parietals and supraoccipital. Outgrowths of this sort provide surface area for muscle attachment.

Large air spaces, or sinuses in communication with the respiratory passages, are present to varying degrees in the skulls of mammals. The most common are the frontal, ethmoid, sphenoidal, and maxillary sinuses.

An ossified derivative of the dura mater, the *tentorium,* in such forms as the dog and cat, becomes secondarily attached to the skull on the inner surface of the parietals. It serves to separate partially the cerebral and cerebellar fossae.

The lower jaw, or mandible, of mammals (Fig. 6.42) differs from that of other ver-

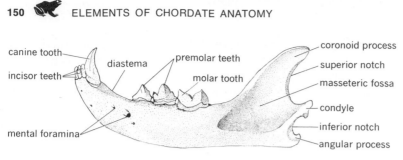

FIG. 6.42 Lateral view of mandible of cat.

tebrates in its articulation with the squamosal rather than with the quadrate and by the fusion and loss of elements, resulting in a single bony structure. The two halves of the mandible join in front at the symphysis. In primates, bats, and members of the order Perissodactyla an inseparable fusion occurs between the two halves of the mandible at the symphysis.

The mammalian hyoid apparatus shows much variation. Typically it consists of a body, or *basihyal,* and two pairs of horns. The *styloid processes* are attached to the otic region of the skull. In some forms the hyoid consists of a chain of bones extending from the basihyal to the otic region. The posterior cornua are usually smaller, each consisting of a single bone which extends from the basihyal to the thyroid cartilage of the larynx. In other mammals the separate bones may lose their identity, being represented only by fibrous bands. The hyoid apparatus serves to support the tongue and furnishes a place for muscle attachment. The basal portion and anterior cornua are derivatives of the hyoid arch, which has also contributed the stapes of the middle ear. The posterior cornua represent remnants of the third visceral arches. The remaining arches contribute to the laryngeal cartilages.

Despite similarities, evident differences distinguish the skulls of monotremes, marsupials, and placental mammals. Monotremes most closely resemble the ancient reptilian stock. The cranial cavity of marsupials is relatively small in comparison with that of placental mammals.

Ribs

Ribs consist of a series of cartilaginous or bony elongated structures attached at their proximal ends to vertebrae. In the primitive condition there is a pair of ribs for each vertebra. In some vertebrates the ribs consist of little more than small cartilaginous tips on the transverse processes of vertebrae. They may be fused permanently with the transverse processes (pleurapophyses), or a joint may appear between the two. In higher forms they may be stout, strongly arched structures surrounding the thoracic cavity and uniting ventrally with a median bony structure, the *sternum,* or *breastbone.*

RIBS OF FISHES Ribs first appear in fishes and are of two kinds, dorsal and ventral. Many fishes possess either one type or the other, but in *Polypterus* and many teleosts both types are present.

Dorsal Ribs The upper, intermuscular, or *dorsal, ribs* (Fig. 6.43) extend out as costal processes from the transverse processes of vertebrae into the lateral skeletogenous septum which separates epaxial and hypaxial muscles. They occur metamerically at points

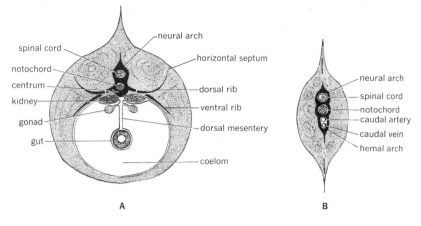

FIG. 6.43 *A,* diagrammatic cross section of typical vertebrate, showing position of dorsal and ventral ribs; *B,* section through tail of fish, showing hemal arch. (*After Parker and Haswell, "Textbook of Zoology," vol. II, used with permission of The Macmillan Company.*)

where the myocommata, or septa between adjacent myotomes, cut across the lateral septum. Some authorities are of the opinion that true dorsal ribs are present only in fishes. Most, however, believe that the ribs of tetrapods are homologous with these dorsal ribs.

Ventral Ribs The ventral, or *pleural, ribs* of fishes are attached at the ventrolateral angles of the centrum. They lie beneath the muscles of the body wall outside the peritoneum and are situated along the lines where the myocommata, or septa, pass toward the coelom. There is general agreement that the ventral ribs represent the two halves of hemal arches of caudal vertebrae which have split ventrally and spread apart. However, this conception may be erroneous, for the caudal vertebrae of some fishes possess both ventral ribs and hemal arches. Additional intermediate bones of riblike appearance are frequently found embedded in the myocommata of fishes.

RIBS OF TETRAPODS Each tetrapod vertebra typically bears a single pair of ribs. Except in the thoracic region, the ribs are usually short and are immovably fused to the vertebrae. In most amniotes the ribs in the thoracic region are modified, separating from the vertebrae, with which they then form movable articulations. These ribs are elongated and usually unite ventrally with the sternum. The ends attached to the sternum are generally cartilaginous and are spoken of as *costal cartilages.*

Tetrapod ribs usually form articulations both with centra and neural arches of vertebrae. In these *bicipital ribs* (Fig. 6.44) the upper head of a rib, or *tuberculum,* articulates with a dorsal process, the *diapophysis,* coming from the neural arch. The lower head, or *capitulum,* joins a projection of the centrum, the *parapophysis.* The bicipital condition is considered to be primitive in tetrapods, since it exists in some of the oldest fossil amphibians and reptiles. It undoubtedly aids in terrestrial locomotion by strengthening the trunk.

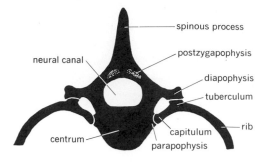

FIG. 6.44 Diagram showing method of articulation of tetrapod rib with vertebra.

In higher forms the diapophysis may be represented by a small *tubercular facet* on the ventral side of the transverse process; the parapophysis in such cases may consist of nothing more than a small facet on the centrum for reception of the capitulum. The latter may articulate at the junction of two centra, in which case each centrum may bear a half facet, or *demifacet,* at either end. Sometimes tetrapod ribs have but a single head. This may represent either the tuberculum or capitulum or a fusion of the two.

In the cervical region, where ribs are generally fused to vertebrae, the space formed between the two articulations of the rib with the vertebra is known as the *foramen transversarium,* or *vertebrarterial canal.* It serves for the passage of the vertebral artery and vein and a sympathetic nerve plexus. In the lumbar, sacral, and caudal regions the ribs are fused with the vertebrae (pleurapophyses), and there is little or no evidence of the primitive bicipital condition.

Ribs in Different Classes of Vertebrates

FISHES In elasmobranchs small ribs are attached to the ventrolateral angles of the centrum (basapophyses). They extend laterally into the skeletogenous septum and probably represent ventral ribs which have moved secondarily into this position. Caudal vertebrae bear hemal arches with which these ventral ribs are homologous.

Most higher fishes possess only ventral ribs, but *Polypterus* and a number of teleosts have both dorsal and ventral ribs.

AMPHIBIANS In some of the fossil labyrinthodont amphibians the ribs were strong structures which extended ventrally around the body for some distance. In modern amphibians they have been reduced and are small and poorly developed. In most urodele amphibians all the vertebrae, from the second to the caudal, bear ribs which are typically bicipital. The strong sacral ribs are attached to the pelvic girdle. Caudal vertebrae lack ribs. Ribs are entirely lacking in anurans or are in the form of minute cartilages fused to the transverse processes. Caecilians have better-developed ribs than other amphibians. Amphibian ribs never form connections with the sternum.

Some extinct amphibians possessed ventral, dermal "ribs," or *gastralia.* These consisted of V-shaped, riblike dermal bones located in the ventral abdominal region. They extended in a posterodorsal direction on either side. Gastralia should not be confused with true ribs, which are endoskeletal structures.

REPTILES Rib development in reptiles has deviated far from the condition observed in amphibians. Ribs may be borne by almost all the vertebrae in trunk and tail regions. Attachment of ribs to the sternum is first observed in the class Reptilia.

Turtles lack ribs in the cervical region. The 10 trunk vertebrae bear ribs which are flat and broad and, like the vertebrae in this region, are immovably fused to the underside of the carapace (Fig. 6.45). The carapace is thus formed partly of dermal skeletal bony

ANTERIOR

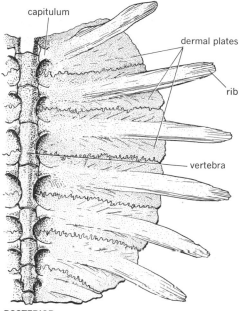

capitulum

dermal plates

rib

vertebra

POSTERIOR

FIG. 6.45 Underside of portion of carapace of young turtle, showing fusion of ribs and vertebrae to dermal bones. The marginal dermal plates in this specimen had not yet become ossified.

plates and partly of endoskeletal structures derived from vertebrae and ribs. Each rib has but a single head, the capitulum, for articulation with the vertebral column. The point of articulation is usually at the boundary of adjacent vertebral centra. The two sacral and first caudal vertebrae are also united to the carapace.

The ribs of *Sphenodon* are extensively developed, being present even in the caudal region. A few of the anterior thoracic ribs join a typical sternum, but most of the remaining ribs join a median, ventral *parasternum* which extends from sternum to pubis. This is an endoskeletal derivative from cartilages which develop in the ventral portions of the myocommata. These ribs should not be confused with gastralia, which are dermal derivatives and which in *Sphenodon* are present in the same region. Each rib is typically composed of three sections: an upper ossified *vertebral section;* a ventral, cartilaginous *sternal section;* and an *intermediate section,* also composed of cartilage. Each rib in *Sphenodon* bears a flattened, curved *uncinate process* which projects posteriorly from the vertebral section, overlapping the next rib and thus providing additional strength to the thoracic wall.

Small cervical ribs occur in lizards, except on the atlas and axis. Geckos have ribs on these two vertebrae as well. In most lizards a few of the anterior thoracic ribs curve around the body to meet the sternum. Each is divided into two or three sections, but only the vertebral section is bony in most cases. Articulation with the vertebra is by the capitulum alone.

In snakes all trunk vertebrae except the atlas and axis bear ribs. They articulate by means of a single head, the capitulum. Since no sternum is present, the ribs terminate freely. The lower ends, however, have muscular connections with ventral scales, which are used extensively in locomotion.

Crocodilian ribs are typically bicipital. All five regions of the vertebral column are rib-bearing, although the ribs are much reduced in the cervical, lumbar, and caudal regions. Even the atlas and axis bear ribs. The two heads of the cervical ribs surround a foramen transversarium. Eight or nine thoracic ribs connect with the sternum. These are composed of vertebral, intermediate, and sternal sections, of which only the vertebral section is completely ossified. Uncinate processes are present on the vertebral sections. The last two or three thoracic ribs are *floating ribs* with no sternal connections. The two pairs of sacral ribs are strong projections which articulate with the ilium of the pelvic girdle. Dermal gastralia, similar to those of *Sphenodon,* are well developed in the ventral abdominal region of crocodilians.

BIRDS In *Archaeopteryx* the ribs resembled those of lizards more closely than those of modern birds. They were slender and articulated by means of a single head, and they lacked uncinate processes. *Archaeopteryx* also possessed gastralia, not encountered in birds of today. In modern birds some of the posterior cervical vertebrae bear movable ribs. Strong, flattened, bicipital ribs connect most of the thoracic vertebrae with the sternum. Each is composed of ossified vertebral and sternal sections. A prominent uncinate process from each vertebral section (Fig. 6.21) overlaps the rib behind, furnishing a place for

FIG. 6.46 Vertebral section of human rib. (*Drawn by G. Schwenk.*)

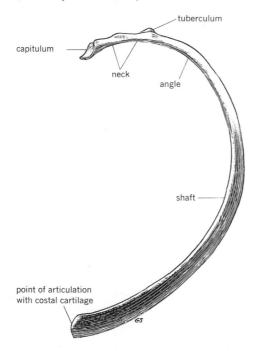

muscle attachment and increasing the rigidity of the skeleton, an important factor in flight. In the posterior thoracic, lumbar, sacral, and anterior caudal regions the ribs are fused to the large synsacrum.

MAMMALS Mammalian ribs are usually bicipital. In the cervical region the transverse processes are composed in part of the remains of ribs which have fused with the vertebrae. Foramina transversaria, present only in cervical vertebrae, represent the original space between points of attachment. Thoracic ribs are well developed in mammals. The tubercular head articulates with the tubercular facet (diapophysis) on the ventral side of the transverse process; the capitulum typically articulates at the point of junction of two adjacent vertebrae, the centrum of each bearing a costal demifacet. In some posterior ribs only the capitular head remains. Each rib consists of two sections: an upper, bony, vertebral section and a lower, usually cartilaginous, sternal section. The latter joins the sternum directly or indirectly and is commonly referred to as the *costal cartilage.* The ribs of monotremes are composed of three sections, as in primitive reptiles. Mammalian ribs do not bear uncinate processes.

The part of the rib between the two heads is called the *neck;* the remainder is the *shaft* (Fig. 6.46). The area where the shaft curves most markedly is the *angle* of the rib. Those ribs which make direct connection with the sternum are called *true ribs. False ribs* are located posteriorly, and their costal cartilages either unite with the costal cartilages of the last true rib or terminate freely. In the latter case they are termed *floating ribs.*

The number of ribs in mammals shows considerable range, from 9 pairs in certain whales to 24 in the sloth. True ribs, however, range in number from 3 to 10 pairs.

Sternum

The sternum, or breastbone, found only in tetrapods, is composed of ventral skeletal elements closely related to the pectoral girdle of the appendicular skeleton and, except in amphibians, with the anterior thoracic ribs. Not all tetrapods have a sternum. It is lacking in some of the elongated salamanders such as *Proteus* and *Amphiuma.* Caecilians, snakes, and some limbless lizards lack a sternum, as do turtles. In the latter the plastron serves as a strengthening support in the absence of a sternum. The sternum provides support and a place for muscle attachment in terrestrial vertebrates, thus aiding in locomotion on land.

The origin of the sternum in evolution is obscure. Several theories have been advanced to explain its origin, none of which is wholly acceptable at the present time. According to one conception, the earliest indication of a sternum is to be found in the salamander *Necturus,* in which it is referred to as the *archisternum.* In this amphibian, cartilages occur in a few of the myocommata in the ventral thoracic region. The largest of these irregular elements is found at precisely the same region in which, in higher salamanders, a sternal plate overlaps the coracoids of the pectoral girdle. Such a theory of sternal origin is plausible, relating it to the *parasternum* of such forms as *Sphenodon,* lizards, and crocodilians. The parasternum is a posterior continuation of the sternum. Fusion of some of the archisternal or parasternal elements may thus possibly account for the origin of the true sternum. If this theory is correct, the connection of ribs to the sternum must be secondary.

According to another theory, the sternum arises as a fusion of the ventral ends of some of the anterior ribs. This type of sternum, found in amniotes, has been spoken of as the *neosternum.* Objection to this theory is based upon the fact that although a sternum is present in most amphibians, in no case are ribs associated with it. It is possible, however, that the ribs have become shortened secondarily in the course of evolution of amphibians and have lost their sternal connections. There is no paleontological evidence for this.

Some authorities are of the opinion that the archisternum of amphibians and the neosternum of amniotes, with its rib connections, are in no way related and are of independent origin. Since, however, evidence from embryology indicates that the ribs of amniotes connect with the sternum secondarily, there is no particular reason for assuming a lack of homology between the two types of sterna.

By using marking techniques it has been shown that the sternum of birds and mammals is a derivative of lateral plate mesoderm. Ribs and vertebrae are clearly somite derivatives. This indicates that the sternum should properly be classified as a portion of the appendicular skeleton rather than of the axial skeleton.

AMPHIBIANS In urodele amphibians the sternum appears for the first time and in its most primitive form. That of *Necturus* has already been referred to. In most salamanders it consists of little more than a small, median, triangular plate, lying behind the posterior median portions of the coracoids of the pectoral girdle.

In anurans the sternum is better developed. In the common frog the anterior clavicles and posterior coracoids of the pectoral girdle are separated from their partners of the opposite side only by narrow cartilaginous strips, the epicoracoids (Fig. 6.47). Anterior to the junction of the clavicles with the epicoracoid cartilages lies the bony *omosternum,* with an expanded, cartilaginous *episternum* joined to it anteriorly. Posterior to the junc-

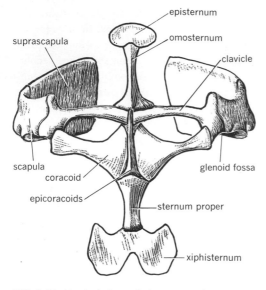

FIG. 6.47 Ventral view of sternum and pectoral girdle of bullfrog, *Rana catesbeiana*. (*Drawn by G. Schwenk.*)

tion of coracoids and epicoracoids lies the *sternum proper*. An expanded cartilage, the *xiphisternum,* is attached to it posteriorly. The sternum is thus considered to be composed of four median elements.

REPTILES Snakes, most limbless lizards, and turtles lack a sternum, but in other reptiles this portion of the skeleton is more fully developed than it is in amphibians.

In *Sphenodon* a few of the anterior thoracic ribs join a typical midventral sternum, but most of the remaining ribs join the median, ventral parasternum, which extends from the true sternum to the pubis. It will be recalled that this endoskeletal structure is derived from cartilages which appear in the midventral portion of the myocommata in this region.

In lizards the sternum usually consists of a large, cartilaginous, flattened plate to which

the sternal sections of several anterior thoracic ribs are united. A dermal membrane bone, the *episternum,* or *interclavicle,* in some cases lies ventral to the sternum. It is *not* part of the sternum.

The sternum of crocodilians is a simple cartilaginous plate which splits posteriorly into two *xiphisternal cornua.* Numerous gastralia of dermal origin are present posterior to the cornua but form no connection with them.

BIRDS If *Archaeopteryx* had a sternum, it is unknown. The sternum of birds is a well-developed bony structure, to the sides of which the ribs are firmly attached. In flying birds and penguins it projects rather far posteriorly under a considerable portion of the abdominal region. The ventral portion is drawn out into a prominent *keel,* or *carina* (Fig. 6.21), which provides a large surface for attachment of the strong muscles used in flight. It is of interest to note in this connection that the extinct flying reptiles, the pterosaurs, had carinate sterna, as do bats among mammals. Running birds of the superorder Paleognathae have rounded sterna which are not carinate.

MAMMALS In mammals the sternum is typically composed of a series of separate bones arranged one behind the other. Three regions may be recognized, an anterior *presternum,* or *manubrium;* a middle *mesosternum,* consisting of several *sternebrae;* and a posterior *metasternum,* or *xiphisternum,* to the end of which a *xiphoid cartilage* is attached (Fig. 6.48). The costal cartilages of the ribs articulate with the sternum at the points of union of the separate bones. In bats the presternum is conspicuously keeled, as is the mesosternum in some cases. The number of sternebrae is variable in mammals, as is the number of true ribs which articulate directly with the

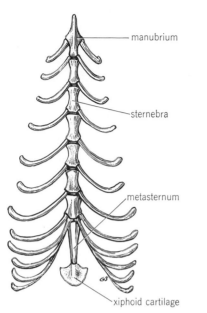

manubrium

sternebra

metasternum

xiphoid cartilage

FIG. 6.48 Sternum of cat, showing relation to costal cartilages.

FIG. 6.49 Sternum of man.

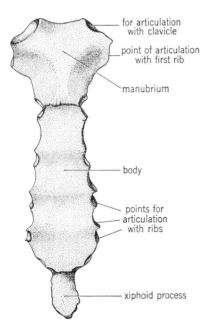

for articulation with clavicle

point of articulation with first rib

manubrium

body

points for articulation with ribs

xiphoid process

sternum. In some cases, e.g., in certain cetaceans, sirenians, and primates, the separate elements may fuse. Thus in man the entire structure, which is flattened (Fig. 6.49), consists only of three parts, an anterior *manubrium,* a *body,* and a small posterior *xiphoid process.* Study of the development of the sternum in man shows it to be composed of separate structures, each derived from separate centers of ossification. These fuse together at various times from puberty to old age. The mammalian sternum arises quite independently of the ribs, which unite with it secondarily.

Median Appendages

In discussing the axial skeleton, certain unpaired, median appendages must be taken into consideration. Only aquatic vertebrates possess median appendages, which are always finlike in character. They are found in cyclostomes, fishes, larval amphibians, some adult urodele amphibians, and cetaceans. Typically they are present in dorsal, anal, and caudal regions. In some forms the median fins are continuous, but most frequently the continuity is interrupted and the various fins are separated by gaps. In some of the newts the dorsal fin of the male becomes developed only during the breeding season. In cyclostomes and other fishes the median fins are supported by skeletal structures (fin rays) and special muscles, but in higher forms thay are merely elaborations of the integument. Dorsal and anal fins are used chiefly in directing the body during locomotion. The caudal fin also helps in this respect, in addition to being the main organ of propulsion.

The skeleton supporting the unpaired dorsal and anal fins in fishes consists of radial, cartilaginous or bony *pterygiophores* supporting slender *fin rays* at their distal ends. Several pterygiophores may unite at the base of the fin to form one or more *basipterygia,* or *basalia.* These are ossified in bony fishes. Pterygiophores and basalia may form secondary connections with the neural spines of the vertebral column. There may or may not be a segmental correspondence between the skeletal elements of the fin and the vertebrae. Distal to the basalia the pterygiophores continue as *radialia.* From the radialia extend numerous fin rays of dermal origin, supporting the greater part of the fin. Fin rays frequently connect directly with the basalia, the radialia being absent. Various types of joints permit the fin to be moved in a complex undulatory manner or merely to be raised or lowered. In the males of many fishes the anal fin and its skeletal elements are modified to form a *gonopodium,* which serves to aid in the transport of sperm from male to female in internal fertilization.

Caudal fins are of several shapes (Fig. 6.50). The most primitive type is the *protocercal* tail of adult cyclostomes. The notochord is straight and extends to the tip of the tail, the dorsal and ventral portions of which are of almost equal dimensions. A second type, the *diphycercal* tail, found in Dipnoi, *Latimeria,* and *Polypterus,* appears at first sight to be similar to the protocercal type. Evidence indicates, however, that the diphycercal tail may really be a secondary modification of a third type, the *heterocercal* tail, in which the skeletal axis bends upward to enter the dorsal flange of the tail. This then becomes more prominent than the ventral flange. A heterocercal tail is typical of elasmobranchs and lower fishes in general. The *homocercal* type of tail is most

FIG. 6.50 Four types of fish tails.

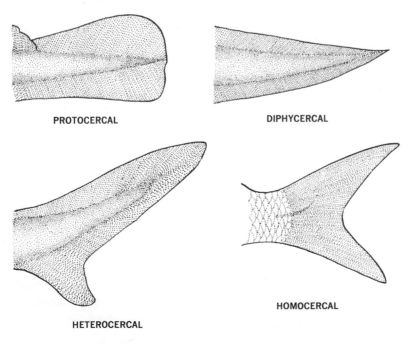

PROTOCERCAL

DIPHYCERCAL

HETEROCERCAL

HOMOCERCAL

common. Superficially it appears to be composed of symmetrical dorsal and ventral flanges. The posterior end of the vertebral column, the *urostyle,* is, however, deflected into the dorsal flange. This type of tail, therefore, is not far removed from the heterocercal type.

In caudal fins the fin rays on the dorsal side may connect with basal pterygiophores or with neural spines directly. On the ventral side, however, they form connections with modified hemal spines which are spoken of as *hypurals.*

APPENDICULAR SKELETON

Speculation about the phylogenetic origin of the paired appendages of vertebrates has long occupied the attention of comparative anatomists. Several theories have been advanced, some of which have been discarded as new paleontological discoveries have been made. Others have been modified from time to time as new lines of evidence have been

brought to light. The limbs of tetrapods undoubtedly arose from the fins of crossopterygian fishes. The *fin-fold theory* of the origin of both unpaired and paired vertebrate appendages seems to be one of the most plausible and in the past was quite generally accepted.

According to one viewpoint, the fin-fold theory (Fig. 6.51) goes back to amphioxuslike animals as a starting point. In this animal the single dorsal fin continues around the tail to the ventral side as far forward as the atriopore. At this point it divides in such a manner that a *metapleural fold* extends anteriorly on either side almost to the mouth region. Gaps appearing in the dorsal fin and in the metapleural folds constitute one explanation for the appearance of median and paired fins. These folds in amphioxus, however, terminate at the atriopore, a structure without homologies in higher forms. Furthermore, amphioxus has no skeletal structures other than a notochord. It is, therefore, improbable that the metapleural folds of amphioxus were in

FIG. 6.51 Diagrams illustrating the fin-fold theory of the origin of paired appendages: *A,* the undifferentiated condition; *B,* how permanent fins may have been formed from the continuous folds. (*After Wiedersheim, "Comparative Anatomy of Vertebrates," copyright 1907 by The Macmillan Company and used with their permission.*)

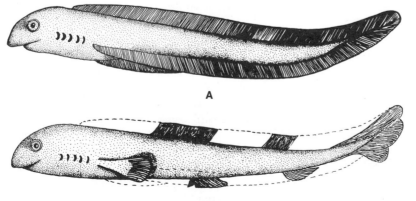

A

B

any way concerned with the origin of the paired appendages of vertebrates. Cyclostomes, lacking paired appendages, are usually dismissed as a specialized, divergent side shoot of the ancestral vertebrate stock.

A second viewpoint holds that the fin folds similar to the metapleural folds of amphioxus were present in some hypothetical ancestral fish but terminated at the anus rather than at the atriopore. In some very primitive acanthodian sharks a row of six or seven spiny fins ex-

FIG. 6.52 Ventral view of acanthodian fish *Euthacanthus macnicoli,* showing paired rows of ventral spines (fins). (*After Watson.*)

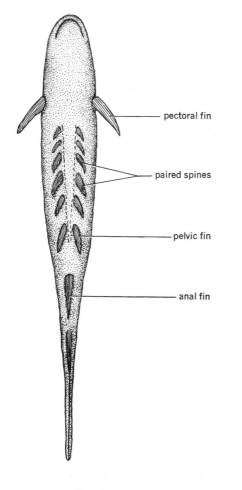

- pectoral fin

- paired spines

- pelvic fin

- anal fin

tended on each side between pectoral and pelvic fins (Fig. 6.52). They have been interpreted as remnants of fin folds. Traces of lateral fin folds may be observed in embryos of certain elasmobranchs as mesodermal proliferations which develop extensively only in the pectoral and pelvic regions as metameric myotomic buds (Fig. 7.9). The pectoral and pelvic fins may thus be regarded as persistent remnants of primitive fin folds. Furthermore, the basic skeletal structure of the paired fins is essentially like that of the unpaired fins, indicating a common origin of the two. The structure of the fins of the extinct shark *Cladoselache* has been cited as providing additional support to the fin-fold theory. Here the fins are broad at their bases and contain numerous parallel pterygiophores, which by their very arrangement suggest a primitive fin-fold origin (Fig. 6.53).

Pterygiophores are primitively metameric. Each is subdivided into several small pieces. Those at the base, the *basalia,* show a tendency to fuse. *Radialia* consist of two or three rows of short cartilages distal to the basalia. Dermal *fin rays* in turn are distal to the radialia. According to one theory, fusion of the anterior basalia with their partners in the midline may have resulted in the formation of a transverse bar, the rudiment of the pectoral or pelvic girdle, as the case may be.

Another and more recent theory of the origin of paired appendages and girdles does not assume a fin-fold origin. According to this concept, the appearance of paired appendages goes back to the ancient ostracoderms, the remote ancestors of existing cyclostomes. Some early ostracoderms had a cephalothorax covered with a shield of plates with posteriorly directed cornua. The plates in some cases are known to have contained bone cells. Certain forms possessed lateral fleshy lobes, projecting from each side, medial to the cornua of the thoracic shield. A bony skele-

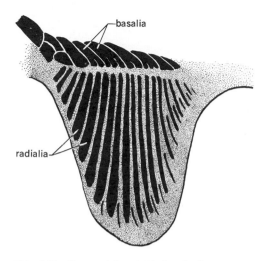

basalia

radialia

FIG. 6.53 Pectoral fin of *Cladoselache.*
(*After Dean.*)

the endoskeleton of the fin may have pushed farther into the body to give rise to the girdles. This concept holds that the paired fins were not originally parts of fin folds but were, from the beginning, separately spaced ridges supported by spines. Much paleontological evidence has been cited in support of the ostracoderm theory.

Paired Fins and Girdles of Fishes

Pectoral Fins and Girdles Certain extinct sharks, such as *Cladoselache,* possessed the simplest fins known to occur in vertebrates (Fig. 6.53). There were several basalia from which radialia extended in a fanlike manner. The basalia of the two sides do not unite.

In existing elasmobranchs the pectoral girdle (Fig. 6.54) is a U-shaped cartilage located just posterior to the branchial region. It is not connected directly with the axial skeleton, except in skates, in which the upper *suprascapular* cartilage joins the vertebral column on each side. The girdle consists of one piece, the *scapulocoracoid* cartilage. The ventral portion is the *coracoid bar,* from which a long scapular process extends dorsally on each side beyond the *glenoid region* where the pectoral fin articulates. Each scapular process may have a *suprascapular* cartilage attached at its free end. The fin itself consists of three basal cartilages (basalia) which articulate with the glenoid region of the girdle. Numerous segmented radial cartilages extend distally from the basalia. They are arranged in rows. From the radialia many dermal fin rays pass peripherally.

In primitive bony fishes certain membrane bones have become associated with the pectoral girdle. There are usually at least two such membrane bones on each side. A pair of *clavicles* meets on the midline close to the coracoid region. A *cleithrum* overlies the scap-

togenous septum passed behind the gill region and pericardium. The fleshy lobes may have become pectoral fins, and the endoskeletal part of the pectoral girdle may have arisen as an ingrowth of basal portions of the pectoral fins. The dermal bones of the pectoral girdle may have been derived from the bony plates of the thoracic shield.

Another group of ostracoderms is known to have possessed a paired row of dermal spines on the ventral side of the body, not unlike the extra spines or fins of acanthodian sharks referred to above. Loss of armor and reduction of spines but their persistence in pectoral and pelvic regions might well account for the origin of paired appendages from ostracoderm ancestors. With the growth of muscles from the metameric myotomes into the bases of the spiny fins, their movement may first have been accomplished. Appearance of radial skeletal elements, the pterygiophores, could have accompanied the segmental muscles as they spread out in a fanlike manner in opposing groups to the peripheral parts of the fin. The basal pieces of

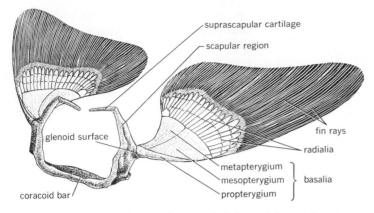

FIG. 6.54 Anterolateral view of pectoral girdle and fins of *Squalus acanthias.*

ula. In *Amia, Lepisosteus,* and teleosts, the clavicle has disappeared. Other membrane bones, *supracleithrum* and *posttemporal,* are present in many forms, the latter forming a connection between the skull and the girdle (Fig. 6.55). The original part of the pectoral girdle may remain cartilaginous, but it usually forms two separate bones on each side, the scapula and coracoid, articulating at the glenoid region. The two halves of the original cartilaginous girdle, unlike those of the elasmobranchs, do not connect ventrally. The membrane bones now make up the greater part of the pectoral girdle, the original components being much reduced.

Much variation is found in the detailed structure of the pectoral fins of bony fishes. In most, the number of separate skeletal elements is reduced, the dermal fin rays taking over much of the supporting function.

The link between the fins of fishes and the limbs of tetrapods is to be sought among the fossil crossopterygians. Study of the fin structure of *Eusthenopteron* (Fig. 6.58A) strongly

FIG. 6.55 Diagram showing the membrane bones of the pectoral girdle of a primitive fish. Membrane bones are stippled; cartilages or cartilage bones are black.

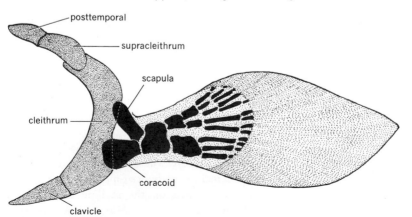

suggests that this group will provide the clue to the true explanation of the origin of tetrapod girdles and limbs.

Pelvic Fins and Girdles The pelvic fins and girdles of fishes are usually more primitive and simple in structure than those of the pectoral region. In *Cladoselache* there is little difference between the two, although the pelvic structures are somewhat reduced. The pelvic girdle of fishes is always free of the axial skeleton. Fusion of the anterior basalia of the two sides occurs in elasmobranchs to form the *ischiopubic bar* (Fig. 6.56), comparable to the coracoid bar of the pectoral girdle. Attachment of the fin is at the *acetabular region* rather than at the glenoid region. A small *iliac process* corresponds to the scapular region of the pectoral girdle. It is rare to find any membrane bones in the pelvic region of fishes comparable to those associated with the pectoral girdle. The pelvic girdle of teleost fishes is reduced to a small element on each side with no connections between the two. It may even be lacking, as in the eels, which have no pelvic fins. In elasmobranchs each pelvic fin typically consists of two basalia from which numerous radialia extend outward. Dermal fin rays complete the skeletal

structure. In male elasmobranchs the skeleton supporting the clasper is a continuation of one of the basalia (metapterygium), which is made up of several segments (Fig. 6.56). Teleosts possess rather degenerate pelvic fins. Usually a single basal bone gives off only a few poorly developed radials, from which the fin rays project. Often the fin rays arise directly from the basal. A curious situation exists in certain advanced fishes, such as the cod, in which the pelvic fins have moved forward to lie *anteroventral* to the pectoral fins and have become attached to the throat region. The fossil crossopterygian *Eusthenopteron* shows a better-developed pelvic fin structure than is to be found in forms living today (Fig. 10.58B). Although somewhat smaller than the pectoral fins, the basic skeletal structure is similar. A strong pelvic girdle is also present.

PAIRED GIRDLES AND LIMBS OF TETRAPODS The appendicular skeletons in all tetrapods show a fundamental similarity in structure. There is a shoulder, or pectoral, girdle, to which the forelimbs are attached and a hip, or pelvic, girdle supporting the hind limbs. The pectoral girdle does not normally form a connection with the vertebral column except through muscles and ligaments. The pelvic girdle, however, attaches directly to the vertebral column in the sacral region. The limbs are typically pentadactyl (having five fingers or toes) or have been modified from the primitive pentadactyl type. Each girdle is composed of two halves, and each half in turn is made up of three bones which show a fairly comparable arrangement in both pectoral and pelvic girdles.

In the pectoral girdle of tetrapods the number of membrane bones has been reduced from the conditions in fishes and only the *clavicle* and interclavicle may persist. The

FIG. 6.56 Pelvic girdle and fin of male *Squalus acanthias.*

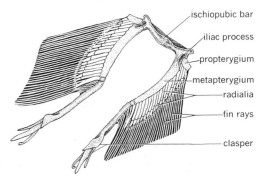

ischiopubic bar

iliac process

propterygium

metapterygium

radialia

fin rays

clasper

cartilage bones of the girdle in tetrapods have become the dominant elements. The bones of the pectoral girdle typically consist of a ventral *coracoid,* which meets the sternum medially; a *scapula* extending dorsally; and a *clavicle,* which lies on the ventral side between scapula and sternum and anterior to the coracoid. In some lower tetrapods an additional cartilage bone, the *precoracoid,* lies anterior to the coracoid and close to the clavicle (Fig. 6.57A). The precoracoid is probably derived from an ossification center in a ventral projection of the scapula. An interclavicle is present in many reptiles, lying ventral to the sternal plate and contacting the clavicles anteriorly. At the junction of scapula and coracoids is a depression, the *glenoid fossa,* which serves as the point of articulation of the forelimb with the pectoral girdle.

The pelvic girdle (Fig. 6.57B) is composed of a ventral *ischium,* comparable to the coracoid; a dorsal *ilium,* similar to the scapula; and an anterior, ventral *pubis* corresponding in position to the clavicle and precoracoid. The two ilia form a union with the sacrum. The two pubic bones usually unite ventrally at the *pubic symphysis.* The ischia also often come together to form an *ischial symphysis.* A depression, the *acetabulum,* located at the junction of the three bones, serves as the point of articulation of the hind limb with the pelvic girdle. The acetabulum is comparable to the glenoid fossa of the pectoral girdle. A fourth and very small *acetabular,* or *cotyloid,* bone often enters in the formation of the acetabulum in mammals. Its homologies are uncertain, but it may possibly be comparable to the precoracoid of the pectoral girdle.

The skeletal elements of the forelimbs and hind limbs are also arranged on the same general plan. In the forelimb there is a single bone, the *humerus,* in the upper arm. The head of the humerus joins the pectoral girdle at the glenoid fossa. Two bones arranged parallel to each other are present in the forearm. They are a lateral *radius* and a medial *ulna.*

FIG. 6.57 Diagram showing skeletal structures of tetrapod limbs and girdles: *A,* pectoral girdle and limb; *B,* pelvic girdle and limb. The clavicle is lightly stippled to indicate that it is a membrane bone.

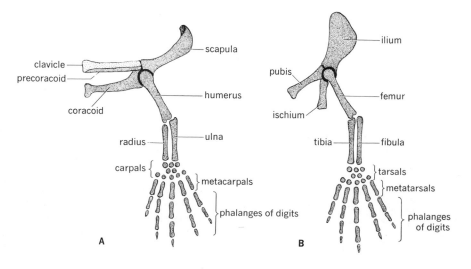

Distal to radius and ulna, in the primitive condition, are nine to ten *carpal,* or wrist, bones; five longer *metacarpals* in the hand; and then a row of small *phalanges* in the fingers. Corresponding bones in the hind limb are the *femur,* which joins the acetabulum, *tibia* and *fibula,* comparable to radius and ulna; nine or ten *tarsal,* or ankle, bones; five *metatarsals;* and a row of phalanges in each toe.

Many remarkable specializations have taken place in the evolution of the limbs of tetrapods. These consist, for the most part, of reductions or fusions of various parts of the primitive, basic, pentadactyl limb skeleton.

Phylogenetic Origin of Tetrapod Girdles and Limbs We have already referred to the origin of tetrapod girdles and limbs from piscine ancestors. Several theories

have been advanced in the past to explain such evolutionary changes. There is now general agreement that connecting links are to be sought among fossil crossopterygians of the extinct order Rhipidistia. The fossil remains of the rhipidistian crossopterygian *Eusthenopteron* rather convincingly indicate that the skeletal parts of the fins of this primitive fish can be closely homologized with those of the limbs of tetrapods.

Each fin of *Eusthenopteron* has a chain of bones along the postaxial side, from which a series of radials comes off, as indicated in Fig. 6.58. It is believed that the first bone at the base of the fin is equivalent to the humerus (femur) and the second to the ulna (fibula). The first radial is comparable to the radius (tibia). The rest of the radials are believed to have finally become carpal (tarsal) bones,

FIG. 6.58 Skeletal structure of *A,* left pectoral, and *B,* left pelvic fins of the crossopterygian fish *Eusthenopteron.* (*After Gregory and Raven, Ann. N.Y. Acad. Sci., by permission.*)

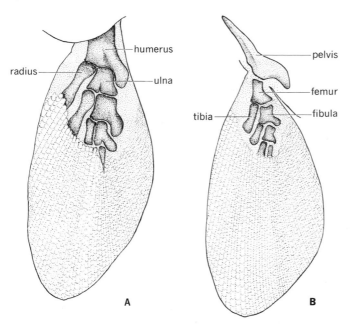

whereas metacarpals (metatarsals) and phalanges have arisen as new distal outgrowths from the margin of the fleshy, muscular portion of the paddlelike fin. *Eusthenopteron* itself is not considered to be the direct ancestor of tetrapods. Some close rhipidistian relative was probably the tetrapod progenitor. Joints are believed to have developed at regions in the appendages where bending was required to produce effective locomotion.

In the evolution of the pectoral girdle of tetrapods, those parts derived from the dermal skeleton have been lost or reduced. The clavicle, however, persists in most forms, although in mammals it is frequently much reduced or absent. A new element, the *interclavicle,* a membrane bone, appears in many tetrapods. The parts derived from the original cartilaginous pectoral arch continue to be of most importance. The scapula and usually the coracoid persist. In most mammals, however, the coracoids are reduced. Except in mammals, precoracoids are usually present but often remain unossified, the clavicles taking their place. In monotremes both precoracoids and coracoids persist, but in higher mammals the precoracoid is lost and the coracoid is represented only by a small *coracoid process* attached to the scapula.

The pelvic girdle, which in fishes is poorly developed and not attached to the veterbral column, is a much more important structure in tetrapods, in which, in most groups, it forms a firm union with the sacrum. It has been suggested that the pelvis of tetrapods could logically have been derived from the primitive pelvic girdle of such a fish as *Eusthenopteron.* This would have involved, among other changes, a shift of the acetabulum to the lateral surface and a union of the dorsally extending iliac processes with the sacral ribs.

Amphibian Girdles and Limbs The pectoral girdle in some primitive amphibians was similar to that of piscine ancestors but for the addition of an interclavicle. In modern amphibians the dermal elements have been lost, except for the clavicle, which persists in anurans. Limbs and limb girdles are much better developed in anurans than in urodeles, but in both groups the basic structural plan is similar. Caecilians lack both limbs and limb girdles.

The limbs and girdles of urodeles (Figs. 6.59 and 6.60) are small and weak. The pectoral girdle is of simple structure and is apt to remain cartilaginous except in the region of

FIG. 6.59 Pectoral girdle of the salamander *Ambystoma jeffersonianum,* shown as though it were flattened out. Actually the suprascapula arches dorsally. (*After Noble, "Biology of the Amphibia," McGraw-Hill Book Company, by permission.*)

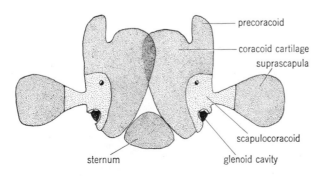

precoracoid

coracoid cartilage
suprascapula

scapulocoracoid

sternum

glenoid cavity

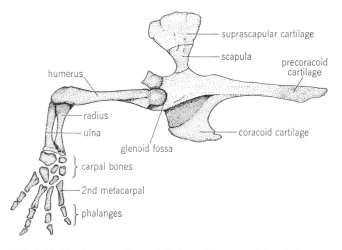

FIG. 6.60 Pectoral girdle and limb of *Necturus,* lateral view.

the glenoid fossa. The coracoids, which are united on each side with the scapula to form a single piece, are broad and overlap medially. A precoracoid cartilage may project anteriorly from each coracoid. Clavicles are absent. Each scapula connects dorsally with a broad suprascapular cartilage.

The urodele pelvis is firm, the irregular puboischia being united at a median symphysis. An iliac process on each side unites with the transverse process of the sacral vertebra.

In the urodele, forelimb, humerus, radius, and ulna are distinct. Carpal bones are reduced in number by fusion. No more than four digits are present, the first, or *pollex* (thumb), probably being the one that has been lost along with its carpal and metacarpal. In the hind limb, femur, tibia, and fibula remain separate and the number of digits is generally reduced.

The pectoral girdle of anurans shows several modifications (Fig. 6.47). In many frogs the two halves are firmly united in the midline and are closely related to the sternum (the firmisternal condition). In toads and some frogs the two halves overlap in the middle (the arciferial condition). Coracoids, precoracoids, clavicles, scapulae, and suprascapulae are present. In frogs, the clavicles are fused to, or cover, the cartilaginous precoracoids, whereas in toads precoracoids and coracoids join their partners medially.

The pelvic girdle of anurans on each side consists of a long ilium with a small ischium and pubis. The ilium is attached anteriorly to the transverse process of the sacral vertebra. All three bones join at the acetabulum. The limbs of anurans are more specialized than those of urodeles (Fig. 6.61). Humerus and femur are typical, but radius and ulna in the forelimb, and tibia and fibula in the hind limb, tend to fuse. The hind limb is pentadactyl; the forelimb usually has only four digits. The tarsal bones are modified and consist of two rows. The proximal row contains two long bones, an inner *astragalus* (*talus*) and an outer *calcaneum.* The distal row of tarsals is reduced to two or three small cartilaginous or bony pieces. Metatarsals and phalanges of the hind limbs are long and webbed. A small additional bone, the *prehallux,* or *calcar,* occurs on the

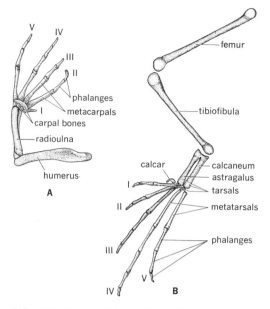

FIG. 6.61 Arrangement of bones in *A,* pectoral, and *B,* pelvic, appendages of the bullfrog *Rana catesbeiana.*

tibial side of the tarsus in many anurans. It probably represents an additional tarsal bone and is not the rudiment of an extra digit.

Reptilian Girdles and Limbs The girdles and limbs of reptiles are well developed except in snakes and limbless lizards. In snakes there is no trace of pectoral girdle or forelimb, but vestiges of a pelvis and hind limb are found in certain primitive families. Pectoral girdles of reptiles usually possess the typical three cartilage bones: coracoid, procoracoid, and scapula. Two membrane bones, clavicle and interclavicle, are usually present. Pelvic girdles are typically composed of the usual three bony elements. The ilium is firmly united with the two sacral ribs. Both pubic and ischial bones usually form median symphyses, but in crocodilians the pubic bones are reduced and only an ischial symphysis is

present. The limbs of reptiles show no very unusual features. They are typically pentadactyl, even in sea turtles, the limbs of which are in the form of paddlelike flippers. A feature of the reptilian hind limb is the presence of an intratarsal joint between the two rows of tarsal bones. Movement occurs here rather than at the junction of the tarsals with tibia and fibula. A *patella,* or kneecap, which is a sesamoid bone, appears in certain lizards for the first time.

In the prehistoric flying reptiles, the pterosaurs, humerus, radius, and ulna were of normal proportions. One heavy metacarpal and three slender metacarpals lay distal to radius and ulna. The slender metacarpals supported the first three digits, which were small and free and bore claws at their tips. The phalanges of what was presumably the fourth digit were enormously elongated. They supported the large integumentary fold which extended outward from the body from shoulder to ankle and was used as a wing. A spurlike sesamoid *pteroid* bone, which was *not* a modified digit, projected toward the shoulder from the base of the metacarpals. It provided additional support to the wing (Fig. 6.62). Flexure of the wing occurred at the junction of the fourth metacarpal and the first phalanx of that digit.

Girdles and Limbs of Birds The appendicular skeleton of birds shows a remarkable uniformity within the group. The pectoral girdle consists on each side of a large coracoid, a thin, narrow scapula, and a slender clavicle. The two clavicles, which are fused medially to a small interclavicle, form the *furcula,* or "wishbone" (Fig. 6.21). In the Paleognathae the pectoral girdle is relatively small, the clavicles being much reduced or absent.

Except in *Archaeopteryx* the three bones of the very large pelvis are fused together to

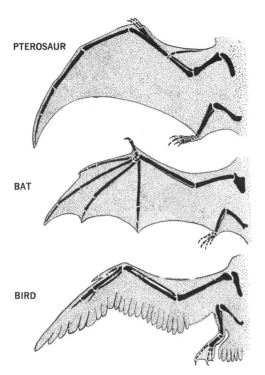

FIG. 6.62 Comparison of skeleton supporting the wing in pterosaur, bat, and bird. (*After Romanes.*)

tionally well developed. Modification of carpals, metacarpals, and phalanges is primarily responsible for the specialized condition of the wing skeleton (Fig. 6.63). The carpal bones are reduced in number to four. These are arranged in two rows of two each. The two distal carpals become fused to the corresponding metacarpals, forming the *carpometacarpus,* thus leaving the proximal carpal bones free. They are referred to as the *radiale* and *ulnare.* Three metacarpals persist in birds, but opinions differ as to which three are actually represented. The general arrangement of the wing bones is indicated in Fig. 6.63. In *Archaeopteryx* claws were present at the ends

FIG. 6.63 Wing bones from right wing of pigeon

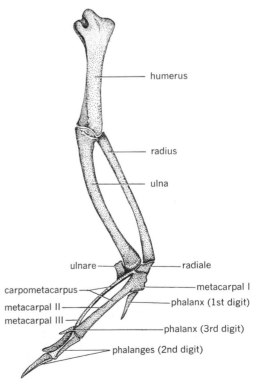

form the *innominate* bone. The ilium projects forward and backward from the acetabulum and is completely fused to the synsacrum. Only two of the vertebrae composing the synsacrum are truly sacral. The ischium extends posteriorly, paralleling the ilium, to which it is fused throughout the greater part of its length. The two pubic bones, projecting posteroventrally, usually terminate freely but may unite with the distal end of the ischium. Only in the ostrich is there a true pubic symphysis. An ischial symphysis occurs in the rhea.

The forelimb (wing) of birds is much modified from the primitive condition. In flying birds, humerus, radius, and ulna are excep-

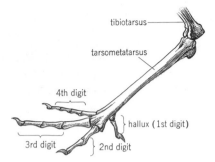

FIG. 6.64 Skeleton of hind limb of chicken.

triches, for example, the first two toes are lacking.

Mammalian Girdles and Limbs The appendicular skeleton of mammals ranges from a primitive reptilelike condition to one of a very high degree of specialization. The monotremes are the most primitive in this respect. In monotremes the cartilage bones of the pectoral girdle consist on each side of scapula, coracoid, and precoracoid (epicoracoid). The coracoids articulate ventrally with the

of all three digits, in which the metacarpals remained separate.

The hind limbs of birds are modified for *bipedal* locomotion. They may be adapted, in addition, for swimming, perching, wading, running, scratching, etc. Although there is much variation, a striking uniformity exists in the skeleton of the hind limbs. A strong femur articulates with the pelvis at the acetabulum. The fibula is usually reduced and may be represented only by a bony splint. The tibia is strong and fused to the proximal tarsal bone to form a *tibiotarsus*. A patella is found in most birds anterior to the junction of femur and tibia. The distal tarsal bones unite with the second, third, and fourth metatarsals, forming a single *tarsometatarsus*. The ankle joint is *intratarsal* in position. The spur of the domestic fowl is a bony projection of the tarsometatarsus (Fig. 6.64). It is covered with a horny cap of epidermis. No more than four digits are present in birds, the fifth metatarsal and the accompanying phalanges being lacking. The first digit (hallux) is usually directed backward, and the other three point forward. The typical number of phalanges included in the first to the fourth digits is two, three, four, five. The terminal phalanges support claws. Occasional variations from the above description are encountered; in os-

FIG. 6.65 Lateral view of left pectoral girdle and forelimb bones of cat.

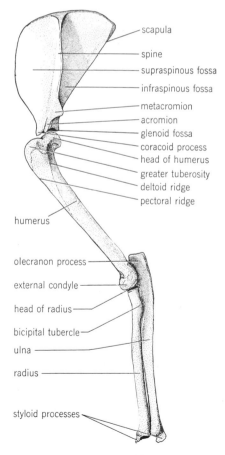

sternum. Precoracoids join a median episternum which lies anterior to the sternum. Clavicles and interclavicle are present, representing the dermal part of the reptilian pectoral girdle. In most mammals, however, interclavicle and precoracoids are lost and the coracoid is reduced to a small *coracoid process* on the scapula adjacent to the glenoid fossa (Fig. 6.65). In some mammals the clavicle persists as a strong bony arch from scapula to manubrium. In others it is lost or remains as a small vestige embedded in muscle.

The pelvic girdle of mammals is made up

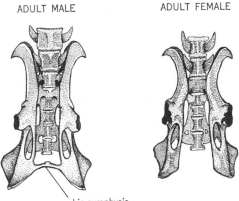

ADULT MALE   ADULT FEMALE

pubic symphysis

FIG. 6.67 Ventral view of pelvis of adult male and female pocket gophers, showing presence and absence, respectively, of the pubic symphysis. (*Original by F. L. Hisaw, J. Exp. Zool.,* **42:**437, *by permission of the Wistar Institute of Anatomy and Biology.*)

FIG. 6.66 Lateral view of left pelvic girdle and hind-limb bones of cat.

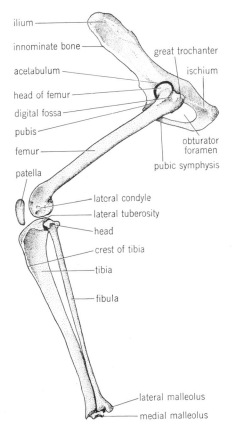

ilium
innominate bone
acetabulum
head of femur
digital fossa
pubis
femur
patella
great trochanter
ischium
obturator foramen
pubic symphysis
lateral condyle
lateral tuberosity
head
crest of tibia
tibia
fibula
lateral malleolus
medial malleolus

of the usual three bony elements, which in most cases are united at the acetabulum. Although the three parts are distinguishable in the young, they are fused in adults to form a single *innominate* bone on each side (Fig. 6.66). Ilium and sacrum are firmly fused. Both pubes and ischia may form symphyses, but in others the ischia do not meet ventrally and only a pubic symphysis is present (Fig. 6.67). Pelvic girdle and hind limbs are lacking in whales and sirenians, although small remnants may persist. In monotremes and marsupials an additional pair of cartilage bones extends forward from the pubes in the ventral abdominal wall. They are the *epipubic,* or *marsupial,* bones, the homologies of which are uncertain.

The forelimbs of mammals, which in many cases are highly specialized, usually deviate less than the hind limbs from the primitive pentadactyl condition. Carpal bones consist of several separate elements, although some fusion may occur. Reduction in number of digits is common in mammals, the tendency

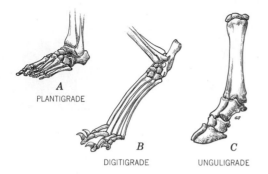

A
PLANTIGRADE

B
DIGITIGRADE

C
UNGULIGRADE

FIG. 6.68 Three types of foot posture in mammals: *A*, man (plantigrade); *B*, cat (digitigrade); *C*, cow (unguligrade).

toward reduction occurring in the following order: one, five, two, four. Many interesting variations are encountered in the forelimbs of mammals. For example, in bats some elements are elongated. The ulna is very much reduced, only the proximal portion being present in most cases. Its distal end is fused to the radius. The first digit is short and free and bears a claw. Metacarpals and phalanges of the remaining digits are greatly elongated and support the web, or wing membrane. The third digit is the longest. Comparison of the wings of bats and pterosaurs is of interest (Fig. 6.62). The same principle is involved in the development of wings in these two groups, but the manner in which it is accomplished differs greatly. The structure of the bird's wing is also of particular interest in this connection. The pentadactyl limbs of cetaceans and sirenians are webbed to form flippers. The skeletal structure differs essentially from other mammals only in the excessive number of phalanges in the central digits.

Mammalian pelvic appendages, in general, show more variation than is found in the pectoral region. A patella is present in most mammals. The tibia is the chief bone of the lower leg; the fibula is usually reduced and may be fused to the tibia as a small splint. Tarsal bones are distinct entities. A projection of one of these, the *calcaneum,* forms the heel. In mammals which show a reduction in number of digits, the hind limbs, in most cases, show a greater reduction than the forelimbs.

Among mammals, three types of foot posture are recognized (Fig. 6.68). The most primitive is the *plantigrade* posture, observed in the hind limb of man, bears, and certain insectivores. The entire foot is in contact with the ground during locomotion. *Digitigrade* mammals such as cats, dogs, etc., place only the digits on the ground, the wrist and ankle being elevated. In *unguligrade* forms (horse, cow, deer, etc.) only the hoof is in contact with the ground. This type of foot posture is the most specialized of the three, since it deviates most from the primitive condition.

The comparative study of mammalian appendages is most interesting. Detailed consideration of the various features cannot be included in a brief volume like this.

SUMMARY

1 All chordates have an endoskeleton which, in the lowest forms, may consist only of a notochord. In a few groups some bony structures are present, in addition, which are referred to as the dermal skeleton. Isolated elements, called heterotopic bones, are present in a few groups of vertebrates.

2   The skeleton is composed of cartilage, bone, or a combination of the two. Bones are of two types: (*a*) membrane bones, which do not go through a cartilaginous stage in development; (*b*) cartilage bones, which do go through this stage. The terms *intramembranous ossification* and *endochondral ossification* are applied, respectively, to the two types of development.

3   Joints are present where bones join, or articulate. There are two main kinds of joints: (*a*) synarthroses, or immovable joints; (*b*) diarthroses, or movable joints. Several types are included under each category.

4   The *dermal skeleton* includes skeletal structures derived from the dermis of the skin. There has been a tendency in evolution to eliminate the dermal skeleton, which is fundamentally a protective device. Among structures included in the dermal skeleton are the scales and fin rays of fishes; the bony plates underlying the epidermal scales of crocodilians, turtles, and a few snakes and lizards; riblike gastralia found in cetain reptiles; the bony armor of the armadillo; and the many membrane bones forming parts of the skull and pectoral girdle.

5   The *endoskeleton* consists of axial and appendicular portions. The axial skeleton includes the spinal column, skull, ribs, and sternum. The appendicular skeleton is composed of the skeletal elements of the limbs and limb girdles.

AXIAL SKELETON   The notochord is the primitive axial skeleton. In amphioxus, cyclostomes, and a few fishes it persists throughout life, but in most chordates it is replaced by the centra of vertebrae.

The *vertebrae* are segmental structures extending from the base of the skull to the tip of the tail. They enclose and protect the spinal cord, give rigidity to the body, furnish a place for attachment of the limb girdles, and provide surfaces to which ribs and muscles are attached. A typical vertebra consists of a solid ventral body, or centrum, dorsal to which is a neural arch enclosing a neural canal in which the spinal cord lies. Various projections from the vertebrae provide for articulation with adjacent vertebrae and ribs as well as for muscle attachment. In lower forms the vertebrae tend to be similar throughout, but higher in the scale they are grouped in several regions, in each of which they show certain peculiar characteristics. Cervical, thoracic, lumbar, sacral, and caudal regions are recognized. Fusion of several vertebrae occurs in many forms but is most pronounced in birds, in which rigidity is important in flight. Otherwise, fusion is most common in the sacral region, to which the pelvic girdle is attached in tetrapods.

The *skull* is derived from three different sources: (1) a chondrocranium, composed of cartilage; (2) a dermatocranium, made up of dermal bones which become attached to the chondrocranium secondarily; (3) the visceral skeleton, or splanchnocranium, derived from the visceral arches which support the gills in lower forms and which also become closely associated with the chondrocranium.

The *chondrocranium* is derived by histological differentiation of cartilage in a mesenchymatous condensation which early surrounds the brain. It first appears as a pair of parachordal cartilages on either side of the notochord ventral to the base of the brain. This is followed by the appearance of a pair of anterior prechordal cartilages. Three pairs of sense capsules next make their appearance in the nasal, optic, and otic regions, respectively. A union of prechordal, parachordal, nasal, and otic cartilages results in the formation of the primitive chondrocranium. The optic capsules do not enter into union with the others. The side portions of the chondrocranium grow dorsally for a short distance on either side of the brain. The posterior, or occipital, portion becomes roofed over by cartilage, but the rest remains open. Ossification centers in various parts of the chondrocranium bring about the formation of the several cartilage bones of the cranium.

The *dermatocranium* consists of membrane bones derived originally from dermal plates of lower forms which have assumed a deeper position in the body. Dermal bones roof over the more anterior part of the chondrocranium to form such bones as the parietals and frontals. Other dermal bones may be closely applied to the lower surface of the chondrocranium, the parasphenoid of bony fishes and lower tetrapods being a notable example. Still other membrane bones become associated with the visceral skeleton.

The *visceral skeleton* is derived originally from the cartilaginous visceral arches which support the gills. The first, or mandibular, arch divides into dorsal and ventral portions which become the upper and lower jaws, respectively. Membrane bones come to invest the original cartilage, which may ultimately disappear, although endochondral ossification in certain regions may result in the formation of certain cartilage bones. The second, or hyoid, arch in lower forms acts as a suspensorium of the jaws. In higher forms it contributes to support of tongue and larynx and to the transmission of sound in the middle ear. The remaining visceral arches are usually lost or reduced.

The skulls of cyclostomes and elasmobranchs do not go beyond the cartilage stage. In other fishes all stages can be observed, from those which are entirely cartilaginous to those which are almost completely bony. Bony fishes have the largest number of skull bones of any vertebrates. In amphibians, the number of bones is notably reduced. The skull articulates with the first vertebra by means of a pair of occipital condyles. Anuran amphibians have better-developed skulls than urodeles. Reptilian skulls show an increased degree of ossification, and several new bones make their appearance. Fusion of various bones is most pronounced in turtles. Articulation with the first vertebra is by means of a single, median occipital condyle. This is also true of birds, in which the skull is light in weight but the various bones are fused to a high degree. In all forms below mammals, articulation of the lower jaw with the rest of the skull is at the quadrate bone. In mammals, however, the lower jaw articulates with another bone, the squamosal. Formation of the malleus and incus bones of the middle ear is the result of this shifting in position. Mammalian skulls have an increased cranial capacity, thus accommodating the larger brain. Two occipital condyles are utilized in articulation with the first vertebra. The visceral skeleton shows a gradual reduction as the evolutionary scale is ascended. The various elements of the vertebrate skull show interesting homologies, there being a surprising uniformity in the basic plan of development and arrangement.

The *ribs* are cartilaginous or bony rods attached at their proximal ends to vertebrae. Primitively there is one pair for each vertebra. There are two kinds of ribs, upper, or dorsal, and ventral, or pleural. The latter represent the spread-out halves of the hemal arch which, anterior to the anus, fail to unite ventrally. They lie just outside the peritoneum. Dorsal ribs appear within the myocommata and lie between hypaxial and epaxial muscle regions. They may be absent or reduced to small tips on the transverse processes of vertebrae. There is disagreement concerning homologies of tetrapod ribs. Many authorities believe that they represent the dorsal ribs of fishes. In most amniotes the ribs in the thoracic region form strong arches which unite ventrally with the sternum. Sacral ribs form a union with the pelvic girdle.

A *sternum* exists only in tetrapods. It is a midventral structure closely related to the pectoral girdle and, except in amphibians, to the ribs. A parasternum, which is a posterior continuation of the sternum in some reptiles, is derived from ossifications in the ventral portions of the myocommata. The sternum typically consists of a row of separate bones, but these may fuse, as in birds and man. The sternum probably should be considered as a portion of the appendicular skeleton rather than of the axial skeleton.

APPENDICULAR SKELETON The appendicular skeleton consists of the anterior, or pectoral, appendages and girdle and the posterior, or pelvic, appendages and girdle. In fishes the appendages are in the form of fins; in tetrapods they are limbs. The limbs of tetrapods are believed to have evolved from the fins of fishes. The connecting links are to be sought among the fossil rhipidistian crossopterygian fishes. Several theories have been advanced to account for the origin of the paired limbs of vertebrates. The "fin-fold" and "ostracoderm" theories offer plausible suggestions.

Fishes Neither the pectoral nor the pelvic girdle in fishes is connected with the vertebral column. Cartilages and cartilage bones, homologous with portions of the girdles of tetrapods, can be identified in the girdles of fishes. Several membrane bones become associated with the pectoral girdle but not with the pelvic girdle. Of these only the ventral clavicles are represented in higher forms. Large cartilaginous or bony plates, the basalia, are located at the base of the typical fin. From these extend numerous radialia which connect distally with dermal fin rays.

Tetrapods The girdles and limbs of all tetrapods show clear-cut homologies in fundamental structure. Pectoral girdles consist of two or three cartilages or cartilage bones. There are a single scapula and usually a coracoid and precoracoid on each side. A glenoid fossa serves as the point of articulation of girdle and forelimb. Membrane bones consisting of clavicle and interclavicle may be added. Coracoids and clavicles unite with the sternum. In higher forms the clavicles may be reduced or lost. The coracoids of mammals, except monotremes, are reduced to a small process on each scapula. The scapula therefore becomes the chief component of the pectoral girdles in mammals. Pectoral girdle and vertebral column have no direct connection.

The tetrapod pelvic girdle is united firmly to the sacral region of the vertebral column. It is composed on each side of three cartilage bones: ilium, ischium, and pubis.

A depression, the acetabulum, usually located at the point of junction of the three bones, furnishes the articular surface for the femur. Pubes and usually the ischia meet their fellows ventrally to form symphyses.

The limbs of tetrapods are fundamentally similar. Forelimbs and hind limbs are often alike. The upper portion of each limb contains a single bone which articulates with the girdle. The lower portion of each limb contains two parallel bones. Next comes the wrist or ankle, as the case may be, with several small carpal or tarsal bones. This is followed by the hand or foot, consisting primitively of five long, narrow bones and of five digits, each of which contains two to five small bones, the phalanges. The tetrapod limb is primitively pentadactyl. Reductions and fusions account for the many variations from the primitive condition that are encountered. Sesamoid bones, which represent ossifications in tendons, are frequently present in various parts of the limb.

BIBLIOGRAPHY

DeBeer, G. R.: The Development of the Vertebrate Skull, Oxford University Press, New York, 1937.

McLean, F. C., and M. R. Urist: "Bone," 3d ed., University of Chicago Press, Chicago, 1968.

Moss, M. L. (ed.): Comparative Biology of Calcified Tissue, *Ann. N.Y. Acad. Sci.,* 109: 1–140 (1963).

Romer, A. S.: "Osteology of the Reptiles," University of Chicago Press, Chicago, 1956.

———: "Vertebrate Paleontology," University of Chicago Press, Chicago, 1966.

7 MUSCULAR SYSTEM

Most movements of the body or parts of the body, at least in the higher forms of animal life, are brought about by muscles. Muscular tissue, more than any other tissue in the body, has developed the power of contractility to a very high degree. Three types of muscular tissue are recognized: (1) *striated, voluntary,* or *skeletal;* (2) *smooth,* or *involuntary;* and (3) *cardiac.* Muscle tissue, with few exceptions, is of mesodermal origin.

MUSCULAR TISSUE Although the property of contractility is possessed by all living protoplasm, it is particularly well developed in muscular tissue. In such tissue the direction of contraction is along definite lines, which correspond to the long axes of the muscle cells. Muscular contraction is responsible in vertebrates not only for locomotion but also for movements of various internal organs, for the beat of the heart, for the propulsion of blood and lymph through vessels, for the passage of food through the digestive tract, and for the passage of glandular secretions and excretory products through ducts. Contrac-

tion of muscle fibers is also of great importance in producing body heat. With the exception of the muscles of the iris of the eye and the myoepithelial cells of sweat glands, lacrimal, salivary, and mammary glands, all of which are ectodermal in origin, muscular tissue is derived from mesoderm.

The distinctions between three main types of muscular tissue are based upon structural and functional differences. *Smooth,* or *involuntary, muscle* makes up the contractile tissue of hollow visceral organs, ducts, and blood vessels. The action is not under conscious control. *Striated, voluntary,* or *skeletal muscle* forms the greater part of the body musculature and is under control of the will. The third type, *cardiac muscle,* represents an intermediate form, since it is striated like skeletal muscle but free of voluntary control.

Smooth-muscle tissue (Fig. 7.1) is composed of long, narrow, spindle-shaped cells each of which bears an elongated nucleus in its central portion. Threadlike *myofibrils,* which are demonstrated with difficulty, run in a longitudinal direction through the cytoplasm, or

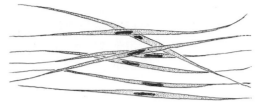

FIG. 7.1 Smooth-muscle cells teased apart.

sarcoplasm as it is called in muscle cells. They are thought to be the actual contractile elements. Smooth-muscle cells, or fibers, may occur individually or in the form of bundles or sheets. Their average length is about 0.2 mm. Individual fibers in close association with collagenous and elastic connective-tissue fibers are found in such regions as the skin about the nipples and the wall of the scrotum and in the wall and villi of the small intestine.

Smooth muscle undergoes slow and rhythmic contractions controlled by the autonomic nervous system which functions independently of the will. Peristaltic and segmenting contractions of the digestive tract, responsible for the propulsion of food down the digestive tract and for mixing it thoroughly with digestive juices and enzymes, are brought about by contraction of smooth muscle. Contraction of the circular layer

narrows the lumen and lengthens the organ; contraction of the longitudinally arranged fibers increases the diameter of the lumen and shortens the digestive tract. In this manner ingested food is propelled down the alimentary tract, in the various parts of which different phases of digestion and absorption occur.

Striated, or *skeletal, muscle tissue,* of which all voluntary muscles are composed, consists of long fibers which are in reality multinucleated cells. A *muscle* consists of bundles of striated-muscle fibers bound together in a sheath of connective tissue. The length of the fibers varies somewhat. In short muscles, fibers may extend the entire length of the muscle, but in longer muscles they are usually attached to tendons or to other muscle fibers. Individual fibers (Fig. 7.2) are independent, show no branching, and course parallel to each other. Each fiber is bonded by a membrane, the *sarcolemma.* Numerous *myofibrils,* which are present within the fiber, are surrounded by sarcoplasm and a sarcoplasmic reticulum, which may be discontinuous. The myofibrils consist of minute, alternating dark and light plates, or discs. These are so arranged within a fiber that all the dark discs lie at the same level, giving the appearance of dark bands within the boundaries of the sarcolemma. The light discs, lying side by side, form light bands in a similar manner.

Attention should be called to the red or white muscles in different vertebrates and even in different regions of an individual animal. They differ in their fine structures, functions, and capillary patterns. Red muscles have more abundant sarcoplasm and are vascular. In pigeons both pectoral and leg muscles are red, whereas in chickens the pectoral muscles are white and the leg muscles are red. In chickens the pectoral muscles are used relatively little, but in the pigeon they are of great importance in flight.

FIG. 7.2 Three striated-muscle fibers teased apart.

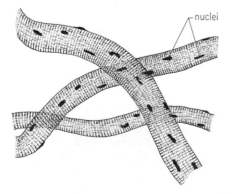

nuclei

Cardiac-muscle tissue is confined to the muscular wall of the heart and to the roots of the large blood vessels joining the heart. It shows some characteristics of both smooth and skeletal muscles. It contracts rhythmically and involuntarily in the manner of smooth muscle, but in structure it more closely resembles skeletal muscle since it is striated. Alternating dark and light bands, or discs, are present in the myofibrils. The nuclei of the fibers, however, are spaced a fair distance apart and lie deep within the fiber in its axial portion, in contrast to those of skeletal muscle, which are peripheral and lie close to the sarcolemma. Moreover, cardiac-muscle fibers branch to form a network, but each fiber tends to course in the same direction as its neighbors. As a result, spaces between the fibers have a slitlike appearance. Endomysium is located in these spaces. Numerous capillaries are inside the endomysium, making close contact with the cardiac-muscle fibers. Lymph capillaries are present in addition. Nerve fibers, which terminate on the cardiac-muscle fibers, are also inside the endomysium.

Peculiar transverse *intercalated discs* are present at intervals in cardiac-muscle fibers

FIG. 7.3 Cardiac-muscle fibers, showing intercalated discs.

(Fig. 7.3). The intercalated discs appear to represent the cell membranes of adjacent cardiac-muscle cells. Each cell therefore is a distinct entity and contains its own nucleus.

DEVELOPMENT It is important to recall that during early development when proliferating mesodermal cells grow outward, between ectoderm and endoderm, they gradually differentiate in such a manner that three different levels, or regions, may be recognized: (1) a dorsal *epimere;* (2) an intermediate *mesomere;* and (3) a lower *hypomere* (Fig. 7.4). It is with the epimeric and hypomeric portions that we are particularly concerned in our consideration of the muscular system. The mesomere is of importance in the development of the excretory and reproductive organs.

The epimere soon becomes marked off by a series of dorsoventral clefts which form in succession from anterior to posterior. Thus a series of blocklike masses of mesoderm, the *somites,* is formed in the epimeric region. Mesomere and hypomere do not usually undergo a similar segmentation. When a somite is first formed it is made up of an almost solid mass of cells, although sometimes a small portion of the coelomic cavity is present in the center. With further differentiation three regions are soon recognizable (Fig. 7.5): (1) a ventromedial mesenchymatous mass, the *sclerotome;* (2) a dorsolateral portion, the *dermatome;* and (3) a dorsomedial region, the *myotome.*

The sclerotome will ultimately give rise to the vertebral column and proximal portions of the ribs; the dermatome contributes to the dermis of the skin.

It is the myotome which gives rise to the greater part of the voluntary musculature of the body. The cells of the myotome proliferate and grow laterally, away from the neural tube but ventral to the dermatome, finally occupying a position parallel and ventromedial to the dermatome (Fig. 7.5*A*).

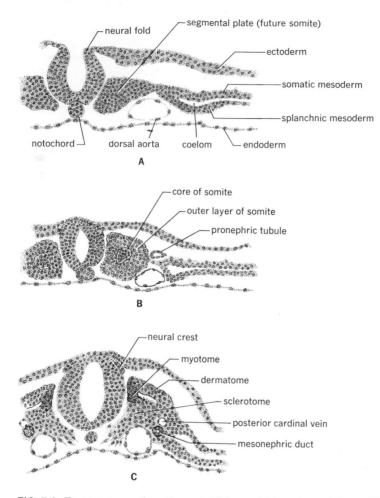

FIG. 7.4 Transverse sections through 33-hour chick embryo at three different levels, illustrating differentiation of somites: *A*, posterior section through segmental plate; *B*, section through middle of somite region; *C*, one of the more anterior somites, showing differentiation of dermatome, myotome, and sclerotome. The formation of neural tube and neural crests is also shown.

The splanchnic layer of the hypomere surrounds the endoderm of the gut. This layer is of importance in forming the smooth muscle of the gut and also cardiac muscle. Furthermore, certain *branchial* or *branchiomeric* (visceral) muscles in the gill region of aquatic forms, which are voluntary and striated, originate from the splanchnic layer of the hypomere rather than from myotomes. They are thus homologous with the smooth muscle of the gut (visceral muscles), despite their striated appearances and the fact that they are under voluntary control. They are used in moving the visceral arches. Derivatives of branchial muscles are present in higher vertebrates. Despite the complexity of the

TABLE 3

Somatic	Visceral
Axial Appendicular	Branchiomeric Smooth (gut and derivatives) Cardiac

muscles within the vertebrates, the most natural groupings of muscle are as shown in Table 3.

The cells of the dorsally located myotomes grow down in the lateral body wall between the outer ectoderm and the somatic layer of hypomeric mesoderm (Fig. 7.5B and C). The sheets of cells thus formed on the two sides almost meet at the midventral line, being separated from each other only by a longitudinal band of connective tissue, the *linea alba*. Connective tissue also comes to occupy the spaces between adjacent myotomes. These partitions are called *myocommata* or *myosepta*. They extend from the spinal column and wall of the coelom (peritoneum) outward to the dermis of the skin. As a result of the foregoing process, the basic musculature of the trunk region of the body is clearly defined. The cells of the original myotome which give rise to muscle tissue are called *myoblasts*. They begin to differentiate and become spindle-shaped and arranged in bundles. Mitotic divisions then occur, resulting in an increase in mass. Further differentiation takes place until the cells give the typical syncytial picture of striated muscle. The muscles derived from myotomes are referred to as *somatic*, or *parietal*, *muscles.* They may be further subdivided into *axial* and *appendicular muscles,* depending upon whether they are confined to the region of the body wall or whether they are associated with appendages. In fishes the axial muscles are of greater importance than the appendicular muscles, since their contraction is responsible for locomotion in these animals. In tetrapods, however, in which locomotion depends largely upon movements of the limbs, the appendicular muscles have assumed greater importance and the axial muscles, although still functional, have taken a minor role.

TERMINOLOGY Smooth, or involuntary, muscles are arranged in continuous sheets, as in the walls of the digestive tract. They are not readily dissected. Voluntary muscles, however, are arranged as separate masses which are easily separated from one another. Each end of a voluntary muscle is attached to a structure of the body.

FIG. 7.5 Series of diagrams of vertebrate embryo, illustrating the epimeric origin of the myotomes and the manner in which they grow ventrally to form the parietal musculature of the body wall. The splanchnic layer of the hypomere surrounds the gut and contributes to the smooth muscle and other mesodermal derivatives of the intestinal wall.

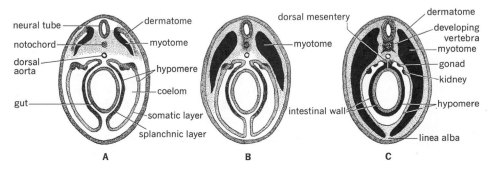

Although muscles are usually attached to bony or cartilaginous skeletal parts, this is not always true. Usually one end is attached to a less movable part than the other (Fig. 7.6). This is the *origin* of the muscle; the other is the *insertion*. Often muscles are attached by means of *tendons,* which are white, fibrous cords of regularly arranged connective tissue, strong and inelastic. The term *aponeurosis* is applied to a broad, flat, ribbon-shaped tendon. In certain regions a small bone may develop within a tendon at a point where the latter moves over a bony surface. Such bones are called *sesamoid* bones. The kneecap, or *patella,* is an example.

Sheets or bands of connective tissue called *fasciae* surround muscles, groups of muscles, and the body musculature as a whole. They tend to bind parts of the body together and in some cases serve as areas of muscle origin and insertion as well.

Muscles are usually present in groups of two. Each group works opposite to the other (Fig. 7.6). Such groups are named according to their action. Thus:

Flexors tend to close or decrease the joint angle.

Extensors tend to straighten or increase the joint angle.

Abductors draw a part away from a median line. Abductors of a limb swing the limb in a direction away from the longitudinal axis of the body; abductors of the digits move the digits in a direction away from the median longitudinal axis of the limb.

Adductors draw a part toward a median line. Adductors of a limb swing the limb toward the median longitudinal axis of the body; adductors of the digits move the digits toward the median longitudinal axis of the limb.

Rotators revolve a part on its axis.

Elevators, or *levators,* raise, or lift, a part, as in closing the mouth by raising the lower jaw.

Depressors lower, or depress, a part.

Constrictors draw parts together or contract a part.

FIG. 7.6 Method of origin and insertion of flexor and extensor muscles.

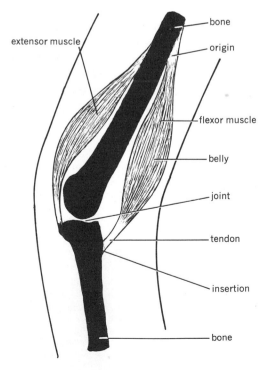

extensor muscle

bone

origin

flexor muscle

belly

joint

tendon

insertion

bone

PARIETAL MUSCULATURE

The slow and rhythmic contraction of smooth-muscle cells is controlled by the autonomic nervous system. Smooth muscle is also responsible for maintaining a condition of sustained contraction, known as *tonus,* by means of which the walls of tubular organs are kept at a relatively constant diameter. This is of great physiological importance in the passage of food through the digestive tract, in the regulation of blood pressure, and the like.

Striated, voluntary, or parietal, muscle acts entirely differently. It may contract extremely rapidly in response to a reflex or the will. It also maintains a partial state of contraction, or tonus, which gives form to the body even when at rest, and is responsible for such phenomena as being able to stand up for prolonged periods, for keeping the head erect, etc.

Carbohydrate, in the form of glycogen, is stored in muscle cells and provides the basic energy released by muscular activity. Energy is released rapidly by a breakdown, within the muscle fibers, of *adenosine triphosphate* to *adenosine diphosphate.* The glycogen store is used in resynthesizing adenosine triphosphate from the diphosphate.

Fundamentally the voluntary, striated, or parietal, musculature of chordates is made up of a linear series of myotomes extending from anterior to posterior ends. Each myotome lies opposite the point of articulation of two adjacent vertebrae and is supplied by the somatic motor fibers of a separate spinal nerve which emerges from the neural canal through an intervertebral foramen (Fig. 7.7). Whether the chordate head is primarily metameric has been the subject of much speculation by comparative anatomists. It is generally agreed that certain muscles of the head, like those of the trunk, are fundamentally metameric and are supplied by somatic motor branches of certain cranial nerves rather than by spinal nerves. In higher vertebrates the metameric condition has been obscured by evolutionary changes, so that it is rather difficult to recognize homologies among the muscles of various vertebrates. In numerous cases certain muscles have been given names which are identical with those of muscles in man because they are located in similar positions. Such muscles may or may not be homologous with the similarly named muscles of man. One of the most reliable criteria in determining muscle homologies is the nerve supply to a muscle. Even this, however, is not infallible. Despite pronounced evolutionary changes, the nerve supply has remained relatively constant. The identification and tracing of small branches of nerves are very difficult, and for that reason the subject of muscle homology is most complex. Only some of the broader aspects of the subject are treated in this volume.

Trunk Musculature

Amphioxus The myotomes of amphioxus, when viewed from the side, are observed to be V-shaped structures with the apex of the V pointing forward. The muscle fibers within the myotome course in a longitudinal direction. The fibers are interrupted from myotome to myotome by the presence of myocommata, to which they are attached anteriorly and posteriorly. The actions of the myotomes of the two sides alternate with each other in amphioxus. When the fibers on one side of the body are contracted, those on

FIG. 7.7 Diagram showing relations of myotomes to sclerotomes of vertebrae and to the somatic motor portions of spinal nerves (see also Fig. 6.11).

ANTERIOR   POSTERIOR

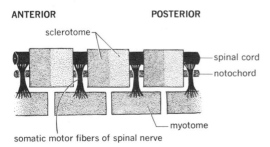

somatic motor fibers of spinal nerve

the other side are relaxed. By contracting fibers on alternate sides of the body in different regions, the animal is able to swim with a wriggling, undulatory motion.

Cyclostomes In cyclostomes, in the absence of paired appendages, the regular segmental arrangement of the myotomes has undergone little change from that of amphioxus. The myotomes, however, instead of being V-shaped, are more nearly vertical. Each is bent forward slightly at its dorsal and ventral ends (Fig. 7.8).

A series of *hypobranchial muscles,* not found in amphioxus, is present ventral to the gill region. These arise from the first few myotomes posterior to the gill region. During early development, outgrowths extend from these postbranchial myotomes in an anteroventral direction to give rise to the hypobranchial musculature.

Fishes Beginning with the elasmobranch fishes, the myotomes of gnathostomes are separated into dorsal, or *epaxial,* and ventral, or *hypaxial,* portions. A longitudinal *lateral septum* of connective tissue separates the two, extending from the vertebral column to the skin at the point where the lateral-line canal is located.

The myocommata now take a zigzag course (Fig. 7.9), but the muscle fibers, particularly those in the epaxial region, continue to course in a longitudinal direction. The epaxial muscles on each side are arranged in the form of two or three *dorsal longitudinal bundles* extending from the base of the skull to the end of the tail. They form rather large muscle masses used primarily in bending the body from side to side in swimming. The hypaxial muscles are divided into *lateral* and *ventral longitudinal bundles.* The lateral longitudinal bundle lies below the lateral septum

FIG. 7.8 Lateral view of several parietal myotomes of lamprey in region just anterior to the first dorsal fin. The skin has been removed to show the muscle fibers beneath.

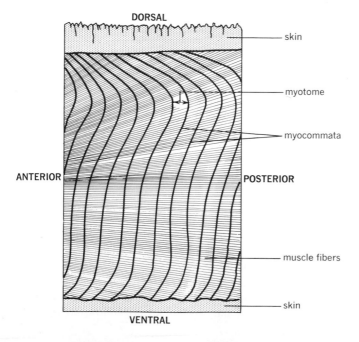

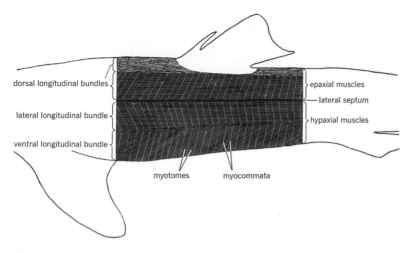

dorsal longitudinal bundles

lateral longitudinal bundle

ventral longitudinal bundle

epaxial muscles

lateral septum

hypaxial muscles

myotomes   myocommata

FIG. 7.9 Lateral view of portion of body wall of dogfish *Squalus acanthias,* showing zigzag arrangement of the myotomes. The skin has been removed to show the underlying musculature.

and is attached anteriorly to the scapular process. Anterior to the pelvic fins its fibers take a slightly oblique course anteriorly and upward, but posteriorly they course in a horizontal direction. The ventral longitudinal bundle, which in turn is divided into two parts, overlaps the lateral bundle to a slight degree. The ventral longitudinal bundle is attached anteriorly to the ventral region of the pectoral girdle. Its fibers are arranged obliquely and course in an anteroventral direction. This is indicative of the direction in which the "pull" caused by contraction of the muscle fibers is being exerted. In some elasmobranchs the portion of the ventral longitudinal bundle on either side of the linea alba is further differentiated into a long but narrow *rectus abdominis* muscle in which the fibers are longitudinally arranged.

Each trunk myotome is supplied by somatic motor fibers of a separate spinal nerve. With the appearance of the lateral septum the epaxial muscles are innervated by the dorsal rami of spinal nerves and the hypaxial muscles by the ventral rami.

The ventral portions of the first few myotomes posterior to the gill region grow forward in an anteroventral direction to form the hypobranchial musculature. Since these muscles are derived from myotomes, they are supplied by somatic motor fibers of spinal nerves. They should not be confused with *branchial* muscles, to be described later, which are derivatives of the anterior visceral musculature and innervated by branches of certain cranial nerves (V, VII, IX, X, and XI).

In fishes certain of the epaxial and hypaxial muscles become specialized, sending portions to the median fins, which thus may be moved in various directions.

The inner epaxial muscles of fishes are attached to the vertebral column. The deeper portions of the hypaxial muscles are attached to myocommata or to ventral ribs. They foreshadow the appearance of intercostal, oblique, and rectus abdominis muscles of higher forms. As yet they do not form distinct muscle masses and have the usual primitive metameric appearance.

Amphibians The axial trunk musculature of tetrapods has become considerably modified from that observed in lower aquatic forms. The original metameric condition becomes more and more obscure as the evolutionary scale is ascended. In most tetrapods the appendicular muscles have developed extensively, and the axial musculature of the trunk, which has been reduced in volume, is of relatively minor significance in relation to body movement.

However, the axial muscle masses do provide components for the tetrapod trunk. These muscle masses and their derivatives can be classified as shown in Table 4.

Urodele amphibians do not have the epaxial muscles so well developed as do fishes, nor are they used to the same extent in locomotion. The lateral septum is more dorsal in position, as are the transverse processes of the vertebrae. Despite the reduction in epaxial musculature, the arrangement of the myotomes in this region is primitive. The number of myotomes corresponds to the number of vertebrae, and the muscle fibers still course from one myocomma to the next, the entire epaxial mass forming the *dorsalis trunci* (Fig. 7.10). The myotomes are vertically arranged and do not exhibit the zigzag condition found in the faster-swimming fishes. The muscle fibers adjacent to the vertebrae are at-tached to them and are referred to as *intersegmental bundles.*

In anurans the epaxial muscles are still further reduced. They are used in bending the vertebral column dorsally rather than laterally. The dorsalis trunci has become further differentiated into *intertransversarial* muscles between the transverse processes and *interneural* muscles between the neural arches. In most frogs the outer fibers of the dorsalis trunci on either side form a long muscle, the *longissimus dorsi,* which passes all the way from the head to near the end of the urostyle. These muscles take the form of a V, the apex pointing posteriorly. The outer fibers are not attached to the vertebrae to any extent. In both urodeles and anurans the anterior region of the dorsalis trunci has split into several muscle masses of various names which are attached to the skull and are effective in turning the head.

The hypaxial muscles of amphibians have undergone greater modification than those in the expaxial region. In larval urodeles the condition is much like that observed in fishes, but in adults these ventral trunk muscles are arranged in four distinct layers, or sheets. Beginning from the outside, there are *superficial* and *deep external obliques,* in which the muscle fibers course in a posteroventral direction, an *internal oblique* layer, in which they

TABLE 4

	Hypaxial muscle mass		
Epaxial muscle mass	**Dorsomedial group**	**Lateral group**	**Ventral group**
Transversospinalis group Longissimus group Iliocostalis group	Subvertebralis	Obliquus abdominis externus Superficialis (supracostals) Obliquus abdominis internus (internal intercostals) Transversus abdominis (subcostals)	Rectus abdominis

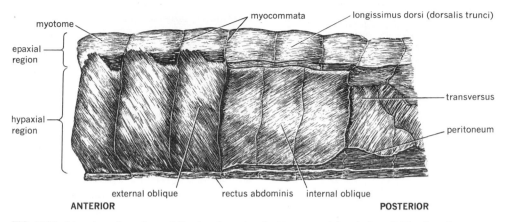

FIG. 7.10 Muscles of portion of the trunk region in *Necturus,* lateral view. In the fourth, fifth, and sixth segments from the left the external oblique layer has been removed to show the underlying internal oblique layer. In the seventh and eighth segments, both external and internal layers have been removed to expose the transversus. The ventral portions of the internal oblique and transversus have been removed to show the underlying peritoneum.

extend in a posterodorsal direction, and a *transversus,* with fibers almost in a vertical position (Fig. 7.10). The transversus lies next to the peritoneum. On either side of the linea alba a rather well-developed *rectus abdominis,* derived from the obliques and with primitive segmentally arranged longitudinal fibers, extends from the head to the pelvic girdle. In some forms this muscle may split into superficial and deep layers.

In certain species of urodeles variations from the above occur, consisting for the most part of a reduction in the number of layers brought about by fusion. Within the group there is also a tendency toward reduction or loss of the myocommata, so that the muscles begin to appear as distinct entities with independent actions.

The more specialized anurans (Fig. 7.11) go even further in reducing the number of layers in the hypaxial trunk musculature, having only an outer superficial external oblique layer and an underlying transversus. The rectus abdominis is now a large muscle ex-

tending from sternum to pubis, its two halves being separated in the midventral line by the linea alba. Myocommata have disappeared from the external oblique and transversus but are retained to some extent in the rectus abdominis as *tendinous inscriptions,* which are transverse in poistion and divide the muscle into segments.

The muscles which move the tongue in amphibians are derived from the hypobranchial musculature.

Reptiles The trunk musculature of reptiles shows a further deviation from the primitive condition, leading to the more complex arrangement found in mammals (Fig. 7.12). A distinct lateral septum separating epaxial and hypaxial muscles is lacking in amniotes. The most conspicuous change occurs in connection with the appearance of well-developed ribs.

The epaxial muscles of reptiles become differentiated into several groups. Immediately next to the vertebral column lie the

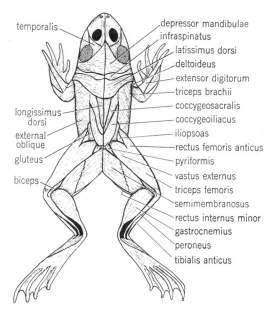

temporalis

depressor mandibulae
infraspinatus
latissimus dorsi
deltoideus
extensor digitorum
triceps brachii
coccygeosacralis
coccygeoiliacus
iliopsoas
rectus femoris anticus
pyriformis
vastus externus
triceps femoris
semimembranosus
rectus internus minor
gastrocnemius
peroneus
tibialis anticus

longissimus
dorsi
external
oblique
gluteus
biceps

FIG. 7.11 Dorsal view of bullfrog, *Rana catesbeiana,* with skin removed, showing the
dorsal musculature.

spinalis and the deeper and laterally situated
semispinalis muscles. These are rather long
muscles originating on the dorsal portions of
vertebrae and inserting on more anterior vertebrae or on the skull. They are sometimes
referred to as the *transverse-spinalis* system of
muscles. The deeper muscles in this region
split into numerous short muscles of various
names, located between adjacent vertebrae or
between vertebrae and ribs. The longissimus
dorsi of amphibians now originates at the
ilium and becomes separated into (1) a *longissimus dorsi proper,* confined to the lumbar
region; (2) a *longissimus capitis,* located on
either side of the neck and extending to the
skull in the region of the temporal bone; and
(3) a laterally situated *iliocostal dorsi,* passing
to the proximal ends of ribs. In the neck
region the iliocostal dorsi passes to the atlas
vertebra and occipital portion of the skull. In
a number of forms the presence of tendinous

inscriptions indicates that these muscles have
originated from segmental myotomes. As
might be expected, the epaxial muscles of
snakes are very well developed, whereas
those of turtles are greatly reduced.

The hypaxial musculature of reptiles
exhibits an even greater change from the
primitive condition. In lizards and crocodilians these muscles, in the abdominal region
at any rate, are much like those of urodele
amphibians, consisting of external and internal oblique and transversus layers. In the
thoracic regions, however, additional layers
are present. Between external and internal
obliques an *intercostal layer,* derived from the
obliques, makes its appearance, with fibers
connecting adjacent ribs. This is subdivided
into *external* and *internal intercostals.* The
fibers of these two sets of muscles course in
opposite directions and are used in respiratory movements. The oblique muscles also
give rise to the *scalene muscles,* which pass
from the anterior ribs to the cervical vertebrae, and to the *serratus muscles,* passing
from the more posterior ribs to the inner surface of the scapula or to the thoracic and
anterior lumbar vertebrae. The rectus abdominis, with prominent tendinous inscriptions, is well developed in lizards and crocodilians.

In snakes the ventral scales are used in
locomotion. Small muscles derived from the
obliques and rectus abdominis pass from the
ribs to the skin underlying the scales. These,
together with certain dermal muscles, are
responsible for movement of the ventral
scales.

The hypaxial muscles of turtles are poorly
developed in association with the rigid shell
which narrowly restricts body movements.

The fact that the number of hypaxial
muscle layers as found in urodele amphibians
is reduced in anurans but increases in reptiles
arouses interesting speculations about evolu-

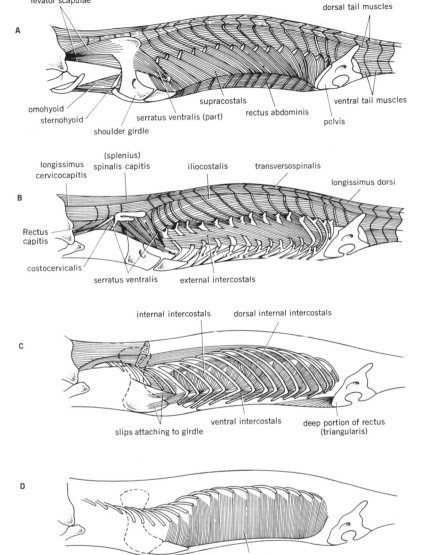

FIG. 7.12 A diagrammatic series dissection of *Sphenodon* showing the anatomy of the trunk muscles: *A*, the thin superficial sheet of the external oblique has been removed; *B*, supracostals, rectus, throat muscles, and superficial muscles to the scapula have been removed; *C*, epaxial muscles have been cut posteriorly, and the internal intercostals and triangularis are indicated; *D*, ribs cut and all muscles removed to show the transversus. (*After Romer.*)

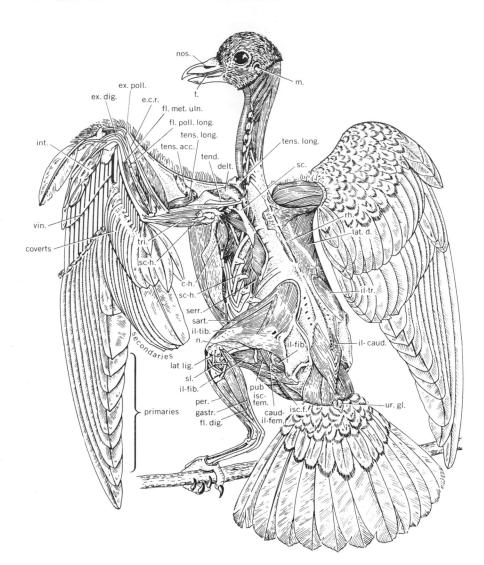

c-h. coraco-humeral
caud-il-fem. caud-ilio-femoralis
delt. deltoid
e.c.r. extensor carpi radialis
ex.dig. extensor digitorum
ex. poll. extensor pollicis
fl. dig. flexor digitorum
fl. met. uln. flexor metacarpi ulnaris
fl. poll. long. flexor pollicis longus
gastr. gastrocnemius
il-caud. ilio-caudalis
il-fib. ilio-fibularis (cut)

il-tr. ilio-trochantericus
int. interosseus
isc. f. ischio-femoralis
lat. d. latissimus dorsi
lat. lig. lateral ligament of knee
m. external auditory meatus
n. sciatic nerve
nos. nostril
per. peroneus
pub. isc. fem. pub-ischio-femoralis
rh. rhomboid
sart. sartorius

sc. scapula
s.c.h. scapulo-humeral (cut)
serr. serratus anterior
sl. sling for tendon of ilio-fibularis
t. tongue
tend. tendon of pectoralis minor
tens. acc. tensor accessorius
tens. long. tensor longus
tri. triceps
ur. gl. uropygial gland
vin. vinculum elasticum

FIG. 7.13 Dissection of the pigeon from the back. (*After J. Z. Young, "Life of the Vertebrates," 2d ed., Oxford University Press, 1962, by permission.*)

tionary trends. In both instances animals have taken to terrestrial existence, but this fact alone would seem to be of minor importance. The increased number of layers in reptiles has developed in accordance with the appearance of ribs, which, of course, are lacking in anurans. This, rather than emergence to life on land, is of importance in considering evolutionary trends.

Birds The trunk muscular system of birds is quite different from that of other forms (Fig. 7.13). This is not surprising when one considers their specialized mode of life. Epaxial muscles are poorly developed. Among the changes to be observed are the absence of the transversus in the abdominal region and reduction of the oblique muscles. A remnant of the transversus is recognized in the *triangularis sterni* on the inner portion of the ribs, near where they are attached to the sternum.

Mammals The trunk muscles of mammals do not deviate markedly from those of reptiles, but the hypaxial musculature is somewhat reduced (Fig. 7.14). With the greater importance of the limbs in locomotion there has been an increase in the musculature of the limbs and limb girdles. Some of these muscles overlie the greater part of the trunk musculature. The original metameric arrangement of the trunk myotomes is now only slightly indicated. The *intercostal, serratus, scalene, intervertebral,* and *rectus abdominis* are trunk muscles in which the original segmental arrangement is retained, at least in part.

The epaxial muscles of mammals show great similarity to the reptilian condition. A bewildering variety of names is applied to the numerous muscles of this region, which, if described in detail, would only add to the confusion of the student. A few of the more obvious muscles will be mentioned. A *rector spinae muscle* originates on the sacrum and spinous processes of the posterior vertebrae.

Anteriorly it splits into three columns on each side, a lateral *iliocostal,* an intermediate *longissimus dorsi,* and a medial *spinalis dorsi.* Each of these continues forward to the cervical region, but only the longissimus dorsi and spinalis dorsi go as far as the skull. A *multifidus* muscle represents the transversespinalis system of reptiles. It fills in the groove on either side of the spinous processes all the way from the sacrum to the axis. The deeper epaxial muscles form *interspinal* and *intertransversarial* muscles between adjacent vertebrae.

The hypaxial muscles of mammals are but little modified from the reptilian condition, at least in the abdominal region. The muscles of the abdominal wall consist of an inner *transversus abdominis,* a middle *internal oblique,* and an outer *external oblique,* which lies beneath the integument. The large aponeurosis of the external oblique interlaces with that of its fellow from the opposite side at the linea alba. A portion of the aponeurosis is continued posteriorly as the *inguinal ligament.* The *quadratus lumborum* is a dorsoposterior derivative of the external oblique. The internal oblique is smaller than the external and best developed in its dorsal portion. Its large aponeurosis also passes to the linea alba. The transversus is a thin muscular layer quite similar to the internal oblique in arrangement. The *cremaster* muscle of the scrotum is continuous with the internal oblique and in some cases with the transversus. The *rectus abdominis* in mammals lies on either side of the linea alba, just as in lower forms. It is enclosed in a sheath formed from the aponeuroses of the oblique and transversus muscles. A few remnants of the rectus abdominis, which originally extended the length of the body, are to be found in mammals in the neck region. They include the *sternohyoid, sternothyroid, geniohyoid, omohyoid,* and *thyrohyoid.* Tendinous inscriptions are clearly indicated on

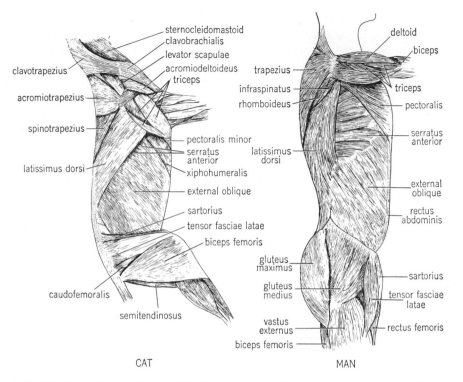

FIG. 7.14 Lateral view of superficial body muscles of cat and man, shown together for comparison.

the rectus abdominis, but only to a minor extent on its anterior derivatives.

In the thoracic region the mammalian hypaxial muscles show a reduction from the reptilian condition. They are almost entirely covered by the large appendicular muscles associated with the pectoral girdles and limbs. *External intercostals* pass from one rib to the next, their fibers coursing posteroventrally. *Internal intercostals* lie beneath the external intercostals. They also pass between adjacent ribs, but their fibers run anteroventrally. The intercostals are used in respiration. Under the internal intercostals lies the transversus, which, in the thoracic region, is referred to as the *transversus thoracis*. It is confined to the inner surface of the anterior thoracic wall. *Scalene* muscles pass from the anterior ribs to

the transverse processes of the cervical vertebrae. The *serratus* muscles pass from the angles of more posterior ribs to the inner surface of the scapula or to thoracic and anterior lumbar vertebrae. These are trunk muscles originally derived from the obliques, their hypaxial origin being indicated by the fact that they are supplied by the ventral rami of spinal nerves.

General Remarks The trunk musculature of all vertebrates is built upon the same fundamental plan. The metameric arrangement of myotomes observed in adults of lower forms becomes less apparent with advance in the evolutionary scale. This variation from the primitive condition occurs as myotomes increase in size and thickness and as the myocommata, between them, disappear. Many of the long trunk muscles have developed in this manner. In some regions a degeneration of portions of certain myotomes has led to the formation of aponeuroses and fasciae. Other myotomes may be drawn out of position or may migrate to new regions, thus obscuring their original arrangement and position. In some cases portions of adjacent myotomes may fuse to form a single muscle.

HYPOBRANCHIAL MUSCULATURE

In cyclostomes, elasmobranchs, and other fishes, as already pointed out, the hypaxial portions of the first few myotomes posterior to the gill region send buds anteriorly and ventrally to form the hypobranchial musculature. These muscles, then, are actually continuations of the ventral longitudinal bundles and are not derivatives of the branchial musculature derived from the splanchnic layer of the embryonic hypomere.

In tetrapods the hypobranchial muscles have been greatly modified. Several muscles in the neck and throat region are hypobranchial muscles which, like all trunk muscles, are innervated by spinal nerves. The forward continuation of the rectus abdominis in the neck region is called the *rectus cervicis.* From it are derived the sternohyoid, thyrohyoid, sternothyroid, geniohyoid, and omohyoid, previously mentioned. These, as well as the intrinsic muscles of the tongue, which are of hypobranchial origin, are, in amniotes, supplied by the hypoglossal nerve (XII), a purely somatic motor cranial nerve. It is generally conceded that the hypoglossal nerve, which is found only in amniotes, represents a union of the first two or three spinal (spino-occipital) nerves such as are found in lower forms and which are somatic motor nerves. The fact that the above-mentioned muscles are supplied by the hypoglossal nerve furnishes additional evidence of their hypobranchial origin.

EYE MUSCLES The hypoglossal nerve is not the only cranial nerve which bears somatic motor fibers. Certain others (III, IV, VI) are somatic motor nerves which supply the muscles, derived from myotomes in the head region, which move the eyeball.

Primitively a series of myotomes existed in the head region from the anterior part of the head to the trunk. Their presence is still indicated in cyclostome and elasmobranch embryos. In the ear region, however, these somites seem to have been crowded out and have disappeared for the most part. Three of the most anterior ones persist and appear uniformly throughout the vertebrate series. They are known as the *prootic somites,* referred to, respectively, as the *premandibular, mandibular,* and *hyoid somites* (Fig. 7.15). The myotomes of these somites contribute to the parietal musculature. As is true of other myotomes of the body, each is supplied by the somatic motor branch of a single nerve. The prootic somites are innervated by cranial nerves III, IV, and VI, respectively, rather than by spinal nerves. The vertebrate eye makes its appearance in cyclostomes. Its structure is essentially similar in all vertebrate groups. With the advent of the eye, six muscles which move the eyeball make their appearance (Fig. 7.16). All six are derived from the three prootic myotomes. The first of these differentiates into four muscles: *superior rectus, inferior rectus, internal rectus,* and *inferior oblique,* all of which are supplied by the oculomotor nerve (III). The second prootic myotome develops into the *superior oblique* muscle supplied by the trochlear nerve (IV). The third, which gives rise to the *external rectus,* is innervated by the abducens nerve (VI).

In the lamprey the abducens nerve appears to supply the inferior rectus muscle as well as the external rectus. The oculomotor, on the other hand, seems to innervate only the supe-

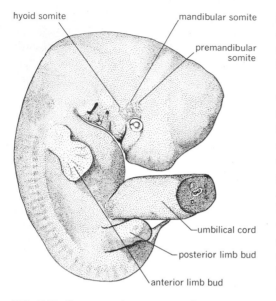

hyoid somite

mandibular somite

premandibular somite

umbilical cord

posterior limb bud

anterior limb bud

FIG. 7.15 Human embryo at an early age, showing position of the prootic somites.

rior rectus, internal rectus, and inferior oblique muscles. This peculiar situation raises questions of homology. It has been demonstrated, however, that in the lamprey the sixth cranial nerve, which leaves the brain rather far forward and close to the oculomotor nerve, actually contains fibers of the third nerve and that the latter are the ones which pass to the inferior rectus.

Several other small muscles associated with the eye in many tetrapods are also derived from the prootic myotomes. They include the *levator palpebrae superioris,* which elevates the upper eyelid, is supplied by the oculomotor nerve, and is derived from the premandibular prootic somite; the *retractor bulbi,* which in the frog pulls the entire eyeball deeper within the orbit, is derived from the hyoid somite and innervated by the abducens nerve. Muscle fibers moving the nictitating membrane, if present, are also supplied by the abducens nerve, indicating their derivation from the third prootic somite.

OTHER MUSCLES OF THE HEAD

The remaining muscles of the head, as well as several in the neck region, are derivatives of the dermal or branchial musculature. It may be recalled that the latter are derived from splanchnic mesoderm of the gut, even though they are of the voluntary, striated type. They

FIG. 7.16 Posteromedial view of left eyeball of *Squalus acanthias,* showing nerve supply to the six eye muscles. (*Drawn by G. Schwenk.*)

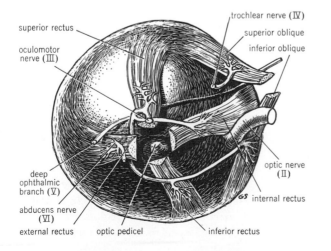

superior rectus

oculomotor nerve (III)

deep ophthalmic branch (V)

abducens nerve (VI)

external rectus

optic pedicel

inferior rectus

trochlear nerve (IV)

superior oblique

inferior oblique

optic nerve (II)

internal rectus

are *not* parietal muscles of myotomic origin and are therefore discussed elsewhere.

DIAPHRAGM The muscular diaphragm, present only in mammals, is derived embryonically from the *septum transversum* and certain other membranes. The portion adjacent to the sternum, ribs, and vertebral column becomes invaded with muscle tissue to varying degrees in different species. This is contributed by myotomes in the cervical region. The fact that the diaphragm is innervated by branches of cervical spinal nerves indicates the embryonic origin of its muscle constituents.

APPENDICULAR MUSCULATURE One important advance in the muscular system of vertebrates takes place in connection with the appearance of paired appendages. During early development when the myotomes are growing ventrally in the body wall, hollow buds grow out from them into the folds from which the paired fins arise (Fig. 7.17). They develop into the muscles which move the fins. These muscles are therefore derivatives of the myotomes of the trunk. Since the muscles become modified to course in different directions, their original metameric arrangement may not be apparent.

Two general types of appendicular muscles are recognized, i.e., *extrinsic* and *intrinsic*. The former attach the girdle or limb to the axial skeleton directly or indirectly. They originate in the axial musculature and are inserted on some skeletal element within the limb. Extrinsic muscles serve to move the entire appendage. Intrinsic muscles have both origin and insertion on parts of the limb skeleton itself, serving to move parts of the limb rather than the appendage as a whole. Actually the distinction between extrinsic and intrinsic muscles is not of very great importance, since the appendicular muscles, at least in fishes, are clearly derived from myotomes.

In various species of elasmobranchs different numbers of buds from the myotomes are concerned in the development of fin musculature. In the electric ray, *Torpedo*, 26 myotomes are involved in forming the muscles of the pectoral fins. Each bud divides into anterior and posterior primary buds. These then give off dorsal and ventral secondary buds. The two dorsal buds of each myotome concerned give rise to the dorsal musculature of the fin; the ventral buds form the ventral fin muscles. The dorsal muscles become the extensors or levators of the fins; the ventral muscles become flexors or depressors.

Attempts to homologize the fin musculature of fishes with conditions in tetrapods have led only to confusion. The limbs of tetrapods originate embryonically from limb buds which develop in pectoral and pelvic regions. These at first consist of swellings of the lateral body wall, covered with superficial ectoderm and containing an undifferentiated mass of mesenchyme. It is from this mesenchyme that the appendicular muscles are differentiated (Fig. 7.18). Only in elasmobranchs, however, have the appendicular muscles been traced to myotomic buds. In higher vertebrates it is practically impossible to trace the origin of mesenchyme so far as any relation to definite myotomes is concerned. Nev-

FIG. 7.17 Developing appendage of elasmobranch fish *Pristiurus,* showing muscle buds arising from the myotomes. (*After Rabl.*)

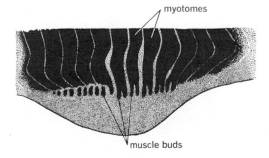

ertheless, the ultimate nerve supply to the appendicular muscles is precisely what one would expect if the muscles were formed from outgrowths of specific myotomes, as in elasmobranchs. It seems as though in tetrapods a certain stage of development has dropped out completely. Every other indication suggests that the appendicular muscles are of myotomic origin.

With emergence to life on land the limb musculature has undergone considerable change from the relatively simple condition observed in fishes. Separate muscles are differentiated from the dorsal and ventral

FIG. 7.18 Anterior limb bud of 16-mm human embryo, showing developing bones and muscle primordia. (*Redrawn from C. K. Arey, "Developmental Anatomy," 7th ed. W. B. Saunders Company, 1965, by permission..*

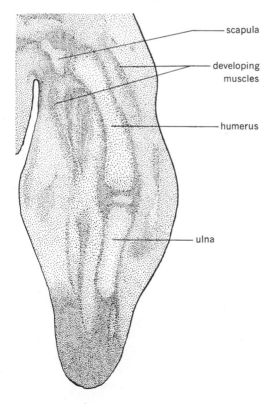

scapula

developing muscles

humerus

ulna

muscle masses. The muscles are larger and sturdier and have broad attachments. A clear division into extrinsic and intrinsic groups is apparent. This provides for the greater freedom of movement in many directions usually required in terrestrial locomotion.

At a certain point in development the muscles begin to differentiate from the mesenchyme of the limb bud. This accompanies the development of the appendicular skeleton, which is also differentiated from limb-bud mesenchyme. Those muscles which arise in the dorsal part of the appendage in most cases become extensors, whereas the ventral derivatives become flexors. Outgrowths of these muscles toward the trunk form abductors and adductors, respectively.

So many modifications are to be found in the appendicular muscles of various tetrapods that it is not feasible to discuss them in a book of this size. The tailed amphibians are the lowest vertebrates to possess limbs terminating in digits (Fig. 7.19). Accordingly, they give a clue to the origin of the appendicular musculature of higher forms. Special treatises on the subject should be consulted by the interested student.

Table 5 provides the usual names given to the large number of muscles found in the limbs of terrestrial vertebrates. The accepted nomenclature is patterned after mammals, but many muscles are not homologous to the names given them. Many of the superficial muscles are illustrated (Fig. 7.14) for mammals.

BRANCHIAL MUSCULATURE

An important feature of elasmobranchs is to be found in the branchial muscles which move the visceral arches and jaws. They are voluntary, striated muscles derived from the splanchnic mesoderm of the hypomere, *not*

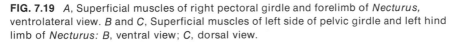

from myotomes. The branchial muscles begin anteriorly at the mandibular arch and run in series to each of the successive visceral arches. Their metameric character is related to the segmental character of the visceral pouches and visceral skeleton, not to that of the axial musculature derived from the metameric somites. Furthermore, they are innervated by *visceral motor* branches of certain cranial nerves. The mandibular branch of the

trigeminal nerve (V) supplies the branchial muscles associated with the mandibular arch; the facial nerve (VII) is the nerve of the hyoid arch; the glossopharyngeal nerve (IX) innervates the branchial musculature of the third visceral arch located between the first and second *typical* gill slits; the accessory nerve (XI) via the vagus nerve (X) supplies the remainder of the branchial musculature.

The outer, or superficial, branchial muscles

FIG. 7.19 *A*, Superficial muscles of right pectoral girdle and forelimb of *Necturus*, ventrolateral view. *B* and *C*, Superficial muscles of left side of pelvic girdle and left hind limb of *Necturus: B*, ventral view; *C*, dorsal view.

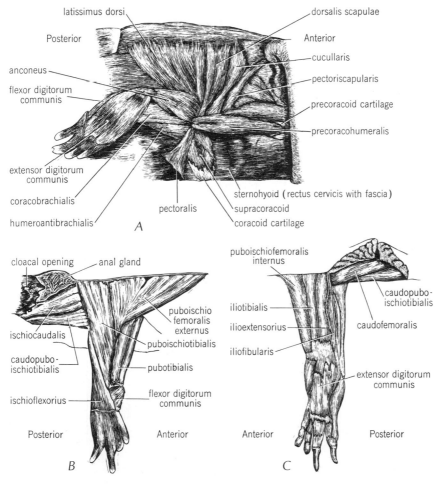

TABLE 5 Muscle Homologies in Various Tetrapod Groups*

Mammal	Reptile	Urodele	Frog	Bird
PECTORAL LIMB, DORSAL MUSCULATURE				
Latissimus dorsi Teres major } Subscapularis	Latissimus dorsi { Subcoracoscapularis { Scapulohumeralis posterior }	Latissimus dorsi Subcoracoscapularis	Latissimus dorsi	Latissimus dorsi { Coracobrachialis posterior { Subcoracoscapularis { Scapulohumeralis anterior }
Deltoideus Teres minor Triceps	Dorsalis scapula Deltoides claviculari Scapulohumeralis anterior Triceps	Deltoides scapularis Procoracohumeralis longus Procoracohumeralis brevis Triceps	Dorsalis scapulae } Deltoideus Scapulohumeralis brevis Anconeus	Deltoideus + propatagialis Scapulohumeralis anterior Triceps
Supinator Brachioradialis Extensores carpi radiales } Extensor digitorum communis } Extensor digiti quinti	Supinator Extensores carpi radiales Extensor digitorum communis	Supinator longus Extensores carpi radiales Extensor digitorum communis	Extensor antibrachii radialis Extensor carpi radialis Extensor digitorum communis	Extensor antibrachii radialis Extensor carpi radialis Extensor digitorum communis
Extensor carpi ulnaris Anconeus	Extensor carpi ulnaris } Anconeus	Extensor carpi manus	Extensor carpi ulnaris } Epicondylocubitalis	Extensor carpi ulnaris Extensor antibrachii ulnaris
Abductores pollicis Extensores digitorum 1–3	Abductor pollicis longus Extensores digitorum breves	Supinator manus Extensor digitorum brevis	Abductor indicis longus Extensores digitorum breves	Abductor pollicis Extensores digitorum breves
Dorsal interossei	Dorsal interossei	Dorsal interossei	Dorsal interossei	
PECTORAL LIMB, VENTRAL MUSCULATURE				
Pectoralis Supraspinatus } Infraspinatus } Biceps brachii Coracobrachiales Brachialis	Pectoralis Supracoracoideus Biceps brachii Coracobrachiales Brachialis inferior	Pectoralis Supracoracoideus Coracoradialis (Not formed) Coracobrachiales Brachialis	Pectoralis Coracohumeralis Coracoradialis (Not formed) Coracobrachialis	Pectoralis { Coracobrachialis anterior { Supracoracoideus Biceps brachii Coracobrachiales Brachialis
Pronator teres Flexor carpi radialis } Flexor palmaris superficialis Palmaris longus Epitrochleoanconeus Flexor carpi ulnaris	Pronator teres Flexor carpi radialis } Flexor palmaris superficialis Epitrochleoanconeus Flexor carpi ulnaris	Flexor carpi radialis Flexor palmaris superficialis Flexor antibrachii ulnaris Flexor carpi ulnaris	{ Flexor antibrachii medialis { Flexores carpi radiales } Palmaris longus { Flexor antibrachii laterales { Flexor carpi ulnaris { Epitrochleocubitalis }	Epicondyloradialis Flexor digitorum sublimis Flexor carpi ulnaris
Flexor digitorum profundus Pronator quadratus	Flexor digitorum profundus } Pronator profundus	Ulnocarpalis Flexor palmaris profundus Pronator profundus	Ulnocarpalis Flexor palmaris profundus Pronator profundus	Flexor accessorius } Pronator profundus
Superficial short digital flexors: palmaris brevis, contrahentes, lumbricales, etc.				
Deep short digital flexors, digital interossei, etc.				

* No attempt has been made to include all the variants present in different members of the groups tabulated. For each group only one common name is given for each muscle; often there are synonyms. In many instances homologies are doubtful.

HIND LEG, DORSAL MUSCULATURE

Mammal	Reptile	Urodele	Frog	Bird
Sartorius Rectus femoris Vasti-complex Gluteus maximus Iliacus Psoas Pectineus Gluteus {medius, minimus}	?Ambiens Iliotibialis Femorotibialis ?Iliofibularis Puboischiofemoralis internus Iliofemoralis	?Iliotibialis Ilioextensorius ?Iliofibularis Puboischiofemoralis internus Iliofemoralis	?Tensor fascia latae Crureus glutaeus ?Iliofibularis Iliacus Pectineus } Adductor longus Iliofemoralis	?Ambiens Iliotibialis Sartorius Femorotibialis ?Iliofibularis Iliofemoralis internus Iliofemoralis externus Iliotrochantericus
Tibialis anterior Extensor digitorum longus Extensor hallucis longus Peroneus tertius Peroneus longus Peroneus brevis	Tibialis anterior Extensor digitorum communis Peroneus longus Peroneus brevis	Tibialis anterior Extensor digitorum communis Peroneus longus } Peroneus brevis }	Extensor cruris brevis Tibialis anticus longus Peroneus	Tibialis anterior Extensor digitorum communis Peroneus longus } Peroneus brevis }
Extensores digitorum breves	Extensores digitorum breves	Extensores digitorum breves	Tibialis anticus brevis	Extensores digitorum breves
Dorsal interossei	Dorsal interossei	Dorsal interossei		

HIND LEG, VENTRAL MUSCULATURE

Mammal	Reptile	Urodele	Frog	Bird
Obturator externus } Quadratus femoris } Obturator internus } Gemelli } Adductores femoris brevis et longus } Adductor magnus	Puboischiofemoralis externus Ischiotrochantericus	Puboischiofemoralis externus Ischiofemoralis	Adductor magnus (pt.) Gemelli } Obturator internus } Obturator externus } Quadratus femoris } ?Adductor magnus (pt.) Caudalipuboischiotibialis	Obturator externus Ischiofemoralis
Crurococcygeus } Pyriformis }	Adductor femoris ?Pubotibialis	Adductor femoris ?Pubotibialis Caudifemoralis longus Caudifemoralis brevis	Pyriformis	Coccygeofemorales
Gracilis	Caudifemoralis longus Caudifemoralis brevis Puboischiotibialis	Puboischiotibialis	Semitendinosus } Sartorius } Semimembranosus } Gracilis }	
Semimembranosus } Semitendinosus }	Flexor tibialis internus ?Flexor tibialis externus	Ischioflexorius		Ischioflexorius Caudilioflexorius
Biceps				
Gastrocnemius medialis } Flexor hallucis longus } Flexor digitorum longus } Tibialis posterior } Popliteus Gastrocnemius lateralis } Soleus } Plantaris }	Gastrocnemius internus } Flexor digitorum longus } Pronator profundus } Popliteus Gastrocnemius externus	Flexor digitorum sublimis } Flexor digitorum longus } Pronator profundus Popliteus Fibulotarsalis (?)	Plantaris longus Tibialis posticus	Gastrocnemius internus } Flexor hallucis longus } Flexor profundus } Tibialis posticus } Popliteus } Gastrocnemius externus }
Interosseus	Interosseus	Interosseus		

Superficial short digital flexors: flexor digitorum brevis, contrahentes, lumbricales, etc.

Deep short digital flexors: flexor short digital flexors, digital interossei, etc.

Source: A. S. Romer, "The Vertebrate Body," 3d ed., Saunders, 1962.

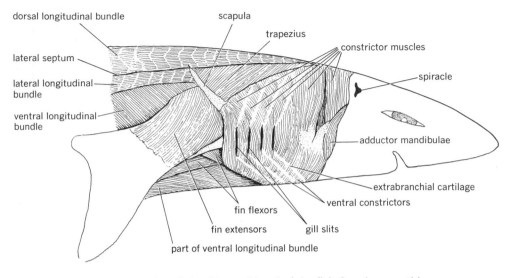

dorsal longitudinal bundle

scapula

trapezius

constrictor muscles

lateral septum

spiracle

lateral longitudinal bundle

ventral longitudinal bundle

adductor mandibulae

extrabranchial cartilage

ventral constrictors

fin flexors

fin extensors

gill slits

part of ventral longitudinal bundle

FIG. 7.20 Superficial muscles of shoulder and head of dogfish *Squalus acanthias.* (*Original by Howell, J. Morphol., 54:401, by permission of the Wistar Institute of Anatomy and Biology.*)

function as constrictors of the pharynx. They tend to close the mouth and gill openings during respiratory movements. They are divided into *dorsal* and *ventral constrictors,* lying above and below the gills, respectively. Each constrictor is separated from the next by a band of connective tissue arranged vertically and known as a *raphe.*

In *Squalus acanthias* (Fig. 7.20) the dorsal constrictor of the first, or mandibular, arch is a small muscle lying anterior to the spiracle. The *adductor mandibulae,* which lies outside the angle of the mouth and acts in closing jaws, is a derivative of the first dorsal constrictor. The first ventral constrictor forms the broad *intermandibularis* muscle. The constrictors of the remaining visceral arches are bounded anteriorly and posteriorly by the gill slits. The dorsal portion of the second constrictor is known as the *epihyoideus muscle.* The ventral portion gives rise to the *interhyoideus,* which is attached to the hyoid arch and lies

just above the intermandibularis. The last four constrictors are simply arranged.

Other branchial muscles, covered for the most part by the constrictors, serve as levators of the visceral arches in *Squalus acanthias* (Fig. 7.21). The first levator is the *levator maxillae,* used in raising the upper jaw. The second is the levator of the hyoid arch. The *trapezius* (*cucullaris*) belongs to the levator group and probably represents the posterior portion of the remaining levators. Other muscles of the levator group include *lateral interarcuales* and *dorsal,* or *medial, arcuales.*

Certain muscles ventral to the gill region are actually hypobranchial muscles derived from the anterior myotomes of the trunk musculature since they are supplied by spinal rather than cranial nerves. They include the *common arcuales, coracomandibular, coracohyoids,* and *coracobranchials.*

Homologs of the branchial muscles as found in higher forms have been determined mostly by tracing the visceral motor branches of the four cranial nerves, which in fishes supply the branchial musculature. The adductor mandibulae gives rise in higher forms to the *masseter, temporal, internal* and *external pterygoids,* and *tensor tympani muscles.* The intermandibularis gives rise to the *mylohyoid* and part of the *digastric* muscle of mammals. The ventral constrictor of the hyoid arch also contributes to the digastric muscle, the double innervation of which, coming from the trigeminal and facial nerves, bears witness to its homologies. Some of the muscles of the face as well as the *platysma* of mammals are believed to be derivatives of the hyoid constrictors. The constrictors of the remaining visceral arches are apparently not represented in higher forms.

The levator muscles of the visceral arches are also imperfectly represented in tetrapods.

The *trapezius* (*cucullaris* of lower forms) and the *sternocleidomastoid,* or any of its variations such as the *sternomastoid* or *cleidomastoid,* may be such derivatives. The voluntary muscles of the pharynx and larynx are probably derived from branchial muscles, but their homologies are uncertain.

The anterior fibers of the eleventh cranial nerve (spinal accessory) of amniotes are so closely associated with the vagus that the muscles which it supplies are considered to be derived from the branchial musculature. The sternocleidomastoid and the trapezius are such muscles. The trapezius in mammals is so obviously a muscle of the pectoral appendage that its inclusion among the branchial muscles might at first sight seem peculiar. This muscle may, however, have a twofold origin, since it is supplied by certain spinal nerves as well as by the spinal accessory (XI).

FIG. 7.21 Deeper muscles of the shoulder and head of the dogfish *Squalus acanthias.* (*Original by Howell, J. Morphol.,* **54:**401, *by permission of the Wistar Institute of Anatomy and Biology.*)

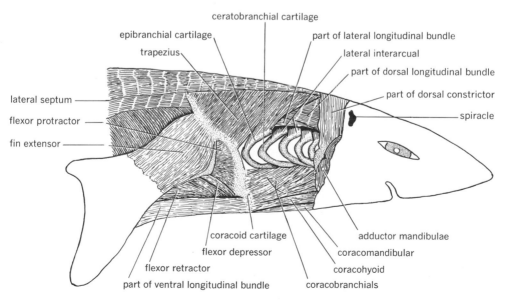

VISCERAL MUSCULATURE

The muscular tissue derived from the splanchnic layer of hypomeric mesoderm is, for the most part, of the involuntary, smooth type. An exception is the branchial musculature, already discussed, which is voluntary and composed of striated fibers. Smooth muscle is found in the walls of the digestive tract and respiratory passages; in the ducts of digestive glands and the urogenital system; in blood vessels; in certain lymphatic vessels; and in the spleen. The internal sphincter muscle of the anus is composed of smooth muscle, as are such sphincters as the pyloric and ileocolic valves.

The peculiar cardiac muscle tissue, found only in the walls of the heart and in the great vessels leaving the heart, is derived from visceral musculature. Its histological structure reveals that it has characteristics of both striated and smooth muscle.

The visceral muscles are more properly discussed in connection with the organs of which they form a part.

DERMAL MUSCULATURE

Integumentary, or dermal, muscles appear, with few exceptions, only in amniotes. Often these muscles are attached to some part of the skeleton and insert on the skin. Usually, however, both origin and insertion are in the integument. Most dermal muscles are derivatives of the parietal musculature, from which they have become separated to varying degrees. Some, however, seem to have split off the branchial musculature.

FISHES Integumentary muscles, as such, are not present in fishes. The myocommata between myotomes are attached to the dermis, and the outermost muscle fibers of the myotomes are closely applied to the dermis.

AMPHIBIANS A few integumentary muscles may occur in anurans. The *gracilis minor* of the hind limb inserts in part upon the neighboring *gracilis major,* but also has an insertion on the skin of the posterior region of the thigh. Another muscle, the *cutaneous*

FIG. 7.22 Semidiagrammatic sketch of inner side of ventral integument of snake, showing costocutaneous muscles extending from ribs to scales. The dermal layer of skin is not shown. (*Modified from Buffa.*)

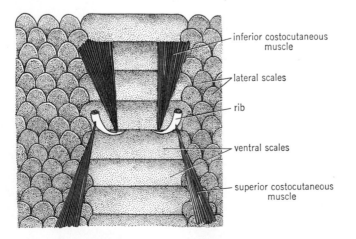

inferior costocutaneous muscle

lateral scales

rib

ventral scales

superior costocutaneous muscle

pectoris, on the ventral side of the anterior part of the body, comes from the body wall and inserts anteriorly on the skin between the forelimbs. Otherwise, integumentary muscles are usually present in amphibians only in the region of the external nares. In urodeles these are of the smooth, involuntary type, serving to control respiratory currents. Rudimentary smooth muscles are also to be found about the external nares of anurans but with few exceptions are unimportant in opening and closing the nares.

REPTILES Intrinsic dermal muscles in reptiles are used in moving certain scales. Locomotion in snakes (Fig. 7.22), aside from the wriggling, undulatory movements of the body proper, is accomplished by alternately elevating and lowering the ventral and ventrolateral scales. Erecting the scales provides a greater friction with the ground, furnishing a temporary anchorage from which the body may be pulled or pushed. *Costocutaneous parietal* muscles, extending from ribs to scales, are also used and are of great importance in progressive locomotion.

BIRDS Movement of individual feathers in birds is accomplished by contraction of cutaneous muscles. The feathers may be made to lie smoothly or may be erected to different degrees, thus enabling the bird to regulate heat radiation from the skin to some extent. The *patagium,* or web of skin which passes from the body to the wing, is provided with dermal *patagial* muscles. Stretching of the patagium increases the resistance of the wing to air during flight.

MAMMALS Integumentary muscles are far more elaborately developed in mammals than in any other vertebrates. In monotremes and marsupials a broad sheet of muscular tissue, the *panniculus carnosus,* almost envelops the entire trunk and appendages. Twitching movements of the skin, used to

dislodge insects and other foreign objects, are brought about by contraction of this muscular sheet. Defensive movements of the European hedgehog and armadillos, i.e., rolling into a ball when in danger, are brought about by contraction of an unusually well-developed panniculus carnosus muscle. Originally derived from the *latissimus* and *pectoralis* muscles, the panniculus carnosus becomes reduced in higher forms. Remnants most often are found in the shoulder, sternal, and inguinal regions. In marsupials the inguinal portion forms the sphincter muscle of the marsupial pouch. The cloacal sphincter of monotremes is also such a derivative. In horses and cattle the skin of the forelegs and shoulders may be twitched to ward off annoying flies and other insects. The posterior part of the body, which can be reached by the swishing tail, is not provided to the same extent with integumentary muscles.

In lower primates both axillary and inguinal remnants of the panniculus carnosus are encountered, but in higher apes and man only a few axillary muscle slips may remain. A *sternalis* muscle, overlying the pectoralis major, appears occasionally in man.

In the head and neck region is another group of integumentary muscles derived from a muscular sheet, the *sphincter colli,* present in turtles and birds. The fact that these muscles are supplied by visceral motor fibers of the facial nerve (VII) indicates that they have been derived from the branchial musculature of the hyoid arch. The *mimetic* muscles, or muscles of facial expression, are derivatives of the sphincter colli. In most lower mammals these muscles are poorly developed, and as a result the facial expression of such forms does not change. In carnivores and primates, however, they are better developed, but only in man do they reflect emotion to a high degree. The original sphincter colli apparently forms two layers, an outer, or superficial, *platysma* layer and an underlying sphincter colli. The muscle fibers of these two layers course in opposite directions. Both layers form slips which serve to move special parts of the face.

MISCELLANEOUS

ELECTRIC ORGANS Electric organs, consisting of modified muscular tissue, are to be found in numerous fishes. Noteworthy are those of the electric ray, *Torpedo,* and the South American electric eel, *Electrophorus electricus.* In *Torpedo* the electric organs lie on either side of the head between the gill region and the anterior part of the pectoral fin. In most fishes with electric organs, however, they are confined to the tail. In *Electrophorus* (Fig. 7.23), which grows to 8 ft, the tail is very elongated, forming approximately four-fifths of the total length of the animal. The electric

FIG. 7.23 Diagrammatic cross section through the posterior body showing the position of the electric organs in the electric eel, *Electrophorus.* (*Adapted from Hildebrand, "Analysis of Vertebrate Structures," John Wiley & Sons, Inc.*)

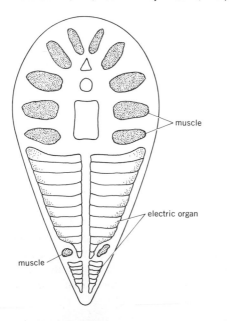

organ is well developed and occupies the position of the ventral longitudinal bundles. In both these fishes electric shocks of considerable intensity may be emitted at will. Voltages of 50 to 600 V at several amperes of current for brief periods have been recorded. They are used by the animals in stunning or killing prey and also in self-defense. Several other fishes, including *Raja* and *Mormyrus,* possess the ability, but the electric shocks they produce are feeble in comparison with those mentioned above. The discharge may be periodical for prey stunning or continuous for object location.

The electric organs of fishes, with the possible exception of *Malapterurus,* are built upon the same general plan, regardless of position. Each organ is composed of many parallel flat-muscle fibers arranged in series. The electroplates are thin and waferlike, the outer surface being a specialized nervous layer and the inner surface, a nutritive layer. Some electric organs develop from muscle or nerve fibers, and others are modified motor end plates. Between adjacent plates lies a thick layer of gelatinous material enclosed in connective tissue, which separates the electroplates into compartments. The nerve fibers leading to the electric plates ramify through the connective tissue to terminate in fine networks on one side of each plate. The nervous layers of all the plates face in the same direction.

In *Electrophorus* the electric plates form longitudinal columns, but in *Torpedo* the columns are vertical. In the former, the electric current passes from the tail to the head; in the latter, from the ventral to the dorsal surface.

How the electric shock or field is produced is well known. Movement of ions across the cell membrane and a difference in the responses of the two faces of the electroplate result in a resting potential difference of about 45 mV. Connecting the electroplates in series produces a storage-battery effect, the discharge of which is under direct nervous control.

The phylogenetic relationships of electric organs are obscure. Since they occur in various parts of the body in different species of fishes, they are not even homologous within the group. This is further borne out by a study of their nerve supply. In *Torpedo* the electric organ is innervated by branches of *cranial nerves* VII, IX, and X, together with a single branch of nerve V. The nerves arise from an enlarged portion of the medulla oblongata spoken of as the *electric lobe*. It is believed that the adductor muscle of the mandible and the constrictor muscles of the visceral arches have given rise to the electric organs of *Torpedo*. In *Electrophorus* and other fishes *spinal nerves* are distributed to the electric organs which are derived from the caudal musculature. More than 200 nerves pass to the electric organ in *Electrophorus*. The high voltage is a result of the polarization of one side of the electroplate and not the other adding in series.

SUMMARY

The muscles of the vertebrate body are derived from the epimeric and hypomeric mesoderm of the embryo. The epimere, which becomes marked off into mesoblastic somites, forms segmentally arranged myotomes which give rise to the voluntary, striated, parietal musculature of the body. The hollow, unsegmented hypomere forms somatic and splanchnic layers enclosing the coelom. The splanchnic layer, which surrounds the gut, gives rise to the smooth, involuntary muscle of the gut and also to cardiac muscle. An exception is to be found in the voluntary, striated, branchial muscles of the gill region, which are derived from hypomeric mesoderm.

PARIETAL MUSCULATURE　The parietal muscles consist primarily of a linear series of muscle segments or myotomes, extending from anterior to posterior ends of the body. Each myotome lies opposite the point of articulation of two vertebrae and is supplied by a separate spinal nerve. Connective-tissue septa, the myocommata, occupy the spaces between adjacent myotomes. The myotomes of the two sides are separated at the midventral line by the linea alba, composed of connective tissue.

Trunk Musculature　In amphioxus, and to a lesser degree in cyclostomes, the myotomes tend to be V-shaped. In fishes they take more of a zigzag course. Beginning with elasmobranchs, and in higher forms to a less conspicuous extent, the myotomes are separated into dorsal (epaxial) and ventral (hypaxial) portions. Each of these regions becomes subdivided into two or more longitudinal muscular bundles. In tetrapods the primitive metameric condition becomes more and more obscure as the evolutionary scale is ascended. This is brought about by a splitting into various layers, disappearance of myocommata, degeneration of portions of some myotomes, and a drawing out of position, a migration, or even fusion of others.

The epaxial musculature becomes reduced in higher forms but shows a tendency to split up into short and long systems of longitudinally directed muscles. The shorter ones connect adjacent vertebrae and are termed, according to position, interneural, intertransversarial, interspinal, etc., muscles.

The hypaxial muscles of tetrapods show greater modification than those of the epaxial region. In urodele amphibians they split into four layers, or sheets, the fibers in each sheet extending in a direction different from that of the others. These layers include the superficial and deep external obliques, an internal oblique, and a transversus. The last lies next to the peritoneum. On either side of the linea alba lies a rectus abdominis, derived from the obliques and with primitively arranged longitudinal fibers. In anuran amphibians the number of layers is reduced, but in reptiles additional layers are to be found in the thoracic region. An intercostal layer, derived from the obliques, makes its appearance. It in turn divides into external and internal intercostals, which are attached to the ribs and are used in respiratory movements. Scalene and serratus muscles are also derived from the obliques. The hypaxial muscles in the thoracic region of mammals are somewhat reduced, being almost entirely covered by the large appendicular muscles of the pectoral girdle and limb.

Hypobranchial Musculature In cyclostomes and fishes, hypobranchial muscles, lying ventral to the gills, are derived from the hypaxial portion of the first few myotomes as buds which grow anteriorly. They are innervated by spinal nerves. In tetrapods these muscles have been greatly modified, but their innervation by spinal nerves bears witness to their origin. The muscles of the tongue, supplied by cranial nerve XII, are considered to be hypobranchial in origin, since this nerve, which is found only in amniotes, represents a union of the most anterior two or three spinal nerves of lower forms.

Eye Muscles The six extrinsic eye muscles of vertebrates develop from three prootic somites supplied by cranial nerves III, IV, and VI, respectively. From the first come the superior, inferior, and internal rectus muscles and the inferior oblique; from the second comes the superior oblique; and from the third the external rectus is differentiated.

Diaphragm The diaphragm of mammals is composed partially of muscle fibers innervated by branches of certain cervical spinal nerves. The muscles, therefore, are believed to originate from cervical myotomes.

Appendicular Musculature In elasmobranchs hollow buds from certain myotomes grow out into the ventrolateral fin folds, accompanied by branches from corresponding spinal nerves. The original metameric arrangement of these buds is soon lost, but they give rise to the dorsal and ventral muscles moving the fins. No similar development of appendicular muscles from myotomic buds has been reported for higher forms, but in all other ways these muscles appear to be derivatives of myotomes. In tetrapods the dorsal and ventral masses differentiate into separate muscles. These are divided into extrinsic and intrinsic groups which provide for the greater freedom of movement required for terrestrial locomotion.

BRANCHIAL MUSCULATURE The striated, voluntary branchial muscles which move the visceral arches of fishes are derived from the splanchnic hypomeric mesoderm rather than from myotomes. They begin anteriorly at the mandibular arch and run in series to each of the successive visceral arches. The outer muscles serve as constrictors of the pharynx and are divided into dorsal and ventral portions. Deeper muscles serve as levators of the visceral arches. Even in fishes many of these muscles have deviated from the primitive condition, but their cranial nerve supply indicates their branchial origin. The mandibular branch of cranial nerve V supplies the muscles of the mandibular arch; the facial nerve (VII) goes to those of the hyoid arch; the third arch is innervated by the glossopharyngeal (IX); and the remaining arches by the vagus (X) and accessory (XI). In higher forms, in which the visceral arches have been greatly modified, the nerve supply to certain muscles of the neck and shoulder regions indicates that they have been derived from branchial musculature.

VISCERAL MUSCULATURE   The smooth muscles of the hollow organs, those of the ducts of digestive glands, etc., as well as the cardiac muscle of the heart, are all developed from the unsegmented splanchnic mesoderm of the hypomere.

DERMAL MUSCULATURE   Integumentary, or dermal, muscles appear, with few exceptions, only in amniotes. They are, for the most part, derivatives of the parietal and branchial musculature, from which they have become separated to varying degrees. In snakes the muscles which erect the ventral scales are dermal derivatives. They are important in locomotion. Movement of feathers in birds is accomplished by integumentary muscles. Greatest development is to be found in mammals. In lower mammals a broad sheet of muscular tissue, the panniculus carnosus, almost envelops the entire body. Its contraction causes twitching movements of the skin. In higher mammals this layer becomes reduced and remnants are found only in the axillary, sternal, or inguinal regions.

Another muscular sheet, the sphincter colli, present in turtles and birds, is most modified in primates, in which it gives rise to numerous muscles of facial expression.

ELECTRIC ORGANS   Certain fishes possess electric organs which give off true electric shocks used to stun or kill prey and in self-defense. These organs are formed from modified muscular tissue. Their phylogenetic relationships and homologies are obscure.

BIBLIOGRAPHY

Alexander, R.: "Animal Mechanics," University of Washington Press, Seattle, 1968.

Bennett, M. V. L.: Comparative Physiology: Electric Organs, *Ann. Rev. Physiol.,* 32:471–528 (1970).

Gray, J.: "Animal Locomotion," Weidenfeld and Nicolson, London, 1968.

_____: Vertebrate Locomotion, *Symp. Zool. Soc. Lond.,* no. 5, London, 1961.

Huxley, H. E.: The Mechanism of Muscular Contraction, *Sci. Am.,* 213(6):18–27 (1965).

Murray, S. M., and A. Weber: The Cooperative Action of Muscle Proteins, *Sci. Am.,* 228(2):57–71 (1974).

Schmidt-Nielsen, K.: "Animal Physiology," Prentice-Hall, Englewood Cliffs, N.J., 1970.

8 RESPIRATORY SYSTEM

The term *respiration* has more than one meaning. The general and widely used definition referring to the exchange of respiratory gases between external environment and the individual cells will not be used in this book. In its restricted meaning the term refers to the oxidation of substances within the cell, which produces energy (ATP), heat, and carbon dioxide and water.

PHASES OF RESPIRATION In vertebrates there are two phases of respiration, external and internal. The term *external respiration* denotes the exchange of gases between the blood and the environment. *Internal respiration* refers to the gaseous exchange between the blood and the tissues or cells of the body.

In vertebrates, the blood serves to bring oxygen from the environment to the cells and carbon dioxide from the cells to the environment. The pigment *hemoglobin,* which gives the vertebrate blood its capacity for carrying oxygen and carbon dioxide, is confined to the red blood cells, or erythrocytes. In reptiles, birds, and mammals a related molecule, *myoglobin* is found in the muscles.

External respiration usually takes place in the capillaries of the gills, skin, or lungs. In gills the capillaries are in almost direct contact with water in which oxygen is dissolved. In lungs they are practically in contact with air that has been taken into the cavities of the lungs. Even though lungs lie *within* the body, the term external respiration is used. During external respiration the blood takes up oxygen and loses most of the carbon dioxide which it is carrying. In internal respiration oxygen is taken from the blood, which in turn gains carbon dioxide given off by the tissues. The principles involved in gaseous exchange follow the laws of gases and are physical rather than biological in character.

Respiratory organs are present in vertebrates, serving to facilitate the external phase of respiration. Certain requirements are demanded of respiratory organs: (1) a large, vascular surface area must be provided so that an ample capillary network can be exposed to the environment; (2) the membranous surfaces through which gaseous exchange occurs must be moist at all times and thin enough to permit the passage of gases; (3) provision must be made for renewing the supply of the oxygen-containing medium (air or water) which comes in contact with the respiratory surface and for removing the carbon dioxide given off from that surface; and (4) blood in the capillary network must circulate freely.

With few exceptions the organs of respiration in vertebrates are related to the pharynx in their development. In some forms, notably amphibians, the skin itself is an important respiratory organ.

The function of the pharynx as part of the digestive tract is essentially to serve as a passageway from mouth to esophagus and to initiate swallowing movements. The part it plays in respiration is of greater importance. The internal gills of aquatic vertebrates and the lungs of air breathers are *pharyngeal derivatives* particularly adapted for respiration. In some forms the vascular epithelium of the pharyngeal wall itself is important. Several structures of pharyngeal origin have no relation to respiration. Among these are the swim bladders of certain fishes, tonsils, middle ear, and Eustachian tube, as well as such glands as the thyroid, thymus, parathyroids, and ultimobranchial bodies.

NASAL PASSAGES

In air-breathing vertebrates there is a close association between the olfactory organs and the organs of respiration. In lower aquatic forms, however, these structures are usually completely divorced from one another. The advantages of a close relationship between the organs of smell and those of respiration should be obvious. When air is drawn into the lungs, volatile substances carried by the air may stimulate the sensory endings of the olfactory nerves situated in the nasal passages. In most of the Sarcopterygii and in the larval forms of some of the more primitive sala-

manders, water enters the mouth through the nares and leaves via pharyngeal gill slits. Chemical substances in the water may be detected by the olfactory apparatus located in the nasal passages.

Comparative Anatomy of Nasal Passages

CYCLOSTOMES In the lamprey an unpaired nasal aperture on the top of the head leads to a blind olfactory sac from which a nasopharyngeal pouch extends ventrally. This complex serves as an olfactory apparatus and has no connection with the pharynx or with respiration. The hagfish has a similar apparatus, but the nasopharyngeal pouch connects with the pharynx and furnishes a passage whereby a current of water reaches the pharynx and gills via the olfactory apparatus.

FISHES In most fishes there is no connection between nostrils and mouth cavity. In some elasmobranchs, however, an *oronasal groove* on each side forms a channel connecting the olfactory pit and mouth. This foreshadows the appearance of a direct connection between the two in higher forms. It is in the fishes of the subclass Sarcopterygii that for the first time a direct connection exists between nasal and mouth cavities in the form of a pair of closed tubes. The openings to the outside are the *external nares;* those opening into the mouth are the *internal nares,* or *choanae.* Internal nares, apparently, were lost secondarily in the coelacanths and are not present in *Latimeria.* Certain teleosts, the stargazers (*Astroscopus*), have well-developed nasal passages and choanae, but these are not homologous with those of the Sarcopterygii or tetrapods. None of the living fishes with internal nares is known to use them to take in atmospheric air. Water is drawn into the mouth through both mouth and nasal passages. In the Sarcopterygii the olfactory epithelium is located in the dorsal parts of the nasal passages.

AMPHIBIANS The nasal passages of amphibians are short and well developed, the internal nares being located just inside the upper jaw (Fig. 8.1). In larval urodeles the current of water passing through the nasal passages is produced either by ciliary action or by movements of the gill apparatus and muscles of the lower jaw. In many urodele larvae and in anuran tadpoles, valves around the internal nares control the direction of flow of water.

At the time of metamorphosis in most urodeles, and with the appearance of lungs, smooth-muscle fibers develop around the external nares, thus providing a means of regulating the opening and closing of the aperture. This is the first instance in which atmospheric air is drawn into the nasal passages. Adult anurans, on the other hand, use a peculiar device for closing their nostrils. A small projection, the *tuberculum prelinguale,* at the tip of

FIG. 8.1 Oral cavity of salamander *Ambystoma texanum.* The angle of the mouth on each side has been cut in order to depress the lower jaw.

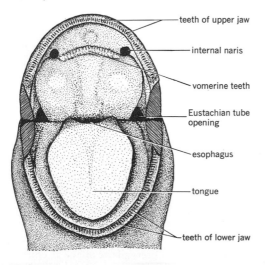

- teeth of upper jaw
- internal naris
- vomerine teeth
- Eustachian tube opening
- esophagus
- tongue
- teeth of lower jaw

the lower jaw is thrust forward and upward, pushing apart the two premaxillary bones in the upper jaw. The rotational movement of these bones changes the position of parts of the nasal cartilages and closes the nasal passages. In a few urodeles a projection into each nasal passage from its lateral wall indicates the first appearance of the *conchae,* which become so highly developed in certain mammals. The nasal passages themselves consist of upper olfactory and lower respiratory regions.

REPTILES Beginning with reptiles there is a tendency toward elongation of the nasal passages. This is brought about by the development of a pair of *palatal folds,* horizontal, shelflike projections of the premaxillary, maxillary, vomer, palatine, and even of the pterygoid bones. Only in crocodilians among reptiles do the processes from the two sides fuse along the median line, thus forming a bony *secondary palate* which separates the nasal passages from the mouth cavity. In some lizards (*Echinosaura*) an attempt at a bony secondary palate is seen. Only the premaxillary, vomer, and maxillary bones are involved. The internal nares communicate with the mouth cavity posterior to the secondary palate. In reptiles a single concha, supported by the maxillary bone, projects into the nasal passage on each side from its lateral wall.

BIRDS The external nares of birds are usually located at the base of the beak, although those of the kiwi are placed almost at the tip. The secondary palate, like that of most reptiles, is incomplete, consisting of a pair of palatal folds which fail to fuse medially. The choanae lie above the palatal folds. The nasal passages of birds are relatively short. Three conchae, supported by *turbinate bones,* are present in birds. Only the most posterior concha is covered with olfactory epithelium.

In numerous marine birds and some reptiles a pair of large *nasal,* or *salt,* glands functions in excreting excess salt from the body. A duct from each leads to the anterior part of the nasal cavity. The salty discharge passes through the external nares and may drip from the end of the beak.

MAMMALS The nasal passages of mammals have been elongated and are larger and more complicated than those of lower forms. Median projections from the premaxillary, maxillary, and palatine bones fuse in the midline to form the *hard palate.* The pterygoids, which are reduced in mammals, do not contribute to palate formation. Extending posteriorly from the hard palate is the *soft palate,* composed of connective tissue covered with epithelium and without any bony foundation. The nasal passages are thus extended posteriorly, and the choanae open into the pharyngeal region.

The presence of a *nose* is characteristic of mammals. In some mammals the nose is excessively developed and forms a *proboscis,* at the end of which the external nares are located. A well-developed proboscis occurs in moles, shrews, tapirs, and elephants. The trunk of the elephant represents the very much drawn-out nose and upper lip.

The nasal passages of mammals are divided into three general regions: vestibular, respiratory, and olfactory. The *vestibular region,* lined with skin, leads from the outside to the inner mucous membrane. Hairs and sebaceous glands may be abundant in this area. Into the *respiratory* and *olfactory regions,* which make up the greater part of each nasal cavity, extends a veritable labyrinth of conchae. These are coiled and scroll-like projections of the *nasal, maxillary,* and *ethmoid bones,* covered with mucous membrane and called the *superior, ventral,* and *lateral nasal conchae,* respectively. The bony projections are called *turbinate bones.*

The respiratory region is lined with respiratory epithelium, a ciliated, pseudostratified, columnar epithelium rich in goblet cells. The lateral and ventral nasal conchae are covered with respiratory epithelium which keeps the nasal passages moist. The mucous membrane lining the nasal cavities is highly vascular. That covering the lateral and ventral conchae has an abundant supply of venouslike spaces which closely resemble those of erectile tissue (see page 334), and which, in some individuals, may even be affected by erotic stimuli, becoming distended with blood and tending to obstruct the nasal passages. The olfactory region occupies the innermost upper recesses. It is covered with olfactory epithelium, which contains nerve endings for the sense of smell. The olfactory epithelium lies over much of the surface of the superior nasal conchae, the sensory area being thus increased. In man, with his rather poorly developed sense of smell, there has been a great reduction of the conchae, compared with many other mammals. The conchae merely project into the nasal passages in the form of shallow ridges. The olfactory epithelium in man is confined on each side to an irregular area measuring about 250 sq mm.

Sinuses, or spaces, within certain bones communicate with the nasal passages. The chief sinuses found in mammals are those of the frontal, maxillary, ethmoid, and sphenoid bones. They are lined with ciliated mucous membrane, but glandular cells are not numerous. The sinuses, which in man are so often the seat of infection and give so much trouble, have no known function. Their presence lightens the weight of the skull and gives resonance to the voice.

JACOBSON'S ORGAN In many tetrapods a *vomeronasal organ* or *Jacobson's organ* is present (Fig. 8.2). It consists of a pair of blind diverticula usually extending from the

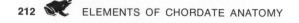

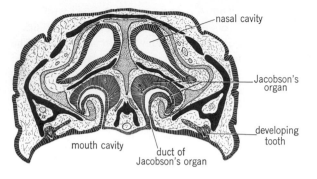

FIG. 8.2 Cross section through head of lizard showing Jacobson's organ. Black areas represent bone; stippled areas represent cartilage. (*After Schimkewitsch.*)

ventromedial portion of the nasal capsule. In amphibians, reptiles, and mammals it receives branches from the terminal (0), olfactory (I), and trigeminal (V) cranial nerves. Jacobson's organ is believed to serve as an accessory olfactory device and aids in the recognition of food.

The vomeronasal organ first appears in amphibians. In frogs it is situated in the anteromedial portion of each nasal cavity, but in urodeles it is more lateral in position. The organ is lacking in *Necturus* and in the cave salamander *Proteus*.

In *Sphenodon*, lizards, and snakes Jacobson's organ is best developed of all. Instead of opening into the nasal cavity, it connects directly with the mouth cavity, near the choanae. In crocodiles and in birds it is merely an embryonic vestige. The tips of the forked tongues of snakes and some lizards, when retracted, are placed in the small, blind pockets of Jacobson's organ. These apparently are able to detect volatile chemical substances that may have adhered to the surface of the tongue when it was thrust out of the mouth. In mammals there is much variation. In some, as in man, it appears only as an embryonic vestige, but in others it persists throughout life as a definite structure. It is best developed in monotremes and is well developed in mar-

supials, insectivores, rodents, the domestic cat, and others. In some rodents its duct penetrates the secondary palate and opens into the mouth cavity.

PHARYNGEAL POUCHES

In the embryos of all chordates a series of pouches develops on either side of the pharynx. These endodermal structures push through the mesenchyme until they come in contact with invaginated *visceral furrows* of the outer ectoderm. The pharyngeal pouches arise in succession in an anterior-posterior direction. They develop in such a manner as to become successively smaller from first to last (Fig. 8.3), and the pharynx, therefore, is funnel-shaped, tapering toward the esophagus. Except in amniotes, perforations usually occur where endoderm and ectoderm come in contact. Even in amniotes, however, temporary openings are occasionally established. The *visceral,* or *pharyngeal, pouches* are then called *clefts.* Primitively the cavity of the pharynx is thus connected by a series of clefts to the outside. The openings connecting the pharynx proper with the clefts are called *internal gill slits;* those connecting the clefts with the outside are *external gill slits.* The number of

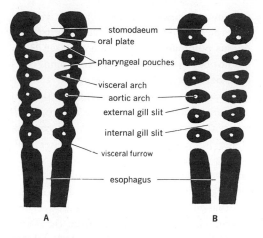

stomodaeum

oral plate

pharyngeal pouches

visceral arch

aortic arch

external gill slit

internal gill slit

visceral furrow

esophagus

A   B

FIG. 8.3 Diagram showing arrangement of pharyngeal pouches: *A*, during development; *B*, after connection with the outside has been established and the oral plate has disappeared.

rior or interior surface. Within each septum lies a cartilaginous or bony barlike structure, the *visceral arch*. This serves to support the septum. Blood vessels called *aortic arches* branch from the ventral aorta and course through the septa, as do branches of certain cranial nerves. The visceral arches, which make up the so-called *visceral skeleton,* are modified in higher vertebrates to form various portions of the skeleton in the head and neck region. The aortic arches also undergo marked changes in the different classes.

The apparently segmented, or metameric, arrangement of structures in the branchial region is a reflection of the arrangement of the visceral pouches or clefts and is unrelated to the metamerism of the trunk myotomes (see page 184).

pharyngeal pouches, or clefts, is greatest in the lowest groups of vertebrates and least in the higher classes. Thus among cyclostomes certain forms (*Bdellostoma*) possess as many as 14 pairs, although the lamprey (*Petromyzon*) has 7 pairs and the hagfish (*Myxine*) but 6. In fishes, amphibians, and reptiles 5 or 6 pairs commonly appear in the embryo, but in birds and mammals the reduced number of 5 or 4 is the typical condition. Pharyngeal pouches are of greatest significance in lower aquatic vertebrates since they bear gills and are directly concerned with respiration. Those of amniotes do not bear gills and generally disappear, except for the first, which becomes the Eustachian tube and middle ear on each side. Remnants of the others persist in the form of certain glandular structures, to be discussed later.

The visceral clefts are separated from one another by septa, which are mesodermal structures covered with epithelium derived either from ectoderm or endoderm, depending upon whether it is toward the exte-

GILLS

Gills are composed of numerous *gill filaments* or *gill lamellae,* which are thin-walled extensions of epithelial surfaces. Each contains a vascular network. Blood is brought very close to the surface, thus facilitating a ready exchange of gases. In their aggregate, gills present a relatively large surface for respiratory exchange.

TYPES OF GILLS Gills are of two general types, external and internal. *External gills* (Fig. 8.4) develop from the integument covering the outer surfaces of visceral arches. They are usually branched, filamentous structures covered entirely with ectoderm and are not related to the visceral pouches. *Internal gills* (Fig. 8.5) are usually composed of a series of parallel *gill lamellae,* although in a few forms they may be filamentous. They may be borne on both sides of an interbranchial septum, but in some cases are present on one side alone. A series of lamellae on one side of

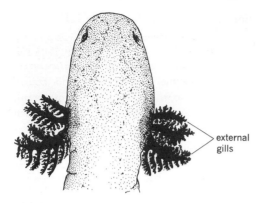

external
gills

FIG. 8.4 Head of salamander larva with external gills.

an interbranchial septum is termed a *half gill* or *hemibranch.* Two hemibranchs enclosing between them an interbranchial septum make up a complete *gill,* or *holobranch.* Two hemibranchs bounding a gill cleft thus belong to different holobranchs. It is generally assumed

that internal gills are covered entirely with endoderm, but there is considerable controversy on this point, and the question of exact origin has not been completely clarified. Some fishes have both external and internal gills.

The function of external gills poses no problems since the filaments are in direct contact with water containing dissolved oxygen. When internal gills are present, water containing dissolved oxygen is generally taken into the mouth, and pumped through the internal gill slits into the gill clefts and out the external gill slits. As the water passes over the gill lamellae, oxygen is taken from the water and carbon dioxide is released. It has been shown that in lampreys and marine teleosts the gills also play an important role in excretion of salt from the blood. Glandular elements in the gill lamellae may, in addition, excrete such nitrogenous wastes as ammonia. The gills are used to absorb salts from the water in freshwater fishes.

FIG. 8.5 Head of bony fish (carp). The operculum has been cut away to expose the internal gills. (*From Storer, "General Zoology," McGraw-Hill Book Company, by permission.*)

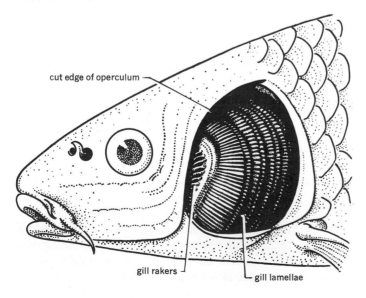

cut edge of operculum

gill rakers

gill lamellae

Comparative Anatomy of Gills

AMPHIOXUS The pharynx of amphioxus is provided with large numbers of vertically elongated gill clefts (Fig. 8.6). Over a hundred may be present. The gill clefts are separated by *primary gill bars*. During development a *secondary,* or *tongue, bar* grows down between two primary gill bars and thus divides each primary gill cleft in two. Later on, small *crossbars* appear connecting one primary bar with the next. There are no filaments or lamellae; the blood vessels course through the pharyngeal bars. A few of the anterior gill slits for a time open directly to the outside. With the development of the <u>ectoderm-lined</u> *atrium,* however, the pharynx no longer forms a direct connection with the outside and all gill clefts then open into the *atrial cavity,* or *peripharyngeal chamber* A single opening, the *atripore,* connects the atrium to the outside. It is located on the ventral side about two-thirds of the way back (Fig. 2.6).

Although much of respiration of am-

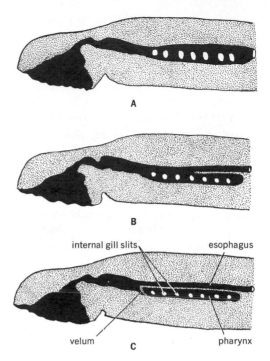

internal gill slits esophagus

velum   C   pharynx

FIG. 8.7 Stages *A, B,* and *C* in the separation of esophagus and pharynx in the lamprey (semidiagrammatic).

FIG. 8.6 Partly dissected amphioxus (*Branchiostoma*) from the left side illustrating the vertical gill clefts.

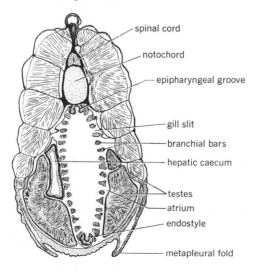

spinal cord

notochord

epipharyngeal groove

gill slit

branchial bars

hepatic caecum

testes

atrium

endostyle

metapleural fold

phioxus takes place through the skin, blood in the gill bars is oxygenated by the stream of water passing from mouth to pharynx, through the gill clefts to the atrium, and leaving through the atriopore.

CYCLOSTOMES The respiratory system of an adult lamprey is highly specialized and at first sight seems atypical. Study of its development in the ammocoetes larva, however, reveals that it is built upon the typical vertebrate plan. Eight pairs of pharyngeal pouches begin to develop, but the first pair flattens out before long and disappears. Seven pairs of pouches remain, each with its internal and external gill slits. At this stage the pharynx connects with the mouth in front and with the esophagus behind. Later a peculiar change occurs. The esophagus and pharynx become separated in such a manner that each has its own connection with the mouth (Fig. 8.7). The esophagus then lies dorsal to the pharynx. The latter becomes a blind pouch, the opening of which is guarded by a *velum*. Seven pairs of internal gill slits open from the pharynx into seven pairs of gill clefts, which are rather large and spherical. The gill lamellae are arranged in a more or less circular fashion, but, nevertheless, each gill cleft is bordered by a hemibranch on both anterior and posterior walls. There are thus 14 hemibranchs on each side but only 6 holobranchs, since the first and last hemibranchs are not parts of holobranchs (Fig. 8.8).

During respiration the lamprey usually forces water in and out the gill clefts through

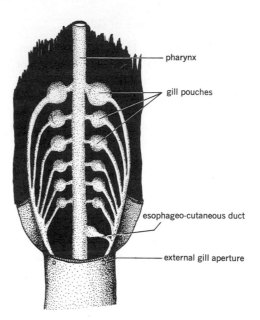

pharynx

gill pouches

esophageo-cutaneous duct

external gill aperture

FIG. 8.9 Diagram showing relation of gill pouches to the hagfish, *Myxine,* to the pharynx and to the single pair of external gill apertures, ventral view. (*After Müller.*)

FIG. 8.8 Frontal section of anterior end of lamprey, showing arrangement of the gills, as seen from below.

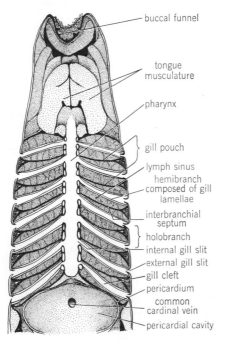

buccal funnel

tongue musculature

pharynx

gill pouch

lymph sinus

hemibranch composed of gill lamellae

interbranchial septum

holobranch

internal gill slit

external gill slit

gill cleft

pericardium

common cardinal vein

pericardial cavity

the *external* gill slits. Thus the gill lamellae are bathed by a constantly replenished supply of water. This is quite different from the method employed by true fishes, in which water enters the mouth and passes over the gills on its way to the outside through the external gill openings. The method used by the lamprey is necessary because when the animal is attached to some object or engaged in feeding, the mouth opening is blocked.

In the hagfish the pharynx connects the mouth and esophagus. Six pairs of gill pouches and internal gill slits are present, but only a single pair of external openings exists (Fig. 8.9). These are situated near the midventral line at some distance from the anterior end. In *Myxine* an *esophageo-cutaneous duct* connects the esophagus with the common exit on the left side. It is similar to a gill cleft but lacks gills.

In cyclostomes of the genus *Bdellostoma,* the number of gill clefts varies from 6 to 7 up to 13 or 14 pairs. The gill clefts connect internally with the pharynx, which is like that of *Myxine* and not a blind pouch like that of the lamprey. They open separately to the outside, however, as in the lamprey. An esophageo-cutaneous duct is also found on the left side in *Bdellostoma.*

FISHES In fishes and tetrapods a series of skeletogenous *visceral arches* encircles the pharynx. In tetrapods these become greatly modified, but in fishes they serve primarily to support the gills. They are located between the gill clefts, one behind the other, at the bases of the interbranchial septa. The first is called the *mandibular arch,* the second the *hyoid arch,* the remainder being referred to by number (3, 4, 5, 6, etc.). The first gill pouch, or cleft, lies between the mandibular and hyoid arches and is often called the hyomandibular cleft. In fishes it is either modified to form a *spiracle* or is closed altogether. The ancestral placoderms had a full-sized hyomandibular gill slit.

Two types of *internal gills* are to be found in fishes. The first, and more primitive, type is found in elasmobranchs (Fig. 8.10*A* and 8.11*A*). In this group the interbranchial septa are well developed and extend beyond the hemibranchs. Each bends posteriorly at its distal end in such a manner that a row of separate *external* gill slits is formed. Small cartilaginous *gill rays* in the form of a single row project from each visceral arch into the interbranchial septum to which they give support. Rigid, comblike *gill rakers* usually project from the gill arches. They prevent food from entering the gill clefts.

The second type of gill is found in the remaining fishes. In these the interbranchial septa are reduced to varying degrees (Fig. 8.10*B* and *C*) and the hemibranchs project into a single *branchial chamber* on each side. Two rows of gill rays may be present. The septum of the hyoid arch is enormously developed and extends caudad as the *operculum*; this covers the branchial chamber, which thus opens to the outside through a single gill aperture. Although in elasmobranchs, chimaeras, sturgeons, and some others a *true* hemibranch receiving unoxygenated blood is located on the posterior side of the hyoid arch, this is lacking in *Polypterus, Amia, Polyodon,* teleosts, and Dipnoi. Tel-

FIG. 8.10 Types of fish gills: *A*, elasmobranch; *B*, chimaera; *C*, teleost.

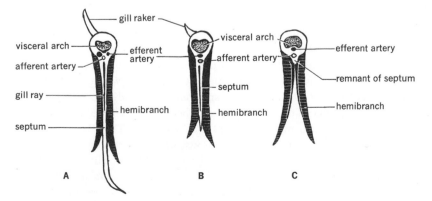

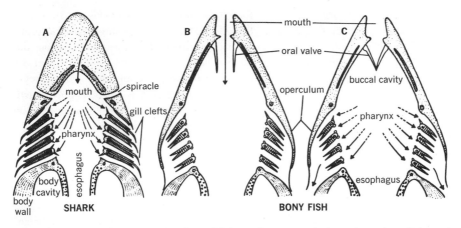

FIG. 8.11 The respiratory mechanism of fishes; diagrammatic frontal sections (lobes of oral valve actually are dorsal and ventral); arrows show paths of water currents. Shark: *A*, water enters ventrally placed mouth, which then closes, and floor of mouth region rises to force water over the gills and out the separate gill clefts. Bony fish: *B*, inhalant, opercula are closed, oral valve is open, cavity is dilated, and water enters; *C*, exhalant, oral valve closes, buccal cavity contracts, water passes over gills in common cavities at sides of pharynx and out beneath opercula. (*From Storer, "General Zoology," McGraw-Hill Book Company, by permission.*)

eosts have four complete holobranchs on each side. No hemibranch is present on the posterior wall of the last gill cleft except in *Protopterus,* in which its homologies are doubtful. Usually a pliant *branchiostegal membrane* extends from the operculum to the body wall. It serves as a one-way valve allowing water to leave the branchial chamber. Other folds form *oral valves* within the mouth opening. These permit water to enter the mouth (Fig. 8.11*B* and *C*).

In most elasmobranchs and in a few other fishes (*Acipenser, Polyodon,* and *Polypterus*) the first gill pouch has become modified and opens to the outside by means of a *spiracle*. Rudimentary gill lamellae may be located on the *anterior* wall of the spiracle. Since the blood supplying these lamellae has been oxygenated, they do not perform a respiratory function. The term *false gill,* or *pseudobranch,* is customarily applied to them. The spiracles open on top of the head posterior to the eyes.

In sharks water passes *out* of the pharynx through the spiracles as well as through the gill slits, but in the bottom-dwelling rays most of the water *enters* the pharynx through the spiracles, and little comes in through the mouth.

Although in many fishes a *true* hemibranch, receiving unoxygenated blood, is lacking on the posterior side of the hyoid arch, a modified *opercular gill,* or *pseudobranch,* receiving oxygenated blood, may be present. Opercular gills of this type are found in *Amia,* the Dipnoi, *Latimeria,* and many teleosts. Such opercular gills probably represent the spiracular pseudobranch, which, upon closure of the spiracle, has shifted its position.

The number of gills and gill clefts varies among fishes (Fig. 8.12). There is some confusion as to just what each represents, since different authors use different numbers in designating them. We shall consider the mandibular arch to be the *first* visceral arch and the hyoid the *second* and shall refer to the remainder as 3, 4, 5, 6, etc. The hyomandibular cleft is the first cleft. However, since it is either modified to form a spiracle or is closed altogether, we shall not consider it as a typical gill cleft. The cleft between the hyoid and third visceral arches (second cleft) is the first *typical* gill cleft and will be so designated hereafter. The shark *Heptanchus* with seven gill clefts has the largest number of any gnathostome.

The presence of external gills is rare among fishes. The larva of *Polypterus* has a single pair of integumentary gills in the region

FIG. 8.12 Diagrams showing number of gill slits and arrangement of hemibranchs in lamprey (*Petromyzon*) and seven other representative fishes. Dotted line indicates gill slit or spiracle which has failed to open to the outside; solid parallel lines indicate lamellae of true hemibranchs; stippled areas indicate pseudobranchs; 1–9, position of visceral arches; I–VII, typical gill slits.

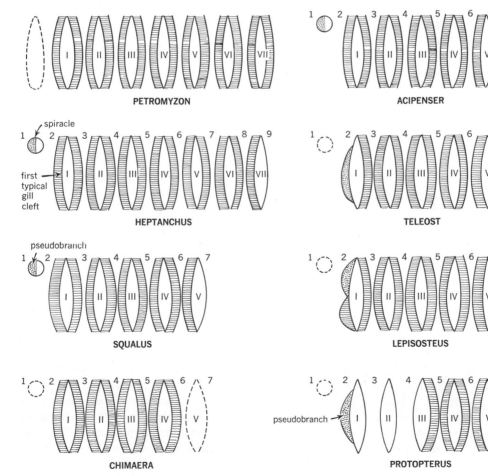

PETROMYZON

ACIPENSER

HEPTANCHUS

TELEOST

SQUALUS

LEPISOSTEUS

CHIMAERA

PROTOPTERUS

of the hyoid arch. Larval dipnoans possess four pairs of external gills, which disappear in adult life although vestiges may remain.

AMPHIBIANS Most amphibians spend their larval life in water and after a period of metamorphosis go to the land. Larval amphibians use external integumentary gills, in addition to the skin, as organs of respiration. In a few urodeles the gills are retained throughout life, but in most amphibians they disappear during metamorphosis. Newly developed lungs usually take over the respiratory function. Cutaneous respiration is important in adults of this class also. Plethodontid salamanders fail to develop lungs when the gills are lost and depend entirely upon cutaneous and buccopharyngeal respiration.

During embryonic development, with few exceptions, five pairs of pharyngeal pouches form in the characteristic manner. The first and usually the last do not become perforated. The first gill pouch in anuran amphibians usually gives rise to the Eustachian tube

and middle ear. In most amphibians all gill slits become closed over at the time of metamorphosis, but in certain urodeles some of them persist throughout adult life.

The external gills of amphibians consist of tufts of filaments with bases on the third, fourth, and fifth visceral arches. The gills are covered with ciliated epithelium. Waving movements of the gills and ciliary action ensure a constant change of water with which they are in contact.

In the larvae of anurans an operculum develops shortly after the external gills appear. It arises from the hyoid region but does not contain any skeletal elements. The operculum grows posteriorly, covering the gills, gill clefts, and the region from which the forelimbs will later develop. It then fuses with the body, behind and below the gill region, in such a manner that the gills are confined to an *opercular chamber* (Fig. 8.13). A small *opercular aperture* connects the opercular chamber with the outside. This may be a single midventral opening or one that is lateral in position. Paired apertures are present in some anuran groups. The external gills soon degenerate, and a new set of gills develops. Even though these are located *within* the opercular cavity and are often referred to as internal gills, their homology with internal gills of fishes is questionable. They are actually covered with ectoderm and should be considered as integumentary derivatives. An opercular fold also forms in urodele amphibians, but it is much reduced in size and consists of a mere crease immediately anterior to the gill region. The gills of larval caecilians are, in general, similar to those of urodeles, a reduced opercular fold also being present in this group.

The gills of anurans are resorbed during metamorphosis. Among other changes which occur at this time, the developing forelimbs push out into the opercular chamber. In the

FIG. 8.13 Diagram showing relations of external gills of anuran larva before (*right*) and after (*left*) the operculum has grown back to enclose the gills in an opercular chamber. (*After Maurer.*)

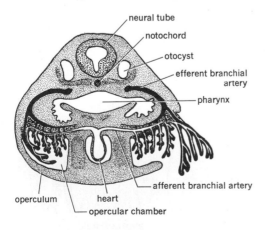

frog, *Rana pipiens,* the left limb passes through the opercular aperture, the right limb pushing through the opercular wall itself.

In urodeles several conditions are exhibited. Three pairs of gills and two pairs of gill clefts persist in *Necturus.* In *Amphiuma* the gills are resorbed, and a single pair of gill slits remains. In most, however, gills and gill slits disappear during metamorphosis. A condition called *neoteny* in certain salamanders refers to retention of certain larval characteristics beyond the normal period. For example, the tiger salamander, *Ambystoma tigrinum,* normally loses its gills at metamorphosis and becomes a terrestrial animal. The western *axolotl,* formerly believed to be another species, is an aquatic, gill-breathing animal even during its adult life. Under certain conditions the axolotl can be induced to lose its gills and becomes a typical lung-breathing salamander identical with *Ambystoma tigrinum.* Several explanations of the meaning of neoteny have been advanced. It seems most probable that neotenic development is a form of evolutionary adaptation. *Necturus,* which retains its gills, cannot be induced to lose them under any condition (page 429).

In many amphibians the gill stage is passed through very rapidly while the embryo is still enclosed within the jellylike egg envelopes. By the time of hatching metamorphosis may already have occurred, and the animal is fitted for terrestrial existence early in life.

REPTILES, BIRDS, AND MAMMALS Five pairs of pharyngeal pouches appear in reptilian embryos, but only four develop in birds and mammals. A fifth pair may appear, but it is rudimentary and attached to the fourth pair. Only occasionally do the pouches break through to the outside. The first pouch persists to become the Eustachian tube and middle ear, but the others become reduced and disappear, except for certain vestiges and derivatives, to be discussed later in the chapter. The homologies of these pouches with those of lower forms should be apparent. If the pharyngeal pouches fail to become obliterated in the normal manner, branchial cysts and fistulae may form. Gills do not develop in association with the pharyngeal pouches of amniotes.

PHYLOGENETIC ORIGIN OF LUNGS AND SWIM BLADDERS

The swim bladder in various fishes shows a rather wide diversity in structure and function. Generally speaking, it is a gas-filled diverticulum which arises from the pharyngeal or esophageal region of the digestive tract. In certain primitive fishes its structure and function are such that it is difficult to make a distinction between swim bladder and lung. In fact the swim bladder is a better lung in some fishes (Dipnoi) than the lungs of most amphibians. The term *lung* is usually applied specifically to the respiratory organ of terrestrial vertebrates. The similarity of lungs of tetrapods to the swim bladders of such fishes makes such a distinction artificial. Some authors refer to the swim bladder as a lung if it has a connection with the ventral surface of the digestive tract. Nevertheless, in the following pages we shall refer to these structures in fishes as swim bladders since in most cases their primary function is not respiratory. A swim bladder may be single or bilobed and may open into the digestive tract dorsally, ventrally, or not at all. It may be large and extend the length of the body cavity or may be so small as to be practically indistinguishable. It usually lies directly beneath the vertebral column, dorsal aorta, and opisthonephric kidneys, but outside the coelom. The peritoneum covers only its ventral surface. Swim bladders which retain the open *pneumatic duct,* or connection with the digestive tract, are said to be *physostomous;* those which are completely closed are *physoclistous.*

In the past, the swim bladder was considered to be the forerunner of lungs of higher forms. The modern point of view, however, is that the presence of lungs was a primitive character and the swim bladder in higher fishes represents a specialized modification of the lung.

The origin of the swim bladder in the phylogenetic series has been obscure. In 1941 it was discovered that lungs were present in the fossil remains of the Devonian placoderm, *Bothriolepis.* The placoderms are considered to have been the oldest and most primitive gnathostomes. The lungs of tetrapods, the lunglike swim bladders of lower fishes, and the modified swim bladders of higher fishes probably came from a common source among the placoderms.

Although in some of the lower fishes the swim bladder functions chiefly as an accessory respiratory organ, it may also serve as a hydrostatic organ and, in certain cases, as an organ for sound production and sound reception as well. The modified swim bladders of higher fishes seem to play little or no part in respiration and serve almost entirely as hydrostatic or equilibratory organs, enabling the fish to swim at different levels with but little effort.

Comparative Anatomy of Swim Bladders

Placoderms The lungs of placoderms, as evidenced by study of *Bothriolepis,* actually consisted of little more than a posterior pair of pharyngeal pouches extending ventrally. They probably functioned as primitive lungs used in conjunction with the gills.

Chondrichthyes No swim bladder is present in the cartilaginous fishes, although a transitory rudiment may appear during development. Opinions differ as to the significance of this rudiment, but it is possible that it represents a lunglike development of some remote progenitor.

Latimeria The swim bladder in *Latimeria* is a tube, only 2 or 3 in. long, prolonged as a filament to the end of the body cavity. It connects to the ventral side of the gut.

Polypterus The most primitive of living ray-finned fishes, *Polypterus,* which is placed in the superorder Chondrostei, is of particular interest because of its swim bladder (Fig. 8.14). It is bilobed and smooth-walled, and its pneumatic duct connects to the ventral side of the digestive tract. The two lobes of the swim bladder are of unequal size, the left being smaller than the right. The arteries which supply the swim bladder of *Polypterus* arise from the last pair of efferent branchial arteries.

Dipnoi The true lungfishes include three surviving genera: *Protopterus, Lepidosiren,* and *Epiceratodus* (Fig. 8.14). Study of the sacculated swim bladders of these forms offers a clue to the origin of the specialized swim bladders found in the teleost fishes. During the dry season some of the Dipnoi estivate in burrows in the mud and use their swim bladders (lungs) in respiration. *Protopterus* and *Lepidosiren,* like *Polypterus,* have bilobed swim bladders which connect with the esophagus on its ventral side. The two lobes, however, are practically of the same size. The pulmonary arteries which supply the swim bladder arise directly from the dorsal aorta posterior to the last pair of aortic arches. Thus, at least partially oxygenated blood is carried to the swim bladder (page 337).

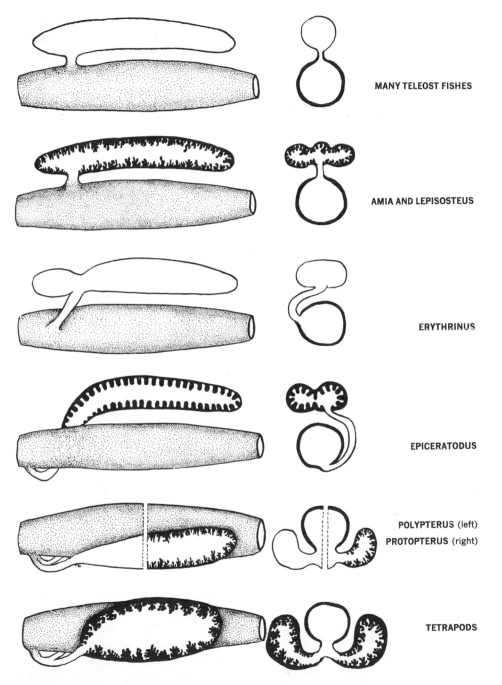

MANY TELEOST FISHES

AMIA AND LEPISOSTEUS

ERYTHRINUS

EPICERATODUS

POLYPTERUS (left)
PROTOPTERUS (right)

TETRAPODS

FIG. 8.14 Various types of swim bladders (or lungs), shown in longitudinal and cross section (diagrammatic). (*After Dean.*)

Pulmonary veins return the blood directly to the left atrium of the heart. Two atria, right and left, appear in the Dipnoi for the first time in the vertebrate series.

The swim bladder of *Epiceratodus* consists of a single lobe. The pneumatic duct courses around the right side of the esophagus and opens on the ventral side. A small diverticulum, present only during embryonic development, represents the left lobe, which is lacking in the adult. The pulmonary arteries arise from the last efferent branchial arteries. Blood going to the swim bladder under normal conditions has been oxygenated in the gills. The pulmonary veins unite to form a single vessel which joins the left atrium. The right and left atria are not so completely separated in *Epiceratodus* as in other dipnoans.

The evolution from the ventrally connecting type of swim bladder to the lungs of tetrapods would seem to be relatively simple and logical. The origin of the physoclistous organs of many teleosts from the single-lobed swim bladder of the type found in *Epiceratodus* would also appear to follow a logical sequence (Fig. 8.14).

Tetrapods The lungs of amphibians, reptiles, birds, and mammals (Fig. 8.14) arise during embryonic development as a bilobed diverticulum from the ventral portion of the pharynx. In most forms there is a tendency for the left lung to be smaller than the right, as in *Polypterus* and some other fishes. The walls of the lungs become subdivided into compartments of varying degrees of complexity. In mammals each lung is commonly made up of two or more large lobes. In structure and development the tetrapod lung so closely resembles the swim bladder of *Protopterus* and certain other fishes that there is little doubt concerning homologies and the phylogenetic origin of the lungs of higher vertebrates.

Holostei Included in the superorder Holostei are the bowfin, *Amia,* and the garpike, *Lepisosteus.* The swim bladder of *Amia* is unpaired, large, and highly vascular (Fig. 8.14). Contrary to the condition in *Epiceratodus,* however, the pneumatic duct connects with the *dorsal* side of the esophagus. In *Amia* a transitory rudiment of the left lobe also appears during development. The arterial supply comes from the last pair of efferent branchial arteries. The venous drainage, like that of *Polypterus,* enters the hepatic veins. The swim bladder of *Lepisosteus* is much like that of *Amia,* but its arteries arise directly from the dorsal aorta. The veins leading from the organ, like those of teleosts, join the hepatic portal and postcardinal veins.

Teleostei The most specialized swim bladders are found among the teleosts (Fig. 8.14). They bear the least resemblance to the primitive structure of *Polypterus.* Some families of teleosts, including such bottom-dwelling forms as the flounder and halibut, lack a swim bladder. The typical teleostean swim bladder is a thin-walled, gas-filled sac which lies dorsal to the digestive tract and extends the entire length of the body cavity. Branches of the celiac artery supply it with blood, and the venous return is through the hepatic portal or postcardinal veins. Teleostean swim bladders, in general, probably play little or no part in respiration, their primary function being concerned with hydrostasis and equilibration. Among the major specializations to be found is the loss of the pneumatic duct, which may atrophy and disappear or persist as a remnant in the form of a solid fibrous cord. The salmon, carp, pickerel, and eel are forms which retain the open pneumatic duct. Some, as in the salmon, are of simple construction. In the carp family the swim bladder is reduced in size and a diverticulum extends forward into the head region.

This is the *anterior chamber*. The *posterior chamber* is joined to the esophagus by a pneumatic duct. The anterior chamber is connected by a chain of four small bones, the *Weberian ossicles,* to the inner ear. These are derived from portions of the anterior four vertebrae and are found in members of the order *Cyprinoformes,* to which the carp, catfish, and sucker belong. In some other teleosts (herring, etc.) extensions of the swim bladder itself make *direct* contact with the inner ear. They are of importance in static and auditory perception.

There is much variation in the vascularity of the swim bladders of teleosts. In some the entire inner surface is uniformly vascular; in others, localized masses of tightly packed capillaries are present. This structural grouping is referred to as a *retia mirabilia*. In certain teleosts with physostomous swim bladders, one or more of these masses become further specialized to form peculiar structures known as *red bodies,* which project into the cavity of the swim bladder near its anterior end. The lining epithelium of the swim bladder is continued over the red bodies, apparently without any modification. The common eel possesses several red bodies, the largest of which lies near the opening of the pneumatic duct. Red bodies are able to remove oxygen, nitrogen, and carbon dioxide from the blood and to pass these gases into the swim bladder.

Completely closed physoclistous swim bladders are found in such teleosts as the perch, toadfish, cod, and haddock. With the disappearance of the pneumatic duct a modification of the red body occurs. The capillaries are still densely arranged, but the epithelium covering the projecting portion of these structures seems to be of a glandular nature with numerous crypts and folds. The red body has now become the *red gland*. A single red gland is found in the cod and haddock, but in other teleosts several may be present. Physoclistous swim bladders show many variations in shape. In some there are constrictions, either forming chambers or giving a sacculated appearance to the structure.

Chondrostei The swim bladders of paddlefish and sturgeons have not been mentioned since they do not fit into the general scheme of things as outlined here. Some sturgeons lack a swim bladder. In others it is oval and lined with ciliated epithelium and opens into the dorsal side of the esophagus near the stomach. In certain forms, gastric glands may even be present in its walls during embryonic development. This group of fishes is considered to be degenerate in other respects, and the loss or degenerate condition of the swim bladder may be nothing more than an association with bottom-dwelling habits.

FUNCTIONS OF THE SWIM BLADDER The physostomous swim bladders of some of the lower fishes serve as an important adjunct to the gills in respiration. The observed behavior of these fishes and the structure and blood supply of the swim bladder support this assumption. Its position in the body and the fact that it contains gas indicate that the swim bladder may also function in hydrostasis, pressure equilibrium, and sound production. It is doubtful whether the physoclistous swim bladders of some higher teleost fishes are of any importance in respiration. They seem to serve chiefly as hydrostatic and equilibratory organs. Those teleosts having Weberian ossicles use their swim bladders in sound reception.

Respiration It is well known that *Lepisosteus,* dipnoans, and other species of fishes are unable to live for any length of time if kept submerged and prevented from coming to the surface of the water to gulp air. Such fishes normally inhabit shallow waters or stay rather close to the surface. Their swim bladders are usually lined with vascular folds providing an excellent respiratory surface. However, blood going to the swim bladder under normal conditions has passed through the gill capillaries, where it has been oxygenated. If the respiratory surface of the gills is inadequate, or if the oxygen content of the water is depleted and the gills cannot obtain enough oxygen, then the ability of the swim bladder to take over the respiratory function is of great adaptive value to the animal. In *Protopterus* (Fig. 8.12) gills are lacking on the borders of the first two gill slits and anterior portion of the third. The anterior aortic arches in *Protopterus* form direct channels from ventral to dorsal aortae, and the pulmonary arteries, which arise from the dorsal aorta posterior to the gill region, supply the swim bladder with blood which has been only partially oxygenated. The exact mechanism by means of which air can be taken directly into the physostomous swim bladder is not well understood, but muscles in the walls of the swim bladder in certain forms, as well as muscles at the entrance of the pneumatic duct, may possibly play a part. In teleosts with pneumatic ducts, liberation of gas by the red bodies is probably of importance. Any excess of gas in the swim bladders of physostomous fishes may be released via the pneumatic duct.

Hydrostasis   The hydrostatic function of the swim bladder has been clearly demonstrated. The organ serves to equilibrate the density of the fish with the surrounding water at any level. The animal literally "floats" in the water. Change in the volume of the contained gas is one method by which this may possibly be effected. Muscular contractions, at least in part, may be used to change the volume. Increasing or decreasing the gaseous content is the more probable explanation. In physostomous teleost fishes passage of gases

in or out of the swim bladder via the pneumatic duct is but one method of changing the volume of gas. Use of the red bodies to increase the amount of gas requires a greater length of time and is probably the method employed when the fish does not have ready access to surface air.

In physoclistous fishes any increase in gaseous content of the swim bladder must be brought about by diffusion of gases from the bloodstream via the red glands or, lacking red glands, via the blood vessels in its walls. Liberation of gas usually occurs at the anterior end of the swim bladder. Absorption of gas and decrease in volume seem to be the functions of a more diffuse rete mirabile at the posterior end. In species having two chambers in the swim bladder, a sphincter muscle separating them regulates the passage of gas from one chamber to another. In certain forms the posterior chamber is reduced to a small vascular pocket, the *oval,* where most of the absorption occurs.

Sound Production   The swim bladder plays an important part in sound production in many fishes. This is confined mostly to the teleosts although *Polypterus, Calamoichthys,* and dipnoans are capable of sound production. Sounds may originate by expulsion of air from the swim bladder via the pneumatic duct or by passage of gas from one chamber to another over the edges of loose vibrating septa.

Sound Reception   The swim bladder plays an important role in the amplification of sound. The gas in the swim bladder is a bubble and is affected by changes in pressure surrounding the fish. In some fishes a connection between the swim bladder and ear, a series of bones derived from the anterior vertebrae called the *Weberian apparatus* (Fig. 8.15), transmits the movement of the swim

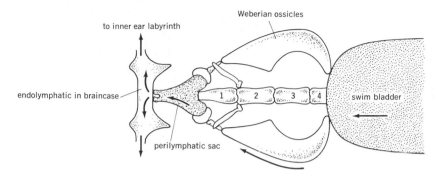

to inner ear labyrinth

Weberian ossicles

endolymphatic in braincase

1   2   3   4

swim bladder

perilymphatic sac

FIG. 8.15 Dorsal view of the Weberian ossicles and their connection between the swim bladder and inner ear. Arrow indicates the direction of transmission. Arabic numbers indicate the vertebrae from which Weberian ossicles are derived. (*After Chvanilov; adapted from Romer.*)

bladder due to pressure changes, to the ear. This movement results in changes in the fluid of the inner ear (see page 462) providing the fish with sensitivity to sound.

LUNGS AND AIR DUCTS

The diverticulum which in the embryo gives rise to lungs grows out from the floor of the pharynx posterior to the last pair of gill pouches. It soon divides into *lung buds,* which give rise to the bronchi and lungs. The lung buds grow posteriorly, invested by an envelope of mesoderm, until they reach their final destination. They may branch to varying degrees, depending on the species. The original unpaired duct which connects the lungs to the pharynx becomes the *trachea,* which serves to carry air to and from the lungs. In most anuran amphibians the duct is so short as to be practically nonexistent. In many tetrapods its anterior end becomes modified to form a voice box, or *larynx,* which opens into the pharynx through a slitlike *glottis,* the walls of which are supported by cartilages. The lower end of the trachea usually divides into two bronchi which lead to the lungs.

LARYNX   The skeletal elements supporting the walls of the larynx are derivatives of certain visceral arches which have become modified. No such structures are to be found in association with the pneumatic ducts of fishes.

Amphibians   Certain urodele amphibians, such as *Necturus,* exhibit the simplest condition. A pair of *lateral cartilages* bounds the glottis. These are derived from the last pair of visceral arches. In other amphibians the lateral cartilages are supplanted by an upper pair of *arytenoid cartilages,* bounding the glottis, and a lower pair of *cricoid cartilages* (Fig. 8.16). The latter may fuse to form a cartilaginous ring located within the walls of the larynx. The muscles which move the arytenoid cartilages are also derived from the gill region. In anurans two thickened ridges of tissue, composed of elastic fibers for the most part, project into the cavity of the larynx. These are the *vocal cords* (Fig. 8.17). They are arranged to parallel the glottis. The lower rim of each vocal cord can vibrate as air passes over it, thus causing the production of sound. Alterations of pitch are influenced by tightening or relaxing the vocal cords to different degrees.

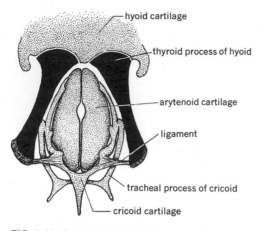

FIG. 8.16 Dorsal view of laryngeal cartilages of frog *Rana esculenta.* Black represents bone; stippled areas, cartilage. (*After Gaupp.*)

Although a few urodeles can produce sounds, they seldom amount to much more than a slight squeak or hiss. In anurans sound production is often well developed, particularly in males. The larynx is larger in males and they also may have *vocal sacs* (Fig. 8.18) which, when inflated, act as resonators. These are ventrolateral diverticula of the mouth cavity.

FIG. 8.17 View of interior of left half of larynx of bullfrog, *Rana catesbeiana,* showing position of vocal cord, attached at its ends and along its length to the wall of the larynx.

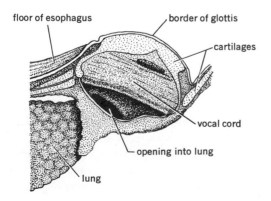

During sound production, air is passed back and forth between the inflated vocal sacs and lungs, the nostrils generally remaining closed. Sounds, therefore, may be produced when the animal is under water as well as on land. Sounds produced by female anurans are of much less magnitude than those of males.

Reptiles The reptilian larynx is generally no better developed than that of amphibians. The skeletal support consists of a pair of arytenoid cartilages and an incomplete cricoid ring. In crocodilians another element, the *thyroid cartilage,* is present. Otherwise a thyroid cartilage appears only in mammals. A fold of tissue anterior to the glottis in certain lizards and turtles may represent the beginning of an *epiglottis.*

Most reptiles lack the ability to produce sounds other than a sort of hissing noise. Some lizards possess vocal cords and make guttural noises. The loud bellowing sounds of male alligators in the mating season can be heard for some distance.

Birds The larynx is poorly developed in birds and incapable of producing sounds. Another structure, the *syrinx,* located at the lower end of the trachea is responsible for sound production. Arytenoid cartilages, sometimes ossified, guard the glottis. In some birds the cricoids have separated into additional elements, the *procricoids.*

Mammals The mammalian larynx is more complicated than that of other goups of tetrapods. Both extrinsic and intrinsic muscles are present. A thyroid cartilage, derived from the fourth and fifth visceral arches, is located on the ventral side of the larynx. It articulates with the hyoid apparatus at its anterodorsal angles. The thyroid cartilage usually consists of two parts which, except in monotremes, are fused along the midventral line. In man this portion forms a prominent protuberance

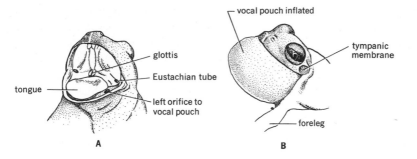

FIG. 8.18 Vocal pouch of American spadefoot toad, *Scaphiopus holbrookii. A*, showing location of openings to vocal pouch; *B*, the vocal pouch inflated. (*From Noble, "Biology of the Amphibia," McGraw-Hill Book Company, by permission.*)

called the *Adam's apple.* The arytenoid cartilages, covered with mucous membrane, form the lateral boundaries of the posterior portion of the larynx; the cricoid makes up the posterodorsal portion. Its middorsal part is sometimes separated from the rest and forms the *procricoid.* Arytenoid and thyroid cartilages articulate with the cricoid. Small *corniculate* and *cuneiform* cartilages are sometimes present in close association with the arytenoids.

An *epiglottis* is characteristically present in mammals. It is composed of elastic cartilage and extends upward from the anteromedial portion of the thyroid cartilage, generally in front of the glottis. During the act of swallowing, the epiglottis stands erect, and the larynx is raised in such a manner that the glottis is sheltered by the epiglottis and root of the tongue. Elevation of the larynx and closure of the glottis itself during the act of swallowing are more important than the presence of the epiglottis in preventing food from entering the respiratory passages. The homologies of the epiglottis are not clear.

The larynx extends from the root of the tongue to the posterior end of the cricoid cartilage, where it is continuous with the trachea. It is divided into two regions by the *true vocal cords.* Each of these is a bandlike fold of yellow elastic tissue stretched between the anterior portion of the thyroid cartilage and the posterior part of the arytenoid. The vocal cords are covered with mucous membrane with a stratified squamous epithelium. The glottis lies between the vocal cords and is bounded by the arytenoid cartilages posteriorly. Sound is produced by vibration of the vocal cords. The degree to which the cords are contracted determines pitch. The portion of the larynx in front of the true vocal cords is called the *vestibule.* A pair of crescentic *false vocal cords* (*ventricular folds*) extends into the vestibule for a short distance. The portion of the vestibule between the true and false vocal cords is called the *ventricle* of the larynx. In the howling monkeys a large diverticulum of the ventricle forms a resonating chamber supported by a greatly enlarged basihyal cartilage.

In man at the age of puberty the larynx of the male becomes much larger than that of the female. The enlarged vocal cords are responsible for the lower pitch of the masculine voice. Growth of the male larynx is controlled by the hormone testosterone secreted by the testes.

The larynx also serves to prevent anything

but air from entering the lungs. A laryngeal spasm, in the form of a cough, results if any irritating substance enters the larynx.

TRACHEA AND BRONCHI The trachea is the main trunk of a system of tubes through which air passes between larynx and lungs. It is a cartilaginous and membranous tube the walls of which are composed of fibrous and muscular tissue stiffened by cartilages which prevent it from collapsing. The trachea is lined with mucous membrane the epithelium of which is composed of columnar-ciliated and mucus-secreting cells.

Amphibians In most anurans the trachea is so short that it can scarcely be said to exist.

In urodeles it is a short but definite structure with small, irregular cartilages in its walls. In caecilians the cartilages are in the form of half rings. The trachea, in any case, divides at its lower end into two bronchi which lead directly to the lungs.

Reptiles The extent of elongation of the trachea in reptiles and other amniotes varies with the length of the neck. It is shortest in lizards and longest in crocodilians and turtles. In some turtles it is even convoluted. Tracheal cartilages are better developed and more complete than in amphibians. Some anterior cartilages may be in the form of complete rings, but the remainder are imperfect bands, deficient on the dorsal side. Sup-

FIG. 8.19 Trachea of jackass penguin, dissected to show duplication which begins about 1 cm past the larynx. (*From a photograph by Dr. Pearl M. Zeek.*)

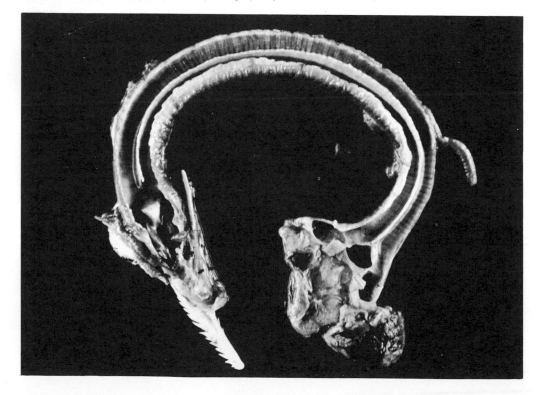

porting cartilages are also present in the walls of the bronchi. In some snakes only one bronchus and lung (the right) are present.

Birds The trachea in birds is often of unusual length and may even exceed that of the neck. In such cases it is convoluted. In swans and cranes a loop lies within a cavity in the modified keel of the sternum, which serves as a resonator. An unusual condition is encountered in the penguins, in which the trachea is double (Fig. 8.19). The duplication begins about 1 cm from the larynx. Tracheal cartilages in birds are usually in the form of complete rings. The bronchi are also supported by cartilages.

The sound-producing organ of birds, the *syrinx,* is located at the lower end of the trachea where division into bronchi occurs. There is much variation in structure, but three general arrangements are recognized: *tracheal, bronchotracheal,* and *bronchial.* They are composed of modified cartilages, vibratile membranes, and complicated sets of muscles.

Mammals In mammals, the length of the trachea varies with the length of the neck. In adult man it is about 4 to 5 in. long. Tracheal cartilages in mammals are usually incomplete on the dorsal side, the esophagus pressing against the gap thus formed.

In most mammals the trachea divides into two main bronchi, but in pigs, whales, and some ruminants a third, the *apical,* or *eparterial, bronchus* arises independently from the right side of the trachea anterior to the main bifurcation. In some others the apical bronchus is a main branch of the right bronchus. Cartilaginous rings support the walls of the bronchi in mammals.

LUNGS The lungs of terrestrial vertebrates are saclike structures, the walls of which in most cases are subdivided. The divisions become more numerous and complex as the vertebrate scale is ascended, reaching the highest degree of branching in mammals. The actual respiratory surface of the lung (except in birds) is composed of minute chambers called *alveoli*. The greater the number of subdivisions, the greater the total respiratory surface.

Amphibians The lungs of amphibians are relatively simple in structure. In most anurans they connect directly with the larynx. In urodeles the lungs consist of a pair of elongated sacs, the left often being longer than the right. In some the lining is smooth, but in others alveoli are present. Often the alveoli are confined to the basal portion. The left lung in caecilians is very short, and alveoli line the entire surface of the right lung. In frogs and toads the lining is more complex since the wall is thrown into numerous *infundibular folds,* which in turn are lined with alveoli.

Amphibians usually develop lungs at the time of metamorphosis, but in some frogs, toads, and salamanders, they appear during larval life. Plethodontid salamanders, in which they never appear, depend upon cutaneous and buccopharyngeal respiration during adult life. Such neotenic salamanders as *Necturus* have lungs in addition to gills. It is doubtful whether they are used very effectively under normal conditions since they are supplied with blood which has already been oxygenated.

Surgical removal of lungs in newts indicates that they are of little or no importance in respiration.

Reptiles The lungs of reptiles exhibit a wide range of complexity, being generally more efficient than those of amphibians but less efficient than those of mammals. In snakes and lizards there is a lack of symmetry between the two sides. In snakes the left lung is smaller than the right or is absent altogether.

The simplest condition is found in

Sphenodon, in which the lungs are simple sacs uniformly lined with infundibula. In snakes the infundibula are usually confined to the basal portion. In lizards, septa, or partitions, divide the cavity of the lungs into chambers which communicate with one another. In some of the higher lizards and in turtles and crocodilians the bronchi divide into smaller and smaller branches which finally terminate in infundibula lined with alveoli. The lungs are of a spongy consistency. Crocodilian lungs most nearly approach the condition found in mammals. In chameleons several thin-walled, saclike diverticula come off the distal portion of the lung. Only the proximal end is spongy. The diverticula seem to foreshadow the structure of air sacs found in birds.

Birds The lungs of birds differ from those of other groups and are probably the most efficient of all. They are small, very vascular, and capable of little expansion, being firmly attached to the ribs and thoracic vertebrae. The lower surface of each lung is covered by a membrane into which are inserted muscles arising from the ribs.

The main bronchus on each side enters the lung on its medial ventral surface and passes to the distal end. The portion within the lung proper is called the *mesobronchus.* A number of *secondary bronchi* arise from it. These then branch into numerous small tubes of rather uniform diameter, called *parabronchi,* which form loops connecting with other secondary bronchi (Fig. 8.20). Surrounding each parabronchus are large numbers of minute tubules, the *air capillaries,* which form anastomosing networks and loops with each other and open into the parabronchi. The lining of the air capillaries makes up the actual respiratory surface. This is a system of intercommunicating tubes rather than a branching system, as in mammals and some others. The mesobronchus and several (usually four) secondary bronchi continue through the walls of the lung and expand into large *air sacs,* which ramify among the viscera and even enter the cavities of several of the bones. The air sacs do not furnish a respiratory surface, since their walls are smooth and have a poor blood supply. Moreover, they are furnished with oxygenated blood. In addition to the one large

FIG. 8.20 Diagrammatic medial view of left lung of chicken, illustrating relations of bronchus, mesobronchus, secondary bronchi, and parabronchi. (*After Locy and Larsell.*)

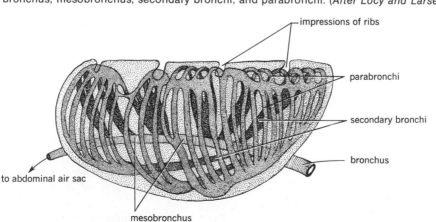

impressions of ribs

parabronchi

secondary bronchi

bronchus

to abdominal air sac

mesobronchus

tube which enters each air sac there are several small tubes, called *recurrent bronchi,* which connect the sac with the adjacent portion of the lung.

Respiration involves the circulation of air under pressure through this complicated intercommunicating system of tubes and sacs. There are no blind alveolar pockets. When the bird is at rest, the intercostal muscles, by raising and lowering the ribs and sternum, provide the means for circulating the air. During flight, when the ribs are kept rigid, movement of air in the air sacs is effected by pressure from surrounding viscera and by movement of the flight muscles and hollow wing bones which contain diverticula of the air sacs. Air is forced back and forth through the respiratory channels. The efficiency of this type of respiration should be obvious.

Air sacs may also play a significant role in temperature regulation by serving as an internal cooling device.

Mammals The original lung buds of mammals divide many times to form primary, secondary, tertiary, etc., bronchi. The smaller bronchi give rise to *bronchioles* of several orders of magnitude. The terminal bronchioles usually divide into two or more *respiratory bronchioles,* which then branch into several *alveolar ducts.* From these arise *alveolar sacs* (infundibula) lined with *alveoli* (Fig. 8.21). Respiratory bronchioles may have a few alveoli projecting from their walls. A single bronchiole and its branches together form a unit called a *lobule.* In some species lobules are separated by connective-tissue septa. In others, such septa are lacking.

The entire respiratory surface is endodermal, being derived from the pharyngeal epithelium. The remainder of the lung is mesodermal and is composed of reticular and collagenous connective tissues, together with a large number of elastic fibers and blood

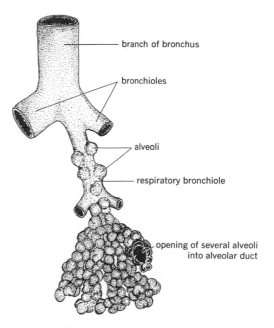

branch of bronchus

bronchioles

alveoli

respiratory bronchiole

opening of several alveoli into alveolar duct

FIG. 8.21 Diagram showing terminal branching of a small bronchus in a typical mammalian lung.

vessels. The epithelium lining the greater part of the "respiratory tree" is of the ciliated pseudo stratified columnar type, but the cilia end a short distance down the respiratory bronchioles. The cilia, which beat in the direction of the trachea, remove dust and other particles from the lungs. Recent studies, using the electron microscope, have revealed details of structure of the interalveolar wall, or septum, which cannot be seen under the light microscope. Despite the thinness of the wall, there are, between the air in an alveolus and the blood in an underlying capillary, the cytoplasm of the epithelial cells lining the alveolus; the basement membrane of the capillary endothelium; and the cytoplasm of the endothelial cells of the capillary.

Cartilaginous rings are present in the main bronchi. Within the lungs they become smaller, and in their place are small, irregular

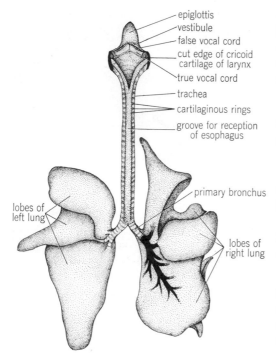

epiglottis
vestibule
false vocal cord
cut edge of cricoid cartilage of larynx
true vocal cord
trachea
cartilaginous rings
groove for reception of esophagus
primary bronchus
lobes of left lung
lobes of right lung

FIG. 8.22 Dorsal view of respiratory organs of cat. The larynx has been split open to show the vocal cords.

cartilages. When the bronchioles are no more than 1 mm in diameter, the cartilaginous supports disappear.

The entire surface of the lung is covered with serous membrane, the *visceral pleura*. The lungs of mammals lie entirely within a pair of airtight pleural cavities which are separated from each other and from the rest of the coelom.

In most mammals the lungs are subdivided externally into lobes, the number varying somewhat in different species. The number of lobes on the right side generally exceeds that on the left. Thus in man there are three lobes on the right and two on the left. The cat has four on the right and three on the left (Fig. 8.22). The left lung of the rat is not lobulated,

even though the right side has three large lobes and one small one. Whales, sirenians, elephants, *Hyrax,* and many members of the order Perissodactyla have lungs showing little, if any, indication of lobulation.

MECHANISM OF LUNG RESPIRATION

The exact mechanism by which air is taken in and forced out of the physostomous swim bladder of the lungfishes is not clear (page 223). In higher forms inhalation and exhalation are accomplished in various ways, many of which are fairly well understood.

AMPHIBIANS The lungs of amphibians lie in that portion of the coelom referred to as the *pleuroperitoneal cavity.* A partition, the septum transversum, has developed posterior to the heart, cutting off a part of the coelom, the *pericardial cavity,* in which the heart lies, from the rest of the body cavity. The lungs of amphibians are located in the anterior lateral portions of the pleuroperitoneal cavity.

In several amphibians the lungs develop before metamorphosis while the larvae are still living in an aquatic environment. Such larvae frequently come to the surface and seem to snap for air. The air bubble thus obtained can be forced through the glottis and into the lungs by elevating the floor of the mouth. This method of obtaining air is sometimes employed by the adult *Necturus,* in which the rather poorly developed lungs serve as an accessory respiratory organ. In such cases expiration is brought about by the pressure exerted when the elastic walls of the lungs return to their original position as the glottis is opened.

When breathing air, both urodeles and anurans keep their mouths tightly closed. Air is sucked into the mouth and forced out again

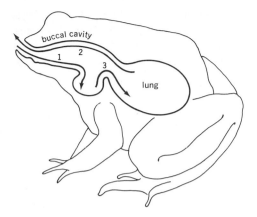

FIG. 8.23 Diagrammatic section through a frog illustrating the path of air through the buccal cavity into and out of the lungs.

through the nares by alternating lowering and raising the muscular floor of the mouth. Smooth-muscle fibers around the external nares of urodeles help to control the respiratory current. Anurans, as mentioned previously, utilize the tuberculum prelinguale in moving the premaxillary bones and thus regulate the opening and closing of the nasal passages. Filling the lungs with air is accomplished by lowering the throat muscles and increasing the volume of the buccal cavity, allowing air to enter through the open nostrils. The glottis is opened, and air from the elastically contracting lungs passes out through the external nares. The nares are closed, the muscles in the floor of the buccal cavity contract and force the fresh air through the glottis into the lungs (Fig. 8.23). The process is then repeated. Sometimes when frogs are making strong respiratory movements, their eyes sink down into the orbit for a moment. Since mouth and orbit are separated only by a thin layer composed of connective tissue and mucous membrane, retraction of the eyes helps compress the air into the lungs. Repetition of these processes en-

sures a slow but adequate supply of oxygen to the lungs. The elasticity of the lungs themselves, possibly aided by pressure exerted upon them by contraction of the muscles of the body wall, is responsible for the escape of air from the lungs into the buccopharyngeal cavity. Even though many amphibians employ these mechanisms in respiration, cutaneous respiration, in most forms, continues to be of prime importance.

REPTILES As in amphibians, the lungs of reptiles lie in that part of the coelom referred to as the pleuroperitoneal cavity, specifically in the anterolateral portions. In many reptiles, respiration is like that of amphibians, utilizing throat movements and opening and closing of valves in the nostrils. However, the presence of ribs and rib muscles helps reptiles carry on respiration more effectively than is possible in the ribless amphibians. Raising the ribs and increasing the size of the pleuroperitoneal cavity reduces the pressure within the cavity. This, in turn, brings about inspiration and expansion of the lungs. Conversely, lowering the ribs results in an increased pressure upon the lungs, and expiration is the result. In certain reptiles, e.g., crocodilians, the pleuroperitoneal cavity is divided by coelomic folds into anterior and posterior portions. The lungs are shut off from the remainder of the coelom by these folds, which are at least functionally comparable to the diaphragm in mammals.

The presence in turtles of a hard, rigid shell, formed in part by fusion of the ribs and vertebrae with the dermal plates of the carapace, makes general expansion and contraction of the body cavity by the usual method impossible. Turtle respiration, therefore, presents some unusual physiological problems. It was formerly thought that these animals employed throat movements, similar to those of amphibians, in respiration, but it has been demonstrated that this is not the case and that turtles utilize a unique method of breathing. Expiration is the result of contraction of paired muscular membranes which enclose the viscera. Inspiration is brought about by the contraction of other paired muscular membranes enclosing the flank cavities. This brings about an increase in volume of the pleuroperitoneal cavity. The glottis remains closed except when movements of expiration and inspiration are in progress. When the glottis is closed, pressure on the air within the lungs, brought about by various degrees of contraction of the paired muscular membranes, varies independently of the atmospheric pressure. Alternation in pressure brings about changes in the diffusion gradient across the lung epithelium and results in an exchange of gases from lung cavities to blood, or vice versa.

BIRDS A membrane on each side, known as the *oblique septum* and sparsely supplied with muscle fibers, extends from the dorsal body wall to the septum transversum. This separates the lungs from the other viscera so that they are enclosed in separate

pleural cavities. The remainder of the coelom is the *peritoneal,* or *abdominal cavity.* The lungs are small and closely adherent to the thoracic vertebrae and ribs. *Costopulmonary muscles,* inserted in the membrane covering the lower surface of each lung, bring about expansion and contraction of these respiratory organs under normal conditions when the bird is not in flight. The respiratory cycle is in two phases (Fig. 8.24). During inspiration air passes through the trachea into the posterior air sacs. At the same time air moves from the lungs into the anterior air sacs. During expiration air flows from the posterior air sacs into the lung, and at the same time air from the anterior air sacs passes out of the body through the trachea. This system provides a continuous flow of air in one direction, through the lungs. During flight, however, the body must be kept as rigid as possible so that the well-developed flight muscles are firmly anchored. At such times contraction and expansion of the air sacs, brought about by movements of viscera, muscles, and bones, cause air to circulate back and forth through the intercommunicating parabronchi and secondary bronchi. It has been shown under experi-

FIG. 8.24 Diagrammatic section through bird showing relationship of lung, air sacs and tracheal-bronchial system.

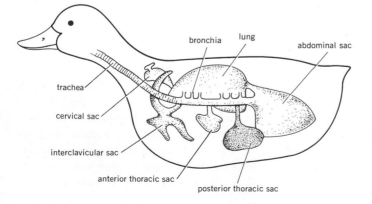

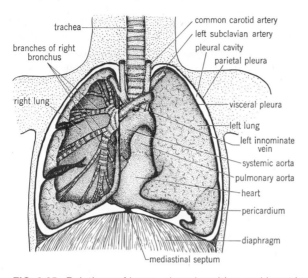

trachea

branches of right bronchus

right lung

common carotid artery

left subclavian artery

pleural cavity

parietal pleura

visceral pleura

left lung

left innominate vein

systemic aorta

pulmonary aorta

heart

pericardium

diaphragm

mediastinal septum

FIG. 8.25 Relations of lungs, pleural cavities, and heart in man, ventral view (semidiagrammatic). The pleural cavities have been exaggerated for clarity, since in life the visceral and parietal pleurae are in apposition.

mental conditions that if the trachea of a bird is occluded and an air sac, such as that in the humerus of the wing, is opened, the bird can continue to breathe without difficulty.

MAMMALS A musculotendinous diaphragm, present only in mammals, separates the pleuroperitoneal cavity into two parts. That portion anterior to the diaphragm contains the lungs. Each lung is enclosed in a separate *pleural cavity,* which is lined with a fibroelastic membrane called the *pleura.* The part of the pleura lining the wall of the pleural cavity is referred to as the *parietal pleura.* The lung itself is invested by a layer of *visceral pleura* reflected over its surface (Fig. 8.25). The outermost layer of each pleura, or that facing the pleural cavity, consists of squamous mesothelial cells. Visceral and parietal pleurae are continuous with each other at the root of the lung, or *hilus,* where the bronchi and blood vessels enter. Toward the median line the parietal pleurae of the

two sides come close together to form a two-layered septum, the *pleural septum,* separating the two pleural cavities. The space between the two layers of the pleural septum is called the *mediastinal space.* It widens as it nears the diaphragm and contains the aorta, esophagus, and posterior vena cava. In the region of the heart the mediastinal membrane passes over the parietal layer of the pericardium. The pleural membrane is rich in capillaries and lymphatic vessels. A serous fluid, present in the pleural cavities, facilitates the movement of the lungs within the cavities by serving as a lubricant.

The pleural cavities are airtight. Air from the outside is pushed in by atmospheric pressure, and additional blood is concomitantly drawn into the blood vessels of the lung and those of the thorax proper. Increasing their size results in an expansion of the lungs. Reduction in the size of the pleural cavities increases the pressure within the lung, forcing the air out. The mechanism of respiration in mammals therefore consists of alternately increasing and decreasing the size of the pleural cavities. This is generally accomplished in two ways. Contraction of the muscles of the arched diaphragm flattens or lowers that structure. This increases the size of the pleural cavities, causing inspiration, and brings about an increased pressure on the abdominal viscera. Contraction of the abdominal muscles causes the abdominal viscera to push against the now relaxed muscular diaphragm, reducing the size of the pleural cavities and resulting in expiration of air from the lungs. This method of breathing is known as *abdominal respiration*. The other method of expanding the pleural cavities, known as *thoracic respiration,* is brought about by contraction of the external intercostal muscles, which pass from the lower border of one rib to the upper border of the rib below. Contraction brings about an elevation of the ribs. How expiration is effected in this type of breathing is not so well understood. In some mammals, e.g., the cat, contraction of the internal intercostal muscles reduced the size of the thoracic cavity. The elasticity of the expanded lung walls, together with the tension of the stretched body wall and ribs, undoubtedly accounts for the greater part of expiratory movement.

MISCELLANEOUS RESPIRATORY ORGANS

Although gills are the chief respiratory organs of aquatic vertebrates and lungs serve terrestrial forms in a similar capacity, there are other structures in certain vertebrates which serve as accessory respiratory organs. The yolk sac and the allantois are of great importance in respiration during embryonic life but are of no significance in adult life.

SKIN In a few fishes the skin may function in respiration. The walking goby, for ex-

ample, may emerge from the water with its head and trunk exposed to the air, but its vascular caudal fin remains submerged and undoubtedly functions as a respiratory organ.

A highly vascular integument is very important in amphibian respiration. The skin must be kept moist in order to function in this respect. In some species of frogs over 70 percent of the carbon dioxide leaves the body through the skin. Filamentous integumentary outgrowths and barbels on the head are sometimes employed in increasing the respiratory surface.

BUCCOPHARYNGEAL EPITHELIUM The buccopharyngeal epithelium is extensively vascularized in many amphibians and some fishes. Passage of air or water over this surface acts as another respiratory mechanism.

RECTUM AND CLOACA A very vascular rectum serves some fishes as an accessory organ of respiration, water being alternately taken in and squirted out of the rectal chamber. Some aquatic turtles have a pair of thin-walled, saclike extensions of the cloaca, which are sometimes referred to as accessory urinary bladders. These are abundantly supplied with blood vessels and are frequently filled and emptied of water via the anus. They may thus serve as accessory respiratory organs, particularly in emergencies when the animal remains submerged for an undue length of time.

OTHER DERIVATIVES OF THE PHARYNX

A number of structures derived from the pharynx but not associated with respiratory activity may conveniently be discussed at this point.

Except for the thyroid gland, the structures

derived from the pharynx arise as modifications of pharyngeal pouches. During development small dorsal and ventral recesses appear at the outer ends of the pouches. Cellular proliferation about these recesses results in the formation of small masses of endodermal origin which represent the primordia of certain glands confined to the neck region. The thymus, tonsils (palatine), parathyroids, and ultimobranchial bodies are derived from such primordia. The history of these pharyngeal derivatives varies somewhat among the vertebrate classes, and there is much confusion, particularly in lower forms, over what the various structures in the neck region actually represent. They are sometimes referred to collectively as *epithelial bodies.*

THYROID GLAND   In lower vertebrates the thyroid gland arises as a midventral diverticulum of the pharynx in the region between the second and fourth visceral pouches. In higher forms it usually develops between the first and second pouches. In some it becomes paired, but in others it remains as a single organ. It has been reported that small paired primordia from the fourth pair of visceral pouches may be added to the median thyroid diverticulum. Since the thyroid gland is an endocrine organ, it is discussed more fully in Chap. 15, Endocrine Organs.

THYMUS GLAND   The thymus was formerly believed to be an endocrine gland, but this is now considered very doubtful. Evidence indicates that it belongs to the lymphatic system rather than to the endocrine system and may ultimately be responsible for some, if not all, of the functions of the body having to do with immunological reactions. In lower vertebrates it is difficult to determine whether certain masses ordinarily referred to as thymus tissue are actually homologs of the thymus gland as we recognize it in higher forms.

In lampreys small primordia arising from the dorsal angles of all seven gill pouches become separated from the pharynx and persist throughout life. If these actually represent thymus tissue, the lampreys exhibit the most generalized condition of all vertebrates in respect to thymus development. Some lobular structures lying posterior to the gill region in the hagfish may be thymus tissue. In fishes there are fewer primordia. Teleosts usually show a fusion of the components on each side into a single lobulated mass. In some forms connections with the pharyngeal epithelium are retained. Except for caecilians, amphibia exhibit a reduction in thymus primordia. In urodeles the gland consists of a lobulated mass of tissue on each side of the neck. In the adult frog it lies behind the tympanic membrane and under the depressor mandibulae muscle. It is lymphoid in character and reported to be smaller in the adult than in the tadpole. There is some variation in origin of the thymus in reptiles and birds. In some reptiles there may be separate elements, but there is a tendency toward fusion. Elongated lobular strands usually lie along the sides of the neck.

Mammals are exceptional in that the thymus arises from the *ventral* portions of pharyngeal pouches rather than the dorsal. Only pouches 3 and 4 are involved. The components are bound together in what appears to be a single mass. It lies along the ventral aspect of the trachea and extends back to the base of the heart. The thymus of some species is of value as food and is sold under the name of neck, or throat, sweetbreads.

Because the thymus gland is considered to belong to the lymphatic portion of the circulatory system, further discussion of its structure and function is included in Chap. 12, Circulatory System (see page 326).

PARATHYROID GLANDS Definite parathyroid glands seem to be lacking in cyclostomes and fishes. They first appear in amphibians. Parathyroids arise from the ventral angles of certain visceral pouches in amphibians, reptiles, and birds. In mammals they come from the *dorsal* angles. The parathyroids are endocrine glands related in function to the calcium and phosphorus metabolism of the body. They are discussed further in Chap. 14, Endocrine Organs.

TONSILS *Palatine tonsils,* found only in mammals, are lymphoid masses derived from the ends of the second pair of visceral pouches. They are located on either side, where mouth and pharynx join. The pouch itself forms the fossa and provides the covering epithelium of the tonsil. *Lingual tonsils,* also present only in mammals, are situated at the base of the tongue. They are lymphoid masses which develop in relation to lingual glands. In reptiles and birds, and occasionally in mammals, *pharyngeal tonsils* are present. These appear in the mucous membrane on the roof of the pharynx and come to lie behind the choanae. They may be homologous with certain lymphoid masses in the roof of the amphibian pharynx. In man, enlarged pharyngeal tonsils are called *adenoids.* Whether tonsils have any function aside from being part of the lymphatic system is doubtful. Palatine tonsils are frequently the site of infection and often must be removed.

MIDDLE EAR AND EUSTACHIAN TUBE The first, or hyomandibular pharyngeal pouch in elasmobranchs, *Polypterus, Polyodon,* and sturgeons, forms the spiracle, discussed elsewhere. In most fishes, however, and in amphibians, reptiles, birds, and mammals, the first pouch fails to break through to the outside or becomes closed secondarily. In anurans, for the first time, the tissue between the pouch and the exterior forms the *tympanic membrane,* or eardrum. The pouch itself becomes the *middle ear.* The narrowed portion between the middle ear and the pharynx proper is the *Eustachian tube.* Middle ear and Eustachian tube function primarily as a pressure-regulating mechanism which keeps the pressure on the two sides of the tympanic membrane equal. In caecilians, urodele amphibians, some anurans, snakes, and a few other reptiles, middle ears and Eustachian tubes have degenerated and are lacking. Detailed description of these structures is reserved for Chap. 15, Sense Organs.

OTHER STRUCTURES Other derivatives of the pharynx include some small, glandular-appearing structures known as postbranchial or ultimobranchial bodies. They arise from, or close behind, the last pair of pharyngeal pouches and come to lie in the vicinity of the thyroid gland. They are frequently confused with the parathyroid glands and may possibly represent accessory parathyroid tissue. Their actual function is unknown. In some forms, e.g., as in the golden hamster, the ultimobranchial bodies contribute to thyroid-gland formation.

The so-called *carotid body,* closely associated with the common carotid artery where it bifurcates into internal and external branches, is conspicuous in frogs and toads and is also found in higher vertebrates. It is composed of masses of epitheliallike cells richly supplied with nerve endings and capillaries of a sinusoidal nature. The carotid body is probably related to the autonomic nervous system and may play a role in controlling circulation in the carotid vessels. Although it has been reported that the carotid bodies of amphibians and lizards are derived from pharyngeal pouches, their origin in birds and mammals is questionable. It is doubtful whether the carotid bodies actually contain any epithelial elements. If not, the pharynx may not in any way be concerned with their formation. Confusion exists because various investigators have assigned the name carotid body to different structures in the neck region which vary from species to species.

The epitheliallike cells of the carotid body, with their abundant nerve supply, seem to respond to variations in the oxygen and carbon dioxide content of the blood. Chemoreceptors in the carotid body send impulses via a branch of the glossopharyngeal nerve to brain centers controlling the heart, arterial vessels, and respiration.

SUMMARY

1   With few exceptions the organs of respiration in vertebrates are formed in connection with that part of the digestive tract known as the pharynx. They are usually covered or lined with epithelia derived from endoderm.

2   In all chordate embryos a series of visceral pouches develops on either side of the pharynx. The number of visceral pouches is greatest in the lowest groups of vertebrates, and least in the higher classes. The pouches push through the mesenchyme to fuse with the outer ectoderm, forming thin, platelike areas. In fishes and larval amphibians perforations occur, forming gill slits which connect the pharynx to the outside. The pouches are then called gill clefts. They are separated from each other by visceral arches composed of skeletal material.

Internal gills supported by visceral arches are vascular, lamellar, or even filamentous extensions of the epithelial surface of the gill pouches. External gills, generally covered with ectoderm, may project from the outer surfaces of the visceral arches as filamentous outgrowths.

3   In lampreys and elasmobranchs each gill cleft opens separately to the outside. In higher fishes an operculum, which is an extension of the hyoid arch, covers the gill chamber. Although most fishes possess internal gills, a few have external gills in addition, at least during larval life.

4   Only external gills occur in amphibians. With few exceptions they disappear upon metamorphosis and the gill slits close over. In reptiles, birds, and mammals the pharyngeal pouches fail to break through to the outside and gills are lacking.

5   In anurans, most reptiles, birds, and mammals the first gill pouch gives rise to the Eustachian tube and middle ear. The original lining of the remaining pouches contributes to such structures as tonsils, parathyroid and thymus glands, and ultimobranchial bodies. The thyroid gland arises as a midventral outgrowth of the pharynx.

6   Lungs develop as a bilobed diverticulum of the floor of the pharynx posterior to the last gill pouch. In higher forms the connection between lungs and pharynx lengthens and becomes the trachea. Lungs show unmistakable homologies to the swim bladders of fishes, which in certain cases are used as lungs but in others have become specialized as hydrostatic organs. Many modifications of the swim bladder occur.

7   The upper part of the trachea becomes modified as a larynx, or voice box, the walls of which are supported by skeletal elements derived from the visceral arches. In birds, a syrinx, located at the lower end of the trachea, is the organ of sound production.

8   The lungs of lower forms are rather simple, vascular sacs, but in higher vertebrates the walls become subdivided into numerous pocketlike air spaces. The divisions become more and more complex as the vertebrate scale is ascended and reach the highest degree of branching in mammals. The lungs of birds are complicated in that they give rise to air sacs which penetrate among the viscera and even enter the hollow bones. Similar structures in certain lizards foreshadow their appearance in birds. The thin-walled, moist, highly vascular lining of the lungs forms an ideal respiratory surface.

9   In mammals the lungs lie in separate pleural cavities partitioned from the abdominal cavity by a muscular diaphragm. In amphibians and reptiles they lie in the anterior part of the coelom, which is known as the pleuroperitoneal cavity. The lungs of birds lie in separate pleural cavities, even though no diaphragm is present.

10   Nasal passages develop in connection with the olfactory apparatus. Blind nasal pits are found in most fishes, but in the Sarcopterygii they form connections between the oral cavity and the outside. This condition is retained in amphibians which, for the first time, employ the nasal passages for intake and outgo of air. In reptiles a secondary palate begins to form, partially separating the nasal and mouth cavities. In crocodilians and mammals the secondary palate becomes complete, so that the two passageways are in communication only in the region of the pharynx. In these forms the internal nasal openings are located relatively far back and are situated near the opening of the trachea.

BIBLIOGRAPHY

Comroe, J. H., Jr.: The Lung, *Sci. Am.,* 217(2):57–68 (1966).

Fange, R.: Physiology of the Swimbladder, *Physiol. Rev.,* 46(2):299–322 (1966).

Hoar, William S.: "General and Comparative Physiology," Prentice-Hall, Englewood Cliffs, N.J., 1966.

Hughes, G. M.: "Comparative Physiology of Vertebrate Respiration," Harvard University Press, Cambridge, Mass., 1963.

Johansen, K. (ed.): Cardiorespiratory Adaptations in the Transition from Water to Air Breathing, *Fed. Proc.,* 29(3):1118–1153 (1970).

Satchell, G. H.: "Circulation in Fishes," Cambridge Monographs in Experimental Biology, Cambridge University Press, London, 1971.

Schmidt-Nielsen, K.: How Birds Breathe, *Sci. Am.,* 225(6):72–79 (1971).

Steen, J. B.: "The Comparative Physiology of Respiratory Mechanisms," Academic Press, New York, 1972.

9 DIGESTIVE SYSTEM

The chief function of the digestive system is to prepare foods for use in growth, structural maintenance, and production of various forms of energy. The main principle involved in digestion is breaking down complex molecules of certain foods, by a series of chemical changes, into molecules of simpler structure. These can then either be absorbed in water-soluble form through the walls of the digestive tract and enter the bloodstream (in the case of proteins and carbohydrates) or be otherwise prepared so that they can enter the small vessels or lacteals of the lymphatic system (if the food is in the form of oil or fat). Water is absorbed unchanged through the walls of the digestive tract, but few kinds of food are available to a vertebrate in such form that they can be absorbed directly without first undergoing digestion.

Food in the digestive tract cannot properly be said to be inside the body. Not until it has been digested and enters the blood or lymphatic systems or is otherwise incorporated within the body tissue can it be spoken of as being within or part of the body. Ingested material that cannot be digested thus never really enters the body proper. It merely passes down the digestive tract and is eliminated.

GENERAL STRUCTURE

In all vertebrates the digestive, or alimentary, tract is complete, having a *mouth* at one end and an *anus* at the other. The greater part of the lining of the alimentary tract is of endodermal origin, but the regions of ingress (mouth, or stomodaeum) and egress (anal canal, or proctodaeum) are lined with epithelium of ectodermal derivation. In the fully developed digestive tract no exact spot indicates the original point of junction of ectodermal with the endoderm. In general, the original stomodaeum ends in the region just posterior to the teeth. The extent of the proctodaeum in different classes is variable and indefinite.

The original body cavity, or coelom, in vertebrates becomes partitioned off to varying degrees in members of the several classes, forming such cavities as the _pericardial cavity_, in which the heart lies, the _pleural cavities_, containing the lungs, and the _peritoneal cavity_ in which the various abdominal viscera are located. In the lower classes only pericardial and _pleuroperitoneal_ cavities are present, and in some forms even these remain connected by a _pericardioperitoneal_ canal.

Inside the large pleuroperitoneal or peritoneal cavity, as the case may be, the body wall is lined with a smooth, shiny membrane, the _parietal peritoneum_. Reflected over the surface of the alimentary tract in this region is the _visceral peritoneum_. The parietal and visceral peritonea are connected along the mid-dorsal line of the body cavity by the _dorsal mesentery_. This is actually a two-layered sheet continuous dorsally with the two halves of the parietal peritoneum. The relationships of these layers are indicated in Fig. 9.1. In reptiles and mammals the dorsal mesentery is continuous, but in amphibians and birds it becomes discontinuous. A _ventral mesentery_, connecting the gut with the ventral body wall, is usually present during developmental stages but soon disappears. In some regions a part of the alimentary tract is connected to an adjacent organ or part by the persistent remnant of the ventral mesentery, referred to as _omenta_ or _ligaments_.

The wall of most of the alimentary tract (Fig. 9.2) is made up of four main cell layers, or coats, serous, muscular, submucous, and mucous. The outer _serous layer_, or visceral peritoneum, almost entirely surrounds those parts of the tract which lie within the coelom. This layer is lacking in the region of the esophagus, which is extracoelomic. The cells of the serous layer rest on loose connective tissue. Beneath this lies the _external muscular coat_, which in turn is composed of two layers, called the outer _longitudinal_ and inner _circular layers_, respectively. They are actually arranged in the form of spirals. Between the two layers lies a nervous network, the _parasympathetic myenteric plexus_, or _plexus of Auerbach_. Next to the circular layer is the _submucosa_, composed mostly of dense connective tissue and elastic fibers. Intestinal glands are located in this layer. Blood and lymphatic vessels are present in the submucosa, in addition to a plexus of nerve cells and fibers, the submucous _plexus of Meissner_. Some ganglionic cells of the para-

FIG. 9.1 Diagram showing relationship of a mesentery and an omentum to parietal and visceral peritoneum.

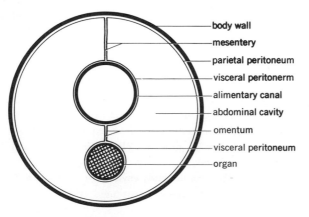

- body wall
- mesentery
- parietal peritoneum
- visceral peritonerm
- alimentary canal
- abdominal cavity
- omentum
- visceral peritoneum
- organ

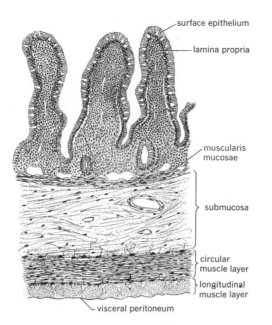

surface epithelium

lamina propria

muscularis mucosae

submucosa

circular muscle layer

longitudinal muscle layer

visceral peritoneum

FIG. 9.2 Part of transverse section of ileum, showing three villi and the layers of the intestine.

sympathetic nervous system are present, but the nerve fibers here are mostly postganglionic fibers of the *sympathetic* nervous system which originate in the superior mesenteric ganglion (page 413). The *mucosa,* or innermost layer of the alimentary tract, is composed of three regions, or layers. Next to the submucosa is the *muscularis mucosae,* a thin layer of smooth-muscle fibers and elastic connective tissue consisting of inner circular and outer longitudinal layers. Internal to the muscularis mucosae is a layer of peculiar connective tissue, the *lamina propria* of the mucosa. The lamina propria is covered with a basement membrane which supports a single layer of simple, columnar, epithelial cells, the *surface epithelium.*

The muscular layers are under the involuntary control of the autonomic nervous system, of which the <u>plexuses of Auerbach and Meiss-</u>

ner are parts. Contraction of the smooth-muscle fibers is responsible for <u>peristalsis,</u> which propels food materials along the intestine, and for segmenting and pendular movements which do not propel food but aid in kneading it and mixing it thoroughly with digestive juices and enzymes. Striated, or voluntary, muscle fibers are found only at the ends of the alimentary canal.

The digestive tract is admirably suited for its functions. Its wall contains muscles which thoroughly mix the contents and propel them along; it is provided with glands which secrete certain fluids and enzymes necessary for proper chemical reactions to take place. The thin cell walls of the lining epithelium are well suited for absorption, and the rich supply of blood and lymphatic vessels not only ensures proper nourishment of the tissue of the digestive tract itself but serves as a pathway by means of which digested substances are transported to the liver, where metabolic modification occurs, and thence to all parts of the body.

In vertebrates the alimentary canal is typically divided into several sections. These are, in order, the *mouth, pharynx, esophagus, stomach, small intestine, large intestine,* and *cloaca* (Fig. 9.3). The cloaca is a common chamber into which the intestine, urinary ducts, and reproductive canals discharge. A cloaca is present in the adult stage of birds, reptiles, amphibians, and many fishes, being notably absent in teleosts, chimaeras, *Polypterus,* sturgeons, and others. Among adult mammals a cloaca is lacking except in monotremes and in the pika. In higher mammals, however, a cloaca is present for a time during embryonic development. A derivative of the embryonic cloaca is the rectum, which leads to the anus. Such digestive glands as the *salivary glands, liver, gallbladder,* and *pancreas* are derived from the alimentary canal (endoderm), with which they are connected by means of ducts.

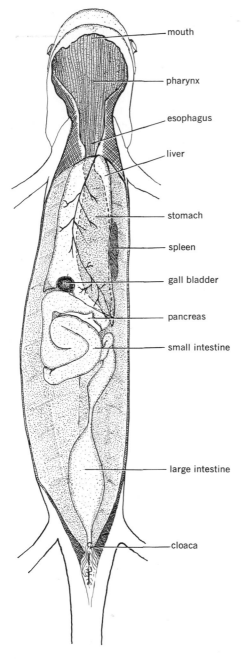

mouth

pharynx

esophagus

liver

stomach

spleen

gall bladder

pancreas

small intestine

large intestine

cloaca

FIG. 9.3 Digestive organs of salamander *Eurycea bislineata*. The other organs have been removed.

MOUTH AND STRUCTURES ASSOCIATED WITH IT

It will be recalled that the primitive mouth, or stomodaeum, which develops early during embryonic life, is lined with ectoderm. It is important to remember, therefore, that the various structures derived from the stomodaeum are, for the most part, ectodermal in origin or are at least covered with material derived from ectoderm.

AMPHIOXUS At the anterior end of amphioxus is a funnel-shaped structure called the *oral hood*. It is bordered by a number of tentaclelike *cirri*. The cavity within the oral hood is the *vestibule*. It is *not* homologous with the mouth cavity of higher forms. The true mouth opening is located at the apex of the vestibule. On the walls of the oral hood are several lobelike structures covered with cilia. These form the *wheel organ*. Posterior to this is a membranous *velum,* which contains the true mouth opening in its center and which also bears several *velar tentacles*.

CYCLOSTOMES In cyclostomes a *buccal funnel* is present at the anterior end. It is bordered by many small *papillae.* In lampreys the lining of the buccal funnel is beset with numerous horny teeth (Fig. 9.4) not homologous with the true teeth of higher forms. At the apex of the funnel is the mouth opening, through which the tongue projects. No jaws are present, and the mouth is permanently open.

GNATHOSTOMES All the remaining vertebrates have both upper and lower jaws and are generally referred to as *gnathostomes.* The size and shape of the mouth opening are varied, being correlated with the type of food eaten and the manner in which it is obtained. *Cheeks* may mark the boundary of the opening of the mouth on either side. Many gnatho-

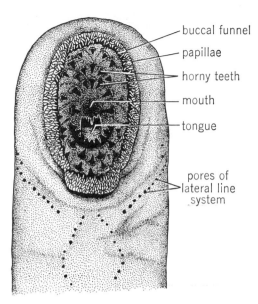

buccal funnel

papillae

horny teeth

mouth

tongue

pores of
lateral line
system

FIG. 9.4 Ventral view of anterior end of lamprey showing buccal funnel and tongue beset with horny, epidermal teeth.

stomes, however, have no cheeks and can open their mouths to an astonishing degree.

LIPS

The upper and lower boundaries of the mouth opening are known as *lips*. In all forms below mammals they are immovable and consist of epithelial folds or pads of connective tissue which meet when the mouth is closed. In turtles and birds the jawbones are covered with hard, horny beaks. Movable lips occur only in mammals. Notable exceptions include the platypus and whales. The bill of the platypus is not horny but is soft, pliant, and extremely sensitive. Whales merely have soft pads of connective tissue which serve to keep the mouth tightly closed.

Mammalian lips are covered with skin on the outside and lined with mucous membrane on the inside.

VESTIBULE

The space between the lips and jaws is the *vestibule.* Lateral to the mouth opening the vestibule may be bounded on the outside by the cheeks and on the inside by the gums. In mammals a *labial frenulum,* composed of a fold of mucous membrane, connects the middle of each lip to the gum, or *gingiva.*

In certain mammals cheek pouches of considerable size are present, but not all are homologous. In some species, e.g., the pocket gopher, they are actually external pouches of the skin, lined with fur and used as temporary storage places for food when these rodents are foraging.

A variety of glands opens into the vestibule. These are mucous glands, for the most part. Labial mucous glands are small structures opening on the inner surface of the lips. In man they may be identified as small lumps when one pushes the tongue against the lips. In some mammals *molar mucous* glands open near the molar teeth. The *parotid duct* (*Stensen's duct*), of the parotid salivary gland, opens into the vestibule opposite the first or second upper molar teeth in man and at a corresponding point in other mammals.

ORAL CAVITY

In fishes the mouth, in addition to serving as a passageway for food, is concerned with the passage of water, which contains the dissolved oxygen used in respiration. In most fishes the nasal cavities are quite independent of the oral cavity, but in the Sarcopterygii they first communicate with the mouth by means of a pair of *internal nares,* or *choanae,* In amphibians the lining of the oral cavity is ciliated. The tissue underlying the epithelium is unusually vascular in amphibians, serving as an important aid to respiration.

The roof of the mouth and pharynx is

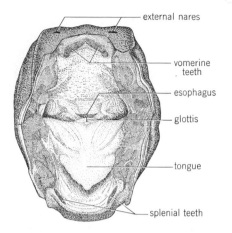

external nares

vomerine
teeth

esophagus

glottis

tongue

splenial teeth

FIG. 9.5 Open mouth of salamander *Siren,* showing the palate and associated structures. The angle of the mouth has been cut in order to depress the lower jaw. (*Drawn by R. Speigle.*)

often referred to as the *palate* (Fig. 9.5). In most reptiles and birds a pair of *palatal folds* grows medially for a variable distance on each side. The palatal folds do not meet in the median line, thus forming a *palatal cleft* through which the nasal and mouth cavities are in communication. In both crocodilians and mammals horizontal shelflike processes of certain bones of the skull grow medially and fuse with their partners of the opposite side, thus forming a *secondary palate*, which effectively separates the nasal passages from the mouth cavity except at the pharyngeal end. The anterior part of the secondary palate is the *hard palate,* since it is reinforced by portions of the premaxillary, maxillary, palatine, and (in some forms) the pterygoid bones. The posterior part in mammals is the *soft palate,* since it has no bony foundation. Abnormal development of the palate in mammals may result in a condition known as *cleft palate.* The nasal passages communicate with the pharynx superior to the soft palate. In man the soft palate ends in a fleshy, pointed elongation,

the *uvula,* which hangs downward and backward. It serves to close off the nasal passageway from the mouth during the act of swallowing. In many forms transverse ridges, the *palatine rugae,* are present on the secondary palate. These are not so prominent in man as in carnivores such as dogs and cats. Both epidermal and true teeth are found with some frequency as palatal developments among various vertebrates.

It is convenient at this point to mention the baleen, or whalebone, found in the mouths of the whalebone whales. It consists of large, often enormous, horny sheets, or plates, which hang in series from the edges of the upper jaw. The lower ends are fringed, serving to strain from the water the microscopic organisms on which these huge animals feed. Baleen is formed from large cornified papillae which become fastened together.

Glands of the Oral Cavity

In cyclostomes a so-called *salivary gland* opens into the mouth cavity on either side just below the tongue. It secretes an anticoagulant called *lamphedrin,* which facilitates the flow of blood from an animal attacked by the lamprey.

Fishes and amphibians that spend their entire lives in the water have no glands other than simple mucous cells opening into the mouth cavity. Glands first appear in the mouths of terrestrial vertebrates, their primary function being to moisten food and render it slippery to facilitate swallowing. In higher forms a specialization of the mouth glands occurs and may play a part in capturing prey and in digestion. In poisonous snakes and lizards certain mouth glands serve as dangerous organs of defense.

AMPHIBIANS In terrestrial amphibians a mucous *intermaxillary,* or *internasal, gland* lies in the nasal septum. The secretion of this gland helps to give the tongue its adhesive properties. It is lacking in caecilians. In frogs and toads a *pharyngeal gland* lies near the choanae, its secretion passing into them. Mucous *lingual glands* are numerous on the protrusible tongues of frogs, toads, and some salamanders. Their secretion aids in the capture of prey. In some frogs a digestive enzyme, *ptyalin,* is secreted by the mouth glands.

REPTILES The oral glands of reptiles show much more distinct grouping than those of amphibians. A *palatine gland* is present which is homologous with the intermaxillary gland of amphibians. In addition, *lingual, sublingual,* and *labial glands* are present.

In poisonous snakes the gland which secretes the venom is apparently a modification of a labial gland in the upper jaw. It may be homologous with the parotid salivary gland of mammals. Its duct opens into the cavity or groove of the poison fang (Fig. 9.6).

In the Gila monster, the only poisonous lizard extant, it is the sublingual gland which forms the poisonous secretion. This passes through four ducts, which penetrate the bone of the lower jaw to emerge in the vestibule in front of the grooved teeth, down which the poison flows.

In marine turtles and in crocodilians, oral glands are poorly developed.

BIRDS Well-developed *sublingual glands* opening in the floor of the mouth are present in birds. An *angle gland,* possibly homologous with the labial glands of reptiles, lies at the angle of the mouth. Numerous groups of small glands open separately on the roof of the mouth. A digestive function has been ascribed to the mouth glands of many birds, the enzyme *ptyalin* being present in their secretion.

MAMMALS Many small mucous glands are located on the palate and tongue of mammals. Encountered for the first time are the large and distinctly grouped *salivary glands.* Only in whales and sirenians are they secondarily reduced or wanting. There are usually three sets of salivary glands, named according to their position. The *parotid gland* lies in the region of the ear, usually beneath and somewhat anterior to the external auditory meatus. The *parotid duct (Stensen's duct)* courses over the masseter muscle of the cheek, to enter the vestibule of the mouth opposite the upper molar teeth. The disease known as mumps, which affects human beings, is marked by an inflammation and swelling of the parotid gland. The *submaxil-*

FIG. 9.6 Head of rattlesnake dissected to show relation of poison gland to hollow fang.

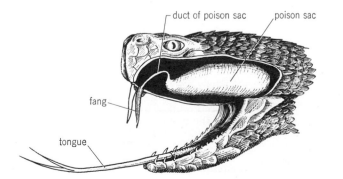

duct of poison sac   poison sac

fang

tongue

lary gland lies in the posterior part of the lower jaw; the *submaxillary duct* opens in front of the tongue near the lower incisor teeth. The *sublingual gland* is smaller than the others. It is composed of a major portion, with its *duct of Bartholin,* and numerous smaller elements, each with its own *duct of Rivinus.* The ducts open separately into the oral cavity along the jaw-tongue groove.

Mucous *molar glands* are well developed in such herbivorous mammals as the Artiodactyla. Their secretion aids in swallowing the coarse vegetation on which these animals feed. The large *orbital glands* of the dog family are mucous glands opening into the mouth near the last upper molar teeth.

TONGUE

The tongues of vertebrates show much diversity, and not all are homologous with the mammalian organ with which we are most familiar.

CYCLOSTOMES The "tongue" in cyclostomes is a very highly specialized structure which is not homologous with the tongues of higher vertebrates. It bears horny teeth on its surface and is used for rasping.

FISHES The primary tongue of fishes is merely a fleshy fold which develops from the floor of the mouth between the mandibular

FIG. 9.7 Head of salamander *Eurycea bislineata* with tongue extended. (*Drawn by J. Dohm.*)

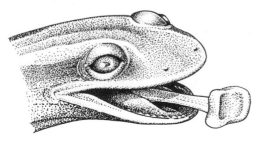

and hyoid arches. The hyoid arch frequently extends into this fold and supports it. The tongue is devoid of muscles and can be moved only within narrow limits by varying the position of the hyoid arch. Sensory receptors may be present. In some fishes the tongue bears small papillae, and in a few teleosts, e.g., the salmon, teeth may even be present on the tongue, which is supported by a *glossohyal bone* uniting the hyoid arches of the two sides.

AMPHIBIANS Four different conditions pertaining to the tongue prevail in amphibians. In the so-called *aglossal toads,* no tongue of any sort is present. This is a degenerate condition. In some urodeles, e.g., *Necturus,* in which the entire life is spent in the water, the tongue shows very little change over that of fishes. Other amphibians have movable tongues which can be thrust out of the mouth and used in capturing the animals on which they feed. Such tongues consist of a basal portion, homologous with the primary tongue of fishes, and an expanded anterior glandular portion well supplied with protractor and retractor muscles. These together make up the *definitive tongue.* In frogs and toads the base of the tongue is attached anteriorly at the margin of the jaw. Its free end is folded back on the floor of the mouth when at rest and is flipped out of the mouth during activity. In urodeles the tongue has a more extensive attachment to the supporting hyoid arch. Some salamanders, e.g., *Eurycea,* can thrust out their boletoid tongue (Fig. 9.7) directly and pull it back again with great rapidity.

REPTILES In turtles and crocodilians the tongue is not protrusible and lies on the floor of the mouth. Snakes and lizards, on the other hand, have well-developed tongues which can be extended and retracted. The tongues of some lizards are used as prehensile organs in capturing prey. It has been reported that the bifurcated tongues of certain lizards and of snakes, when retracted, are placed in the blind pockets of Jacobson's organ (page 211), which opens on either side into the mouth cavity near the choanae. This seems to be an accessory olfactory device. Volatile chemical substance particles which cling to the surface of the tongue may thus be detected by the animal.

The reptilian tongue represents a higher degree of development than that of amphibians. It includes a fold over the hyoid arch and a homolog of the glandular tongue of amphibians. The latter arises between the basihyal bone and the lower jaw. A pair of *lateral lingual folds* above the mandibular arch also contributes to tongue formation. The tongue is thus composed of four portions which have fused together.

BIRDS The tongue of birds is practically lacking in intrinsic muscles. It is usually cov-

FIG. 9.8 Tongue of cat, showing arrangement of papillae.

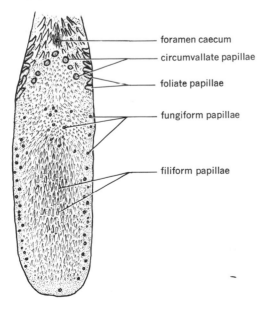

- foramen caecum
- circumvallate papillae
- foliate papillae
- fungiform papillae
- filiform papillae

ered with horny material. The only way in which the tongue can be moved is by altering the position of the hyoid apparatus, which supports it at its base. It has lost the lateral lingual folds possessed by reptiles.

MAMMALS The mammalian tongue is best developed of all. It is movable except in the whales and contains well-developed intrinsic muscles. These striated muscles are arranged in bundles coursing among one another in three planes, thus accounting for the high degree of mobility of the mammalian tongue. In the anteaters it is highly specialized and its muscles may extend back as far as the sternum.

In mammals the tongue is derived from five separate portions which have united. These include the unpaired tuberculum impar, paired lateral lingual swellings from the mandibular arch which make up the *body* of the tongue, and paired fleshy ridges contributed by the hyoid arch and the third and fourth visceral arches, which form the *root* of the tongue. The thyroid gland originally arises from the midline at the point where the anterior and posterior portions come together. A small depression, the *foramen caecum,* represents the point of origin of the thyroid gland.

The mucous membrane under the front of the mammalian tongue forms a median fold, the *lingual frenulum,* which is attached to the floor of the mouth.

The tongue bears papillae of various kinds (Fig. 9.8) which may or may not be associated with taste buds. Among the most common in mammals are *filiform, fungiform, foliate,* and *circumvallate (vallate) papillae.* Filiform papillae are small, conical projections making up the greater part of the plushy surface of the tongue. In the cat family (Felidae) the epithelium covering the filiform papillae is highly cornified. Fungiform papillae are mushroom-shaped structures scattered over the surface of the tongue. Foliate papillae are broad and leaflike. Their numbers vary in different species, but they are generally situated near the base of the tongue. They are lacking in man. Circumvallate papillae are few in number but large. Each is surrounded by a trench. These papillae are arranged in the form of an inverted V with the apex pointing toward the base of the tongue. The foramen caecum lies just behind the apex of the V.

The tongue of the dog is highly important in keeping the body temperature constant. Elimination of excess heat is accomplished by panting. The moist, highly vascular tongue swells and hangs from the mouth. Rapidly expired air, passing over the surface, evaporates the moisture and produces a cooling effect.

TEETH

Teeth are primarily employed by animals in cutting, grinding, or crushing food but may, in addition, serve as structures used in attack or defense. Vertebrates use their teeth only to hold captured prey. Among vertebrates, two types of teeth are found, *epidermal teeth* and *true teeth*.

EPIDERMAL TEETH These consist of hard pointed, cornified epithelial projections (Fig. 9.9). Only a few vertebrates have them, and they are not homologous with true teeth. Perhaps the best-developed epidermal teeth are those of cyclostomes. In lampreys they are definitely arranged on the inner surface of the buccal funnel and on the tongue.

Among amphibians, epidermal teeth are located on the edge of the jaws of larval anurans and are used in scraping off algae on which the larvae feed.

Epidermal teeth in the form of horny plates are found in the platypus. This animal possesses true teeth during development, but they are lost even before birth. Horny crushing plates are also present in sirenians, which possess true teeth as well.

TRUE TEETH All other teeth found in vertebrates are built upon the same general plan, although variations exist in the composition of the hard, shiny, translucent covering of the exposed surface. In mammals this outer layer is made up of *enamel,* derived from ectoderm. In fishes and some amphibians possessing true teeth the hard covering layer is composed of *vitrodentine,* which is claimed to

FIG. 9.9 Section through epidermal tooth of lamprey. (*After Warren.*)

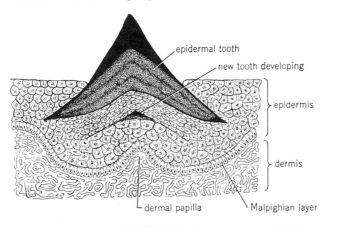

be of mesodermal origin. The structure of the latter type of tooth is, in general, rather similar to that of the placoid type of scale found in elasmobranch fishes (page 85).

An opinion has been held that true teeth derived phylogenetically from the placoid type of scale. The modern view is that both are modified remnants of bony dermal plates which formed the armor of such ancestral forms as ostracoderms and placoderms and are therefore homologous. The true type of tooth covered with vitrodentine accordingly deviates less from the ancestral type than the mammalian tooth with its covering of enamel.

Structure The portion of the true tooth that is exposed and called the *crown* is covered with vitrodentine or enamel, as the case may be. Under this lies the somewhat softer *dentine* (ivory), composed of a peculiar type of calcified connective tissue. The dentine, in turn, surrounds a *pulp cavity* filled with soft connective tissue and supplied with small nerves and blood vessels. The embryonic origin of the mammalian type of tooth is twofold. Enamel is derived from stomodaeal ectoderm; the rest of the tooth from mesoderm. The vitrodentine-covered type of tooth is thought to be entirely mesodermal in origin but is formed under the organizing influence of an enamel organ which comes from the stomodaeal ectodermal epithelium. In mammals a layer of peculiar histological structure, the *cementum,* forms around the *root* of the tooth, or that part below the crown. At the tip, or apex, of the root is a small opening through which nerves and blood vessels enter the pulp cavity. It is called the *apical foramen.*

Replacement of Teeth Except in mammals and a few others, teeth may be replaced an indefinite number of times (*polyphyodont*). In most mammals only two sets of teeth develop (*diphyodont*), but in a few there is only one set (*monophyodont*). The persistent molar teeth of man have no successors and actually belong to the first set of teeth.

Location of Teeth Teeth may be located almost anywhere in the oral or pharyngeal walls where stomodaeal ectoderm is present and where there is cartilage or bone to support them. In addition to their usual location on premaxillaries, maxillaries, and mandible, in certain forms they are found on the vomer, palatine, pterygoid, parasphenoid, and splenial bones as well. In some fishes, teeth are located on the tongue and attached to the hyoid arch which supports it. In others they may even be found on the gill arches.

Shape of Teeth Most vertebrates below mammals have teeth of uniform shape (*homodont*). In most mammals and in several kinds of fishes the teeth are arranged in groups which differ in shape and function (*heterodont*). Such mammals as the toothed whales have homodont dentition, but most are heterodont, the teeth being differentiated into *incisors, canines, premolars,* and *molars*.

Attachment of Teeth The manner in which teeth are attached at their bases also varies. In some fishes, crocodilians, and mammals the teeth are set in sockets in the

FIG. 9.10 Diagram showing three methods of tooth attachment. (*Drawn by G. Schwenk.*)

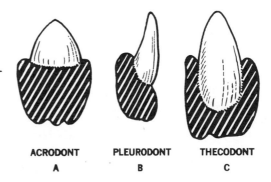

ACRODONT   PLEURODONT   THECODONT
A   B   C

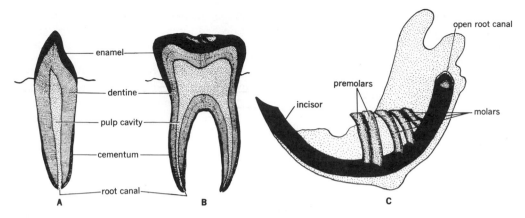

FIG. 9.11 *A*, section through incisor tooth; *B*, section through premolar tooth; both with closed pulp cavities. *C*, lower jaw of rodent *Geomys*, showing large incisor tooth with open pulp cavity. (*C, after Bailey, modified from Wiedersheim, "Comparative Anatomy of Vertebrates," copyright 1907 by The Macmillan Company and used with their permission.*)

jawbones (*thecodont*) (Fig. 9.10). Numerous fossil reptiles had thecodont dentition, as did *Archaeopteryx* among birds. In mammals there are two kinds of roots, open and closed (Fig. 9.11). In the first type the pulp cavity, or root canal, is wide open, and by addition of new layers of dentine from beneath, such teeth may continue to grow throughout life. The tusks of elephants and the incisors of rodents and lagomorphs are examples of this type of tooth. In the closed type, which is the usual form, the root canal is very small and serves only for the passage of nerves and blood vessels. These teeth fail to grow after they have reached their definitive size.

In other vertebrates the method of tooth attachment is more simple. In sharks, skates, and rays, the teeth are not directly attached to the cartilaginous jaws but are embedded in the tough, fibrous, mucous membrane which covers the jaws. In most vertebrates the teeth are fused to the underlying bone (*acrodont*) (Fig. 9.10). In some fishes and snakes certain of the teeth may be *hinged*. Hinged teeth yield to pressure and bend backward, only to

snap back to the upright position when the pressure is released. Hinged teeth are held in position by dense, fibrous ligaments. Another method of attachment is the *pleurodont* type. Here the teeth are joined not only to the bone below but also along a shelflike indentation on the inside of the jawbone (Fig. 9.10). In teeth of the acrodont or pleurodont types, where there are no roots, nerves and blood vessels enter the pulp cavity at the base.

Comparative Anatomy of Teeth

CYCLOSTOMES Only epidermal teeth are present in cyclostomes.

FISHES A few species of fishes are naturally toothless, at least as adults. Among these are sturgeons, sea horses, and pipefish. Teeth in fishes are all of the true type. Dentition in most is polyphyodont, acrodont, and homodont, but many teleosts and some elasmobranchs have heterodont dentition. In such species as the haddock, garpike, barracuda, and several others, the thecodont method of attachment is utilized. A primitive type of *labyrinthine* tooth occurs in fossil crossopterygians and in the living coelacanth *Latimeria*. Fish teeth usually consist of little more than conical projections, but in some they are vertically flattened triangles. In chimaeras and dipnoans several teeth have fused to form platelike structures used for crushing. In different species teeth may be located in various parts of the mouth or pharynx wherever bony or hard parts are situated.

AMPHIBIANS Save for the epidermal teeth of larval anurans, the teeth of amphibians are all true teeth. Teeth are lacking in toads, genus *Bufo,* and in *Pipa*. In frogs, teeth are lacking on the lower jaw except in members of the genus *Amphignathodon*. Some fossil amphibians had conical labyrinthine teeth, but in modern forms they are simple peglike structures. Amphibian teeth are confined to the jawbones, palatines, and vomers and, in a few, to the parasphenoid bone. They are polyphyodont, homodont, and acrodont. Characteristic of modern forms is a weak point of calcification near the base of the tooth. The upper part of the tooth is usually lacking in prepared skeletons, only the base of the teeth being present. These teeth are referred to as *pedicilate* teeth.

REPTILES Turtles are the toothless representatives of the class Reptilia. The horny beaks of these animals are sharp and strong and serve them in lieu of teeth. Some prehistoric pterosaurs also lacked teeth.

In other reptiles the teeth are situated on the jawbones, but in snakes and lizards they may also occur on the pterygoids and palatines. *Sphenodon* has vomerine teeth, an unusual feature among reptiles. *Sphenodon* is monophyodont, but most reptiles have polyphyodont as well as homodont dentition. Variations occur in the method of tooth attachment. Snakes and lizards are either acro-

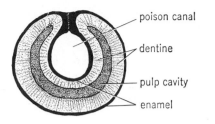

poison canal

dentine

pulp cavity

enamel

FIG. 9.12 Section through poison fang of rattlesnake.

dont or pleurodont, but in several ancestral fossil reptiles as well as living crocodilians thecodont dentition occurs. The extinct cynodonts, or mammallike reptiles, had heterodont teeth.

The poison fangs of certain snakes are specialized maxillary teeth bearing either a groove or poison duct for the passage of venom (Fig. 9.12). The lower front teeth of the Gila monster bear similar grooves on their anterior surfaces.

The teeth in lizards vary from a simple conical type to biconodont or triconodont to large flattened molariform structures. It is not unusual to find ontogenetic changes in crown-tooth form among lizards.

A so-called *egg tooth* is present on the tip of the upper jaw in the embryos of lizards and certain snakes. It projects beyond the other teeth and helps the young in emerging from the leathery eggshell. It is lost soon after hatching. In *Sphenodon,* turtles, and crocodilians, a *horny epithelial* egg tooth at the tip of the the snout of the embryo serves the same function. This is also a transitory structure, and although analogous to the egg tooth of snakes and lizards, is not homologous with it.

BIRDS Modern birds lack teeth. *Archaeopteryx* had thecodont teeth embedded in sockets. In *Ichthyornis* the conical teeth were set in shallow sockets, whereas in *Hesperornis* they were set in a continuous groove. Many birds have a transitory, horny, epithelial egg tooth which aids in breaking the shell at time of hatching.

MAMMALS Only a few toothless mammals are known. The adult platypus has epidermal teeth but no true teeth. A single set of true teeth appears during development, but these are soon shed. In *Echidna* no teeth are present at any time. Whalebone whales lack teeth in the adult, as do certain anteaters. Although other anteaters, armadillos, and certain other forms possess teeth, they are imperfect structures which lack roots and enamel.

The thecodont condition is the rule in mammals. Most members of this class are diphyodont, having but two sets of teeth. The first set, known as the baby teeth (milk teeth, lacteal teeth, deciduous teeth) is lost and then replaced by a permanent set. In bats and guinea pigs the milk teeth are lost even before birth. A few mammals are monophyodont. The platypus, toothed whales, sloths, and sirenians are examples. Marsupials are an intermediate group since they retain all their milk teeth except the last premolars.

In certain marsupials a prelacteal dentition has been observed, and occasionally in man a postpermanent dentition has been recorded. Thus it seems that in mammals there may originally have been four sets of teeth, a complication not so far removed from the polyphyodont condition of lower vertebrates.

Among mammals only the toothed whales are homodont, the number of teeth ranging from 2 to 200. With this exception the heterodont condition prevails in mammals. Four types of teeth are commonly identified in the upper jaw. *Incisors* are in front, their roots being embedded in sockets, or alveoli, in the premaxillary bones. On either side the incisors are followed by a single *canine,* which is the most anterior tooth in the maxillary bone. It usually differs from the remaining teeth in

having a single, pointed crown and a single root. Following the canines are the *premolar* teeth, which are more complex in form and have two roots. Lastly come the *molars,* with several roots. There is but a single set of molars, which should therefore really be considered as belonging to the milk dentition but differing from the others in not being replaced. The mandible, or lower jaw, bears teeth corresponding to those just described. A natural space between two types of teeth is referred to as a *diastema*. In rodents and lagomorphs, which lack canine teeth, a diastema is present between the incisor and cheek teeth on each side. In the cat there is a diastema between the canine and first premolar on each side of the lower jaw.

Only in mammals is there a fixed number of teeth characteristic of each species. The maximum number of teeth in *placental* mammals with heterodont dentition is 44. Some marsupials, however, possess a greater number. There are usually fewer than this because of reduction in number of one or more types. The number of teeth in any particular species is surprisingly constant.

In various mammals the shapes of the molar and premolar teeth vary according to feeding habits and the type of food that is eaten. The cheek teeth in terrestrial carnivores have sharp, cutting crowns. This condition is known as *secodont*. A so-called *carnassial tooth* in each jaw, a premolar above and a molar below, is specially developed and used for shearing. In man and some other mammals the flattened cheek teeth with their small tubercles are of the *bunodont* type and used for grinding. In the horse, ruminating mammals, and some others, the cheek teeth are characterized by the presence of vertical, crescent-shaped folds of enamel enclosing softer areas of dentine. Cementum may fill the interstices. This condition is known as *selenodont* (Fig. 9.13*A*). Because the softer

FIG. 9.13 *A*, grinding surface of selenodont tooth from lower jaw of a sheep; *B*, grinding surface of lophodont molar tooth of African elephant. (*B, after Tomes.*)

dentine wears away more rapidly, a rasping surface, well adapted for grinding, is maintained. Elephants have enormous grinding teeth, which may measure as much as 1 ft in length and 4 in. in width. There is an intricate folding of the enamel and dentine to form transverse ridges. Such teeth are of the *lophodont* type (Fig. 9.13*B*).

Certain unusual teeth found in mammals deserve special mention. The tusks of elephants are incisor teeth with open root canals. They grow continuously throughout life. The gnawing, chisellike incisor teeth of rodents and lagomorphs are also examples of teeth with open pulp cavities which grow continuously. Only the anterior face of this type of tooth is covered with enamel. The tusks of the male wild boar are greatly enlarged canine teeth with large persistent pulp cavities. Both upper and lower canines curve upward. Tusks

of the walrus, occurring in both sexes, are also enlarged upper canine teeth.

In a number of mammals rather striking sexual differences are apparent in the dentition. In the apes, for example, the canines and first premolars of males are decidedly larger than in females.

Embryonic monotremes, being the only mammals to develop within an eggshell outside the body of the mother, are the only mammals which have a horny egg tooth at the end of the snout. This serves the young in escaping from the eggshell.

The origin of the rather complex cheek teeth of mammals from the simple conical type found in most of the other vertebrates has long been a subject for conjecture. The single-rooted incisor and canine teeth show so little modification that they need not enter the discussion. The theory that is generally

FIG. 9.14 Theoretical development of tooth cusps from a single pointed tooth. (*After Osborne.*)

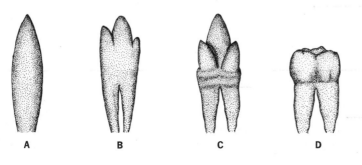

held, and for which there is considerable pale-ontological evidence, states that, starting with a primitive conical tooth, two additional pro-jections, or buds, developed, giving rise to the so-called *triconodont shape*. Later these cones shifted in position, to give rise to separate tubercles, or cusps, arranged in a triangle. This has been called the *tritubercular position* (Fig. 9.14). Still later other parts may have developed from these three original tubercles, to form additional cusps, ridges, and folds; thus the many and varied types of mammalian cheek teeth found in living forms may have arisen. For further details and elaborations of this theory of tooth evolution the reader is referred to the Bibliography.

ANTERIOR AND INTERMEDIATE LOBES OF THE PITUITARY BODY

The pituitary body can under no circum-stances be considered as part of the digestive system. However, since its anterior and inter-mediate lobes arise from the ectodermal epithelium of the primitive mouth cavity, they should be mentioned here as stomodaeal de-rivatives. For a more complete description of this important endocrine organ the reader is referred to Chap. 14, Endocrine Organs, page 419.

PHARYNX

The pharynx, lined with endoderm, is the portion of the alimentary tract which is directly continuous with the ectoderm-lined mouth. Its function as part of the digestive system is merely to serve as a passageway between the mouth and esophagus. It is in the walls of the pharynx that the muscles which initiate swallowing movements are located. The pharynx is particularly important in con-nection with respiration. Derived from its

walls are internal gills, lungs, and several other structures. These are better described in relation to the respiratory system (Chap. 8). The occasional presence of teeth in the pharynx is an indication that stomodaeal ec-toderm has invaded this region.

ESOPHAGUS

The esophagus is that portion of the ali-mentary canal immediately following the pharynx. It joins the stomach at its other end. In some of the lower vertebrates it is difficult to distinguish the point at which the esoph-agus ends and the stomach begins.

The histological structure of the esophagus is similar, in general, to that of the remainder of the digestive tract, except that, since it does not lie within the coelom, it is not surrounded by a layer of visceral peritoneum. Further-more, the mucous membrane lining the esoph-agus is of the stratified squamous type of epithelium rather than of the simple colum-nar type. In the upper part of the esophagus the muscle fibers, in most cases, change grad-ually from the striated, voluntary type to the smooth, involuntary type. There are excep-tions to this, particularly in the ruminating, or cud-chewing, mammals, in which striated fibers extend throughout the entire length of the esophagus. Contraction of the esophagus in these forms is under voluntary control.

CYCLOSTOMES   In the adult lamprey a very specialized condition occurs which has no parallel in other vertebrates. From the mouth cavity two tubes extend posteriorly (Fig. 9.15): a dorsal esophagus and a ventral pharynx which ends blindly and is entirely concerned with respiration. At the entrance to the pharynx is a valvelike *velum*. The esophagus is lined with numerous folds. In the hagfish the condition is much as in other

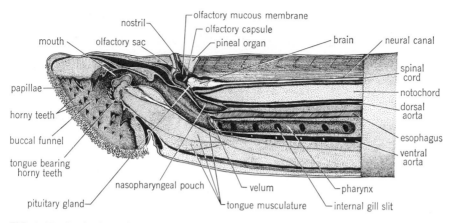

mouth
nostril
olfactory sac
olfactory mucous membrane
olfactory capsule
pineal organ
brain
neural canal
spinal cord
notochord
dorsal aorta
esophagus
ventral aorta
papillae
horny teeth
buccal funnel
tongue bearing horny teeth
pituitary gland
nasopharyngeal pouch
velum
tongue musculature
pharynx
internal gill slit

FIG. 9.15 Sagittal section of anterior end of lamprey.

vertebrates, and the esophagus merely extends posteriorly from the pharynx.

FISHES In fishes the esophagus is <u>very short</u>, its junction with the stomach being almost imperceptible. In such elasmobranchs as *Squalus acanthias* (Fig. 9.16) numerous backward-projecting papillae line the esophagus and first part of the stomach. The esophagus commonly bears longitudinal folds which permit a considerable degree of distention.

AMPHIBIANS The esophagus in amphibians is <u>extremely short</u> and consists of little more than a <u>constricted</u> area of the alimentary tract (Fig. 9.3). In most amphibians the lining of the esophagus as well as of the mouth is ciliated. By this means small food particles are carried to the stomach. Secretory cells in the esophageal epithelium of the frog are reported to have a digestive function.

REPTILES In reptiles the esophagus is <u>generally longer</u> than in lower forms. <u>Longitudinal folds</u> in the walls <u>permit</u> considerable <u>distention,</u> which is of special use in snakes, all of which are capable of swallowing large objects. The lining of the esophagus of a cer-

tain marine turtle is covered with cornified papillae which point posteriorly (Fig. 9.17).

BIRDS In many birds the esophagus is likewise lined with horny papillae. Its junction with the stomach is abrupt. In grain-eating birds, as well as birds of prey, either the esophagus forms a large sac, or a ventral, pouchlike outgrowth, the *crop,* is present (Fig. 9.18). The crop is primarily useful in permitting the bird to <u>secure and store</u> an abundance of <u>food</u> in a <u>short period.</u> It also enables the bird to compete with others for a limited amount of food. Although it is doubtful whether any digestion occurs here, such food particles as grain may swell and are thus rendered more capable of being broken down and digested later. The crop releases portions of its contents at intervals. The food then passes on to the stomach. The *furcula* (wishbone) supports the crop.

In pigeons a peculiar adaptation is found in the <u>so-called crop "glands,"</u> which are two modified areas in the walls of the crop in both sexes (Fig. 9.19). These are <u>really not glands</u> since they are cytogenic, or cell-forming, structures. Upon proper stimulation the crop glands enlarge, and a very rapid cellular prolif-

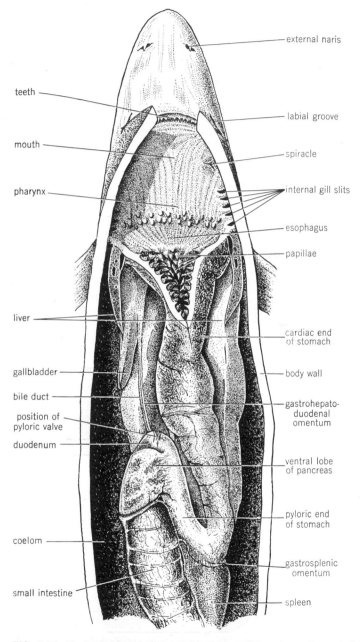

external naris

teeth

labial groove

mouth

spiracle

pharynx

internal gill slits

esophagus

papillae

liver

cardiac end
of stomach

gallbladder

body wall

bile duct

gastrohepato-
duodenal
omentum

position of
pyloric valve

duodenum

ventral lobe
of pancreas

pyloric end
of stomach

coelom

gastrosplenic
omentum

small intestine

spleen

FIG. 9.16 Ventral view of abdominal viscera of the elasmobranch fish *Squalus acanthias.*

FIG. 9.17 Cornified papillae lining esophagus of a marine turtle. (*Drawn by G. Schwenk.*)

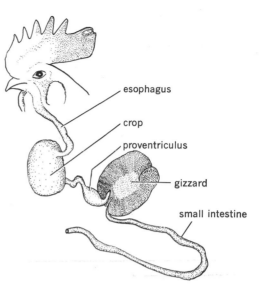

esophagus

crop

proventriculus

gizzard

small intestine

FIG. 9.18 Relation of crop and stomach regions of the domestic fowl.

FIG. 9.19 Crops of pigeons cut open to show crop "glands." *A*, during resting, or inactive, period; *B*, from a bird previously injected with lactogenic hormone. (*Drawn by J. Dohm.*)

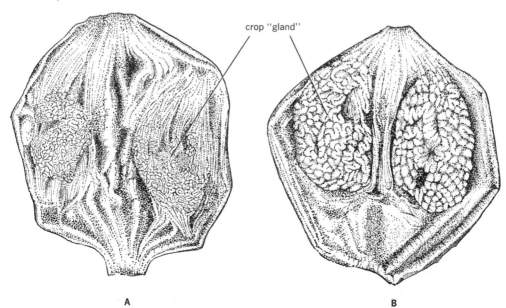

crop "gland"

A

B

eration occurs. A nutritious, cheesy material called *pigeon milk* is sloughed into the cavity of the crop. This is regurgitated into the throats of the young squabs and serves as their food. The activity of the crop glands is controlled by *prolactin,* which is a hormone from the pituitary body. Activity of the crop glands seems to be stimulated via the pituitary body by the act of incubating the eggs, in which both male and female participate. After the young are reared, the crop glands return to the inactive condition.

MAMMALS In mammals there is a clear distinction between stomach and esophagus. The length of the esophagus varies with the length of the neck, the giraffe having the longest esophagus of all. On the way to the stomach it must pass through the diaphragm. That portion below the diaphragm is covered by visceral peritoneum, which is lacking from the upper portion. In ruminant mammals the first three parts of the stomach are modified regions of the true stomach.

STOMACH

The stomach is basically a dilatation of the digestive tract for the temporary storage of food. Only when its lining epithelium contains gastric glands (with certain special exceptions) is it properly called a true stomach. Its digestive function is apparently a secondary acquisition. The shape of the stomach is related to the shape of the body. In such elongated creatures as snakes, it extends longitudinally, but in those with wider bodies it occupies a more transverse position. The end of the stomach which connects to the esophagus is the *cardiac end.* The main portion is called the *body.* The *pyloric end* connects to the intestine and terminates at the *pylorus,* or *pyloric valve.* This consists of a fold of the lining mucous membrane surrounded by a thick, involuntary sphincter muscle which regulates the passage of the contents of the stomach into the intestine.

In most vertebrates the stomach has assumed a transverse position and is U-shaped or J-shaped. It has become twisted so that the cardiac end is on the left side of the body and the pylorus on the right. Two faces or curvatures are thus formed, the *lesser* and *greater curvatures,* respectively (Fig. 9.20). The lesser curvature is actually ventral and the greater curvature dorsal. The expansion at the cardiac end of the stomach, formed by the greater curvature, is the saclike *fundus.*

The muscular walls of the stomach, particularly of the pyloric end, contract in such a manner as to mix, or churn, the contents of the stomach thoroughly. Peristaltic waves originate near the body of the stomach and pass through the pyloric end down to the intestine, forcing the contents along. The rate of peristalsis and the strength of the contractions vary with the type and amount of food eaten.

CYCLOSTOMES The cyclostome stomach is very poorly developed and consists of little more than an almost imperceptible enlargement at the posterior end of the esophagus.

FISHES There is practically no distinction between esophagus and stomach in fishes, and the longitudinal folds of the former may extend for some distance into the stomach. A considerable variety of stomach shapes may be observed. Some are simple, straight tubes without any digestive function, as in Dipnoi, chimaeras, and a number of teleosts. In others, e.g., *Polypterus,* the cardiac and pyloric limbs have fused along the line of the lesser curvature so that the stomach appears much as a blind pouch.

In elasmobranchs the J-shaped stomach is

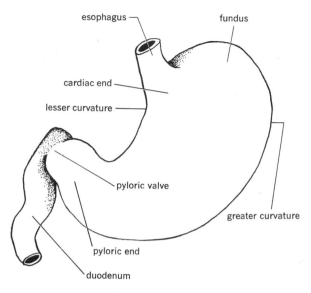

FIG. 9.20 Outline of human stomach, showing regions (ventral view).

FIG. 9.21 Ventral view of digestive organs and lungs of *Necturus*.

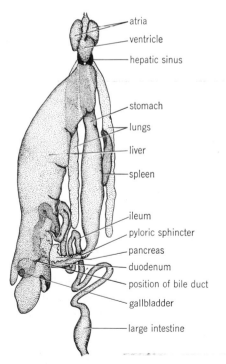

typical. The pyloric end is smaller than the cardiac portion. Many other fishes have similarly shaped stomachs. Teleosts exhibit a great variety of stomach shapes; some even have a ciliated lining.

AMPHIBIANS In frogs the cardiac end of the stomach is wide, there is no fundus, and the pyloric end is short and narrow. In certain salamanders the stomach is straight (Fig. 9.21). All amphibian stomachs have a digestive function.

REPTILES No striking deviations are to be observed in the stomachs of reptiles. Snakes and lizards have long, spindle-shaped stomachs in correlation with their elongated and narrow body shape. There is a clear-cut line of demarcation between stomach and esophagus, however. Of all the reptiles crocodilians have the most specialized gastric organs. Part of the stomach is modified into a gizzardlike muscular region.

BIRDS In accordance with the lack of teeth and the type of food eaten by birds, the stomach of most of them has been modified greatly for trituration. (pulverize) It has become differentiated into two regions (Fig. 9.18). The first, which is continuous with the esophagus, has a glandular lining which secretes gastric juices. It is known as the *proventriculus.* Food which is mixed with the digestive fluid in the proventriculus passes into the much-modified and highly muscular *gizzard,* which represents the pyloric portion of the stomach. The muscles form a pair of disclike areas with tendinous centers. The glandular cells lining the gizzard secrete a tough, horny layer which in some cases bears bumps, or tubercles, on its surface. These aid in the grinding process. In grain eaters, pebbles are taken into the gizzard. They help reduce the contents to a pulp. The gizzard is best developed in grain-eating birds. It is less well developed in insectivorous forms and in birds of prey, in which little differentiation is noticeable.

MAMMALS Many modifications exist in the transversely arranged stomachs of mammals, although most of them are of the typical vertebrate type. In monotremes the lining epithelium lacks glands, and the pouchlike structure, which serves merely for the storage of food, is therefore not considered to be a *true* stomach. In the platypus the two limbs are fused along the lesser curvature, so that the organ appears as a wide sac. In some monkeys, rodents, and others, a constriction marks off cardiac and pyloric regions. Such a stomach is often referred to as an *hourglass stomach.* It occasionally occurs in man.

The gastric organs of herbivorous mammals are larger than those of carnivores. They are frequently divided into two or more compartments. Ruminating mammals (those which chew their cud) have the most complex stomachs of all. The stomach in cattle consists

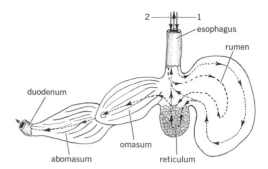

FIG. 9.22 Diagram of stomach of ruminant, showing functional relationships: 1, initial swallowing; 2, path taken by food upon reswallowing. (*Modified from Kingsley, "Comparative Anatomy of Vertebrates," McGraw-Hill Book Company, by permission.*)

of four separate chambers: the *rumen, reticulum, omasum,* and *abomasum* (Fig. 9.22). The rumen is a large sacculated compartment which is filled fairly rapidly with grain, grass, or other herbage. In cattle, fluid is added, and by muscular contractions, the food in the rumen is churned about. The food undergoes bacterial and protozoan breakdown (specifically cellulose processing) during its stay in the rumen. While this is going on, the animal may lie down and chew its cud at leisure. Food in the rumen passes by degrees into the reticulum or directly to the esophagus and, together with a fair amount of fluid, is then regurgitated into the mouth. It is referred to as the *cud* or *bolus.* The voluntary musculature of the esophagus makes this possible. When the animal begins to chew its cud, it first swallows most of the fluid, which passes back into the rumen. When the cud has been well masticated and mixed with secretions of the salivary glands, it is swallowed once more, and again passes into the rumen. A new bolus is then formed, regurgitated, and the entire process repeated. Most of the food which has been thoroughly masticated goes from the rumen into the reticulum and then passes directly to the omasum, a small chamber with many well-defined longitudinal folds in its walls, and thence to the abomasum, the only chamber of the stomach provided with gastric glands.

In camels the stomach is not quite so complicated, since an omasum is lacking. Pouchlike diverticula called *water cells* arise from both rumen and reticulum. Their openings are guarded by sphincter muscles. The water cells actually contain pure water, but this is generally metabolic water drawn from other parts of the body. Much comes indirectly from the breakdown of glycogen stored in the muscles and of fat stored in the hump. As much as 8 gal of water may be obtained by the animal as a result of processes involved in the breakdown of fat stored in the hump alone.

The stomachs of whales and hippopotamuses are divided into several compartments. In the kangaroo the pyloric portion has many peculiar sacculated folds in its walls. The pyloric portion of the stomach of the vampire (blood-lapping) bat *Desmodus* is elongated into a cecumlike structure which fills with blood when the animal is engaged in feeding.

Details of the histological structure and secretory activity of the stomach are of considerable interest but beyond the scope of this volume. Recent books on histology and physiology should be consulted.

INTESTINE

The portion of alimentary tract following the stomach is the intestine. Here the acid *chyme* from the stomach is mixed with alkaline bile from the liver, pancreatic juice from the pancreas, and the digestive secretions of great numbers of small glands in the walls of the intestine itself. These secretions finally convert food into such a form that it can be absorbed through the wall of the intestine into the blood or lymphatic systems, which then distribute it to all parts of the body. The intestine usually consists of two main parts, a long, but narrow, *small intestine* and a short, but wider, terminal *large intestine,* or *colon.* The part of the small intestine which immediately follows the stomach is called the *duodenum.* It is here that the ducts from the liver and pancreas open.

Many modifications are to be found in the intestines of vertebrates. They serve, for the most part, to increase the surface area of the intestinal epithelium for secretion of digestive fluids and for the absorption of digested foods. The intestine may be elongated and coiled; its diameter may be increased; saclike diverticula called *ceca* may be present; the lining is often thrown into longitudinal folds; circular folds of various types may develop; there may be great numbers of tiny projections, or *villi,* all over the lining surface. In some, a fold, the *spiral valve,* extends throughout the length of the small intestine. Alternate contractions of circular and longitudinal muscles in the intestinal wall are responsible for three types of contractions that may be observed. These include peristalsis, segmenting movements, and pendular movements. Together they result in a churning and mixing of the contents and the propulsion of materials toward the posterior end.

The length of the intestine is related to the feeding habits of the animal. It is relatively short in carnivorous forms and long in herbivores. This is unusually well shown in the frog. During its existence as a tadpole it feeds on vegetation and the intestine is long and coiled. After metamorphosis the animal becomes carnivorous and the intestine is not only relatively but actually shorter than that of a tadpole half its size.

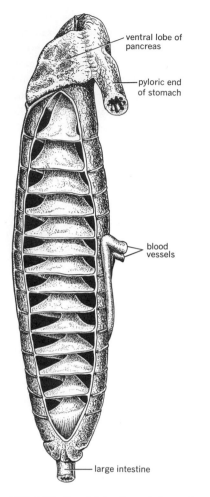

ventral lobe of pancreas

pyloric end of stomach

blood vessels

large intestine

FIG. 9.23 Small intestine of elasmobranch fish *Squalus acanthias,* cut open to show spiral valve. (*Drawn by J. Dohm.*)

CYCLOSTOMES The cyclostome intestine is straight. At its posterior end it enlarges slightly to form a rectum, which terminates in an anus. This opens into the anterior end of the cloacal depression. A longitudinal fold, the *typhlosole,* which takes a somewhat spiral course, projects into the cavity of the intestine.

FISHES In elasmobranchs the small intestine is shorter than the stomach. It is wide but rather straight and contains a well-developed spiral valve (Fig. 9.23). Spiral valves are also found in the small intestines of chimaeras, Dipnoi, *Latimeria,* members of the superorder Chondrostei, *Amia, Lepisosteus,* and others. Traces have been found in certain teleosts. The presence of a spiral valve in some fossil placoderms is an indication of its primitive character. In elasmobranchs the large intestine, a short passageway between the small intestine and anus, bends slightly and then opens into the cloaca. A long, slender, *rectal gland* connects to the intestine by means of a duct near the point where small and large intestines join. The rectal gland secretes a highly concentrated solution of NaCl, thus ridding the blood of excess salt.

In the Dipnoi a cloacal cecum is present. Many fishes have *pyloric ceca* coming off that part of the intestine immediately following the pylorus (Fig. 9.24). In *Polypterus* there is but one. In the mackerel as many as 200 may be present. Sometimes the pyloric ceca are bound together in a compact mass by connective tissue.

A circular valve often separates the small and large intestines. In many fishes the anus opens separately to the outside. This is true of teleosts, chimaeras, sturgeons, *Polypterus,* and some others. In other fishes a true cloaca is usually present. When no cloaca exists, the posterior part of the large intestine is known as the *rectum.*

AMPHIBIANS In caecilians the intestine is not differentiated into large and small regions and shows but a slight degree of coiling. In salamanders a greater degree of coiling is evident, and in anurans this tendency is much more marked. The large intestine in urodeles and anurans is short, straight, and plainly marked off from the small intes-

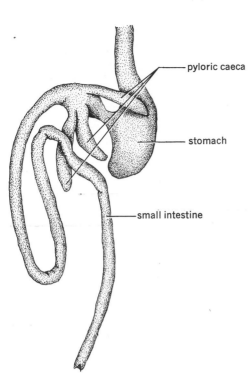

pyloric caeca

stomach

small intestine

FIG. 9.24 Digestive tract of perch, showing pyloric ceca. (*After Wiedersheim, "Comparative Anatomy of Vertebrates," copyright 1907 by The Macmillan Company and used with their permission.*)

tine. It opens into a cloaca. A ventral diverticulum of the amphibian cloaca gives rise to the *urinary bladder*. It is difficult to discern exactly from which germ layer, ectoderm or endoderm, the lining of the amphibian cloaca is derived. It is most probably endodermal, and the bladders of amphibians are thus homologous with those of higher forms. Frequently an *ileocolic valve* is present between small and large intestines. Villi first become evident in certain members of this class, although they are lacking in the common leopard frog and others. Circular folds, the *valvulae conniventes*, are also present in the small intestines of some amphibians.

REPTILES The reptilian small intestine is elongated, coiled, and of fairly uniform diameter. The large intestine is generally straight, of greater diameter, and opens into a cloaca. An ileocolic valve is located at the junction of small and large intestines and at this point, except in crocodilians, a *colic cecum* arises. Reptiles, then, are the first vertebrates to have true colic ceca.

BIRDS A tendency toward greater length is evident in the small intestine in this group. The large intestine is straight, relatively short, and terminates in a cloaca. A colic cecum is lacking in parrots, woodpeckers, and a few others, but most birds have one or two such structures. In certain birds (ducks, geese, turkeys, ostriches, etc.) the colic ceca attain a very large size and the walls may even bear villi. The relatively enormous cecum of the ostrich contains a spiral fold not found in other birds.

MAMMALS The intestines of mammals are more elaborately developed than those of other vertebrates. The coiled small intestine is made up of three regions, *duodenum, jejunum,* and *ileum*.

A pouchlike structure, known as *Meckel's diverticulum,* is sometimes found projecting from the ileum. It represents a remnant of the embryonic yolk stalk which has failed to degenerate in the normal manner. A Meckel's diverticulum is found in about 2 percent of all human adults and may give rise to serious complications.

It is generally stated that the average length of the small intestine in man is 22½ ft. There is actually much variation. The large intestine, or *colon,* is much shorter than the small intestine but of considerably greater diameter. In man it averages about 4 to 5 ft in length. The colon terminates in the rectum, which opens to the outside through the anus.

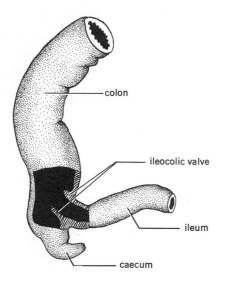

FIG. 9.25 Junction of ileum and colon of cat, showing cecum and ileocolic valve.

Among mammals only the monotremes and the pika (a lagomorph) possess a cloaca. In herbivorous mammals the intestine is very long and may be 20 to 28 times the length of the body. In the cow the average length is 165 ft and in the horse approximately 95 ft. Carnivorous forms have an intestine only 5 to 6 times the length of the body.

At the junction of ileum and colon lies the *ileocolic* (sometimes called the *ileocecolic*) *valve,* which regulates the passage of material from small to large intestine (Fig. 9.25). Coming off the colon at this point of juncture is a single *colic cecum,* found in almost all mammals. Some edentates have two, but this is exceptional. Great variations occur in the form and size of the cecum. It is small in bats, carnivores, some edentates, and certain whales. In marsupials, herbivores, and some rodents the cecum is relatively enormous and may even exceed the length of the body. The cecum provides additional space for colonic function and is a reservoir for bacteria. In

man its distal end has degenerated, the remnant being represented by the *appendix* (Figs. 9.26 and 9.27). This is situated on the abdomen about 2 in. from the right anterior superior spine of the ilium on a line between the latter and the umbilicus, or navel. An appendix is also found in certain monkeys, civets, and a few rodents. The exact function of the appendix is not clear.

The myriads of intestinal glands are of two main types. Their cells secrete certain hormonal substances (*secretin, pancreozymin,* and *cholecystokinin*) as well as numerous enzymes of importance in digestion.

In addition to the great numbers of small intestinal glands, the epithelium of the small intestine bears quantities of villi of various shapes. Those of man vary from 0.2 to 1 mm in height. They give a velvety appearance to the lining of the small intestine. In a human embryo villi are also present in the large intestine, but they gradually disappear, so that at birth the lining epithelium is relatively smooth. Villi contain blood vessels and lymphatics (lacteals), which collect the products

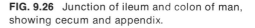

FIG. 9.26 Junction of ileum and colon of man, showing cecum and appendix.

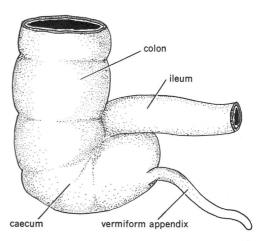

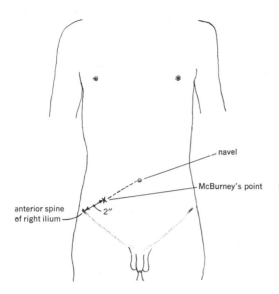

FIG. 9.27 Diagram showing location of appendix at McBurney's point.

of digestion after they have been absorbed. The free surfaces of the villi bear innumerable microvilli, which increase the absorptive surface immeasurably.

Nodules of lymphoid tissue, called *Peyer's patches,* are large oval areas in the small intestine, most numerous in the region of the ileum. They are sometimes present in the duodenum and jejunum. Villi are lacking from the epithelium which covers these structures.

In mammals several types of folds are found in the intestinal lining. In man these include (1) *plicae circulares,* or *valvulae conniventes,* which are circular folds found only in the jejunum and ileum; (2) *plicae semilunares,* or internal transverse folds of the colon; (3) *plicae transversales* in the lining of the rectum. Bulges of the colon between the plicae semilunares are known as *haustra.* In the region of the anal canal several longitudinal folds called *rectal columns* are present.

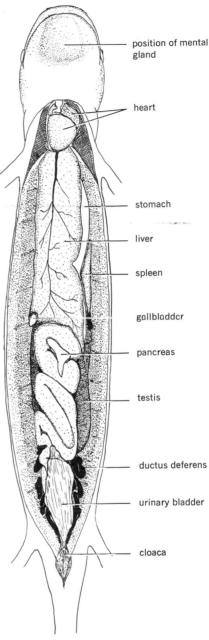

FIG. 9.28 Internal organs of male salamander *Eurycea bislineata,* showing relation of urinary bladder to cloaca.

PROCTODAEAL DERIVATIVES

Derivatives of the proctodaeum are few in number when compared with those of stomodaeal origin. It is very difficult to determine the point of transition between the endoderm lining the cloaca and the ectoderm lining the proctodaeum. Although some authorities disagree, it is generally conceded that the urinary bladder of amphibians, which opens directly into the cloaca (Fig. 9.28), is lined with *endoderm* derived from the cloaca, *not* with *proctodaeal* ectoderm. The *bursa Fabricii, present in most young birds,* is another endodermal cloacal derivative often thought to be derived from the proctodaeum. It forms a secondary connection with the proctodaeum during development and comes to lie in the body cavity between the spinal column and posterior part of the large intestine. It apparently functions as a lymphoid organ during early life but degenerates as the animal approaches sexual maturity. In this way it somewhat resembles the thymus gland, another lymphoid organ. The bursa Fabricii seems to be basically responsible for the formation of cells involved in antibody production and in reactions against invading bacteria. Certain *cloacal glands,* used primarily for defense or sexual allurement, are probably derived from the ectodermal, proctodaeal invagination.

LIVER

The liver is an important digestive gland, although its major function is in the treatment of food material after digestion and absorption into the body. Proteins are synthesized and fats altered in composition by the liver. It is the largest gland in the body. During embryonic development it develops as an outgrowth of the endodermal wall of the primitive gut. The region from which the liver arises becomes the duodenal portion of the small intestine. The liver proper consists of great numbers of *lobules* composed of branched columns of cells, the liver *trabeculae,* which are separated from one another by small blood spaces, or sinusoids. Narrow *bile capillaries* course between the columns of cells, anastomosing with one another. Larger ducts drain the bile capillaries, and these finally unite to form the *hepatic duct.* A fairly typical arrangement of those parts is to be found in man (Fig. 9.29). Two main ducts, one from the right lobe and one from the left, drain the liver. These unite to form the single hepatic duct. The *gallbladder,* in which bile, secreted by the liver, is stored, is an enlarged, pear-shaped saccular structure, its narrow neck connecting with a *cystic duct,* which in turn joins the hepatic duct to form the *common bile duct,* or *ductus choledochus,* which opens into the duodenum. The main pancreatic duct (ventral pancreatic duct; duct of Wirsung) joins the common bile duct shortly before it enters the duodenum. The common orifice of the two ducts opens at the tip of a *duodenal papilla.* The short segment common to the bile and pancreatic ducts is dilated, forming the *ampulla of Vater.* The preampullary portion of the bile duct is surrounded by a strong sphincter *muscle of Boyden.* Another, but weaker, sphincter, the *muscle of Oddi,* surrounds the ampulla as well as the distal portion of the ventral pancreatic duct.

Blood in the sinusoids is contributed by the hepatic portal vein, which drains the digestive tract and spleen, bringing the products of digestion, particularly those of carbohydrates and proteins, to the liver. Blood in the hepatic portal vein is unoxygenated. The usual type of oxygenated blood, which supplied all organs, reaches the liver through a *hepatic artery,* a branch of one of the main vessels coming off the dorsal aorta. The capil-

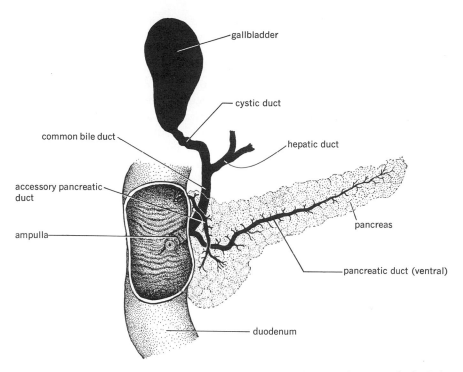

FIG. 9.29 Fairly typical arrangement of gallbladder, bile ducts, and pancreatic ducts in relation to duodenum, as found in man (semidiagrammatic.)

lary network of the hepatic artery anastomoses with the hepatic portal vein, the oxygenated blood entering the sinusoids.

Many variations from the arrangement just described are found in different vertebrates. Some are discussed below. The shape of the liver varies with the shape of the animal and is typically divided into right and left lobes. It is generally larger in herbivorous forms than in carnivores. The liver is held firmly in place by ligamentous folds of peritoneum. Its main attachment is to the *septum transversum,* a structure which in mammals contributes to the diaphragm.

Comparative Anatomy of the Liver

AMPHIOXUS Although no true liver exists in amphioxus, the presence of such a

structure in higher chordates may be foreshadowed in amphioxus by a hollow, forward-projecting, ventral saclike structure which comes off the intestine just posterior to the branchial region. The lining of this pouch is ciliated glandular epithelium, and it may have some important metabolic and secretory function. A system of veins coming from the intestine breaks up in capillaries in this structure, thus presaging the appearance of the hepatic portal vein of higher forms.

CYCLOSTOMES The liver of cyclostomes is unusually small. In the lamprey it appears to be single-lobed, but in the hagfish it consists of two halves, this representing the most typical condition. The ducts from these two parts in the hagfish open separately into the gallbladder. Neither gallbladder nor bile duct is present in an adult lamprey. The am-mocoetes larva, however, possesses both these structures, which degenerate during metamorphosis.

FISHES The livers of fishes are lobed and relatively large. Some are of economic value because of the oil and vitamins which they contain. A gallbladder is almost uni-formly present, though lacking in a few species of sharks.

AMPHIBIANS Amphibian livers are large in proportion to the size of the body. They are lobed, and a gallbladder is present.

REPTILES No important deviations in liver structure occur in reptiles. In snakes the liver consists only of one elongated lobe. All reptiles possess gallbladders.

BIRDS Lobed livers are the rule in birds, but in many a gallbladder is lacking. In the pigeon, which lacks a gallbladder, there are two bile ducts opening separately into the duodenum. One comes from each main liver lobe. It has been reported that a transitory gallbladder is present in the pigeon for a short time during embryonic development.

MAMMALS More variations occur in the lobulation of the mammalian liver than in any other group studied. The two main lobes are usually subdivided into smaller units ar-ranged in various ways. In some mammals there may be as many as six or seven lobes. In man there are five. Many mammals lack gall-bladders. These include certain rodents, whales, conies, some of the Artiodactyla, and all the Perissodactyla.

PANCREAS

The pancreas is the second largest of the digestive glands. It arises from the endoderm of the primitive gut in the same general region from which the liver develops. A single dorsal, and usually one or two ventral, pancreatic diverticula develop in most ver-tebrates. The proximal parts of the diverticula form the pancreatic ducts, and the distal parts, by an intricate process of budding, give rise to the main mass of pancreatic tissue. The ducts undergo various degrees of fusion or reduc-tion so that the number and arrangement show much diversity. Usually only one or two ducts persist. They may open independently into the duodenum or may join the bile duct. This duct joins the bile duct, the two having a common opening into the duodenum. A dorsal accessory *duct of Santorini* opens into the duodenum independently. The pancreas is a single organ of irregular shape and may con-sist of several lobes. It generally lies in the loop between stomach and duodenum. The pancreas of calves and sheep is sold on the market under the name "stomach sweet-breads."

The pancreas plays a dual role in the body, serving as both an exocrine and an endocrine gland. The exocrine secretion is called *pancreatic juice.* That part which is responsible for the endocrine secretion consists of small groups of cells called the *islands* (*islets*) *of Langerhans* (Fig. 9.30), which in most forms are scattered throughout the gland. They are not concerned with digestive processes.

Comparative Anatomy of the Pancreas

AMPHIOXUS No pancreas is present in amphioxus. Cells with pancreatic character-istics are located in the anterior end of the in-testine.

CYCLOSTOMES In the adult lamprey no well-defined pancreatic organ is evident. Recognizable pancreatic tissue is present, however, embedded within the liver and in the wall and typhlosole of the intestine. The hagfish possesses a small pancreas located

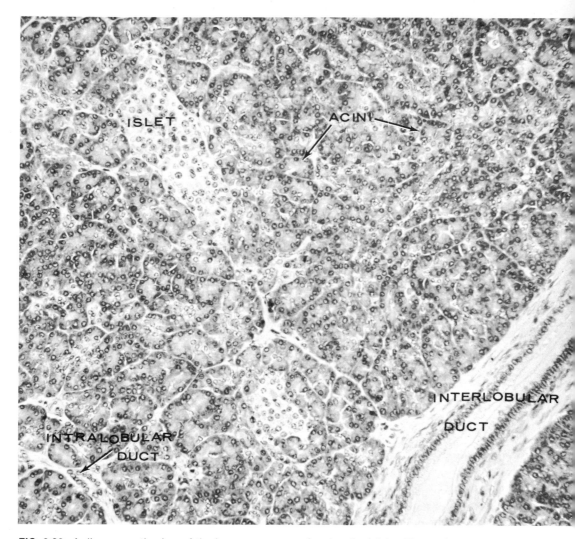

FIG. 9.30 A diagrammatic view of the human pancreas showing the islets of Langerhans. (*From R. O. Greep, "Histology," 3d ed., McGraw-Hill Book Company, New York, 1973, by permission*).

near the bile duct into which several pancreatic ducts open independently.

FISHES In certain fishes (Dipnoi and many teleosts) the pancreas is so diffuse as to be almost unrecognizable. In some, the endocrine portion is separated from the remainder of the pancreatic tissue and is spoken of as the *principal island.* Elasmobranchs have a well-defined pancreas consisting of dorsal and ventral lobes connected, in many cases, by a narrow isthmus. A single duct enters the duodenum.

AMPHIBIANS, REPTILES, AND BIRDS No noteworthy features are exhibited by the pancreas in members of these three classes of vertebrates. One duct or several may be present, and these may open directly into the duodenum or indirectly through the bile duct.

MAMMALS Generally two pancreatic ducts are present in mammals. The ventral duct opens into the bile duct near the ampulla (Fig. 9.29). The dorsal duct enters the duodenum directly, although its position varies in different forms. Both ducts are functional in the horse and dog. In the cat, sheep, and man the ventral duct only is usually functional, although the dorsal duct may persist in reduced form. In the pig and ox only the dorsal duct persists. A *pancreatic bladder* is sometimes found in the cat. Pancreatic juice may accumulate within it, and then be released into the intestine in much the same manner as bile is released from the gallbladder. The alkaline pancreatic juice contains numerous enzymes of great importance in digestion.

SUMMARY

The digestive tracts of vertebrates are built upon the same fundamental plan and function in a comparable manner. Many variations are to be found which may be considered as adaptations to meet particular needs. The signs of evolution are not so apparent in a study of the digestive system as in some of the other organ systems, but the adaptive radiations from a simple plan are strikingly brought out.

Among the more important advances that are to be noted within the group are:

1 Separation of mouth and nasal passages
2 Change to heterodont condition of teeth
3 Development of muscles in lips and tongue
4 Beginning of digestive function on part of mouth glands
5 Distinct separation of esophagus and stomach
6 Digestive function of stomach added to that of mere storage
7 Complicated stomachs in many forms, providing for proper utilization of type of food eaten
8 Differentiation of intestine into distinct regions
9 Tendency toward lengthening of alimentary canal as a whole
10 Provisions for increasing surface for absorption and secretion
11 Differentiation of cloaca so that openings of digestive and urogenital systems are separate

BIBLIOGRAPHY

Edmund, A. G.: Tooth Replacement Phenomena in the Lower Vertebrates, *R. Ont. Mus., Toronto, Life Sci. Div., Contrib.* 52, 1960.
Peyer, B.: "Comparative Odontology," University of Chicago Press, Chicago, 1968.
Rouiller, C.: "The Liver," Academic Press, New York, 1962.
Symposium of Evolution and Dynamics of Vertebrate Feeding Mechanisms, *Am. Zool.,* 1:177–234 (1961).

10 EXCRETORY SYSTEM

The word *metabolism* is used to designate the different chemical changes that take place in the cells of the body upon which heat production, muscular energy, growth, and maintenance of life processes depend. Some products of metabolism cannot be further utilized as sources of energy. They are called the end products of metabolism or *wastes* and must be eliminated. Among the waste products of metabolism are carbon dioxide, urea, ammonia, uric acid, creatinine, various pigments, inorganic salts, and water. Carbon dioxide is eliminated, for the most part, through the gills in aquatic forms, through the skin and lungs in amphibians, and through the lungs in terrestrial vertebrates. The remaining substances are excreted almost entirely through what are called the *excretory,* or *urinary, organs.* In many fishes nitrogenous and other ionic wastes may be excreted in considerable quantity by way of the gills. In man small amounts may be eliminated through the sweat glands. It should be recalled at this point that although feces may contain substances excreted through the bile or by the cells of the intestine, the greater bulk is not made up of true excreta but is indigestible material which has never entered the body proper.

Most excretory substances are in solution in water. Water itself is not considered to be a waste product, but any excess is eliminated along with substances dissolved in it. The final product is known as *urine.* The organs which in vertebrates are involved in urine formation are called *kidneys.*

All parts of the excretory system in vertebrates, except certain structures near the terminal duct openings, are mesodermal in origin and derived from the embryonic mesomere.

TYPES OF KIDNEYS

The excretory tubules of amphioxus (Fig. 10.1) are of an entirely different type from those of vertebrates. A series of such tubules opens into the atrium, or peribranchial space. Each lies on the outer dorsal side of a second-

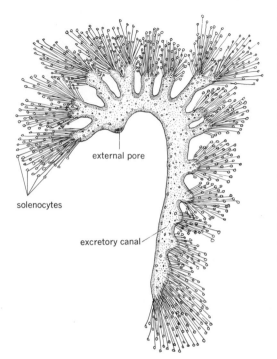

external pore

solenocytes

excretory canal

FIG. 10.1 One of the numerous excretory tubules of amphioxus. (*After Goodrich.*)

ary gill bar. They are apparently of ectodermal origin, have no connection with the coelom, and are composed of numerous flame cells called *solenocytes,* which collect wastes.

It has been difficult to work out homologies among the various types of excretory organs found in vertebrates. Variations that are encountered are correlated with problems with which vertebrates have had to cope in the past in adapting themselves to the many different environmental conditions. Whether animals remained aquatic or became terrestrial posed important problems in connection with excretion. Whether aquatic forms lived in fresh or salt water made a great difference in how they adapted themselves to changes in osmotic pressure, variations in salt concentration, and the like. Comparative anatomists have long speculated concerning the evolu-

tion of the various types of vertebrate kidneys. Gradually a fairly logical sequence of events has been postulated. There has been, and still is, much confusion in this field, particularly in regard to terminology. Investigators who studied kidney development in embryos of birds and mammals in the past have attempted to homologize these structures with excretory organs of lower forms. It now appears that certain of these speculations were erroneous. The student should be wary in accepting certain interpretations of homologies that appear in the older literature on the subject. Often the ideas advanced were correct, but the terminology employed has confused the issue.

ARCHINEPHROS

It is now generally held that the primitive vertebrate ancestor possessed an excretory organ referred to as an *archinephros* or *holonephros* (Fig. 10.2). This was composed of a pair of *archinephric ducts* located on the dorsal side of the body cavity and extending the length of the coelom. Each duct was joined by a series of segmentally arranged tubules, one tubule to each segment. At its other end the tubule opened into the coelom through a ciliated, funnel-shaped aperture called the *nephrostome.* Closely associated with each tubule was a small knot of capillaries interposed within the course of an arteriole. A thin layer of peritoneum was reflected over its surface. Tissue fluid, exuded at these *glomeruli,* passed into the coelom and thence through nephrostomes into the archinephric tubules and finally through the archinephric ducts to the outside. From this type of kidney the various kidneys of present-day forms may originally have been derived. Even today the larval form of the hagfish and the larvae of some caecilians possess kidneys resembling this archinephric arrangement.

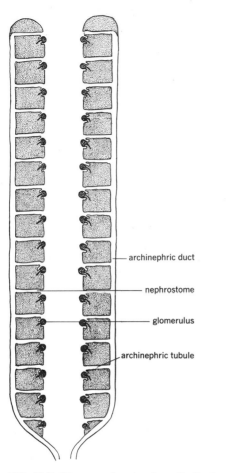

— archinephric duct

— nephrostome

— glomerulus

— archinephric tubule

FIG. 10.2 Diagram showing hypothetical structure of archinephros.

THE ANAMNIOTE KIDNEY

The anterior portion of the archinephric kidney has persisted in only a few vertebrates, and even in the adult hagfish it is modified. It appears in the embryos of most vertebrates as a transitory structure, usually referred to as the *pronephros* (Fig. 10.3). In the few amniote vertebrates in which it persists in the adult stage it is called the *head kidney*. The remainder of the kidney posterior to the

pronephric region is known as the *opisthonephros.*

There has been confusion over the correct term to be applied to the kidney duct. Both *archinephric duct* and *pronephric duct* have been used. We shall use the term *archinephric duct* here in referring to this primitive kidney duct as it appears in cyclostomes, fishes, and amphibians (*Anamniota*).

PRONEPHROS The pronephros in anamniotes consists of several anteriorly located *pronephric tubules* together with a pair of archinephric ducts. The tubules and ducts lie in the dorsolateral mesoderm on either side of the base of the mesentery that supports the gut. The ducts extend posteriorly, usually opening into the cloaca. One end of each of the segmentally arranged tubules connects with the archinephric duct near its anterior end; in fact, the duct is actually formed by successive tubules bending posteriorly and fusing with adjacent tubules. The other end of the tubule opens into the coelom by means of a ciliated nephrostome. In a few forms, an *external glomerulus* (Fig. 10.3), coming from segmental branches of the dorsal aorta, projects into the coelom near the nephrostomes. A thin layer of peritoneal epithelium is reflected over the projecting surface of each external glomerulus. Most forms, however, possess *internal glomeruli*. These are small knots of interarterial capillaries, each surrounded by a double-walled structure, called *Bowman's capsule,* the two together being known as a *renal* or Malpighian corpuscle (Figs. 10.4 and 10.5). Blood is brought to the glomerulus by an *afferent arteriole* and leaves through an *efferent arteriole*. The latter may break up into true capillaries along the course of a pronephric tubule, and the blood is ultimately returned to the heart through one of the postcardinal veins. Pronephric tubules retain a connection with the coelom. Whether glomeruli are of the internal or external type is of no particu-

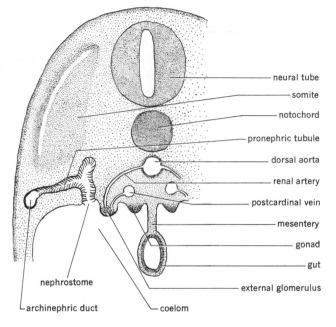

FIG. 10.3 Diagrammatic section through part of a vertebrate embryo, showing relation of external glomerulus to pronephric tubule and nephrostome.

FIG. 10.4 Diagrammatic section through part of a vertebrate embryo, showing relation of internal glomerulus to pronephric tubule and archinephric duct.

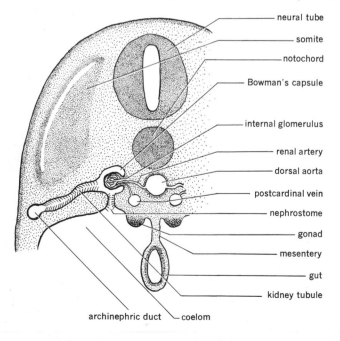

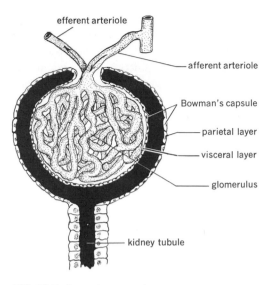

efferent arteriole

afferent arteriole

Bowman's capsule

parietal layer

visceral layer

glomerulus

kidney tubule

FIG. 10.5 A renal corpuscle

lar significance in their relation to the pronephros. A protein-free filtrate of blood plasma passes from the glomerulus into the coelom or into the cavity of Bowman's capsule and thence into the tubule. Cells in the wall of the tubule probably excrete nitrogenous wastes into the lumen, where they are added to the fluid filtered from the glomerulus. Selective reabsorption of water and certain other constituents may occur as the fluid passes down the tubule, so that a smaller volume of fluid with a higher concentration of wastes in solution finally enters the archinephric duct.

Sometimes several glomeruli unite to form a larger *glomus.* Pronephric tubules may expand to form *pronephric chambers,* or several tubules may fuse to form a large chamber.

The importance of the pronephros lies chiefly in the part it plays in forming the archinephric duct, which persists even though the pronephros may disappear.

In members of the class Chondrichthyes the pronephros degenerates soon after it is formed. In many fishes and in larval amphibians, however, it becomes modified and func-

tions for a time as an excretory organ. Only in the hagfish and a few teleosts does the head kidney with its peritoneal connections persist in adult life. The nephrostomes of the hagfish open into the pericardial cavity, and the fluid which passes down the tubules enters an adjacent vein rather than the archinephric duct. In the hagfish pronephros and opisthonephros become completely separated by degeneration of the portion between them.

OPISTHONEPHROS Since the pronephros in most cases is but a transient structure, the opisthonephros is actually the more important part. It serves as the adult kidney in lampreys, most fishes, and amphibians.

Frequently the term *mesonephros* is used in describing what we have called the opisthonephros. The opisthonephros, however, is not quite comparable to the mesonephros of embryonic amniotes, even though the two are structurally similar. We shall reserve the term mesonephros for the structure which appears *during embryonic development* in reptiles, birds, and mammals. The reason for the distinction lies in the fact that the three types of kidneys, pronephros, mesonephros, and metanephros, which appear in amniote embryos represent developments from different levels of the primitive archinephros which appear in succession in an anterior-posterior direction. The opisthonephros of cyclostomes, fishes, and amphibians actually extends over a region which in amniotes forms both mesonephric and metanephric kidneys.

The opisthonephros differs from the pronephros in several respects. In males the archinephric duct may be taken over almost entirely by the reproductive system, in which case accessory urinary ducts are formed which carry away waste materials. A main distinction between pronephros and opisthonephros is that in the opisthonephros the segmental arrangement of the tubules no longer exists and several tubules lie within a given seg-

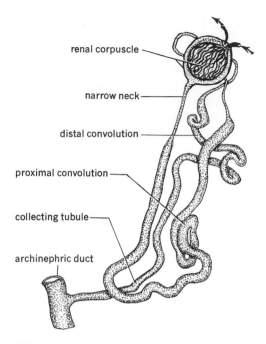

renal corpuscle

narrow neck

distal convolution

proximal convolution

collecting tubule

archinephric duct

FIG. 10.6 An amphibian kidney tubule of the opisthonephros, showing the renal corpuscle and secretory and collecting portions.

ment. This is most apparent toward the posterior end of the opisthonephros. Furthermore, there are rarely any connections between kidney tubules and coelom. Renal corpuscles with internal glomeruli are typically present. In some forms the anterior tubules of the opisthonephros lie in the same segments as posterior pronephric tubules, thus indicating the transitional nature of the two kinds of kidneys.

A typical opisthonephric tubule (Fig. 10.6) consists of a narrow neck adjacent to the renal corpuscle, followed in turn by secretory and collecting portions. The collecting portion joins the archinephric duct. The secretory part forms two loops named the *proximal* and *distal convoluted tubules* in relation to their proximity to the renal corpuscle. The collecting ends of several collecting tubules may

unite to form a ureterlike duct which opens into the archinephric duct or may establish an independent connection with the cloaca. Several such accessory ducts may be present.

Only internal glomeruli are present in the opisthonephros. In a few cases peritoneal funnels connect the tubules with the coelom, but this is unusual.

Comparative Anatomy of the Opisthonephros

CYCLOSTOMES In the larval stage of the hagfish the kidney is of the archinephric type. In the adult, however, as has been pointed out, the anterior end has become modified to form a persistent head kidney, or pronephros. The remainder of the kidney, however, which is separated from the head kidney, becomes an opisthonephros. It differs but little from the original archinephros except that the posterior tubules lose their peritoneal connections.

The opisthonephros of the lamprey consists on each side of a long, strap-shaped body without any peritoneal connections in the adult. The kidneys lie on either side of the middorsal line, from which each is suspended by a mesenterylike membrane. The archinephric duct courses along the free edge of the kidney (Fig. 10.7). In *Petromyzon marinus* a vestige of the part of the archinephric duct, which was associated with the degenerated pronephros, extends forward from the opisthonephros. The ducts from the two sides unite posteriorly to open into a *urogenital sinus* which leads to the outside through an aperture at the tip of a small *urogenital papilla*. Two slitlike openings, the *genital pores*, connect the urogenital sinus with the coelom. Reproductive cells (eggs or spermatozoa) leave the body cavity via genital pores, urogenital sinus, and urogenital aperture. The condition is similar in both sexes.

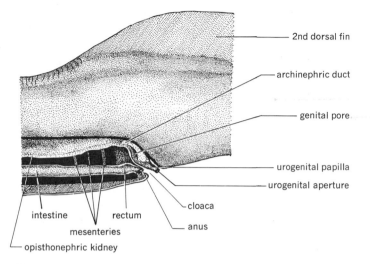

2nd dorsal fin

archinephric duct

genital pore

urogenital papilla

urogenital aperture

cloaca

anus

intestine   rectum

mesenteries

opisthonephric kidney

FIG. 10.7 Position and arrangement of left genital pore of lamprey. It is through the genital pores that the reproductive elements leave the coelom to enter the urogenital sinus.

FISHES The opisthonephric kidneys of fishes show great variation in shape but are fundamentally similar in structure. They are dorsal in position in all species. In some they extend almost the entire length of the coelom. The two sides may show various degrees of fusion. In some they are short and confined to the posterior part of the body cavity. Peritoneal funnels are retained in only a few forms, notably *Amia,* sturgeons, and certain elasmobranchs. In some marine teleosts no external or internal glomeruli are present, such kidneys being referred to as *aglomerular* kidneys.

In general the opisthonephric kidneys of male fishes are longer than those of females, the anterior ends in males having been appropriated by the reproductive system. In the males of some groups small modified kidney tubules, called *efferent ductules,* connect the testes with the archinephric duct. The latter then becomes the *ductus deferens,* which serves primarily for sperm transport. There is a marked tendency in such cases for the posterior portion of the opisthonephros to assume the greater part of the excretory function, with one or more accessory ducts developing and carrying wastes directly to the cloaca or to the outside. Usually the connection of the testis and archinephric duct occurs at the anterior end of the opisthonephros (selachians, chondrosteans, and some others). In teleosts there is no connection between the testes and opisthonephric kidneys. The ducts from the testes either join the archinephric ducts near their posterior ends or open independently to the exterior. The term *ductus deferens* is properly used only in reference to the archinephric duct when it is used for sperm transport.

In female fishes the posterior ends of the archinephric ducts enter a common *urinary sinus* inside a small *urinary papilla.* The latter enters the cloaca in dipnoans and elasmobranchs, but in most other fishes it opens directly to the outside, a cloaca being absent.

Dilatations of the archinephric ducts may form bladderlike enlargements for the temporary storage of urine. In forms in which the archinephric duct is used as a ductus deferens, enlargements called *seminal vesicles* and *sperm sacs* may develop, serving as temporary storage places for spermatozoa.

AMPHIBIANS The primitive archinephric type of kidney found in the larval stage of the hagfish also occurs in larval caecilians in which there is a distinct metameric arrangement of kidney tubules, renal corpuscles, and nephrostomes. In an adult the opisthonephros extends the greater part of the length of the coelom and is lobulated.

FIG. 10.8 Urogenital organs of female salamander, ventral view. The ovary on the right side and the oviduct on the left side have been removed for clarity.

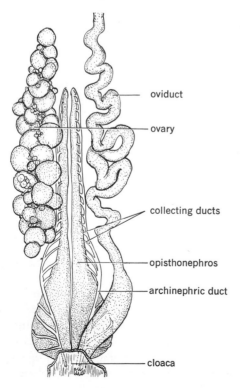

- oviduct
- ovary
- collecting ducts
- opisthonephros
- archinephric duct
- cloaca

Although a small head kidney with peritoneal connections is present in many larval amphibians, it does not persist in the adult stage. It is of interest, however, that adult frogs have ciliated nephrostomes on the ventral surfaces of the kidneys. They are not usually connected with kidney tubules but have become secondarily connected with the renal veins.

Urodele amphibians have opisthonephric kidneys much like those of elasmobranchs. They consist of two regions: an anterior portion, which in males is concerned more with genital than urinary functions, and a posterior expanded region, which makes up the main part of the opisthonephros. The archinephric duct courses along the lateral edge of the kidney a short distance from the kidney proper. Numerous collecting ducts or tubules leave the opisthonephros at intervals to join the archinephric duct. These are less well developed in females than in males (Figs. 10.8 and 10.9). The archinephric ducts in both sexes open on either side of the cloaca at the apex of a small papilla. In *Necturus* peritoneal connections with some of the kidney tubules persist throughout life.

The opisthonephric kidneys of anurans show a more posterior concentration of tubules and are confined to the posterior part of the abdominal cavity. They are dorsally located, retroperitoneal, and flattened in a dorsoventral direction. There is no clear-cut distinction between anterior and posterior ends, as in urodeles. An adrenal gland of a yellowish-orange color is located on the ventral side of the kidney, to which it is closely attached. The kidneys of female anurans have no relation to the reproductive system, but in males an intimate connection exists (Fig. 10.10). Certain anterior kidney tubules have become modified as efferent ductules connecting the testis with the kidney and archinephric duct which serves to transport spermatozoa as well as urinary wastes. The archinephric duct, contrary to the condition in urodeles, is located *within* the kidney along its lateral margin. It leaves the opisthonephros near the posterior end and passes to the cloaca.

A thin-walled urinary bladder connects with the amphibian cloaca a short distance beyond the openings of the archinephric ducts. It is bilobed. There is no direct connection of the ducts with the bladder, so that the thin, watery urine first passes into the cloaca. It has already been mentioned that the amphibian bladder originates from the cloaca. Most authorities are of the opinion that it is

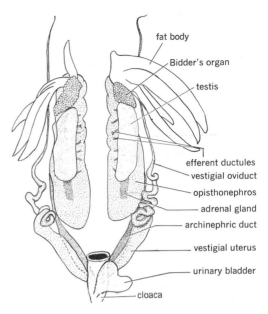

FIG. 10.10 Urogenital organs of male toad *Bufo americanus,* ventral view. (*Modified from Witschi, in Allen, "Sex and Internal Secretions," 2d ed., The Williams & Wilkins Company, Balitimore, by permission.*)

of endodermal origin and homologous with the urinary bladders of higher forms.

THE AMNIOTE KIDNEY

In reptiles, birds, and mammals three types of kidneys are usually recognized: <u>pronephros, mesonephros, and metanephros.</u> These appear in succession during embryonic development, but only one, the metanephros, persists to become the adult kidney. Mesonephros and metanephros actually represent different levels of the opisthonephros of anamniotes, the metanephros being the equivalent of the posterior portion.

In all forms an anteriorly located pronephros appears during very early stages of development, but it soon degenerates, and the more posterior mesonephros then develops.

FIG. 10.9 Urogenital organs of male salamander, ventral view. The collecting ducts on the right side are shown detached from the cloaca and spread out for clarity.

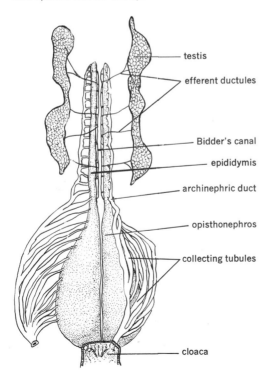

testis

efferent ductules

Bidder's canal

epididymis

archinephric duct

opisthonephros

collecting tubules

cloaca

The duct of the pronephros, however, persists to become the duct of the mesonephros. This is actually the same as the archinephric duct, to which considerable reference has already been made. Since most accounts refer to this duct as the *mesonephric,* or *Wolffian, duct,* and to avoid further confusion, we shall use the term <u>Wolffian duct</u> here in referring to the <u>archinephric duct</u> as it appears in amniotes.

The mesonephros persists for a time and then degenerates. In the meantime the metanephros has begun to develop from the region posterior to the mesonephros. Portions of the mesonephros may persist to contribute to the reproductive system in the male or to remain as vestigial structures without any apparent function.

MESONEPHROS The mesonephros is sometimes called the *Wolffian body.* Mesonephric tubules differentiate in the same manner as opisthonephric tubules (page 280). Some of the anterior tubules may form peritoneal connections, but this is unusual. In some forms the mesonephric kidneys become rather voluminous structures temporarily occupying an extensive portion of the body cavity. On the contrary, in the embryonic stages of the mouse and guinea pig the mesonephros is so poorly developed and appears so early that it can scarcely be regarded at any time as a functional structure. In reptiles, *Echidna,* and certain marsupials, the mesonephros may even persist for a time after birth. Well-developed peritoneal funnels associated with the mesonephros have been described for the monotremes.

As long as the mesonephros functions, the Wolffian duct serves as the urinary passage; but as soon as the metanephros becomes functional, this duct degenerates in the female, although it persists in the male. Remnants of the mesonephros which are left after the main part has degenerated include por-

tions which become associated with the reproductive system. Masculine structures <u>derived from mesonephric components</u> include the *epididymis, ductus deferens, seminal vesicles, paradidymis,* and *ductus aberrans.* Rudiments in the female include the *epoophoron, paraoophoron,* and *canal of Gärtner.*

METANEPHROS The metanephric type of kidney, <u>found only in amniotes</u>, arises posterior to the mesonephros on each side. It is a more compact organ than the mesonephros. It comes from a level which corresponds to the posterior portion of the opisthonephros of anamniotes. The metanephros is made up of essentially the same parts as the mesonephros and contains renal corpuscles, secretory tubules, and collecting tubules. No nephrostomes are present. Each kidney has a twofold origin. A diverticulum, the *ureteric bud,* from the posterior end of the Wolffian duct, grows forward into the so-called *metanephric blastema,* which lies posterior to the mesonephros. The blastema is continuous with the nephrogenous tissue from which the mesonephric tubules were derived. The diverticulum of the Wolffian duct is destined to form the *ureter* and the collecting portion of

FIG. 10.11 Manner of branching of diverticulum of Wolffian duct during early development of metanephros. (*After Huber.*)

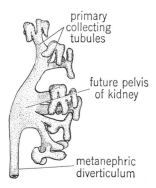

primary collecting tubules

future pelvis of kidney

metanephric diverticulum

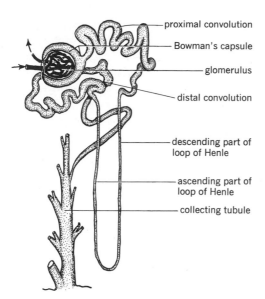

FIG. 10.12 A mammalian metanephric tubule, showing the renal corpuscle as well as secretory and collecting portions.

proximal convolution

Bowman's capsule

glomerulus

distal convolution

descending part of loop of Henle

ascending part of loop of Henle

collecting tubule

the kidney. The ureteric bud branches and rebranches a varying number of times, ultimately to form large numbers of fine *collecting tubules.* At least in mammals, at a point where the bud undergoes its primary divisions, an expanded region may be noted. This becomes the *pelvis* of the kidney (Fig. 10.11).

Condensations in the mesenchyme of the adjacent blastema soon give rise to hollow *secretory tubules* which become S-shaped. One end of each secretory tubule establishes connection with a collecting tubule. The other end expands and is soon invaded by a glomerular tuft from a small branch of the renal artery, so that a typical renal corpuscle is formed. The differentiation of a secretory tubule of the metanephros in mammals, and to a lesser extent in birds, is somewhat more elaborate than that of a mesonephric tubule (Fig. 10.12). Each tubule, as it leaves Bowman's capsule, consists of a *proximal convoluted tu-*

bule, a long *loop of Henle* with its *descending* and *ascending portions,* and a *distal convoluted tubule.* Secretory tubules with their Bowman's capsules are very numerous.

The metanephros functions in a manner similar to that of the mesonephros. Wastes are carried from the kidneys by the ureters, which enter the cloaca or urinary bladder, as the case may be. In reptiles and birds resorption of water occurs in the cloaca, with precipitation of the organic matter as a chalky mass of urea or uric acid.

Comparative Anatomy of the Metanephros

REPTILES The kidneys of reptiles are restricted to the posterior half of the abdominal cavity and are usually confined to the pelvic region. They are generally small and compact, with a lobulated surface. The posterior portion narrows down on each side, and in some lizards the hind parts may even fuse. The degree of symmetry varies and is most divergent in snakes and limbless lizards, which have excessively lobulated, long, narrow kidneys in correlation with the shape of the body. One kidney may lie entirely posterior to the other.

A urinary bladder is lacking in snakes and crocodilians. Most lizards and turtles have well-developed and usually bilobed bladders which open into the cloaca. Except in turtles the ureters open separately into the cloaca. In turtles they are connected to the bladder. In some turtles a pair of *accessory urinary bladders* is also connected with the cloaca. They function as accessory organs of respiration. In females they may be filled with water, which is used to soften the ground when a nest is being prepared.

BIRDS In all birds the kidneys are situated in the pelvic region of the body cavity, the posterior ends frequently being united. They are lobed structures with short ureters which open independently into the cloaca.

Except for the ostrich, birds have no urinary bladders. Urinary wastes, chiefly in the form of uric acid, are eliminated via the cloaca along with the feces.

MAMMALS The typical mammalian metanephric kidney (Fig. 10.13) is a compact, bean-shaped organ attached to the dorsal body wall. It is retroperitoneal. The ureter leaves the medial side at a depression called the *hilum* or *hilus*. At this point a *renal vein* also leaves the kidney, and a *renal artery* and nerves enter it. The kidney is surrounded by a *capsule* of connective tissue, under which lies the *cortex*. The renal corpuscles and the convoluted portions of the secretory tubules lie entirely in the cortex. Immediately beneath the cortex is the medulla of the kidney. It is

FIG. 10.13 Sagittal section of metanephric kidney of man (semidiagrammatic.)

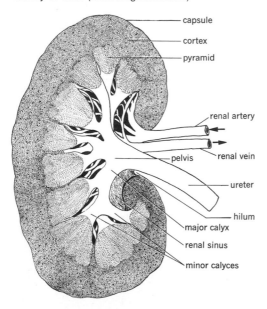

partially composed of large areas known as *renal pyramids.* The outer borders of the pyramids are subdivided into smaller units called *lobules.* The collecting tubules lie in the pyramids but may extend well up toward the cortex. The inner portion of each pyramid, in the form of a blunt papilla, projects into an outpocketing of the pelvis known as a *minor calyx.* Several minor calyces join together to enter *major calyces,* which in turn open into the pelvis. The pelvis leads to the ureter, which empties into the bladder (cloaca in monotremes). Urine, which is stored temporarily in the bladder, passes to the outside through the *urethra.*

The electron microscope has revealed that the surfaces of the cells of the proximal convoluted tubule facing the lumen are covered with exceedingly numerous *microvilli,* each about 1 μm (1 micrometer = 10^{-6} meter) in length. Estimates indicate that in man the total surface of the microvilli in all the proximal convoluted tubules represents an area of 50 to 60 sq m. As much as two-thirds of the water of the glomerular filtrate is resorbed here. The electron microscope has also shown that the detailed structure of the renal corpuscle is much more complicated than was formerly believed.

An enzyme, renin (rē′nin), is produced by certain cells in the kidneys under special conditions. It acts upon a protein, *hypertensinogen,* derived from the liver, to form a substance called angiotensin I. Another enzyme acts upon this to form angiotensin II, which has a profound effect in causing constriction of arterioles and thus elevating the blood pressure. Another factor, *erythropoietin,* presumably of kidney origin, is believed to stimulate the production of erythrocytes, or red corpuscles.

The metanephros of mammals shows marked lobation in the embryo. In many forms this condition is retained throughout life. In others the lobation is not superficially apparent in the adult and the kidney surface is relatively smooth.

Urinary bladders, present in all mammals, are muscular sacs derived from the ventral cloacal wall and possibly from a portion of the allantois. The bladder narrows down and connects to the outside by the urethra. The lower ends of the ureters, except in monotremes, open directly into the bladder on its dorsal posterior surface, usually near the urethral end. In monotremes they open into the urethra through small papillae near the base of the bladder. At the junction of bladder and urethra there is an abundance of elastic tissue. Much of the bladder musculature in the form of bundles continues down into the urethra. Urine flow ceases when the urethra is lengthened and the diameter of the lumen is reduced. In males the urethra passes through the penis to open at the tip of that organ through the *external urethral orifice,* or *meatus.* In females the condition varies. In some, as in the rat and mouse, the urethra opens independently to the outside, passing through the clitoris; in others, it enters a urogenital sinus, or vestibule, which is the terminal part of the genital and urinary tract.

EFFECT OF ENVIRONMENT ON KIDNEY STRUCTURE AND FUNCTION

The importance of the kidney is not confined to waste elimination. It is of great importance in maintaining a fluid environment in the body which keeps the body cells in satisfactory condition (1) by regulating the amount of water given off in the urine and (2) by controlling the amounts of sodium chloride and other electrolytes eliminated through the urine. Vertebrates occupy very diversified habitats in which water may not be available (marine or desert environment) or may be in abundance (freshwater environment). The structure of the kidney reflects, to a degree,

FIG. 10.14 Evolution of renal unity in vertebrates: *A,* freshwater fishes, elasmobranch, amphibians; *B,* marine actinopterygians, reptiles; *C,* birds and mammals. (*After A. S. Romer, "The Vertebrate Body," W. B. Saunders Company, Philadelphia, 1970.*)

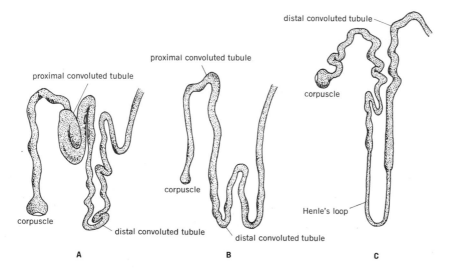

the environmental condition in terms of water availability and has evolved structural means to handle the problem of water balance.

The glomeruli of kidneys (Fig. 10.14) are the initial filters, the size of which determines the amount of filtrate removed from the blood. An increase in the glomerulus size represents an increase of filtrate (water) moving out of the blood into the kidney tubules. Since this filtered fluid, like the blood, has a high concentration of chemicals which are essential for life, it is important that they be retained by the body. The kidney tubules, with their ability to reabsorb these chemicals selec-

tively, have evolved as a means of effecting such a conservation.

Kidney tubules show tremendous variation in vertebrates (Fig. 10.15). All possess a proximal convoluted tubule. In the aglomerular kidneys of the marine teleosts, the proximal convoluted tubule is the only portion that is functional, yet the urine of these fishes does not differ basically from urine of fishes with glomeruli. The proximal convoluted tubule in such cases is undoubtedly excretory. Its excretory activity has been clearly demonstrated, under experimental conditions, for glomerular kidneys.

FIG. 10.15 Variation in the components of vertebrate nephron tubules. *A*, hagfish; *B*, skate; *C*, sculpin; *D*, catfish; *E*, toadfish; *F*, frog; *G*, turtle; *H, I*, chicken; *J, K*, rabbit. (*Modified from Prosser; adapted from Romer, "The Vertebrate Body," W. B. Saunders Company, by permission.*)

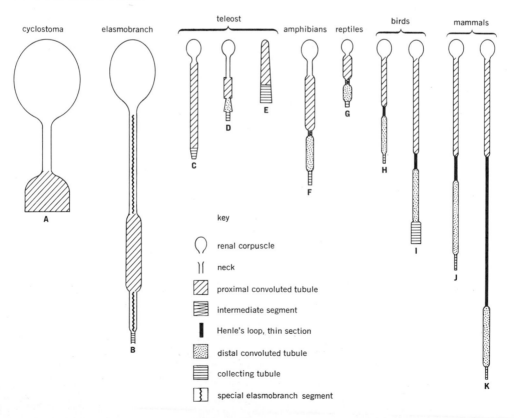

The activities of the distal convoluted tubule include ionic and water reabsorption. In amphibians, a degree of selective reabsorption takes place here, particularly of chloride and bicarbonate.

In reptiles, birds, and mammals the tubules of the kidney are present but their development depends on the amount of water available to the animal. In desert forms, in which it is scarce, the distal convoluted tubules are very long and are capable of resorbing nearly all the water, producing a very concentrated urine. Animals which live where water is easily accessible have shorter tubules and their urine is less concentrated.

Vertebrates in a freshwater habitat have a relatively high concentration of salts in the blood. The osmotic pressure of the body fluids is considerably higher than that of the water in which the animal lives. Under such conditions water is constantly entering the body because of osmotic inflow. Ridding the body of this water while maintaining ionic balance is essential to keep the constituents of the blood constant. The glomeruli of the kidney are well developed, providing for a high output of water. The distal tubules are also well developed and function as the primary absorbent of ions from the watery urine.

Marine vertebrates are faced with the relatively high salt concentration and osmotic pressure of seawater. Osmotic extraction tends to deplete the body of its water, but water must be retained so that the kidneys can form dilute urine for the elimination of wastes. One way in which this is accomplished is by taking seawater into the digestive tract and absorbing most of the water and salts. The salt is then excreted through the gills and rectal gland. It has been shown that the rectal gland of the elasmobranch fish *Squalus acanthias* secretes a solution of sodium chloride, the concentration of which may be greater than that of seawater and is about twice that of blood plasma.

In certain marine teleosts there is no need to remove water from the body since the osmotic pressure of the blood is lower than that of seawater. Nitrogenous wastes, as well as some ionic salts, are eliminated via the gills. In these marine teleosts, glomeruli appear during early developmental stages but later degenerate, the adults having an aglomerular kidney. Such kidneys are of functional importance in the regulation of ions in the blood via tubular excretion.

Another method of absorbing water occurs in the elasmobranchs and coelacanth fishes, in which a high concentration of urea in the blood under normal conditions means that the osmotic pressure of the blood is higher than that of the marine environment. Water is absorbed directly through gills and other membranes. The permeability of the gills and membranes of the mouth to urea is diminished in these fishes, a factor of importance in retaining a high concentration of urea in the blood. The elasmobranch fishes have well-developed kidneys with numerous glomeruli. Since the osmotic pressure of the blood is high, as in freshwater fishes, water must be constantly removed by glomerular filtration.

Kidney tubules of elasmobranchs differ from those of other vertebrates in having two peculiar portions, or segments. The first is situated between Bowman's capsule and the proximal convoluted tubule; the second follows the proximal convoluted tubule. It has been known for some time that elasmobranch fishes differ from all other vertebrates in the large percentage of urea present in tissues, blood, and other body fluids. The urine is particularly low in its concentration of urea, indicating that most of the urea filtered through the glomeruli is reabsorbed. The unique segments of the elasmobranch kidney tubule seem to be specially adapted for selective reabsorption of urea. Not all species of elasmobranchs have these segments. They are notably absent in the electric ray and the bat ray.

In these species, however, the proximal convoluted tubules show a differentiation into two regions, based on cell structure and thickness. These serve a function similar to that of the elasmobranch segments.

Terrestrial forms have much the same problems as marine fishes. They must maintain the water available to them by forming relatively concentrated urine. The glomeruli are reduced in reptiles and birds. In mammals and to some extent in birds an increase in the length of the distal convoluted tubules forming Henle's loop has evolved. Henle's loop is specialized for water absorption and generally bears an inverse relationship of length to the amount of water available to the animal.

In addition to water absorption in the kidney, the bladder in most vertebrates acts as a storage area where additional water may be reabsorbed.

Salt Glands

Terrestrial vertebrates that have invaded the marine habitat, removing themselves from any freshwater source, face an increased problem of removing excess salt from the tissues. Such forms as marine turtles, lizards, and marine birds have an accessory salt-secreting organ on the head termed a *nasal* or *orbital gland.* It secretes a watery fluid containing sodium and potassium chlorides. The concentration of the fluid may be 50 percent higher than seawater and is many times higher than the organism's blood. The gland is located in the corner of the eye in marine turtles and birds, the secretion being shed as tears in the former and flowing down a canal in the beak to the tip in the latter. In the marine iguana the secretion empties via ducts into the nasal passage and is forcefully expelled during exhalation.

A similar structure, the nasal gland, is present in desert reptiles. Its function is the same as for marine forms. A salt gland has been described in the desert roadrunner but is undescribed for other desert birds.

A salt gland (the *natrial gland*) described for marine snakes is located in the palate.

The method of salt regulation in marine mammals is not clear. The kidneys are presumably the major organ of salt balance, excess salt being secreted in the urine.

SUMMARY

1   The excretory organs of vertebrates consist of paired kidneys together with their ducts. Several types of kidney are recognized within the group.

2   The various kinds of kidneys have all been derived from a primitive structure referred to as the archinephros or holonephros. This consisted of a pair of archinephric ducts which extended the length of the coelom and which were joined by segmentally arranged tubules, one pair to each segment. The free end of each tubule opened into the coelom by means of a ciliated, funnel-shaped nephrostome. Associated with each tubule was a small knot of interarterial capillaries known as a glomerulus. The larval stages of the hagfish and caecilians exhibit an archinephros condition. The various types of vertebrate kidneys may be regarded as successive stages which have evolved in a craniocaudal direction from the original archinephros.

3   In adult anamniotes (cyclostomes, fishes, and amphibians) the anterior portion of the primitive archinephros usually becomes modified or degenerates. In the embryo, however, it appears as a transitory structure called the pronephros. In a few lower vertebrates the pronephros persists in the adult stage and is called the head kidney. A head kidney is found in the adult hagfish and in certain teleosts.

4   The remainder of the anamniote kidney, the opisthonephros, retains the original archinephric duct but differs from the pronephros in that several kidney tubules may be present in each segment and the tubules usually lose their peritoneal connections.

5   There is a general tendency for concentration of kidney tubules toward the posterior end of the opisthonephros. The anterior end loses its significance as an excretory organ and in the male is appropriated by the reproductive system, the archinephric duct becoming the ductus deferens. Accessory urinary ducts may then form by a fusion of two or more kidney tubules and serve to drain the urinary wastes.

6   In amniotes (reptiles, birds, and mammals) three types of kidneys are recognized, pronephros, mesonephros, and metanephros. These appear in succession in a craniocaudal direction during embryonic development, but only the metanephros persists to become the adult kidney. Mesonephros and metanephros represent different levels of the opisthonephros of the anamniota, the metanephros being the equivalent of the posterior portion. The archinephric duct is now called the Wolffian duct.

7   The pronephros of amniotes, with its metamerically arranged tubules with peritoneal connections, is merely a transitory structure of the embryo and has little functional importance. Its duct, however, persists as the Wolffian duct.

8   The mesonephros appears only in the embryos of amniotes. Several kidney tubules are present in each segment and most or all peritoneal connections are lost. This kidney may function for a time but then degenerates and disappears. Vestiges of the mesonephros and Wolffian duct persist. In males the latter gives rise to the epididymis, ductus deferens, and certain other parts of the reproductive system.

9   The metanephros has no peritoneal connections. Its origin is twofold. The collecting portion, including the ureter, is derived from a diverticulum of the Wolffian duct; the secretory portion arises from nephrogenous tissue, the metanephric blastema, which lies caudal to the mesonephros.

10   The important factors in kidney function involve glomerular filtration at the renal corpuscle, selective reabsorption by the secretory tubules, and excretion of wastes by cells of the secretory tubules.

11   Ducts from the kidneys lead to the cloaca in most forms. In teleost fishes they open directly to the outside. In mammals they enter the urinary bladder (monotremes are exceptions). Urinary bladders, when present, represent ventral outpocketings of the wall of the cloaca. In mammals, the bladder opens to the outside through a urethra, which in males of all forms, except monotremes, is also utilized by the reproductive system.

12   The important factors in the evolution of the basic structure and function of the vertebrate kidney appear to have been associated with body-fluid regulation. This involves the maintenance of a constant water and salt content of the body and is modified by the type of environment.

13   In marine reptiles and birds an accessory gland, the nasal or orbital gland, assists the kidney in regulating salt balance. A watery fluid is secreted which contains high concentration of salts. A similar structure in desert reptiles helps to regulate ionic concentration.

BIBLIOGRAPHY

Bentley, P. J.: Adaptations of Amphibia to Arid Environments, *Science,* 152:619–623 (1966).

Dunson, W. A., and A. M. Taub: Extrarenal Salt Excretion in Sea Snakes *(Laticanda),* *Am. J. Physiol.,* 213:975–982 (1967).

Fox, H.: The Amphibian Pronephros, *Q. Rev. Biol.,* 38:1–25 (1963).

Fraser, E. A.: The Development of the Vertebrate Excretory System, *Biol. Rev.,* 25:159–187 (1950).

Schmidt-Nielson, K.: Organ Systems in Adaptation: The Excretory System in D. B. Dill, E. F. Adolph, and C. G. Wilber (eds.), "Handbook of Physiology," sec. IV, Adaptation to the Environment, American Physiological Society, Washington, 1964.

———, C. Barker Jorgenson, and H. Osaki: Extrarenal Salt Excretion in Birds, *Am. J. Physiol.,* 193:101–107 (1958).

11 REPRODUCTIVE SYSTEM

Sexual, or *gamic,* reproduction is the rule in the phylum Chordata. It involves the union of *gametes,* or reproductive cells, of male and female to form a *zygote,* or fertilized egg. Under proper conditions and by a series of complicated processes the zygote develops into an embryo.

REPRODUCTIVE ORGANS Egg cells, or *ova,* and sperm cells, or *spermatozoa,* are formed in the *primary reproductive organs,* which are spoken of collectively as the *gonads.* The gonads in the male are known as *testes;* those of the female are the *ovaries.* Besides forming gametes, both ovaries and testes give off endocrine secretions, or *hormones,* which pass into blood or lymph and are transported to other parts of the body, where they may bring about profound effects. A discussion of the endocrine activity of the gonads is included in Chap. 14.

The gonads are paired structures, although in such forms as cyclostomes, certain fishes, and female birds of most species, what seems to be an unpaired gonad is the result either of fusion of paired structures or of unilateral degeneration. Only in the primitive amphioxus among chordates do the gonads show evidence of metamerism.

The gonads are mesodermal derivatives. In very early stages of development sexual differences are difficult to recognize. The ovaries and testes come to be attached to the dorsal body wall by mesenterylike bands of tissue, the *mesorchium* in the male and the *mesovarium* in the female.

The gametes formed in the gonads of both sexes must be transported to the outside of the body either as ova or spermatozoa. For this purpose, in most vertebrates, ducts are utilized, those of the male being known as *deferent ducts* and those of the female as *oviducts.* In a few forms, as in cyclostomes, no ducts are present in either sex, and eggs and sperm escape from the body cavity through *genital pores* (Fig. 10.7). The deferent ducts of the male are usually the archinephric, or the Wolffian, ducts, which may also serve to transport urinary wastes from the opisthonephric or mesonephric kidneys in those animals in which these kidneys function either during adult or embryonic life. In amniotes, in which the metanephros is the functional adult kidney and in which the mesonephros degenerates, the Wolffian duct of the male on each side persists to become the ductus deferens.

When the reproductive ducts first develop, in most vertebrates they open posteriorly into the cloaca, in both sexes. This relationship persists throughout life in many vertebrates, but in others modifications in the cloacal region occur and the reproductive ducts either open separately to the outside or, at least in the male, join the excretory ducts to emerge by a common orifice.

In many aquatic vertebrates fertilization is external. However, in all terrestrial forms except anuran amphibians and even in many aquatic species, internal fertilization is the rule. The transport of spermatozoa from male to female in some is brought about by apposition of the cloacae of the two sexes. More commonly, however, copulatory organs are present in the male which are employed in an intromittent manner in depositing spermatozoa in the reproductive tract of the female. A variety of types of copulatory organs is found among vertebrates.

In both sexes all structures or organs which serve to bring the germ cells, or products of the primary sex organs, together are known as *accessory sex organs.* They include the reproductive ducts, associated glands, and intromittent organs. *Secondary sex characters,* which are indirectly concerned with sex, play a part in the reproductive scheme. Sexual differences in such secondary sex characters as ornamental plumage, body size and strength, and vocal apparatus are but indirectly related to reproduction. The development, maintenance, and function of the accessory sex organs and secondary sex characteristics are controlled, at least in part, by the endocrine secretions of the gonads.

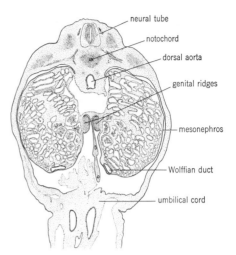

neural tube
notochord
dorsal aorta
genital ridges
mesonephros
Wolffian duct
umbilical cord

FIG. 11.1 Cross section through 12-mm pig embryo, showing relation of genital ridges to mesonephros.

It is generally agreed that, at least in amphibians, birds, and mammals, the *primordial germ cells* which give rise to eggs and sperm originate in the embryo in close association with the yolk-sac splanchnopleure. The cells migrate to areas called *genital ridges* (Fig. 11.1) located along the medial sides of the opisthonephric or mesonephric kidneys, close to the base of the dorsal mesentery. The genital ridges develop into the gonads (Fig. 11.2).

The sex of an individual depends upon the chromosomes present in the nucleus of the zygote. The nuclei of the germ cells contain two kinds of chromosomes, *autosomes* and *sex chromosomes.* In the human female, for example, each ovum contains 22 autosomes and a single sex, or X, chromosome. Sperm cells are of two types, both containing 22 autosomes but with an X or a Y sex chromosome. When an egg and sperm unite to form a zygote, the latter has 46 chromosomes: either 44 autosomes and 2 X chromosomes or 44 autosomes and an X and a Y chromosome. The former will develop into a female; the latter into a male.

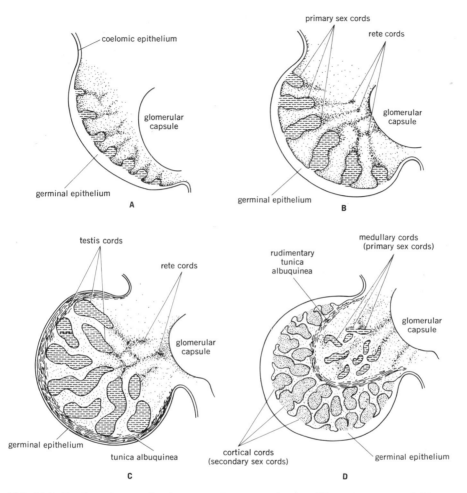

FIG. 11.2 Testis and ovary development of a mammal: *A*, undifferentiated genital ridge, with incipient primary sex cords growing inward; *B*, gonad still in the indifferent state; primary sex cords well developed; cords are developing which will become the rete testis if gonad becomes a testis; *C*, development of testis; *D*, early development of the ovary; primary cords reduced, rete cords rudimentary; great development of the secondary cords in which the eggs will develop. (*Adapted from A. S. Romer, "The Vertebrate Body," W. B. Saunders Company, Philadelphia, 1970, by permission.*)

FEMALE STRUCTURES

OVARIES A typical mammalian ovary is a solid structure with an inner *medullary* region composed of connective tissue, blood vessels, lymphatic vessels, smooth muscle, and nerve fibers. Indistinctly separated from the medulla is the outer *cortex,* in which the cytogenic and endocrine elements are situated. The cortex in a mature ovary is usually irregular in shape because of *ovarian,* or Graafian, *follicles* developing there. An ovum, or *egg cell,* with each follicle receives nourishment from the *follicular cells* surrounding it. After growing to a certain extent and at the proper time certain follicles push out to the surface of the ovary. Either the follicle ruptures, liberating the ovum* into the coelom (ovulation), or the follicle and its contents are reabsorbed. The ruptured follicle is reabsorbed in most lower vertebrates, but the follicle may persist in some vertebrates (always in placental mammals) and fill with a yellow material, from which the name *corpus luteum* is derived. This secretes a hormone, progesterone. Follicles which fail to rupture and undergo degeneration are called *atretic follicles.*

In certain fishes and in amphibians, snakes, and lizards, the ovaries are hollow and said to be of the saccular type. In teleosts the cavities are actually closed-up portions of the coelom into which ripe ova are liberated. This is *not* the case in amphibians, snakes, and lizards, the ovarian cavities of which are not homologous with those of teleosts. Ripe ova escape into the coelom through the *external* surface of the ovaries. It is a general rule in vertebrates that ova are shed into the coelom. New ova are formed periodically by mitotic activity of oocytes present in the *germinal epithelium,* a narrow layer of cells located on the outer border of the cortex.

The size of an ovarian follicle differs greatly in various vertebrates. It depends mostly upon the volume of the ovum characteristic of each particular species and upon the season of the year when the ovaries are examined. The mature ova of most forms below mammals are relatively large because of the yolk they contain. They range in size from those of certain fishes, in which the ova are just visible to the unaided eye, to the

* The term *ovum*, as used here, refers to the female gamete regardless of the stage of oogenesis at the time it is liberated. This varies among vertebrates.

enormous "yolk," or ovum, of the ostrich and mackerel shark. In vertebrates below mammals in the evolutionary scale, the ovum completely fills the cavity of the follicle. The relation of the microscopic mammalian ovum to the ovarian follicle is discussed on page 299. In most annual-breeding vertebrates there is a seasonal fluctuation in the size of the ovaries which is maximum at the breeding season.

Amphioxus The ovaries of amphioxus are metameric structures consisting of approximately 26 pairs of gonads projecting into the atrium, or peribranchial space, from the inner surface of the body wall. The most anterior gonads are located at about the middle of the pharyngeal region. Each gonad has a layer of atrial epithelium reflected over its free surface. Under this is a double layer of coelomic epithelium which almost completely surrounds the gonad. Each ovary therefore projects into a closed pocket of the coelom, the *gonocoel.* When eggs are ripe, the coelomic pouches and atrial epithelium rupture. Eggs are forced into the peribranchial space, from which they pass directly to the outside through the atriopore.

Cyclostomes The adult female lamprey has a single gonad, representing a fusion of two, which runs the length of the body cavity (Fig. 11.3) and is attached to the middorsal body wall by a single mesovarium. At the height of the breeding season the ovary occupies the greater part of the abdominal cavity. In the ammocoetes stage two ovaries are present, but they later fuse into a single organ.

The hagfish is *hermaphroditic,* the anterior part of the single gonad being ovarian in nature and the posterior portion testicular. Usually one or the other region matures and becomes predominant. Only the left gonad develops fully.

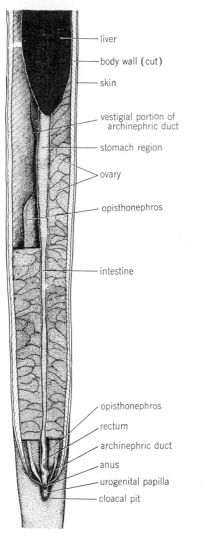

liver

body wall (cut)

skin

vestigial portion of
archinephric duct

stomach region

ovary

opisthonephros

intestine

opisthonephros

rectum

archinephric duct

anus

urogenital papilla

cloacal pit

FIG. 11.3 Ventral view of coelomic viscera of female lamprey. Portions of the ovary have been removed to show details of kidney structure.

In cyclostomes ripe eggs are shed directly into the coelom, and pass out of the body through a genital pore located at the posterior end of the body cavity.

Fishes The ovaries of fishes are paired,

although in some cases they appear to have fused into a single organ. In a number of elasmobranchs only the right ovary becomes fully developed, and the left ovary is degenerate in the adult.

Eggs of elasmobranchs are discharged from the anteriorly located ovaries directly into the body cavity. Few eggs are produced at a time. In correlation with the small number, the ova contain relatively enormous amounts of yolk. It has been reported that in both oviparous and ovoviviparous elasmobranchs certain bodies called *corpora lutea,* presumably having an endocrine function, are formed from the ovarian follicles after ovulation. Some sharks are viviparous.

In most teleosts the ovaries are of the saccular type. During development peritoneal folds form in connection with each ovary and enclose part of the coelom (Fig. 11.4). The anterior portion ends blindly, but in most cases continuations of the folds at the posterior end form an oviduct. Teleost ovaries are very large during the breeding season. Ripe ova, which are sometimes numbered in the millions, are discharged into the central ovarian cavity, from which they pass down the oviduct directly to the outside. In adult teleosts a cloaca does not exist, the oviducts having separate openings from the urinary and digestive systems. It would seem that the condition in teleosts is an exception to the rule that ova are liberated into the coelom. However, since the cavity within the ovary is really a portion of the coelom which has been cut off from the main portion, the exception is only apparent. Although most teleosts are oviparous, there are many ovoviviparous species. Corpora lutea have been identified in the ovaries of numerous forms following ovulation. In some, eggs are even fertilized and undergo development while inside the ovarian follicles; in others, development of the young takes place within the cavities of the ovaries.

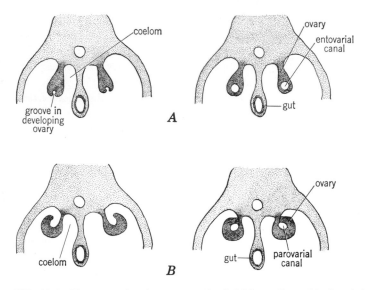

FIG. 11.4 Diagrams showing two methods of formation of hollow teleost ovaries: *A*, entovarial canal formation; *B*, parovarial canal formation.

Amphibians Although amphibian ovaries are saccular structures, ova escape into the coelom through their external walls. The cavity within each ovary becomes lymphoid in character and does not correspond morphologically or functionally to the cavity of the teleost ovary. The shape of the ovaries varies with the shape of the body. They are long and narrow in caecilians; elongated, but to a lesser degree, in urodeles (Fig. 11.5), and decidedly shortened and more compact in anurans. *Fat bodies* are closely associated with amphibian ovaries. They apparently serve for the storage of nutriment and undergo profound changes during the year, being smallest at the approach of and during the breeding season.

A peculiar structure in the male toad, known as *Bidder's organ,* may under certain conditions develop into a true ovary. Corpora lutea have been reported to be present during pregnancy in the ovaries of a certain ovoviviparous African toad, *Nectophrynoides.*

Reptiles Snakes and lizards possess saccular ovaries of the amphibian type, whereas turtles and crocodilians have solid ovaries. In snakes and lizards the ovaries are elongated and are not symmetrically disposed. This is very pronounced in snakes. Only the "yolk" of the reptilian egg is formed in the ovaries, and this represents the true ovum.

In certain ovoviviparous snakes and lizards corpora lutea form from the ruptured follicles after ovulation. It seems likely that these structures, as in mammals, secrete a hormone (progesterone or some other progestin) necessary for the maintenance of pregnancy.

Birds Although both ovaries are present during embryonic development, in most birds (except in many birds of prey) the right ovary degenerates and the left becomes the functional adult gonad. This may be in keeping with the elimination of unnecessary weight observed in connection with other structures

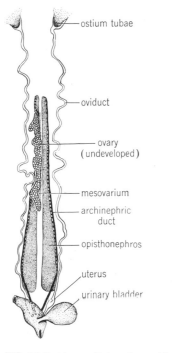

ostium tubae

oviduct

ovary
(undeveloped)

mesovarium

archinephric
duct

opisthonephros

uterus

urinary bladder

FIG. 11.5 Urogenital system of female urodele amphibian *Necturus* during the nonbreeding season, ventral view. The left ovary has been removed to show the opisthonephros above.

of birds. Mature ova of the polylecithal type escape from the ovarian follicles into the coelom through a preformed, nonvascular band, the *stigma,* or *cicatrix,* located on the surface of the follicle (Fig. 11.15). Only the "yolk" of the egg of the bird represents the true ovum formed in the ovary.

Experiments have shown that if the functional left ovary is removed from the domestic fowl, the rudimentary right gonad will develop into a testislike organ. In many birds an increase in the number of hours of daylight stimulates ovarian activity. Even brief exposure to intense light during sleep increases egg production in the domestic fowl. The effect is probably mediated via the hypothalamus and the pituitary gland.

Mammals Cortical and medullary portions of the mammalian ovary are more clearly marked off from each other than in most other vertebrates. The ovarian follicles lie in the cortex and form its most significant components. The relationship of the small mammalian ovum to the ovarian follicle differs somewhat from conditions in other vertebrates. Follicles vary in size according to their state of development and may be

FIG. 11.6 Section of portion of cortex of rat ovary, showing one primary follicle, two growing follicles, and a single large mature follicle.

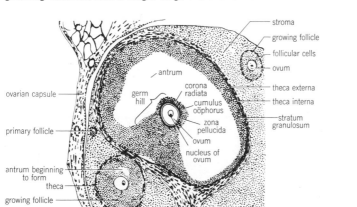

stroma

growing follicle

follicular cells

ovum

antrum

corona
radiata

ovarian capsule

germ
hill

cumulus
oöphorus

theca externa

theca interna

zona
pellucida

stratum
granulosum

primary follicle

ovum

nucleus of
ovum

antrum beginning
to form

theca

growing follicle

grouped into *primary, growing,* and *mature follicles* (Fig. 11.6).

A young, or primary, follicle consists of a centrally located ovum surrounded by a single layer of epithelial *follicular cells. Stroma* separates follicles from one another. In a growing follicle the follicular cells have increased greatly in number, forming several layers of cells about the ovum. A sort of capsular arrangement of cells known as the *theca* now surrounds the follicle. The theca is arranged in two layers, an outer *theca externa,* the cells of which are long and spindle-shaped, and an inner, vascular *theca interna,* made up of more rounded cells. It is highly probable that the sex hormone *estradiol* (page 437) is secreted by the cells of the theca interna. The ovum, which has been growing up to this point, now ceases to enlarge. It is surrounded by a clear, thick, transparent envelope, the *zona pellucida,* derived from both the follicular cells and the ovum. Microvillous processes from the ovum interdigitate with similar processes from the follicular cells within the zona pellucida. These are of importance in the transfer of metabolic substances between the ovum and the surrounding cells. The follicular cells making up the first layer around the zona pellucida are more columnar in shape than the others and form the *corona radiata.* So far the follicle is solid, but it is soon converted into a fluid-filled vesicle by intercellular vacuolation and rearrangement of the follicular cells. The cavity thus formed is the *antrum,* or *follicular cavity.* It is filled with a semiviscous *follicular fluid.* A small hillock of follicular cells, the *cumulus oophorus,* or *germ hill,* within which lies the ovum surrounded by the zona pellucida and corona radiata, protrudes into the antrum are several layers in thickness and collectively called the *stratum granulosum.*

Many follicles in the mammalian ovary begin to grow, but most of them degenerate before long and become atretic. In a mature female, at periodic intervals typical of each species, one or more follicles grow to maturity, the number being closely correlated with the number of offspring cast in a single litter. When these follicles have become fully developed, the cumulus with its contained ovum separates from the remainder of the follicle; the follicle ruptures; and the ovum with its surrounding layers is extruded into the coelom. It is of some interest that in the rabbit, ferret, cat, mink, shrew, and possibly others, ovulation is not spontaneous and will not take place unless the animal copulates.

In some mammals the antrum of the follicle fills with a clot of blood. Such a body is called a *corpus hemorrhagicum.* In other mammals a corpus hemorrhagicum fails to form. In either case the cells of the stratum granulosum and theca interna undergo a transformation. They enlarge and push into the antrum, which is thus gradually obliterated. Finally, a relatively large, rounded mass of cells is formed which is known as a *corpus luteum* (Fig. 11.7). If pregnancy does not occur, the corpus luteum persists only for

FIG. 11.7 Section through human corpus luteum of pregnancy.

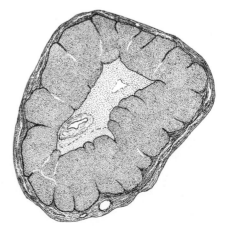

a short time. If pregnancy does ensue, it grows larger to become a *corpus luteum verum.* In some species the corpus luteum persists and functions throughout pregnancy; in others it seems to function only during the first half of pregnancy. In such forms the placenta undoubtedly takes over the secretion of progesterone or other essential steroids. When a corpus luteum finally degenerates, all that remains is a small, whitish remnant of connective tissue, the *corpus albicans.* The corpus luteum is of primary importance as an endocrine gland. The action of its hormone, *progesterone,* is discussed in Chap. 14. A corpus luteum in the ovary of a blue whale is enormous, measuring approximately 9 in. in diameter.

Ovaries in mammals are located in the lumbar or pelvic regions and are usually small in relation to the size of the body.

OVIDUCTS Oviducts, except in teleosts and some other fishes, are modifications of the Müllerian ducts. The latter are formed in either of two ways during embryonic development. In elasmobranchs and urodele amphibians the archinephric duct splits longitudinally, one part remaining as the kidney duct and the other forming the Müllerian duct. In other vertebrates the usual method of Müllerian duct formation begins by the appearance of a groove in the peritoneum covering the ventrolateral part of the opisthonephros or mesonephros, as the case may be, near the archinephric, or Wolffian, duct (Fig. 11.8). The edges of the groove come together and fuse, thus forming a tube, the Müllerian duct. The anterior end does not close, the opening becoming the *ostium tubae abdominale.*

The Müllerian ducts of the male usually degenerate, and only a few functionless remnants persist. However, in certain fishes and amphibians the ducts persist as rather prominent vestigial structures. The oviducts in most

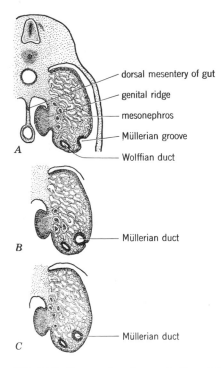

A — dorsal mesentery of gut
— genital ridge
— mesonephros
— Müllerian groove
— Wolffian duct

B — Müllerian duct

C — Müllerian duct

FIG. 11.8 Successive stages, *A, B,* and *C,* in the formation of the Müllerian duct.

vertebrates become differentiated into regions, particularly the posterior portion, which expands to form a uterus. This may serve as a temporary storage place for eggs and in many vertebrates affords a site in which the young develop.

Amphioxus No reproductive ducts exist in amphioxus. Ova which have been discharged into the peribranchial space are normally carried by the respiratory current out the atriopore into the surrounding seawater. Fertilization is external.

Cyclostomes Reproductive ducts are lacking in cyclostomes. Ova make their way from the coelom through the genital pores and out the urogenital papilla (Fig. 10.7). External fertilization also occurs in this group.

Fishes Structures for transport of ova show a great diversity among fishes. In some teleosts (family Salmonidae) and a few others, eggs escape from the coelom through modified abdominal pores. The two Müllerian ducts in elasmobranchs fuse at their anterior ends so that a single ostium tubae opens into the coelom (Fig. 11.9). A narrow but distensible oviduct leads from the ostium on either side. An enlargement called the *shell gland* is present in each oviduct. A short distance beyond the shell gland the Müllerian duct

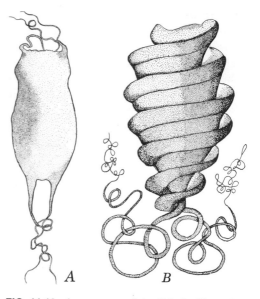

FIG. 11.10 *A*, egg case of dogfish *Scyllium; B*, egg case of bullhead shark, *Heterodontus galeatus*. (*After Waite.*)

FIG. 11.9 Ventral view of urogenital organs of mature, pregnant female dogfish. One ovary is shown only in outline, and part of one uterus has been removed in order to expose structures lying dorsal to them.

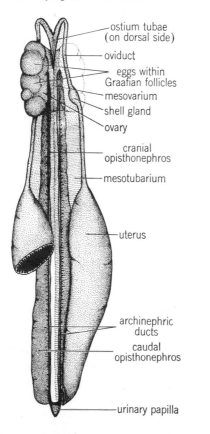

ostium tubae (on dorsal side)

oviduct

eggs within Graafian follicles

mesovarium

shell gland

ovary

cranial opisthonephros

mesotubarium

uterus

archinephric ducts

caudal opisthonephros

urinary papilla

enlarges to form a uterus which opens into the cloaca.

In all elasmobranchs fertilization takes place internally. There are oviparous, ovoviviparous, and viviparous species. Oviparous forms lay eggs encased in elaborate horny shells formed by the shell gland (Fig. 11.10). The shell formed about the ova in ovoviviparous species is very much reduced, and the shell gland is rather poorly developed in such forms. In viviparous forms the shell fails to form around the egg. The egg is kept in a specialized region of the oviduct, the ovisac, where the embryo is sheltered and nourished until its birth.

Amphibians Oviducts in amphibians are of the same general pattern throughout the class. They are paired, elongated tubes with ostia situated well forward in the body cavity. Posteriorly each Müllerian duct is slightly enlarged to form a short uterus, which in most

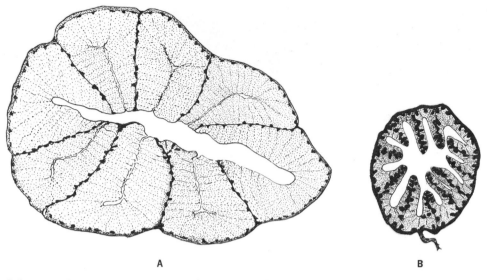

FIG. 11.11 Cross section of oviduct of salamander *Eurycea bislineata: A,* during the breeding season (April 11); *B,* after the breeding season is over (June 21).

species opens independently into the cloaca. In certain toads, however, the oviducts unite before entering the cloaca by a common orifice. The uteri in most amphibians serve only as temporary storage places for ova that are soon to be laid. The lining of the oviducts is glandular. Prior to the breeding season the ducts become greatly enlarged (Fig. 11.11) and markedly coiled, and the glandular lining epithelium begins to secrete a clear, gelatinous substance. As the eggs pass down the oviducts, several layers of this jellylike material are deposited about each ovum. This swells when the egg enters the water (Fig. 11.12).

In most anurans fertilization is external. The male grasps the female in a process called *amplexus,* and as the eggs emerge from the cloaca, spermatozoa are shed over them. No copulatory organs are present. Nevertheless, one group of African toads (*Nectophrynoides*) is apparently ovoviviparous.

Internal fertilization occurs in most uro-

deles, but no copulatory organs are present. In these amphibians the males deposit *spermatophores,* which are actually small packets of spermatozoa held together by secretions of the cloacal glands. One spermatophore is placed in pond or stream (on land in certain species). The structure of the spermatophores shows some variation (Fig. 11.13). In most species the spermatozoa are gathered together in a ball-like structure at the tip of the spermatophore. A complex mating ritual performed by the male stimulates the female, who is led by the male to the spermatophore. The female picks up the packet of sperm by muscular movements of the cloacal lips. A dorsal diverticulum of the cloaca, the spermatheca (Fig. 11.14), serves as the receptacle for the spermatozoa, which are thus available for fertilizing the ova as they pass down the oviducts to the cloaca. Fertilization is external in *Cryptobranchus,* Asiatic land salamanders, and also, apparently, in the Sirenidae. A few salamanders are ovoviviparous.

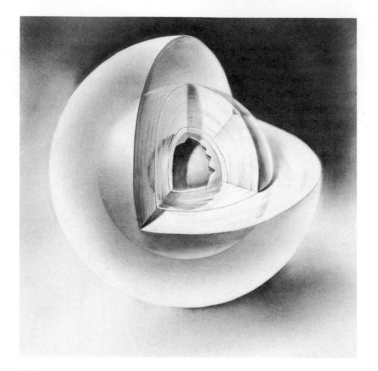

FIG. 11.12 Stereogram showing fertilized frog ovum surrounded by three layers of "jelly." (*Drawn by G. Schwenk.*)

FIG. 11.13 Three types of salamander spermatophores: *A, Triturus (Diemictylus) viridescens; B, Desmognathus fuscus; C, Eurycea bislineata. (From Noble, "Biology of the Amphibia," McGraw-Hill Book Company, by permission.)*

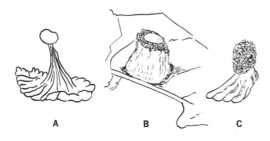

A   B   C

Internal fertilization is the rule in caecilians. The eversible cloaca of the male is considered by some authorities to serve as a copulatory organ. Caecilians are oviparous or ovoviviparous, with some forms retaining the developing embryos in the oviduct, feeding on the lining of the oviduct.

Reptiles The oviducts of reptiles open into the coelom by means of large, slitlike ostia. Each oviduct is differentiated into regions which mediate different functions in forming the envelopes deposited about the ova prior to laying. In *Sphenodon,* turtles, and crocodilians, glands in the upper part of the oviduct secrete albumen about ovum. Eggs of snakes and lizards lack albumen, and corresponding glands are missing in these forms.

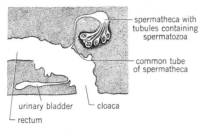

spermatheca with
tubules containing
spermatozoa

common tube
of spermatheca

urinary bladder — cloaca

rectum

FIG. 11.14 Diagram showing relation of spermatheca to cloaca in the salamander *Eurycea bislineata.*

The shell is deposited in the uterus, or shell gland. The uteri enter the cloaca independently. The eggshell in most oviparous reptiles is of parchmentlike consistency, but in some lizards and crocodilians it is hard and rigid like that of a bird's egg.

The size of the oviducts varies with the season and is at its maximum during the breeding period.

In all reptiles fertilization is internal due to the presence of intromittent organs and occurs in the upper part of the oviduct. Most reptiles are oviparous, but some snakes and lizards furnish good examples of ovoviviparity since their eggs are retained in the oviducts until the young are born. In some forms nourishment may pass from mother to young.

Birds The right Müllerian duct as well as the right ovary in most birds is degenerate, and only the left oviduct is functional. Both oviducts are formed in the embryo at the same time, but that on the right side fails to develop further and only vestiges remain. *Both* ovaries and oviducts are functional in certain birds of prey.

The left oviduct is long, coiled, and made up of several regions (Fig. 11.15). The ostium is bordered by a fimbriated funnel through which the egg enters the oviduct. This is followed by a *glandular portion* (*magnum*) in which albumen is secreted. Then follows a short *isthmus,* in which *inner* and *outer shell* membranes, made up of a fibrous meshwork, are deposited around the albumen. The isthmus leads to a dilated *uterus,* or *shell gland,* where the hard calcareous shell of the bird's egg is formed. The uterus opens into the *cloaca.*

Internal fertilization, which is characteristic of birds, is accomplished in most species by apposition of the cloacae of the two sexes in a cloacal kiss. Intromittent organs in the form of *penes* are present in male ostriches, swans, geese, ducks, and certain others. Fertilization takes place in the upper end of the oviduct.

Mammals Paired Müllerian ducts are present in all mammals. In this class various degrees of fusion occur between the two sides, and the ducts become differentiated into regions. Fertilization is internal and generally takes place in the upper part of the oviducts.

The most primitive condition found in mammals is that of monotremes, in which the Müllerian ducts remain separate and terminate independently in a urogenital sinus anterior to ureters and bladder (Fig. 11.16). Each duct consists of a narrow anterior *Fallopian tube* and a posterior expanded *uterus.* Although both oviducts are present in the platypus, only the left side seems to be functional. A very thin layer of albumen is formed about the small ovum in the Fallopian tube. A horny shell is deposited when the ovum reaches the uterus. The monotreme egg is unique in that it continues to grow after the shell has been deposited. This is because of absorption of uterine fluids and the plasticity of the shell. Calcium salts are later deposited in the shell, which thickens and continues to increase in size. At the time of laying, the embryos are well along in their development and at about the same stage as a 36-hour chick embryo.

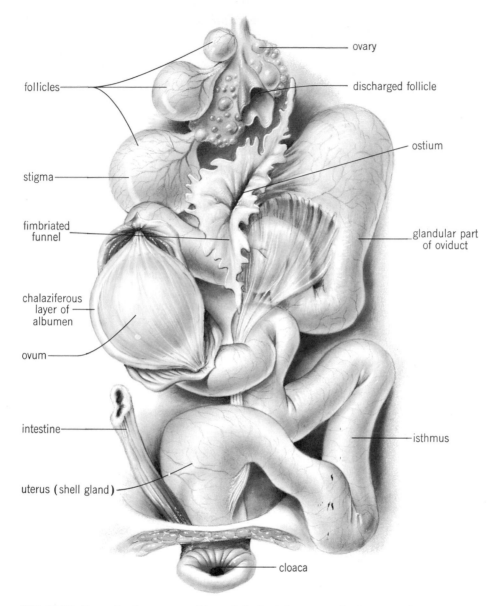

follicles

ovary

discharged follicle

ostium

stigma

fimbriated
funnel

glandular part
of oviduct

chalaziferous
layer of
albumen

ovum

intestine

isthmus

uterus (shell gland)

cloaca

FIG. 11.15 Reproductive organs of hen. Only the left ovary and oviduct are functional. A portion of the oviduct has been cut away to expose a descending ovum. Actually only one ovum is in the reproductive tract at any one time. (*Drawn by G. Schwenk, after Duvall.*)

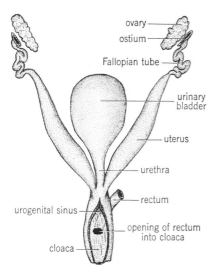

FIG. 11.16 Reproductive tract of female monotreme (semidiagrammatic).

structure directly to the urogenital sinus, a temporary opening being established.

Eutherian mammals are characterized by possessing a single vagina which represents a fusion of two. Except in the pika, a lagomorph, a cloaca is present only in the embryo. The uterine portions of the Müllerian ducts may fuse to varying degrees, resulting in different types of uteri. The *duplex uterus* (Fig. 11.18), found in many rodents, elephants, some bats, conies, and the aardvark, is the most primitive of these types, with two sepa-

FIG. 11.17 Dorsal view of reproductive tract of female kangaroo which had recently given birth. A single sucking young, 9 cm long (crown to rump), was in the marsupial pouch at the time of the mother's death. The urogenital sinus region has been slit longitudinally, and the edges have been spread apart.

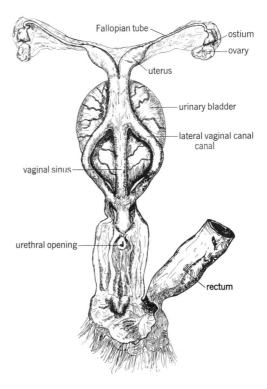

In other mammals the Müllerian ducts become differentiated into three distinct regions. The upper, narrow Fallopian tube, which is often spoken of as the oviduct, is sharply marked off from the expanded uterus. The oviduct opens into the coelom by a funnel-shaped aperture, the margins of which are sometimes fimbriated. The uterus leads to a terminal vagina which serves for the reception of the penis of the male during copulation. The lower part of the uterus is usually telescoped into the vagina to a slight degree. This portion is called the *cervix*. Its opening into the vagina is the *os uteri*. Marsupials retain the primitive paired condition of the Müllerian ducts, and two vaginae open into a urogenital sinus. In some, the vaginae fuse at their upper ends so as to form a *vaginal sinus* which extends posteriorly as a blind pocket (Fig. 11.17). This cecumlike structure may even open independently into the urogenital sinus and is sometimes called a third vagina. During the birth process the young pass through this

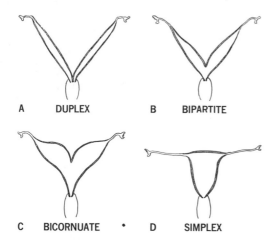

A DUPLEX   B BIPARTITE

C BICORNUATE   •   D SIMPLEX

FIG. 11.18 Diagram illustrating degrees of fusion of the uterine portions of the two Müllerian ducts in four types of mammalian uterus.

rate openings into the vagina. In the *bipartite uterus* (Fig. 11.18) of most carnivores, pigs, cattle, a few bats, and some rodents, the two sides fuse at their lower ends and open by a single os uteri. The *bicornuate uterus* (Fig. 11.18) of sheep, whales, insectivores, most bats, some carnivores, and many ungulates is the result of a still greater degree of fusion. In the *simplex uterus* of apes and man (Fig. 11.18) the fusion of the two sides is complete, only the bilaterally disposed Fallopian tubes indicating the paired origin of this type of uterus. Anomalous uteri of the duplex, bipartite, and bicornuate types occasionally occur in the human being.

The external part of the female reproductive system is called the *vulva.* In primates two folds of skin, the *labia minora,* are located about the margins of the opening of the vestibule. In some apes and in the human female (Fig. 11.19) two additional outer folds, the *labia majora,* make up part of the vulva. They correspond to the scrotal swellings of the male (page 312).

On the ventral wall of the vestibule a small *clitoris* is situated. It bears a rudimentary *glans clitoridis* at its tip, which is homologous with the glans penis of the male but much smaller. It differs from the penis, however, in that it has no connection with the urethra except in a few animals, as in rats and mice.

Certain glands, as well as rudiments remaining from the degenerated mesonephros and its duct (page 284), are associated with the female reproductive system. The glands are generally homologs of similar structures present in the male reproductive system. *Glands of Bartholin* correspond to Cowper's glands (bulbourethral glands) of males. They open into the vestibule and secrete a clear, viscid fluid under sexual excitement. This serves as a lubricant during copulation. *Paraurethral glands (of Skene),* corresponding to the prostate glands of males, are occasionally encountered in females.

The epithelium lining the Fallopian tube consists of ciliated cells and others of a glandular nature. The beat of the cilia is away from the ovaries and toward the uterus. This, together with peristaltic muscular contractions, serves to transport ova down the Fallopian tube. In most species of mammals it takes 3 or 4 days for fertilized ova to pass down the Fallopian tube to the uterus. Unfertilized ova die and usually disintegrate even before they reach the uterus.

The wall of the uterus is made up of three layers, a thin outer serous membrane, the *visceral peritoneum;* a thick middle layer of smooth muscle, the *myometrium;* and an inner layer of mucous membrane, the *endometrium.* Numerous *uterine glands* are present in the endometrium.

The state of development of the endometrium with its uterine glands varies during the reproductive cycle and shows a definite periodicity correlated with changes taking place in the ovaries. The ovarian hormones *estrogen*

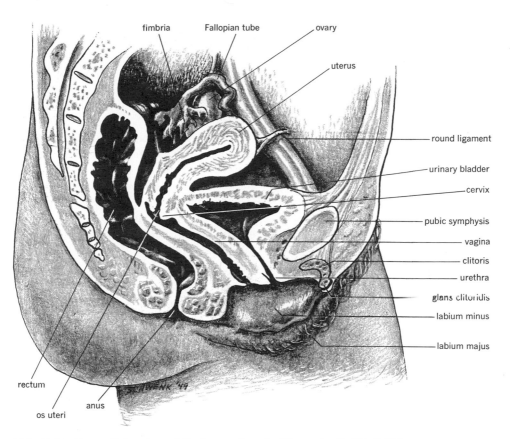

FIG. 11.19 Urogenital system of the human female. (*Drawn by G. Schwenk.*)

and *progesterone* have been shown to control these endometrial changes. In primates, at fairly regular intervals, a phenomenon known as *menstruation* occurs. This involves a degeneration and sloughing of the superficial portion of the endometrium, accompanied by hemorrhage from blood vessels in the endometrium. This process lasts from 3 to 5 days on the average. Further details are discussed in Chap. 14.

During pregnancy, menstruation is held in abeyance. Profound changes occur in the uterus. The fertilized ovum becomes embedded, or implanted, in the endometrium, where it undergoes development. The

placenta forms at the implantation site. This is an organ which establishes communication between the mother and the developing embryo by means of the umbilical cord. In animals that normally have several young at a time there is a fairly even spacing of the embryos along the horns, or cornua, of the uterus.

A phenomenon referred to as *delayed implantation* occurs in a number of mammals, including certain marsupials, bears, mink, the armadillo, and others. Fertilized eggs develop to the blastocyst stage prior to implantation. In delayed implantation the blastocysts lie free but inactive in the uterine lumen for extended periods before being implanted. Endocrine mechanisms are responsible for this phenomenon.

The vaginal epithelium is of stratified squamous type. It undergoes periodic changes in correlation with the stage of the sexual cycle.

MALE STRUCTURES

TESTES   The typical testis is a compact organ the shape of which shows much variation in members of different vertebrate classes. In all except a few low forms each testis is composed of numbers of *seminiferous ampullae,* or *tubules,* which connect by means of ducts to the outside. Two types of cells make up their walls. The first of these are large *Sertoli cells,* which are supporting and nutrient cells. The others are sex cells (spermatogonia) which, through the process of mitosis followed by meiosis and finally a metamorphic change (spermiogenesis), give rise to spermatozoa. At a certain stage of spermatogenesis the developing sperm cells become embedded in the Sertoli cells. Through the use of radioactive material it has been determined that in the rat it takes about 48 days to complete all the phases of spermatogenesis.

In addition to producing sperm the testes of vertebrates are endocrine organs which produce the male hormone *testosterone.* The actual tissue in the testes which secretes testosterone has not been definitely determined in the lower classes of vertebrates. In mammals, groups of cells lying among the seminiferous tubules undoubtedly represent the endocrine elements of the testes. They are called the *interstitial cells* or *cells of Leydig.*

In most vertebrates the testes are located in the dorsal part of the body cavity, where they first appear during embryonic development. In the majority of mammals, however, they undergo, at least seasonally, a descent to a position outside the coelom proper and come to lie in a special pouchlike structure called the *scrotum,* an extension of the coelom and ventral abdominal wall.

In annual breeders, for the most part, the size of the testes, like that of the ovaries, fluctuates with the seasons. In these forms they are largest just before the breeding period, but after spermatozoa have been discharged and the breeding season has past, they shrink to only a fraction of their former size. Spermatogenesis begins once more, following an inactive period, and the testes return to their fully mature condition.

Amphioxus   The gonads of the male amphioxus are similar in arrangement to those of the female. They are metameric structures which consist of about 26 pairs of testes. With the rupture of the wall of the mature gonad, spermatozoa pass into the peribranchial space and thence to the outside via the atriopore.

Cyclostomes   The gonad of the male lamprey differs but little in general appearance from that of the female but is never so voluminous. It is a single structure, representing a fusion of two. Spermatozoa are discharged through the outer wall of the testes and are shed directly into the coelom, from which they escape through genital pores (Fig. 11.20).

Fishes   In elasmobranchs the testes are paired, symmetrical structures situated at the anterior end of the coelom. As in other vertebrates, each is suspended from the middorsal body wall by a *mesorchium.* The testes are roughly oval or slightly elongated, but in most other fishes they are elongated and often lobulated as well. The male gonads vary greatly in size during the year and are largest at the breeding season.

Amphibians   The shape of amphibian testes shows a rough correlation with body

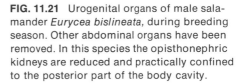

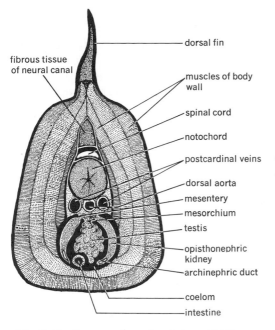

FIG. 11.20 Cross section of male lamprey at level of the first dorsal fin.

Labels (Fig. 11.20): dorsal fin; fibrous tissue of neural canal; muscles of body wall; spinal cord; notochord; postcardinal veins; dorsal aorta; mesentery; mesorchium; testis; opisthonephric kidney; archinephric duct; coelom; intestine

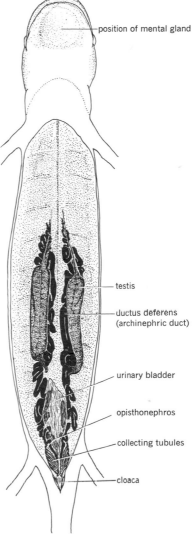

Labels (Fig. 11.21): position of mental gland; testis; ductus deferens (archinephric duct); urinary bladder; opisthonephros; collecting tubules; cloaca

FIG. 11.21 Urogenital organs of male salamander *Eurycea bislineata,* during breeding season. Other abdominal organs have been removed. In this species the opisthonephric kidneys are reduced and practically confined to the posterior part of the body cavity.

shape. In caecilians each is an elongated structure which appears like a string of beads. The swellings consist of masses of seminiferous ampullae which are connected by a longitudinal collecting duct. In urodeles the testes are somewhat shorter and irregular in outline (Fig. 11.21); in anurans they are more compact and of an oval or rounded shape (Fig. 10.10). A pronounced difference in size is apparent during breeding and nonbreeding seasons. Fat bodies are also associated with the gonads of male amphibians. Their size fluctuates with the seasons.

Reptiles The testes of reptiles are compact structures of an oval, rounded, or pyriform shape. Seminiferous tubules within testes are long and convoluted. In snakes and lizards there is a tendency for one testis to lie farther forward in the body cavity then the other. Periodic fluctuations in size according to season are clearly indicated in most reptilian testes.

Birds The round or oval shape of the bird's testes is characteristic. In a few birds such as the domestic fowl the testes are functional throughout the year, and no periodic variations in size are to be noted. In others, e.g., the sparrow and junco, which have a limited breeding period, the testes enlarge conspicuously at the approach of the mating season. Increase in number of hours of daylight has a stimulating effect upon spermatogenesis in certain birds, and hence brings about testicular enlargement, which therefore generally occurs in the spring. This reaction is undoubtedly mediated through the pituitary gland via the hypothalamus (page 421).

Mammals The smooth, oval mammalian testes are enclosed in a thin but tough fibrous envelope, the *tunica albuginea*. Each testis is divided internally into a number of lobules which show occasional intercommunications. Several seminiferous tubules lie within each lobule. In all mammals except monotremes the testes move from their place of origin to the pelvic region of the body cavity; they either remain permanently in this region or descend further into a *scrotum*. In some mammals the testes are located in the scrotum only during the breeding period and are withdrawn into the body cavity after this period has passed. Such testes shrink and become inactive for spermatogenesis but still secrete testostrogen. At the approach of the next breeding period they enlarge and again descend into the scrotum. In several species a relation between the number of hours of daylight and testicular activity has been demonstrated.

Among mammals with permanently abdominal testes are elephants, whales, sirenians, conies, rhinoceroses, some insec-

tivores, many pinnipedians, and most edentates. Bats, most rodents, a few carnivores (otters), llamas, aardvarks, and some insectivores show periodic descent and withdrawal. Those with permanently scrotal testes include marsupials, primates, some edentates and insectivores, as well as most members of the orders Carnivora, Artiodactyla, and Perissodactyla. Eared, or fur, seals of the order Pinnipedia are also included in this category.

The scrotum actually represents a fusion of two *scrotal pouches* which develop during embryonic life. The two halves are separated internally by a partition, the *septum scroti*. Externally the line of fusion is represented by a scarlike *raphe*. Each scrotal pouch contains a diverticulum of the peritoneum known as the *vaginal sac* or *process*. The testis lies under the peritoneum and as it descends into the scrotum pushes against the peritoneum of the vaginal process in such a way that it becomes partly covered by a reflected fold which comes in close contact with the tunica albuginea. This peritoneal fold is called the *tunica vaginalis*. It may become entirely separated from the peritoneal lining of the remainder of the coelom. Thus, the testis lies entirely outside the coelomic cavity but within the scrotum (Fig. 11.22). The descent is brought about primarily by unequal rates of growth of the body proper and certain structures closely related to the developing testis, e.g., the ligamentum testis and the scrotal ligament, which together form the gubernaculum, a cord of fibrous tissue which extends from the fetal testis to the scrotal swelling on each side and which guides the testis during its descent. No muscle fibers are present in the gubernaculum, and it does serve as a contractile unit. It appears first in an embryo as a mass of mesenchyme and is of significance only during development stages. After testicular descent has occurred, and by the time

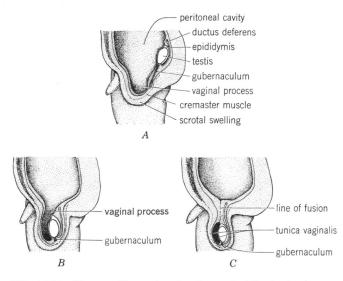

peritoneal cavity
ductus deferens
epididymis
testis
gubernaculum
vaginal process
cremaster muscle
scrotal swelling

A

vaginal process

gubernaculum

B

line of fusion

tunica vaginalis

gubernaculum

C

FIG. 11.22 Diagram illustrating the descent of the testes in man.

of birth, it ceases to exist as a definite structure, becoming blended into the connective tissue forming part of the scrotal wall.

The wall of the scrotum consists of layers of fascia, muscle, and skin. The integument is thin, folded, and often rather deeply pigmented. Just beneath the skin lies the *dartos tunic.* This is a thin layer of smooth-muscle fibers closely connected to the dermis. It reacts to variations in temperature so that under the influence of warmth the scrotum is relaxed and flaccid but under the influence of cold it becomes contracted, thick, and corrugated and pushes the testes up close to the body. A layer of muscle fibers and connective tissue, called the *cremasteric fascia,* lies beneath the dartos tunic. It has been established that the scrotum serves as a temperature regulator, producing an environment for the testes which is several degrees lower than that of the body. A slightly lower temperature seems to be a requirement for spermatogenesis in most mammals. A proper balance of gonado-

trophic hormones and an adequate supply of vitamin E are also essential.

Birds, the only other homoiothermous vertebrates, have relatively high body temperatures and intra-abdominal testes. However, there is some indication that even in birds a lowered body temperature may be a requirement for full spermatogenesis to occur. In sparrows, for example, greatest mitotic activity occurs at night, when the body temperature usually drops several degrees.

Those mammals with intra-abdominal testes would appear to be exceptions to the rule that a lowered temperature is a requirement for spermatogenesis. In addition, many reptiles, e.g., lizards and snakes, have intra-abdominal testes, and yet spermatogenesis occurs at a temperature as high as in mammals or higher.

In marsupials the scrotum is situated anterior to the penis, but in other mammals it lies posterior to the organ. In a few (rhinoceros, tapir, etc.) the scrotum is not typically pendulous and the descended testes lie in recesses in close proximity to the integument. The scrotum of male marsupials is homologous with the marsupial pouch of the female.

SPERMATOZOA The size of the spermatozoon is extremely small compared with that of the ovum. Spermatozoa show a great variety in form (Fig. 11.23). In vertebrates they all possess long, filamentous, flagellalike tails. These are used in a whiplike manner in moving through the fluids in which they are deposited.

The number of sperm cells produced by the male is enormous. In man a single ejaculate of semen measures about 4 ml in volume. An average number of approximately 300 million spermatozoa is present in this quantity of seminal fluid. The volume of a single ejaculate of a boar is enormous, measuring as much as half a liter.

MALE DUCTS The ducts which in most vertebrates serve to transport spermatozoa to the outside of the body are the archinephric

ducts, or the Wolffian ducts, formed in connection with the developing kidneys. It will be recalled that the term *archinephric duct* is applied to the kidney duct in anamniotes. The name *Wolffian duct* is given to the duct in amniotes which forms in connection with the pronephric and mesonephric kidneys. These are really different names for the same thing (page 284). The original function of these ducts is elimination of urinary wastes. In a number of fishes and amphibians certain modified kidney tubules are employed in carrying spermatozoa from the testis to the archinephric duct. They are known as *efferent ductules,* and the archinephric duct then becomes the *ductus deferens.* Even in amniotes, in which the mesonephros degenerates, its duct persists to become the ductus deferens and *epididymis (ductus epididymidis),* the latter establishing connections with the testis via efferent ductules, which in this case are modified and persistent mesonephric tubules.

Reproductive ducts are lacking in amphioxus and cyclostomes.

FIG. 11.23 Various types of vertebrate spermatozoa.

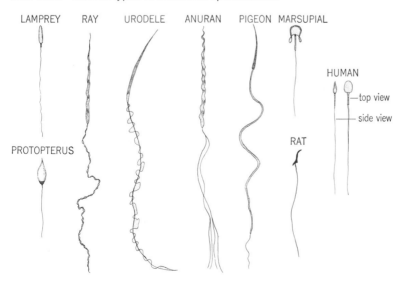

Fishes A variety of conditions characterize the reproductive systems of male fishes. In elasmobranchs (Fig. 11.24) efferent ductules, leading from the testis, course through the mesorchium and connect with certain anterior kidney tubules along the medial border of the opisthonephros. Spermatozoa are thus conveyed from the testis through the efferent ductules and kidney tubules into the archinephric duct, which now serves almost entirely as a ductus deferens. It courses along the ventral side of the opisthonephros and in young specimens is a straight tube with a urinary function. In older specimens it be-

FIG. 11.24 Ventral view of urogenital organs of mature male dogfish *Squalus acanthias.*

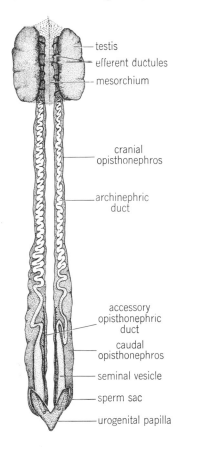

testis

efferent ductules

mesorchium

cranial opisthonephros

archinephric duct

accessory opisthonephric duct

caudal opisthonephros

seminal vesicle

sperm sac

urogenital papilla

comes convoluted. The posterior portion is markedly dilated to form a *seminal vesicle.* The two seminal vesicles open into a common *urogenital sinus,* which in turn communicates with the cloaca through an aperture at the tip of a *urogenital papilla.* A pair of blind sperm sacs passes forward from the ventral wall of the urogenital sinus. The sperm sacs are possibly remnants of the Müllerian ducts which persist in the male.

Although in elasmobranchs the connection of testis and archinephric duct occurs at the anterior end of the opisthonephros, in other fishes conditions may be quite different. For example, in *Polypterus* only the posterior part of the opisthonephros becomes involved. The glomeruli of the posterior kidney tubules frequently degenerate. A *sperm duct,* running the length of the testis, connects by means of posteriorly located efferent ductules to a longitudinal canal within the kidney. This is *not* the archinephric duct. The two, however, join posteriorly to open into a common urogenital sinus (Fig. 11.25).

In *Protopterus* a longitudinal sperm duct also extends the length of the testis. Each sperm duct leaves its testis at the posterior end, and those from the two sides enter the cloaca through a median genital papilla (Fig. 11.25). No relation to the archinephric duct is apparent even though the sperm duct, which is not considered a true ductus deferens, may pass through the posterior end of the opisthonephros. Only when the archinephric duct is utilized for sperm transport is it appropriate to use the term ductus deferens.

The male ducts of teleosts (Fig. 11.25) in many cases are entirely different in origin from most fishes. Their relation to the gonads is similar to that of the oviducts in females. A connection with the archinephric duct is established posterior to the opisthonephros, but the two ducts often have independent openings.

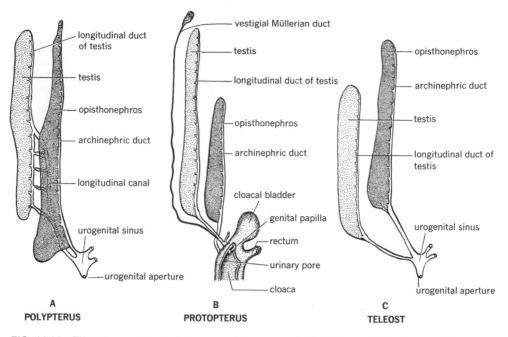

FIG. 11.25 Three types of male fish urogenital organs: *A, Polypterus; B, Protopterus; C,* teleost. (*Modified from Goodrich.*)

The anterior connection of the testes to the archinephric ducts, as observed in elasmobranchs, seems to be the most primitive condition in fishes. The posterior connections found in *Polypterus* and the gradual separation of the testis duct and archinephric duct as observed in *Protopterus* and many teleosts represent deviations and specializations from primitive conditions.

Amphibians The relationship of the reproductive and excretory systems in male amphibians is closer than in most fishes. It resembles the condition found in elasmobranchs. Efferent ductules usually join a longitudinal canal inside the testis or along its medial border. The efferent ductules course through the mesorchium, enter the anterior part of the opisthonephros on its medial side, and may connect directly to the archinephric duct or join certain kidney tubules, which in turn connect to the archinephric duct.

In urodeles, efferent ductules join a narrow longitudinal *Bidder's canal* which courses along, outside the *medial* edge of the kidney, but within the mesorchium. Bidder's canal connects by a number of short ducts to kidney tubules in the narrow anterior part of the opisthonephros. Certain kidney tubules emerge from the *lateral* edge of the opisthonephros and join the archinephric duct which courses posteriorly (Fig. 10.9). The anterior part of the archinephric duct, or ductus deferens, is concerned primarily with transport of spermatozoa, but the posterior end serves for elimination of urinary wastes as well. The archinephric ducts enter the cloaca independently.

Conditions in anurans (Fig. 10.10) are somewhat similar to those of urodeles, but minor variations exist. Efferent ductules enter the anterior end of the opisthonephros along its medial edge. In some forms they connect directly to the archinephric duct, but in others join Bidder's canal, which in these animals lies *within* the opisthonephros close to its medial border. Spermatozoa are then conveyed from Bidder's canal through kidney tubules to the archinephric duct, which also courses within the opisthonephros but along its *lateral* border. A Bidder's canal is present in the opisthonephros of the female, but its function, if indeed it has any, is obscure. The archinephric duct emerges from the kidney near its posterior end and passes to the cloaca. In males of several species a dilatation of the archinephric duct near the cloaca forms a *seminal vesicle* in which spermatozoa may be stored temporarily. These structures are at the height of their development during the breeding season and then undergo regression. Their rise and wane is in all probability controlled by seasonal changes in the secretion of testosterone, the hormone secreted by the testes.

The ductus deferens proper, particularly in urodeles, also varies greatly in size and is larg-est just prior to the breeding season (Fig. 11.26). After the breeding period is over it is reduced to a mere thread. Cloacal glands of male urodeles also show marked periodic changes.

Reptiles When the embryonic mesonephros degenerates in male reptiles, its duct persists as the male reproductive duct. The portion of the Wolffian duct near the testis becomes highly convoluted and forms the epididymis. Certain persistent mesonephric tubules become modified to form efferent ductules which form a connection between the seminiferous tubules of the testis and the epididymis. In some reptiles the epididymis is even larger than the testis. The Wolffian duct continues posteriorly as the ductus deferens, which sometimes is straight but more often is convoluted. In most reptiles the ductus deferens on each side joins the ureter of the metanephros, the two entering the cloaca through a common aperture.

Müllerian ducts, reduced to vestiges, commonly persist in male reptiles. In the European lizard *Lacerta viridis,* however, those of the male are as well developed as those of the female.

The epididymides and deferent ducts of seasonal breeders show periodic fluctuations and are under endocrine control. In many lizards and in certain other reptiles some of the posterior metanephric tubules become modified, producing an albuminous secretion which contributes to the seminal fluid.

Other accessory genital organs in male reptiles include glandular structures in the walls of the cloacae of snakes and lizards. Their secretion passes into a groove formed by the hemipenes (page 321). In chelonians and crocodilians the deferent ducts open at the proximal end of the groove which conveys spermatozoa to the free end of the single penis (page 321).

FIG. 11.26 Cross section of ductus deferens of male salamander *Eurycea bislineata: A,* from specimen captured March 21, just prior to the breeding season; *B,* from specimen obtained May 16, several weeks after the breeding season.

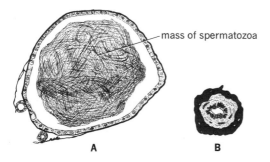

mass of spermatozoa

A   B

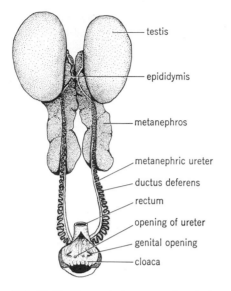

FIG. 11.27 Urogenital system of male fowl.

Birds Efferent ductules connect to a small epididymis, which in turn leads to a highly convoluted ductus deferens. The deferent ducts in birds open independently into the cloaca and are not related to the metanephric ureters (Fig. 11.27). There are no accessory glands. In the few birds possessing copulatory organs (page 321) a groove on the upper surface of the single penis carries spermatozoa to its apex.

Mammals Efferent ductules, which are persistent and modified kidney tubules, are homologous with the similarly named structures of lower forms. They connect the seminiferous tubules of the testis to a compactly coiled epididymis which joins the ductus deferens (Fig. 11.28). During their passage through the epididymis, spermatozoa undergo a physiological maturation necessary for development of full fertilizing power. The ductus deferens joins the urethra a short distance below the bladder. In mammals with scrotal testes the ductus deferens enters the pelvic region of the body cavity, crosses in

FIG. 11.28 Diagram showing relations of mammalian testis to epididymis and ductus deferens.

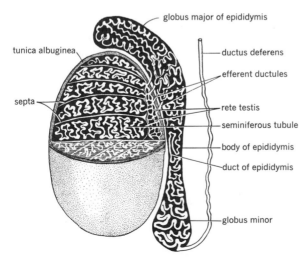

front of the ureter, loops over that structure (Fig. 11.29), and then courses posteriorly for a short distance before joining the urethra.

In several mammals there is an enlargement of the ductus deferens near its posterior end termed the *ampulla*. This may function as a temporary storage place for spermatozoa. An *ampullary gland* may also connect to the ductus deferens at this point. Another glandular structure, the *seminal vesicle,* often comes off the ductus deferens near its junction with the urethra. Its secretion forms part of the seminal fluid. Seminal vesicles are lacking in monotremes, marsupials, carnivores, and cetaceans. Mammalian seminal vesicles, in general, are not used for sperm storage. The portion of the ductus deferens between the seminal vesicle and urethra is called the *ejaculatory duct.*

The urethra, which extends throughout the length of the penis, opens at the tip of that structure through the small *external urethral orifice.* Accessory sex glands are also associated with the urethra. These include (1) the *prostate gland,* which contributes to the seminal fluid; (2) the *bulbourethral glands of*

FIG. 11.29 Urogenital system of human male.

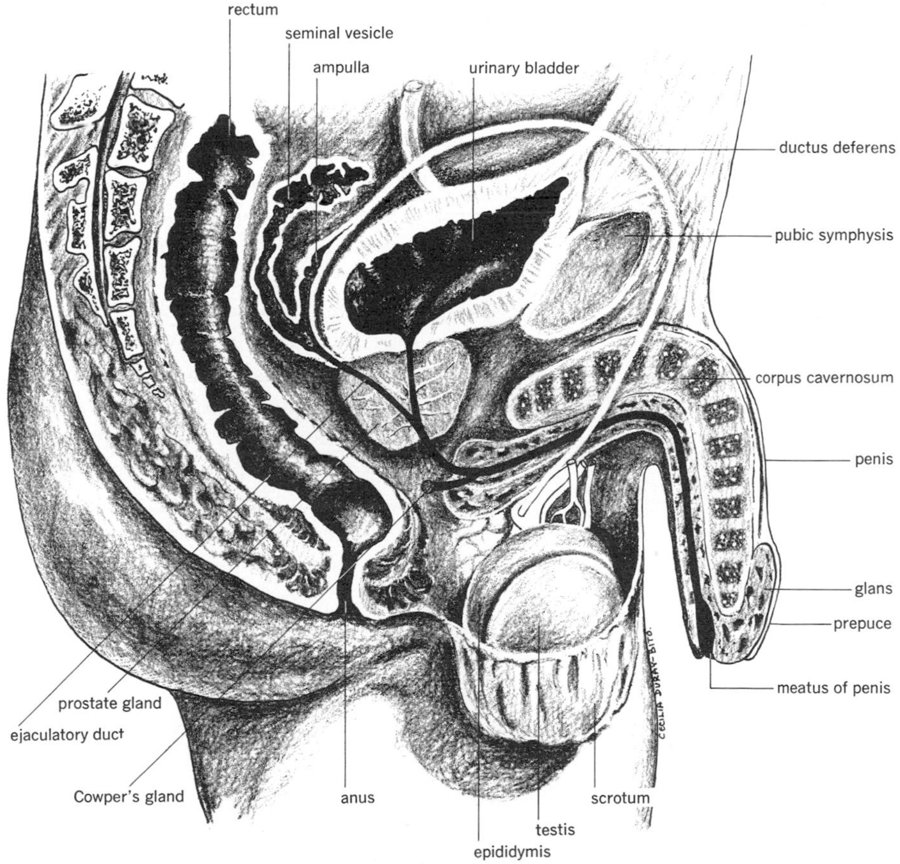

Cowper, which during sexual excitement secrete a clear, viscid fluid with lubricating properties; (3) small *urethral glands,* which are mucus-secreting glands. The normal functioning of the accessory sex organs in mammals is controlled by the hormone secreted by the testes. Monotremes, marsupials, edentates, and cetaceans lack a prostate gland. In monotremes, Cowper's glands are the only accessory reproductive glands.

Vestiges of the mesonephros associated with the reproductive system of male mammals include the *paradidymis, aberrant ductules,* and the *appendix of the epididymis.* Persistent homologs of the Müllerian duct include the *appendix of the testis* and the *prostatic utricle.*

COPULATORY ORGANS Fertilization always takes place in a fluid environment. Spermatozoa, which swim by a whiplike movement of the long flagellumlike tail, must have some medium in which they can travel in order to gain access to the ova. In most aquatic forms external fertilization takes place, and water furnishes the medium through which spermatozoa reach the eggs. In terrestrial forms a liquid environment must be provided. Internal fertilization is the rule in such animals, the necessary fluids being furnished by both male and female. In a number of terrestrial vertebrates spermatozoa are transferred from male to female by cloacal apposition, but in most terrestrial forms, and even in many aquatic species, intromittent, or copulatory, organs are present in the male by means of which spermatozoa suspended in seminal fluid are deposited directly within the reproductive tract of the female. A number of types of copulatory organs exists among vertebrates. Not all are homologous.

Fishes Among fishes, copulation with internal fertilization occurs only in elasmobranchs, holocephalians, and some tel-

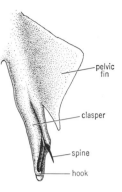

FIG. 11.30 Pelvic fin of male dogfish *Squalus acanthias,* showing clasper in detail. (*After Petri.*)

eosts. In elasmobranchs copulation is accomplished by means of *clasping organs* (Fig. 11.30), which are modifications of the medial portions of the *pelvic fins* of males. Each clasper is provided with a medial groove, the middle portion of which actually serves as a closed tube, since the edges overlap like a scroll. The anterior opening into the clasper tube is the *apopyle;* the posterior exit is the *hypopyle.* Spermatozoa enter the apopyle, which is situated close to the cloaca. In sharks and dogfish a sac, the *siphon,* with heavy muscular walls, lies beneath the skin in the posterior ventral abdominal region. Its anterior portion ends blindly, but posteriorly it communicates with the apopyle. The siphon sac is used with some force in the ejection of spermatozoa. Under normal conditions the apopyle is open and seawater is drawn into the siphon. The clasper tube gradually fills with spermatozoa. During copulation the clasper is bent so as to close the apopyle. Contraction of the siphon forces water down the clasper tube, ejecting the sperm already there, into the uterus of the female. It seems probable that during copulation only one clasper is inserted into the cloaca of the female. Not all elasmobranchs have hallow siphons. In skates

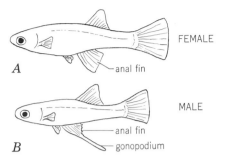

FIG. 11.31 *A*, anal fin of female poeciliid fish; *B*, how the anterior portion of anal fin of male is modified to form the gonopodium.

and rays the space usually occupied by the siphon is filled with a *pterygopodial gland* of uncertain function.

In those teleosts with internal fertilization the anterior border of the *anal fin* of the male may be elongated posteriorly to form an intromittent organ, the *gonopodium* (Fig. 11.31).

Amphibians Copulatory organs are lacking in urodeles and anurans. Internal fertilization in urodeles has already been discussed (page 302). In some caecilians the muscular cloaca is protrusible and serves as a sort of intromittent organ when the cloacae are in apposition.

Reptiles The only reptile lacking copulatory organs is *Sphenodon*. In other reptiles two types of structures are recognized, one being typical of snakes and lizards and the other of turtles and crocodilians.

In snakes and lizards, paired *hemipenes* are employed in an intromittent manner. They are saclike structures (Fig. 11.32), devoid of erectile tissue, normally lying under the skin adjacent to the cloaca. They are everted during copulation by propulsor and retractor muscles and filling of blood sinuses in the hemipenes. Spermatozoa pass down grooves into the cloaca of the female. Reptilian hemi-

penes are *not* homologous with the penes of higher forms.

A single penis is present in turtles and crocodilians. It is derived from paired thickenings, or ridges, in the anterior and ventral walls of the cloaca and is made up of connective and erectile tissues. The paired masses of erectile tissue are called *corpora cavernosa*. The penis can be extruded and retracted. A groove along the dorsal surface provides for the passage of spermatozoa. During the act of mating the corpora cavernosa are filled and distended with blood, and the penis is firm and enlarged. It is then said to be *erect*.

Birds A penis is present in only a few birds such as ducks, geese, swans, and ostriches. It is a single structure built upon the same plan as that of turtles and crocodilians. In the remaining birds sperm is transmitted during a cloacal kiss.

Mammals A single penis is typical of mammals. In monotremes, under normal conditions, the penis lies in the floor of the cloaca. It is similar to the organ in turtles, crocodilians, and birds, except that the groove on the dorsal side has become a closed tube. Moreover, the tube is surrounded by a single mass of erectile tissue, the *corpus spongiosum,* separate from the paired corpora cavernosa.

FIG. 11.32 Hemipenes of lizard *Platydactylus.* (*After Unterhössel.*)

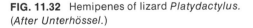

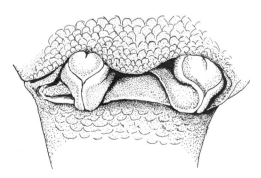

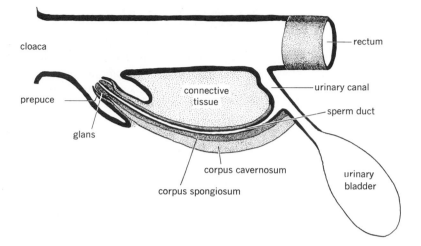

FIG. 11.33 Diagrammatic longitudinal section through cloacal region of male mono-treme, showing penis in retracted position. (*After Boas, based on Wiedersheim, "Comparative Anatomy of Vertebrates," copyright 1907 by The Macmillan Company and used with their permission.*)

The canal in monotremes is believed to carry only spermatozoa, since the urethra has a separate opening into the cloaca (Fig. 11.33). This differs from the condition in any other mammal. The end of the penis is enlarged to form a sensitive swollen *glans* containing erectile tissue continuous with the corpus spongiosum. The glans in monotremes and other mammals is covered with a fold of skin, the *prepuce,* or *foreskin.* In the platypus the glans is bifurcated; in *Echidna* it is divided into two double-lobed knobs (Fig. 11.34).

Marsupials lack a cloaca in the adult stage. The marsupial penis, therefore, does not lie within a cloaca as in monotremes. It is covered by a sheath and opens to the outside just ventral and anterior to the anus. It may be protruded and retracted. The scrotum lies *anterior* to it. The urethra in this and subsequent groups carries both urine and seminal fluid. The two corpora cavernosa are separated by a septum; the corpus spongiosum surrounds the

urethra. In several marsupials the tip of the penis is bifurcated (Fig. 11.34). The only accessory sex glands present are the bulbo-urethral glands, of which there are usually three pairs.

In eutherians there is a tendency for the penis to be directed forward, and in all forms possessing a scrotum the penis lies anterior to it. The penis is situated along the midline of the abdomen, usually in a horizontal position. In most forms it lies within a sheath, from which it may be protruded and retracted. In primates, however, it is permanently exserted, and a preputial sheath, or foreskin, covers and protects only the sensitive glans. A dartos tunic (page 312) is also present in the dermis of the skin covering the penis. In a number of mammals a penis bone develops in the septum between the corpora cavernosa. This helps to increase the rigidity of the penis, which normally becomes erect by distention of the erectile tissue with blood. Penis

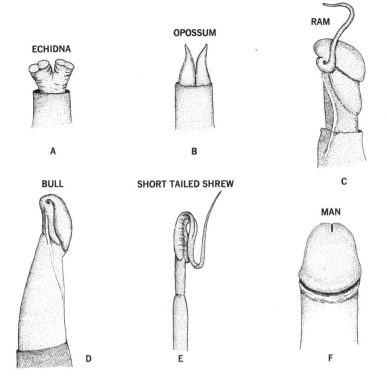

FIG. 11.34 Various types of mammalian glans penis. *(C and D, after Böhm, from Sisson, "The Anatomy of the Domesticated Animals," W. B. Saunders Company, by permission.)*

bones are present in members of the orders Rodentia, Carnivora, Chiroptera, Cetacea, and lower primates.

The glans is really the enlarged distal end of the corpus spongiosum covered with thin and delicate skin. It is supplied with numerous sensory nerve endings and is extremely sensitive to certain stimuli. Many variations in shape and structure are observed (Fig. 11.34).

SUMMARY

1   The reproductive system consists of primary and accessory sex organs. The primary organs are the paired gonads, testes in the male and ovaries in the female. Spermatozoa are formed in the testes and ova in the ovaries. The accessory sex organs are the ducts and glands which provide for the transport of eggs or spermatozoa to the outside of the body. The archinephric duct (in anamniotes) or the Wolffian duct (in amniotes) becomes the reproductive duct, or ductus deferens, in the male. The Mül-

lerian duct, or oviduct, in female elasmobranchs and urodele amphibians arises by a splitting of the archinephric duct. In other forms a groove in the peritoneum covering the opisthonephros or mesonephros, as the case may be, closes over to form the Müllerian duct. Ducts are lacking in amphioxus and cyclostomes.

2   Eggs develop in ovarian follicles within the ovaries. When fully developed they break out of the ovaries into the coelom. Each oviduct generally opens into the coelom by a funnel-shaped ostium. The oviducts of teleosts are derived in a different manner and may not be homologous with those of other vertebrates.

3   In forms below mammals the paired oviducts are separate and usually open independently into a cloaca. In higher mammals (except the pika) a cloaca is no longer present in the adult. Each oviduct differentiates into three regions, Fallopian tube, uterus, and vagina. Various degrees of fusion occur from distal to proximal ends, and different types of uteri result. The Fallopian tubes always remain paired. In higher mammals, vagina and urethra usually have a common opening to the outside. In a few groups, however, separate openings are present. The young in viviparous and ovoviviparous species develop within the uterus.

4   The Wolffian duct in amniotes disappears or is nonfunctional in the female, but its remnants, together with those of the degenerated mesonephros, are frequently found associated with the female reproductive system.

5   Certain accessory glands associated with the female reproductive tract secrete mucus which serves to lubricate the vagina during copulation. Fluids secreted by the Fallopian tubes and uterus aid in the passage of ova down the reproductive tract and, in forms with internal fertilization, help in the passage of spermatozoa to the upper part of the oviduct, where fertilization takes place.

6   In males, spermatozoa are formed within seminiferous ampullae, or tubules, in the testis. Each ductus deferens establishes connections with the testis generally by means of an epididymis and persistent kidney tubules called efferent ductules. In anamniotes the epididymis and ductus deferens consist of portions of the archinephric duct, which may serve in some cases as an excretory duct as well as for the passage of the spermatozoa. In amniotes, however, the epididymis and ductus deferens are formed from the persistent Wolffian duct, which is entirely dissociated from the ureter of the metanephric kidney.

7   The testes are located in the abdominal cavity, except in a number of mammals. In most mammals they are either temporarily or permanently situated in a scrotum outside the body proper. The scrotum serves as a temperature regulator.

8   Functionless remnants of the Müllerian duct persist in males of many species and are found associated with the reproductive organs.

9   Accessory glands, particularly in mammals, secrete seminal fluid in which spermatozoa are suspended and which are necessary not only for the viability of the spermatozoa but for their transport through the reproductive tracts of both male and female.

10   In most vertebrates with internal fertilization, copulatory organs are used in transferring spermatozoa from male to female. In fishes these usually consist of mod-

ifications of fins. In snakes and lizards, paired hemipenes are present; but in turtles, crocodilians, certain birds, and all mammals, a single penis serves this purpose. The penis consists largely of erectile tissue which, when distended with blood, causes the organ to become firm and erect. Only then can it serve as an intromittent organ. In some forms a penis bone contributes to the rigidity of the organ. The urethra, coming from the urinary bladder, passes through the penis in all mammals except in the monotremes. It thus serves for the passage of both urinary and seminal fluids.

11 The parallel development of male and female reproductive organs is striking. The various structures in one sex have definite homologies with those of the opposite sex.

12 The close structural relationship between excretory and reproductive organs is very apparent in vertebrates, but the evolutionary trend is toward a complete separation of the two systems.

BIBLIOGRAPHY

Breder, C. M., Jr., and D. E. Rosen: "Modes of Reproduction in fishes," The Natural History Press, New York, 1966.

Bullough, R. K., Jr.: "Vertebrate Reproductive Cycles," Wiley, New York, 1961.

Nalbandov, A. V.: "Reproductive Physiology," 2d ed., Freeman, San Francisco, 1964.

Parkes, A. S. (ed.): "Marshall's Physiology of Reproduction," vols. I–III, Longmans, London, 1952–1966.

Perry, J. S., and I. W. Rowlands, (eds.): "Biology of Reproduction in Mammals," Blackwell, Oxford, England, 1969.

van Tienhoven, A.: "Reproductive Physiology of Vertebrates," Saunders, Philadelphia, 1968.

12 CIRCULATORY SYSTEM

Circulatory systems of various types have developed in the evolution of higher forms of animal life in response to demands created by the increasing complexity of their bodies. In vertebrates fluids are carried by a closed circulatory system which transports substances to and from cells in all parts of the body. The fluids must circulate, for only by this means can supplies of food, water, ions, and oxygen be replenished and harmful waste products be removed. Hormones must be carried from one part of the body to another to be effective.

The various cells of the body are not actually in *direct* contact with parts of the circulatory system. The fluid surrounding, or bathing, the cells is called *interstitial fluid.* It serves as an intermediary in transporting nutritive materials, oxygen, and waste substances between cells and the minute subdivisions of the circulatory system. Tissue fluid is associated with various intercellular substances, its relationship to them varying in different parts of the body and depending upon the type of intercellular substance in ques-

tion. Tissue fluid is actually composed of water in which nutrients of various kinds are in solution. It usually contains some protein material which has leaked out of the circulatory system.

The circulatory system in vertebrates is highly complex and is made up of elaborately branched tubes which ramify throughout the entire body and convey fluids indirectly to all living cells. This is the *blood-vascular system.* A second system, the lymphatic system, conveys fluids away from the tissues. The blood-vascular system is the better defined and more obvious of the two.

BLOOD-VASCULAR SYSTEM The blood-vascular system is composed of continuous tubes of various sizes known as *blood vessels,* through which *blood,* a type of fluid connective tissue, is pumped to all parts of the body. This system consists of a muscular contractile organ, the *heart,* situated near the anterior end of the body and toward the ventral side. *Arteries* carry blood away from the heart. They divide and subdivide into vessels of smaller caliber, the *arterioles*. These in turn branch into *capillaries,* which are extremely small vessels averaging from 4 to 12 μm in diameter and 1 mm or less in length. *Venules* collect blood from the capillaries. They combine to form *veins,* which return the blood to the heart. Irrespective of the type of blood carried by the vessels, it should be clearly understood that *arteries carry blood away from the heart* and *veins carry it toward the heart*. The arteries and their branches make up the *arterial system,* and the veins with their tributaries form the *venous system*. The blood-vascular system is a closed system of blood vessels having no direct connection with any other system of the body except the lymphatic system, with which it is closely associated.

The arterial system in higher vertebrates is separated into *pulmonary* and *systemic divisions,* the former carrying blood to the lungs, the latter carrying it to all other parts of the body. The venous system likewise has pulmonary and systemic divisions, but in addition two or three *portal systems* may be present. A portal system is a complex of veins which begins in capillaries and ends in capillaries before the blood which courses through it is returned to the heart. All vertebrates have a *hepatic portal system,* in which blood collected from the digestive tract and spleen passes through capillaries (sinusoids) in the liver before entering the heart. Adults of the lower vertebrate classes and embryos of higher forms have a *renal portal system* in addition, in which blood from the posterior part of the body passes through capillaries in the opisthonephric or mesonephric kidneys on the way to the heart. Another small but important *hypophysio-portal system* is found in association with blood vessels of the pituitary gland.

LYMPHATIC SYSTEM The lymphatic system is composed in part of delicate tubes, the *lymphatic vessels*. They have their origin in very fine *lymph capillaries,* which are blind at their free ends and drain the tissue spaces. Lymphatic vessels may form connections in various parts of the body with relatively large spaces called *lymph sinuses*. Some join certain veins in the region of the heart, so that the connection between the blood-vascular and lymphatic systems is direct. The lymphatic system is a closed system, since the walls of even the lymph capillaries are composed of a delicate membrane made up of endothelial cells. The lymphatic system carries a fluid, *lymph,* the composition of which is, in general, rather similar to that of blood except that red corpuscles are lacking and the protein content is very much lower than that of blood plasma. In amphibians pulsating *lymph hearts* are present at points where lymphatic vessels enter certain veins. These hearts are usually few in number, but in Caecilians there are 100 in a paired series. Anurans and urodeles have paired lymph hearts in the pelvic region, and these are reported in some reptiles and birds. They aid in propelling lymph into the blood-vascular system. In most forms, however, pulsations of adjacent blood vessels and movements of the viscera, as well as other parts of the body, propel the lymph, which courses slowly through the vessels. Movement is in one direction; numerous valves prevent backflow in birds and mammals. In mammals, and to a lesser extent in birds, the lymphatic vessels are interrupted at certain places by *lymph nodes*. These are composed of a meshwork of connective-tissue fibers which support large masses of *lymphocytes*. Also present in lymph nodes are certain phagocytic reticuloendothelial cells, as well as others known as *fixed macrophages*. The lymphatic vessel breaks up into fine branches within the lymph node. The branches recombine on the other side of the node. Lymph filters slowly through the lymph nodes but in considerable quantity. Bacteria and other foreign substances are engulfed by the fixed macrophages and reticuloendothelial cells. Lymphocytes pass into the lymphatic stream and ultimately enter the blood-vascular system. Some of the macrophages become free, wandering cells. The lymphatic system is of importance also in connection with the regulation of both the quantity and quality of tissue fluid. It furnishes a means by which colloidal substances can get into the circulatory system, since the endothelial walls of lymphatic vessels are somehow permeable to colloids, whereas those of blood capillaries are not. The lymph nodes serve to defend the body against invasion by bacteria and other injurious agents which might enter via the lymphatic pathway. They are also of importance in producing plasma cells and lymphocytes. Plasma cells are of significance in antibody formation.

BLOOD AND LYMPH

Blood and lymph are classified among the connective tissues even though the cellular elements are free and the intercellular substance is fluid. Both cells and fluid course through the body in vessels which are lined with endothelium. Certain of the cellular elements, the white cells, or leukocytes, are constantly passing back and forth between connective tissue proper and the blood or lymph, migrating en route through the walls of capillaries and small venules. It is therefore difficult to separate these two types of tissue categorically. The behavior and even the appearance of the leukocytes may differ significantly in these different kinds of environment.

The fluid part of the blood, known as *plasma,* is composed mostly of water but is a solution of both colloids and crystalloids. Plasma is the medium which carries nutritive

substances, waste products, hormones, and even gases (to a minor extent) to all parts of the body where they are utilized or eliminated, as the case may be.

Tissue fluid is actually derived from that part of the blood plasma capable of diffusing through the endothelial walls of capillaries. The walls are permeable to solutions of crystalloids, which then can pass out of the blood and make up the tissue fluid. The cellular elements of the blood and the colloids in the plasma do not normally pass through the endothelium and are retained in the blood vessels. Tissue fluid, therefore, differs considerably from plasma in that, for the most part, proteins are lacking.

The cellular, or formed, elements of the blood include red corpuscles, or *erythrocytes;* white corpuscles, or *leukocytes;* and *blood platelets,* or *thrombocytes,* which are found only in mammals but which have a function similar to that of *spindle cells* found in the blood of certain lower vertebrates.

Although erythrocytes in various vertebrates differ in structure and appearance, their function in all is to carry oxygen to the tissues and to help in the removal of carbon dioxide. It is the pigment hemoglobin, a protein contained within the erythrocytes, which is responsible for the oxygen-carrying prop-

FIG. 12.1 Some types of formed elements in the blood of mammals.

BASOPHIL

EOSINOPHIL

NEUTROPHIL

ERYTHROCYTE

LYMPHOCYTE

MONOCYTE

erty of blood. Erythrocytes are confined to blood vessels under normal conditions. They are absent in lymph.

Leukocytes are of several types (Fig. 12.1) but may be separated into two main groups, based on whether they possess granules in the cytoplasm. *Granular leukocytes* are of three main types, *basophiles, eosinophiles,* and *neutrophiles.* The *nongranular,* or *agranular, leukocytes* are of two kinds, *lymphocytes* and *monocytes.*

The composition of blood in the various vertebrates differs rather widely, both in chemical constituents and in formed elements. The blood of amphioxus is practically colorless and contains few red corpuscles. Apparently no leukocytes are present. In vertebrates the hemoglobin of the blood is confined to the red blood corpuscles. The erythrocytes of vertebrates in all classes except mammals are oval and flattened (Fig. 12.1). Each contains a conspicuous nucleus which causes it to bulge in the center. Considerable variation in size occurs, the erythrocytes of the salamander *Amphiuma* being the largest erythrocytes of all vertebrates. Among mammals only camels and llamas possess oval red corpuscles. Those of other mammals are round and biconcave. The erythrocytes of elephants are the largest among mammals, but they do not exceed those of man by a great deal.

Mammals are unique in that the *erythroblasts,* cells which give rise to erythrocytes, extrude their nuclei before becoming mature erythrocytes. It is not surprising that the red cells, lacking nuclei, should be relatively short-lived. It has been estimated that red blood cells in man live in the bloodstream anywhere from 100 to 120 days before they break down. About 10 million red cells are destroyed every second in the human being, and, of course, a comparable number must be formed to replace them. Since the normal number in man is about 5 million per cubic millimeter, the rate of destruction and formation is not so tremendous as it might seem.

Blood-forming Tissues

Tissue which gives rise to various types of blood corpuscles is called *hemopoietic tissue.* There are two varieties. In the adults of higher vertebrates, erythrocytes, granular leukocytes, and thrombocytes trace their origin to red bone marrow, to which the name *myeloid tissue* has been applied. Lymphocytes and monocytes, the nongranular types of leukocytes, are formed for the most part in *lymphoid tissue.* It is in anuran amphibians that red bone marrow first becomes hemopoietic tissue. In these forms lymphocytes originate, in part, in lymphoid tissue in the intestinal wall. Lymph nodes, which also first appear in amphibians, produce great numbers of lymphocytes, monocytes, and plasma cells, the latter being of special importance in antibody formation. During embryonic development in higher forms, before spleen and bone marrow develop, mesenchyme and liver are the chief tissues in which red corpuscles are formed. Later the spleen and bone marrow take over this function, but after birth the red bone marrow is the chief site of erythrocyte formation. The intestinal wall and lymph nodes retain their role of producing lymphocytes. In many fishes and in urodele amphibians granular leukocyte formation takes place in the opisthonephros and even in portions of the gonads. All these structures should therefore be included among hemopoietic tissues.

In vertebrates below mammals the blood contains *spindle cells,* or *thrombocytes,* which are believed to play an important role in blood clotting. These cells are of approximately the same size as small lymphocytes and are biconvex discs. They have a spindle-shaped appearance when viewed edgewise. The cytoplasm is clear, and the nucleus, which stains deeply, is oval or rounded. A nucleolus is absent. In mammals there are numerous small bodies in the circulating blood which are not cells but which appear to be fragments of cytoplasm. These are called *blood platelets.* Unfortunately they are also referred to as *thrombocytes,* thus causing some confusion with the spindle cells, mentioned above, which are entirely different in structure even though their function may be similar. Platelets are even smaller in size than red corpuscles. No nucleus is present. They are fragments of pseudopodia of certain giant cells, called *megakaryocytes,* found in the bone marrow. Platelets in man number between 250,000 and 500,000 per cubic millimeter of blood. It is difficult to determine their exact number since they disintegrate when exposed to air or to rough surfaces. Blood platelets are lacking in other vertebrates, which, instead, possess nucleated spindle cells, or thrombocytes, believed to serve the same function. The platelets play an important part in the clotting of blood.

CIRCULATION

ORIGIN The vessels of the blood-vascular system arise in embryonic development from mesenchyme *in situ* and are included among mesodermal derivatives. The first indication of blood-vessel formation is the appearance of small clumps or cords of cells, called *blood islands,* in certain parts of the embryo (yolk sac, body stalk, etc.). In a chick embryo, for example, at first the blood islands are solid, but they soon hollow out to form a thin, flat endothelium enclosing a fluid-filled space (Fig. 12.1). Some loose cells float about in the fluid and become blood cells, or corpuscles. The fluid, which is apparently secreted by the cells of the blood

island, is blood plasma. The scattered, hollow vesicles derived from the blood islands flow and coalesce to form an irregular, anastomosing network of small blood vessels. The vessels of the main part of the embryo appear as clefts in the mesenchyme. The endothelium of the original vessels proliferates, and the vascular area expands. After a primitive blood-vascular system is once established, the formation of new vessels depends upon growth and branching of previously established channels. Every blood vessel, including the heart, has an endothelial lining.

STRUCTURE At first, during early development, there is no structural difference between arteries and veins, and the early blood vessels are histologically similar to the capillaries of later stages. From the mesenchyme surrounding the endothelium of these primitive vessels are derived the coating layers found in arteries and veins. The nature of the accessory coats differs in these two types of vessels.

Arteries Every artery (Fig. 12.2) has a wall composed of three coats: (1) an inner layer, the *tunica intima,* composed of the endothelium, and an *internal elastic membrane,* the structure of which differs somewhat in arteries of different caliber; (2) a thick intermediate coat, the *tunica media,* consisting of smooth-muscle cells arranged in a circular manner, as well as an elastic-fiber network, also coursing circularly; (3) an outer layer, the *tunica externa,* or *adventitia,* of varying thickness in vessels of different caliber, made up of loosely arranged connective tissue as well as longitudinally arranged collagenous and elastic fibers.

The thick muscular and elastic walls of arteries are of great physiological significance. Blood forced into the arteries by contraction of the heart distends these vessels. Since backflow is prevented by certain valves in the

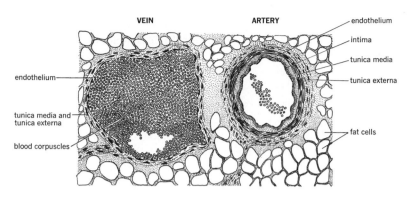

FIG. 12.2 Section through adipose tissue adjacent to ovary of rat, showing structure of a vein and an artery.

heart, tension of the elastic walls of the larger arteries tends to force the blood farther along the vessel. This accounts for the greater part of the movement of blood as well as for the diastolic blood pressure* within the arterial system. The walls of the smaller *arteries,* which are predominantly muscular, are of most importance in regulating the *amount* of blood supplied to a given area. Arterioles, with their thick muscular walls and small lumina, are largely responsible for maintaining the high degree of blood pressure within the arterial system.

Veins The walls of veins (Fig. 12.2) are much thinner than those of arteries but consist, nevertheless, of three similar coats. In some cases the tunica media is difficult to distinguish, particularly in the larger veins. Elastic and muscular elements are poorly developed in veins, but connective tissue is more clearly defined than in arteries. In mammals with long hind legs, the muscular

portion of the tunica media in the superficial subcutaneous veins of the legs is exceptionally well developed, providing necessary strength and resistance to hydrostatic pressure. The tunica externa makes up the greater part of the wall of the vein. In birds and mammals, particularly those with long extremities, many of the medium-sized veins of the limbs are provided with paired semilunar valves on the inner walls (Fig. 12.3). They prevent backflow of blood in the veins and are generally located distal to the point where a tributary enters a larger vessel. The valves are formed by folds of the tunica interna and

FIG. 12.3 Diagram of portion of a vein and its tributaries, showing arrangement of semilunar valves.

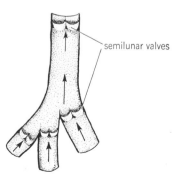

semilunar valves

---

* The term *systolic blood pressure* refers to arterial blood pressure brought about by contraction of the heart. It is somewhat more than half again as high as *diastolic blood pressure,* which exists in the arterial system between contractions of the heart.

contain elastic fibers. In general, the veins take the same course in the body as their corresponding arteries and usually are given similar names. A vein is always of greater diameter than its corresponding artery, since its walls are thinner, not so rigid, and more capable of being distended.

The walls of both arteries and veins of 1 mm or more in diameter are supplied with nutrient blood vessels of their own, since they do not receive nutrient material and oxygen from the blood that is coursing through them. The small nutrient blood vessels are referred to as the *vasa vasorum* (vessels of the vessels). They arise from the same vessel or a neighboring vessel some distance from the point where they are distributed.

Nerve fibers are abundant in blood vessels, particularly in arteries. They are of two types, medullated sensory fibers and sparsely medullated autonomic motor fibers.

Capillaries The endothelium of the capillary is its only component. Since it is at the capillary bed (Fig. 12.4) that exchange of gases, food materials, and wastes takes place between the blood and the extravascular fluids, thinness of the capillary wall is an obvious advantage. Diameter of the capillaries is relatively constant for each species and is related to the size of the red corpuscles which pass through them in single file. In man the average diameter is about 8 μm. They are seldom more than 1 mm long. Since they lack muscular and elastic coats, it has been a point of conjecture whether they are able to expand and contract of their own accord. It is possible that the endothelial cells themselves may be

FIG. 12.4 Diagram of a capillary bed: *a*, arterioles; *v*, venules.

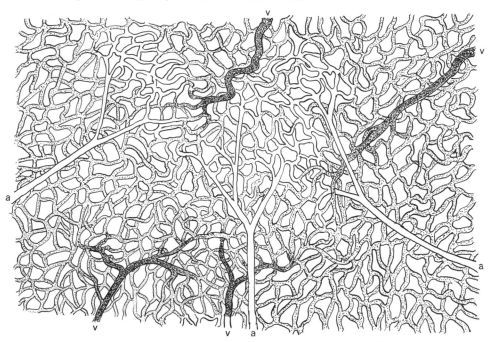

contractile. However, it is more likely that changes in caliber may be passive rather than active and a reflection of alterations in internal and external pressures.

Capillaries are so numerous that it is practically impossible to prick the skin in any part of the body without drawing blood. There are approximately 2,000 in a cubic millimeter of human muscle. It has been estimated that if all the capillaries in a man were stretched out in a single line, they would measure nearly 150,000 miles.

The velocity of blood in arteries is high, but when these vessels break up into arterioles and great numbers of capillaries, the increase in the vascular area brings about a considerable reduction in the velocity of the bloodstream. This plays an important role in the exchange of nutrient materials, gases, and wastes between blood and tissue fluid in the capillary bed.

Variation in the hydrostatic pressure and change in osmotic pressure of the blood in the capillaries and the surrounding tissue fluids determines whether fluid and/or ions pass in or out of the capillaries. Two main exchange processes are involved, bulk filtration and diffusion.

In addition electron-microscopic studies have shown that the capillary wall is only one cell thick and that no more than two cells surround the lumen at a given point. Extremely small vesicles are present in the cytoplasm of these cells; some are clustered along the inner cell membrane and others against the outer cell membrane (Fig. 12.5). They are apparently tiny invaginations of the cell membranes, from which they are capable of separating themselves, moving to the opposite side where they release their contents. It is possible that tissue fluid is absorbed or replenished by means of these vesicles. The process is called *pinocytosis*. The vesicles are

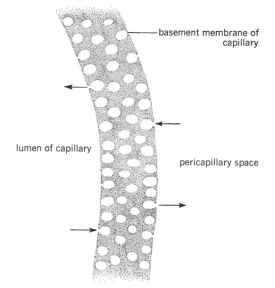

basement membrane of capillary

lumen of capillary

pericapillary space

FIG. 12.5 Diagrammatic representation of portion of cytoplasm of an endothelial cell of capillary wall, illustrating pinocytosis as suggested by electron-microscopic studies. Cytoplasmic vesicles may transport fluid from capillary lumen to the outside of the cell or vice versa.

too small to handle the large molecules of colloids and, therefore, confine their activity to the transport of water and crystalloids in solution.

Other Vessels   In some regions there may be direct connections between arteries and veins without the intervention of a capillary network. Such vessels, called *arteriovenous anastomoses,* are to be found in the toes of birds, ears of rabbits, and such areas in man as the terminal phalanges of the fingers and toes, nail bed, lips, tip of tongue, nose, and eyelids. At ordinary temperatures they are closed, but when critical temperatures are reached, dilatation occurs. These vessels, according to one theory, serve in two capacities: (1) at high temperatures an increased volume of blood circulates at the surface, thereby permitting excess heat to be removed from the body by radiation; (2) at low temperatures increased circulation through these vessels prevents harmful effects which might be caused by cold. Small arteriovenous anastomoses, known as *glomera,* are present in the dermis of the skin of the fingers and toes. Another theory holds that arteriovenous anastomoses may furnish a mechanism which regulates blood pressure and circulation in various parts of the body. Their actual function is not completely understood.

The *sinusoid* is another type of connection between arteries and veins. Sinusoids are relatively large, irregular, anastomosing vessels which are not lined with a continuous endothelial layer. They are to be found in such organs as the liver, adrenal glands, bone marrow, parathyroids, pancreas, and spleen.

In some parts of the body arteries or veins break up into small vessels of capillary dimensions, but no capillary network is formed. The efferent and afferent capillaries are closely packed together so as to provide the maximum surface exposure to one another. They are referred to as *retia mirabilia* and involve a countercurrent exchange principle. Examples of such structures are to be found in the red bodies and red glands in the swim bladders of certain fishes and in the glomeruli of kidneys.

In the penis and clitoris a peculiar type of tissue called *cavernous,* or *erectile, tissue* is to be found. It consists of large, irregular vascular spaces interposed between arteries and veins. They are supplied directly by arteries. *Erection,* or distention of the structure containing this type of tissue, is caused by a filling of the spaces with blood under high pressure and a partial closure, or compression, of the venous outflow.

HEART

The hearts of chordates differ from those in members of the lower phyla in their ventral, rather than dorsal, location. Blood is pumped anteriorly through arteries and then is forced to the dorsal side. The greater volume courses posteriorly, where the arteries terminate in capillaries in various parts of the body. The blood finally returns to the heart through veins.

In the lowest chordates the heart consists of little more than a pulsating vessel, which is sometimes referred to as a one-chambered heart. In cyclostomes and fishes the heart is divided into a series of compartments, which, in a posterior-anterior sequence, are called *sinus venosus, atrium (auricle) ventricle,* and *conus arteriosus,* respectively (Fig. 12.6). The latter leads into a *ventral aorta.* Although some authors refer to a heart of this type as a four-chambered heart, we shall consider the hearts of cyclostomes and fishes (except dipnoans) to be two-chambered, referring to atrium and ventricle as *true* (in the sense of persistent) *chambers,* and the sinus venosus and conus arteriosus as accessory chambers. This is because only the atrial and ventricular portions serve as chambers throughout the entire vertebrate series. In higher forms the sinus venosus is incorporated into the atrial wall and the conus arteriosus becomes modified to form the major arterial trunks.

In some forms an enlargement, the *bulbus arteriosus,* is present at the proximal end of the ventral aorta. It does not contain cardiac muscle tissue and is actually a part of the ventral aorta rather than of the conus arteriosus. Sometimes the term *truncus arteriosus* is used as a synonym for ventral aorta.

Dipnoans, amphibians, and most reptiles have three-chambered hearts with two atria

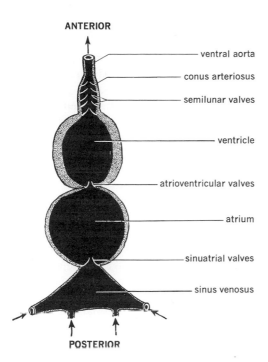

ANTERIOR

ventral aorta

conus arteriosus

semilunar valves

ventricle

atrioventricular valves

atrium

sinuatrial valves

sinus venosus

POSTERIOR

FIG. 12.6 Diagram of two-chambered heart as found in cyclostomes and most other fishes, showing atrium and ventricle with accessory chambers, the sinus venosus and conus arteriosus.

and a single ventricle. The sinus venosus is usually somewhat reduced and joins the right atrium. In crocodilians, birds, and mammals, the heart finally becomes four-chambered, with two atria and two ventricles. The sinus venosus in adult birds and mammals no longer exist as a separate structure.

In animals having two-chambered hearts the venous blood from all parts of the body is collected by the sinus venosus, which in turn enters the atrium. In those having three- or four-chambered hearts unoxygenated blood is returned to the heart through the sinus venosus, or directly to the right atrium, as the case may be. Oxygenated blood, coming from the lungs, enters the heart through the left atrium.

In lower forms the heart is located far forward in the body, but there is a gradual backward shifting as the vertebrate scale is ascended. The heart lies in a *pericardial cavity,* which is surrounded by an investing membrane, the *pericardium.* The pericardial cavity is a portion of the coelom which has become cut off from the remainder of the body cavity. In lower forms such as elasmobranchs the separation is incomplete and the two portions of the coelom are connected by a *pericardioperitoneal canal.* A thin serous membrane, the *epicardium,* covers the surface of the heart. It is similar to, and continuous with, the lining of the pericardium. Pericardium and epicardium correspond respectively to the parietal and visceral pleurae of the pleural cavities and to the parietal and visceral peritoneal membranes of the peritoneal cavity.

In evolutionary terms the heart is a highly modified blood vessel. Like other vessels it is lined with endothelium, the *endocardium.* Between epicardium and endocardium is a thick muscular layer, the *myocardium,* composed of cardiac muscle.

Just as the walls of blood vessels of greater caliber than 1 mm are supplied with blood via the vasa vasorum, the tissues of the heart itself are furnished with blood vessels which bring oxygen and nutrient materials to them and remove wastes. These vessels (arteries and veins) make up the *coronary circulation.* The blood within the chambers of the heart is too far removed from most of the heart tissues to be available to them. In addition, the heart is lined with an essentially impermeable lining.

In the embryos of all vertebrates the heart, when first formed, is a simple tube. The appearance of constrictions, folds, and partitions of various kinds results in the formation of two-, three-, or four-chambered hearts.

Comparative Anatomy of the Heart

Amphioxus A single, median contractile vessel lies ventral to the pharynx in amphioxus. Some consider it to be a one-chambered heart. Usually, however, it is spoken of as the *branchial artery, endostylar artery,* or *truncus arteriosus.* The muscles in its walls contract more or less rhythmically from posterior to anterior ends. On either side of the vessel numerous lateral branches are given off which course through the primary gill bars. At the base of each lateral *afferent branchial artery* is a small contractile enlargement, the *bulbillus* (Fig. 12.28). A single contraction of the "heart" followed by contraction of the bulbilli is sufficient to force blood through the entire body. Since the primitive embryonic condition of the heart in vertebrates is that of a single tube, the "heart" of amphioxus is considered to correspond to an early stage of development of the heart of higher forms.

Cyclostomes The two-chambered heart of the lamprey consists of a large, thin-walled atrium which communicates with the ventricle by means of a small aperture guarded by an *atrioventricular valve.* The ventricle is smaller than the atrium and has thick muscular walls. The lining of the ventricle is irregular and is provided with tough cords, the *chordae tendineae,* which connect to the atrioventricular valve. The cords prevent the valves from being pushed into the atrium when the muscular ventricle contracts. Blood is forced from the ventricle through a poorly developed conus into the ventral aorta, which distributes it to the gills. A single set of two semilunar valves in the conus region prevents any backflow of blood. A small, thin-walled

sinus venosus, which lies in the crevice between atrium and ventricle, opens into the atrium through a slitlike aperture guarded by a pair of *sinuatrial valves.* The heart, which lies posterior to the last pair of gill pouches, is located in a pericardial cavity surrounded by a thick, tough pericardium. In the ammocoetes stage the pericardial cavity communicates with the rest of the coelom, but the connection disappears in the adult.

Only unoxygenated blood passes through the cyclostome heart, which sends blood to the gills, where exchange of gases takes place. This is known as the *single type of circulatory system* (Fig. 12.7), since but one stream of blood passes through the heart.

In hagfishes a secondary caudal "heart" assists in pumping blood forward.

Fishes The hearts of most fishes are essentially similar in structure to the two-chambered heart of the lamprey. The single type of circulation is the rule. Variations in the relative positions of atrium and ventricle occur but are of minor significance. The arrangement of veins entering the sinus venosus also varies. The feature which perhaps deserves the greatest attention is the number and arrangement of valves in the conus arteriosus (Fig. 12.8). These semilunar valves, which prevent backflow of blood into the heart, are most numerous in elasmobranchs and in the ganoid fishes. In these forms the valves are arranged in three longitudinal rows, one dorsal and two ventrolateral in position. Several valves are present in each row. In *Epiceratodus* eight sets of valves are present. Many elasmobranchs have six sets. The valves of the anterior set are the largest. There is a tendency among fishes toward a reduction in the number of valves in the conus. Teleosts are characterized by having a single set, although in a few species two sets are retained.

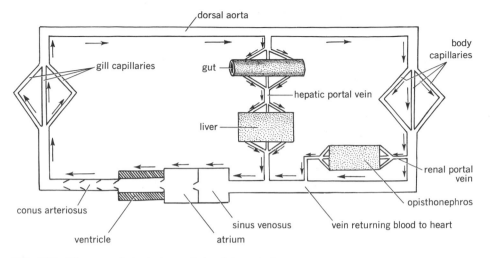

FIG. 12.7 Diagram of single type of circulatory system.

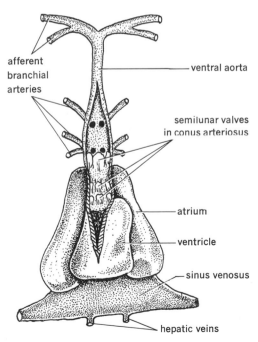

FIG. 12.8 Ventral view of heart of dogfish *Squalus acanthias.* The ventricle and conus arteriosus have been cut open to show the arrangement of the semilunar valves in the conus.

A coronary circulation, supplying and draining the walls of the heart itself, is particularly well developed in elasmobranchs. Similar vessels have been described in certain teleosts and other fishes, but in some a special coronary circulation does not seem to be evident.

The first advance from a two- to a three-chambered condition is seen in the Dipnoi, or lungfishes. In this group the atrium has become partially divided into right and left halves by an incomplete partition, or septum. In *Protopterus* unoxygenated blood from the body enters the right chamber via the sinus venosus, but oxygenated blood coming from the swim bladder (lung) enters the left atrium. The ventricle possesses pocketlike cavities in its walls. These, together with an incomplete interventricular septum made up of fibrous and muscular tissues, prevent mixing to a large extent. In *Protopterus* the conus has become divided longitudinally into two channels so that the two streams of blood, one oxygenated and the other unoxygenated, leave the heart. This is known as the *double type of circulation.* The unoxygenated blood is sent to the posterior gill region and swim bladder, whereas the oxygenated stream goes to the dorsal aorta. It will be recalled that gill lamellae are lacking in the anterior gill region of *Protopterus.*

Amphibians A double type of circulation very similar to that of *Protopterus* is characteristic of adult amphibians. Two streams of blood, one oxygenated and the other partially oxygenated, enter the heart (Fig. 12.9). The oxygenated stream coming from the lungs (or

FIG. 12.9 Heart of bullfrog, enlarged. (*After Storer, "General Zoology," 5th ed., McGraw-Hill Book Company, New York, by permission; drawn by R. Speigle.*)

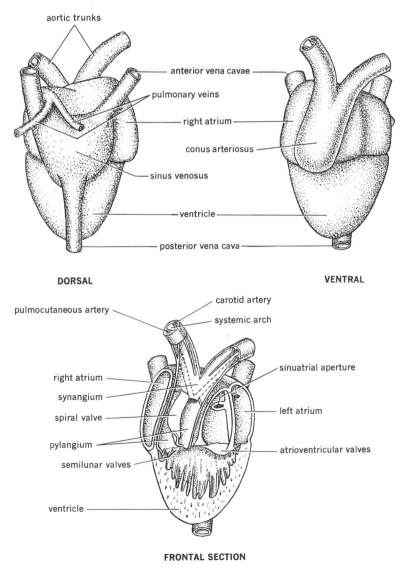

aortic trunks

anterior vena cavae

pulmonary veins

right atrium

conus arteriosus

sinus venosus

ventricle

posterior vena cava

DORSAL

VENTRAL

carotid artery

pulmocutaneous artery

systemic arch

right atrium

sinuatrial aperture

synangium

spiral valve

left atrium

pylangium

semilunar valves

atrioventricular valves

ventricle

FRONTAL SECTION

from the gills via the lungs in certain urodeles) enters the left atrium (via the sinus venosus) and thence passes to the left side of the posterior gill region and swim bladder, composed of unoxygenated blood from most of the body, mixed with oxygenated blood from the skin, enters the sinus venosus, from which it passes to the right atrium and thence to the right side of the ventricle.

The system of vessels coming from the lungs to the heart, and those leading from the heart to the lungs, is referred to as the *pulmonary circulation*. That which is distributed to the body in general and then returned to the heart is called the *systemic circulation*.

The *interatrial septum* is a thin membrane of connective tissue covered with endothelium. Perforations are frequently present in the interatrial septum of urodeles, but little mixing of the two bloodstreams occurs. The ventricle is not partitioned, but its lining is spongy and thrown into many pockets by muscular bands which, to a large extent, influence the blood flow, keeping the blood coming from the right atrium separate from that coming from the left. Chordae tendineae fastened to the atrioventricular valves prevent blood from being regurgitated into the atria when the ventricle contracts.

Despite the presence of a single ventricle, the two bloodstreams mix only to a slight degree. This, however, is not very important, since the systemic blood entering the right side of the ventricle has, to a considerable extent, been oxygenated in the skin and lining of the mouth. Both these areas in amphibians are important, since the animals utilize buccopharyngeal and cutaneous respiration in addition to pulmonary respiration.

Many amphibians have developed a rather complicated system of valves and partitions which ostensibly serve to keep the two bloodstreams well separated upon leaving the ventricle.

The conus arteriosus bears cardiac muscle fibers in its walls and is capable of contraction. In frogs it is made up of two regions. The part next to the ventricle is the *pylangium*. It is more muscular than the distal portion, which is referred to as the *synangium*. When viewed externally the anterior end of the synangium appears to divide into two trunks, each of which in turn separates into three arteries. The most anterior is the *carotid artery*, going to the head region; the second is the *systemic artery*, or *arch*, which gives off a few branches before the two systemic arches join each other posteriorly to form the *aorta*, which in turn distributes blood to the rest of the body; the third, or *pulmocutaneous artery*, leads to the lungs and skin. The latter actually connects with the pylangium.

The internal structure of the conus is rather complicated. Two sets of semilunar valves are present, one set at the base of the pylangium, the other at the junction of pylangium and synangium. In most amphibians one of the latter has become modified to form the *spiral valve*, which plays a role in separating the two bloodstreams as they leave the heart.

A spongy mass, the *carotid body* (sometimes incorrectly referred to as the carotid "gland"), located at the base of the carotid arch, aids in keeping the blood in the carotid vessels leading to the head at a rather constant pressure and gaseous content.

In such perennibranchiate salamanders as *Necturus*, blood is sent to the gills for aeration. Blood going to the lungs through pulmonary arteries has already been oxygenated in the gills. This would indicate that the lung has been reduced as a respiratory organ in such salamanders and may be used only in times of emergency.

Variations from the conditions described above are to be found in many amphibians. The interatrial septum may be incomplete or furnished with numerous small openings or a single, large aperture. This occurs in lungless salamanders and in those aquatic forms in which the lungs have been reduced in size and function. The sinus venosus may open into both right and left atria, and the left atrium may be reduced in size. The spiral valve may be reduced or is wanting in these forms in which no precise separation of the two bloodstreams occurs. Biologists are generally of the opinion that in its primitive condition (Paleozoic forms) the amphibian heart was three-chambered and capable of accommodating two separate bloodstreams. Other conditions, such as those mentioned above, are considered to be specializations.

A special coronary circulation, supplying and draining the muscular wall of the heart itself, is apparently lacking in many amphibians. In the frog the conus arteriosus receives a branch from the carotid artery bearing oxygenated blood. Blood is returned through two small veins entering the left brachiocephalic and anterior abdominal veins, respectively.

Reptiles Reptiles are the first of the chordates to become truly terrestrial. Except for certain aquatic turtles which at times utilize cloacal respiration, lungs are the only respiratory organs. In this group, as well as in birds and mammals, an efficient pulmonary circulation is a necessity. With its development, further changes have occurred in the structure of the heart.

Although a large sinus venosus is present in certain reptiles (turtles), it has, in general, been greatly reduced. Much of it is incorporated within the wall of the right atrium. Valves which are present where veins enter the right atrium represent vestiges of the sinus venosus. A complete interatrial septum separates the oxygenated blood in the left

atrium from the unoxygenated blood in the right.

All reptiles have a three-chambered heart except the crocodilians, which have a four-chambered heart. Even in the three chambered structure, however, the ventricle is partially divided by an incomplete *interventricular septum,* which extends from the apex toward the center. The conus arteriosus no longer exists as such. Its distal portion, as well as the ventral aorta, has split into three main trunks (Fig. 12.22), each of which has a row of semilunar valves at its base. One trunk is the *pulmonary trunk,* or *pulmonary aorta,* which gives off two pulmonary arteries going to the lungs. The pulmonary trunk leaves the right side of the ventricle. The two remaining *systemic trunks* are called the *left* and *right aorta,* respectively. The left aorta leads from the right side of the ventricle and crosses to the left side; the right aorta leads from the left side of the ventricle and crosses to the right side. A small aperture, the *foramen Panizzae,* is located at the point where right and left aortae cross each other and are in contact, so that their cavities are in communication. Even though the ventricle is only partially divided in most reptiles, an incomplete interventricular septum separates the two bloodstreams passing through the heart more effectively than in amphibians. In crocodilians, in which a complete interventricular septum appears for the first time, a true four-chambered heart is found. Some mixing of systemic blood occurs at the foramen Panizzae, however, as well as at the point where right and left aortae unite to form the dorsal aorta.

In reptiles the right aortic arch, carrying oxygenated blood from the left side of the ventricle, gives off a large *brachiocephalic artery,* which distributes blood to the anterior part of the body. At its base arise small coronary arteries, which pass to the walls of the heart itself. A coronary vein returns blood to the right atrium.

Birds A *complete* double circulation (Fig. 12.10) occurs in birds for the first time, since at no point is there any opportunity for oxygenated and unoxygenated blood to mix. The sinus venosus has disappeared, and three large veins, two precavae and one postcava, enter the right atrium directly (Fig. 12.35). Pulmonary veins return oxygenated blood from the lungs to the left atrium. The heart in birds is relatively much larger and more compact than in forms previously discussed. Both atria are thin-walled. The ventricles are completely separated, as in crocodilians, the muscular wall of the left ventricle being much heavier than the right. A single valve sepa-

rates right atrium from right ventricle. Two valves, however, known together as the *bicuspid valve,* are present at the left atrioventricular aperture. Chordae tendineae, attached to the atrioventricular valves, are anchored at their other ends to the lining of the ventricle by heavy projections called *papillary muscles.*

The main advance shown by the heart of the bird over the four-chambered crocodilian heart lies in the elimination of the left aorta. Only two vessels leave the heart in birds: a pulmonary trunk from the right ventricle and a systemic aorta, corresponding to the right aorta of reptiles, from the left. A single set of three semilunar valves is present at the base of each.

In birds the circulating blood passes from the left ventricle to all parts of the body. It is

FIG. 12.10 Diagram of double type of circulatory system in a vertebrate having a four-chambered heart, ventral view.

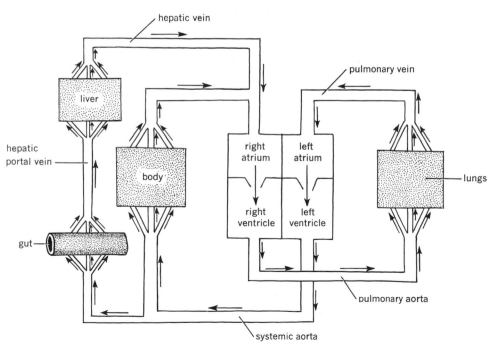

collected by the two precavae and the single postcaval vein and enters the right atrium. From here it passes to the right ventricle, which sends it to the lungs for aeration. Oxygenated blood is returned to the left atrium, which sends it past the bicuspid valve into the left ventricle.

A well-developed coronary system is present in birds. Coronary arteries arise from the systemic aorta. A venous coronary sinus enters the right atrium near the entrance of the postcava.

Mammals The four-chambered mammalian heart (Fig. 12.11) is essentially similar to that of birds, the two sides of the heart being completely separated from each other by interatrial and interventricular septa. A thin

FIG. 12.11 Diagram showing internal structure of four-chambered mammalian heart, ventral view.

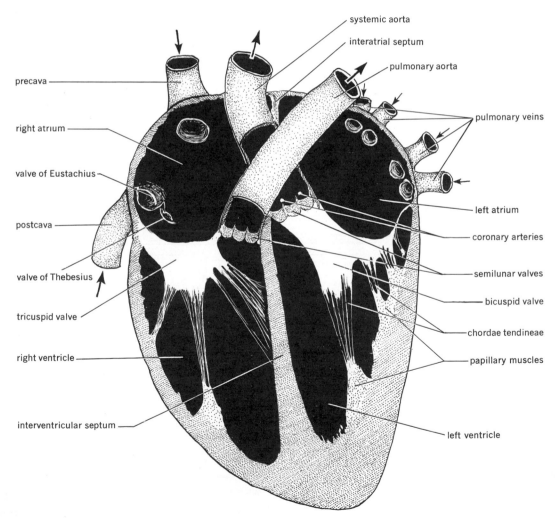

area, the *fossa ovalis,* in the interatrial septum represents the position of an opening, the *foramen ovale,* which is present during fetal life. A sinus venosus, present only during early embryonic development, is lacking in the fully formed mammalian heart, having been incorporated into the wall of the right atrium.

When the atria are in a contracted state, a flaplike projection of each may be observed extending for a short distance over the ventricle. These are the *auricular appendages.* Although the inner lining of the greater part of each atrium is smooth, that of the auricular appendages is ridged by muscular bands, the *musculi pectinati.*

Some variation exists among mammals in regard to the number of systemic veins entering the right atrium. A single postcava guarded by a *valve of Eustachius* is present in all, but there may be either a single precava, as in man and the cat, or two precavae, as in the rabbit, rat, and others. Pulmonary veins, returning oxygenated blood from the lungs, enter the left atrium. Their number varies in different species.

As in birds, a *bicuspid valve,* consisting of two membranous flaps, prevents blood in the left ventricle from being regurgitated into the left atrium. Monotremes are exceptional in that the left atrioventricular valve is tricuspid. In most mammals, a *tricuspid valve,* composed of three somewhat irregular flaps, is located between the right atrium and the right ventricle. Since a single valve is present in birds at this point, the presence of the tricuspid valve is a distinguishing feature of the mammalian heart. Chordae tendineae, attached to the irregular borders of the bicuspid and tricuspid valves, are connected at their other ends either directly to the inner walls of the ventricles and the interventricular septum, or indirectly to papillary muscles which are continuations of the mus-

cular ridges (*trabeculae carneae*) lining the inner surfaces of the ventricles. All are covered with endocardium.

In mammals, contrary to the condition in birds, the right aorta has been eliminated, so that only the left aorta persists. It arises from the left ventricle and distributes blood to all parts of the body. It is called the *systemic aorta.* The *pulmonary aorta* from the right ventricle carries unoxygenated blood to the lungs. A set of three semilunar valves is located at the bases of both pulmonary and systemic trunks.

A well-developed coronary system is present in the mammalian heart. Right and left coronary arteries arise from the systemic aorta as it leaves the left ventricle, just distal to the semilunar valves. During diastole there is a rebound of blood in the aorta. Semilunar valves prevent the blood from returning to the left ventricle. Some of the blood, however, is forced into the coronary arteries, which distribute it to the tissues of the heart itself. Deoxygenated coronary blood is returned through several vessels which converge to enter the right atrium through the *coronary sinus* guarded by the *valve of Thebesius.* Other small openings, the *foramina of Thebesius,* are openings of small veins, the *venae cordis minimae,* which return some blood directly from the heart muscle to the right atrium.

One form of heart failure, known as *coronary occlusion,* is caused by a blood clot or some other fragment blocking a coronary vessel. Death often results when the myocardium fails to receive its normal supply of metabolic substances.

Just as the walls of blood vessels are supplied with blood by the vasa vasorum, and the walls of the heart by the coronary vessels, the tissues making up the framework of the lungs, bronchi, and even the walls of the pulmonary vessels themselves receive a supply of blood distinct from that inside the pulmonary arteries and veins. *Bronchial arteries* and *veins,* branches of certain systemic vessels close to the heart, function in this manner.

Rings of dense fibrous connective tissue furnish support and prevent excessive dilatation of certain regions of the heart when contraction occurs. Since the free ends of cardiac muscle fibers may insert on this tissue, it is sometimes called the "skeleton" of the heart. In some mammals cartilaginous or bony tissue may be associated with the fibrous rings.

RHYTHMICITY OF THE HEART-BEAT The "pacemaker" which regulates or initiates the rhythmic beat of the heart is a bundle of atypical muscle fibers located in the wall of the sinus venosus. It is called the *sinoatrial node.* In higher forms, in which the sinus venosus has been incorporated into the wall of the right atrium, the sinoatrial node lies embedded in the atrial wall. Another mass of atypical muscle fibers, the *atrioventricular node,* in the interatrial septum of the four-chambered heart, also acts as a pacemaker under experimental conditions when the sinoatrial node is destroyed or prevented from functioning. A bundle of tissue, the *bundle of His,* is distributed from the atrioventricular node to the ventricular walls via the *Purkinje fibers.* The two nodes, the bundle of His and the Purkinje fibers, are responsible for the rhythmic sequence of the various phases of the heartbeat. Except in hagfishes, the autonomic nervous system reaches the heart and may affect the rate of contraction.

ARTERIAL SYSTEM

Although the arterial systems of various adult vertebrates appear to be different in arrangement, a study of development reveals that all are built upon the same fundamental plan. The increasing complexity of the heart from the simple two-chambered structure of lower forms to the four-chambered organ of crocodilians, birds, and mammals is associated with certain variations to be found in the blood-vascular system.

During early development the anterior end of the ventral aorta divides into two *aortic arches,* which course dorsally in the mandibular region. Dorsal to the pharynx these are continued posteriorly, where they are known as the *paired dorsal aortae.* Additional pairs of aortic arches then appear in an anterior-posterior sequence, forming connections between ventral and dorsal aortae on each side. Each courses through the tissue between adjacent pharyngeal pouches. The typical number of aortic arches to form in vertebrates is six pairs (Fig. 12.12), although there are certain discrepancies among lower

FIG. 12.12 Diagram illustrating the typical condition of aortic arches in vertebrate embryos, ventral view. Six pairs of arches connect ventral and dorsal aortae.

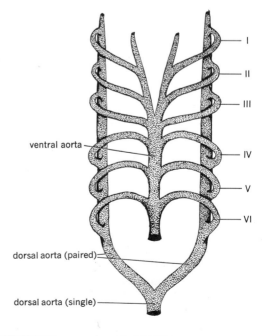

ventral aorta

I

II

III

IV

V

VI

dorsal aorta (paired)

dorsal aorta (single)

forms. The first is known as the *mandibular aortic arch;* the second is the *hyoid aortic arch;* the remainder are referred to as the third, fourth, fifth, and sixth aortic arches, respectively. Each lies anterior to the visceral cleft bearing the corresponding number. The two dorsal aortae soon fuse posterior to the pharyngeal region, so that only a single dorsal aorta ultimately is present. It is continued into the tail region as the *caudal artery.* Various paired and unpaired vessels arise along the length of the dorsal aorta, supplying all the structures of the body posterior to the pharyngeal region. Anterior continuations of the unpaired ventral aorta and of the paired *radices* (singular, *radix*) of the dorsal aorta supply the head and anterior branchial regions. Although the manner in which the dorsal aorta branches is fairly uniform throughout the vertebrate series, the aortic arches undergo profound modifications in different forms, the changes being similar in members of a given class. Blood, which is pumped anteriorly by the heart, passes through the ventral aorta to the aortic arches. These vessels carry the blood to the paired dorsal aortae, from which it goes anteriorly to the head or posteriorly to the single dorsal aorta, which distributes it to the remainder of the body. Veins return blood to the sinus venosus, at least during early stages of devel-

opment, or to the right atrium, as the case may be.

Changes in the Aortic Arches

Amphioxus   The "heart" of amphioxus, sometimes called the *ventral aorta,* is a single contractile vessel lying ventral to the gill region. Lateral branches, the *afferent branchial arteries,* are given off on both sides alternately. They extend up into the primary bars of the pharynx. In an adult specimen 60 or more pairs of afferent branchial arteries may be present. Each bears a contractile *bulbillus* at its base. The afferent branchial arteries give off small branches, which connect with vessels in the secondary gill bars. The latter vessels have no direct connection with the ventral aorta, but those in both primary and secondary bars connect to the paired dorsal aortae by means of *efferent branchial arteries* (Fig. 12.13). The two aortae unite behind the pharynx to form a single median vessel which courses posteriorly. Oxygenation of the colorless blood takes place during its passage through the gill bars.

In amphioxus the aortic arches are much more numerous than in higher chordates.

Cyclostomes   The ventral aorta in cyclostomes continues forward from the heart for

FIG. 12.13   Diagram of portion of pharyngeal region of amphioxus, showing arrangement of aortic arches.

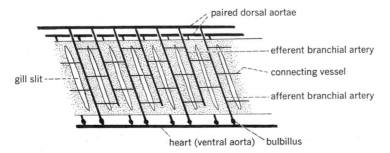

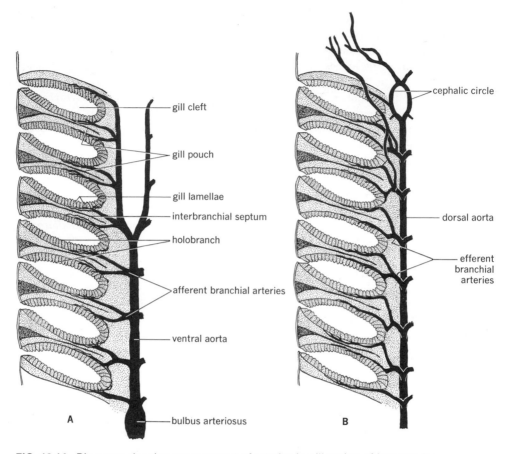

FIG. 12.14 Diagrams showing arrangement of arteries in gill region of lamprey *Petromyzon marinus: A*, ventral aorta and afferent branchial arteries; *B*, dorsal aorta and efferent branchial arteries.

some distance. The number of aortic arches which it gives off varies with the species and depends upon the number of gill pouches. In the lamprey *Petromyzon marinus,* which is representative, seven pairs of gill pouches are present. The ventral aorta leaves the heart as a single vessel which bifurcates at the level of the fourth gill pouch (Fig. 12.14*A*). Four afferent branchial arteries arise from each of the paired anterior extensions, and four pairs are given off by the unpaired portion. The most anterior of the eight pairs on each side

supplies the gill lamellae of the anterior hemibranch. The last furnishes blood to the most posterior hemibranch. Each of the remaining vessels arises at the level of an interbranchial septum and divides almost immediately to supply the lamellae on either side of the septum. Thus each holobranch is furnished with the two branches of an afferent branchial artery. Efferent branchial arteries, corresponding in position to the afferent vessels, collect blood from the gill lamellae (Fig. 12.14). They join the single, median dorsal aorta. The anterior end of this vessel is paired for a short distance, but the two portions come together again, thus forming a *cephalic circle.* From this arise arteries which supply brain, eyes, tongue, and various parts of the head.

The salient features of the aortic arches in cyclostomes are (1) a reduction in number as compared with amphioxus; (2) the manner in which they break up into interarterial capillaries between afferent and efferent vessels.

Fishes   Much variation is to be observed in the aortic arches of fishes. In general there is a reduction in number within the superclass as the evolutionary scale is ascended. The greatest number occurs in certain primitive sharks, in which the number is directly related to the number of gill pouches. Although in most fishes and in members of the higher classes the number of aortic arches is reduced or otherwise modified, practically all pass through a stage in embryonic development in

FIG. 12.15   Diagram showing arteries in the left gill region of dogfish *Squalus acanthias* as seen from the dorsal side. The length of the arteries at the anterior end has been exaggerated for clarity. Capillary connections between afferent and efferent arteries have been omitted. (*Modified from Daniel.*)

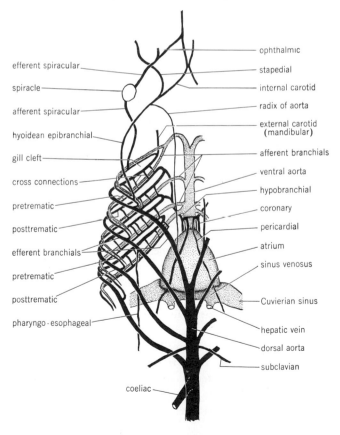

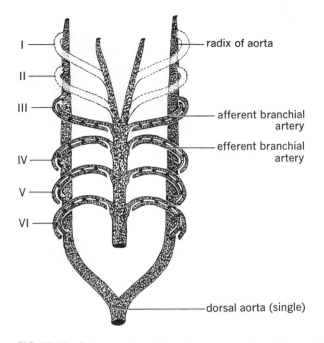

I — radix of aorta

II —

III —

afferent branchial artery

efferent branchial artery

IV —

V —

VI —

dorsal aorta (single)

FIG. 12.16 Diagram of aortic arch region as found in most teleost fishes, ventral view. Arches I and II have degenerated. Each of the remaining arches is divided into afferent and efferent branchial arteries connected with each other by gill capillaries.

which six pairs of aortic arches connect dorsal and ventral aortae (Fig. 12.12). Six, then, should be thought of as the primitive number for vertebrates, whose ancestors undoubtedly possessed a greater number. The anterior continuations of the paired dorsal aortae (*radices*) give rise to the *internal carotid arteries,* supplying the brain. The jaws and face are supplied by an artery running ventroanteriorly from the efferent gill artery. In most fishes each aortic arch except the first, or mandibular, consists of afferent and efferent branchial portions with an interarterial capillary network interposed between. It is in the capillary network of the gill lamellae that aeration occurs. The afferent branchial arteries, unlike those in cyclostomes, course *through* the interbranchial septa. In cyclostomes the afferent vessels divide and pass

between the pouchlike structure, sending a branch to the adjacent halves of the neighboring pouches. In *Protopterus,* which is exceptional, the third and fourth aortic arches pass directly to the dorsal aorta without interruption.

In most sharks only five aortic arches persist, the first having been modified. Although five afferent branchial arteries are present on each side, there are usually but four pairs of efferent vessels (Fig. 12.15).

In teleosts and most other fishes only the last four pairs of aortic arches remain, numbers I and II having been modified or reduced to small branches of the third (Fig. 12.16). In *Polypterus* and dipnoans, a pulmonary artery arising from each sixth arch (or from the dorsal aorta) carries blood to the swim bladder.

Amphibians In amphibians and in the remaining vertebrate classes there is a further reduction in the number of aortic arches, together a greater modification of the entire complex of vessels in the pharyngeal region. The aortic arches do not break up into afferent and efferent portions, since in these higher forms internal gill lamellae do not develop. To be sure, amphibians possess external gill filaments, at least during early development, but these are not homologous with the internal gill lamellae of fishes, nor are they supplied with blood in the same manner (pages 220 and 351).

In anurans (Fig. 12.17), aortic arches I, II, and V disappear. The radix between arches III and IV on each side is greatly reduced and may degenerate completely. The anterior continuations of the ventral aorta become the external carotid arteries. The third arch, together with the anterior portion of the radix on that side, becomes the internal carotid artery. The portion of the ventral aorta from which the internal and external carotids arise becomes the common carotid. The fourth aortic arches persist to become the systemic arches, which unite posteriorly to form the dorsal aorta proper (Fig. 12.18). Arch VI on each side sends a branch to the developing lung and to the skin, thus becoming the pulmocutaneous artery. The portion of arch VI present at first between the pulmonary artery and the radix is the *ductus arteriosus*. It disappears at the time of metamorphosis.

Slight differences from the above are to be found in urodeles (Fig. 12.19). In some salamanders the fifth arch may persist in very reduced form. Frequently the radix between

FIG. 12.17 Diagram showing modification of aortic arches as found in anuran amphibians, ventral view.

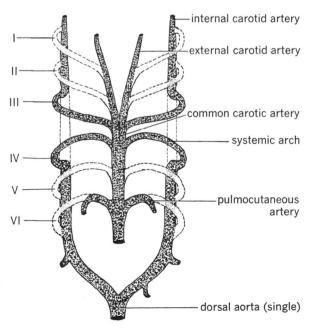

arches III and IV fails to degenerate completely. The ductus arteriosus also persists in urodeles.

The external gills of larval urodeles and early anuran larvae are supplied with vascular loops connected to aortic arches (Fig. 12.20). The loops themselves are composed of afferent and efferent vessels connected by capil-

FIG. 12.18 Diagram of arterial system of adult frog, ventral view.

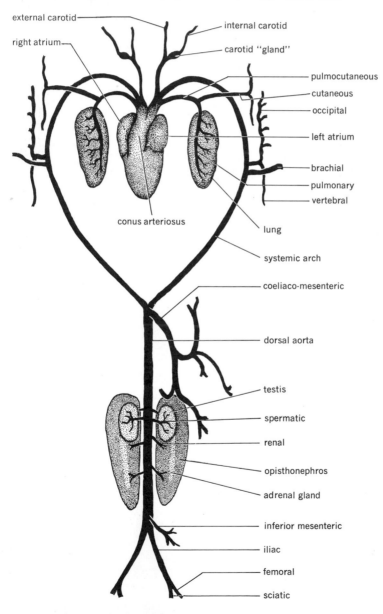

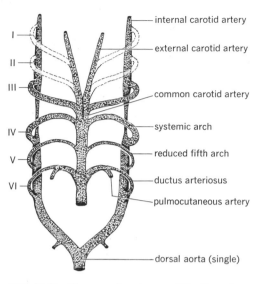

internal carotid artery

external carotid artery

I

II

common carotid artery

III

systemic arch

IV

reduced fifth arch

V

ductus arteriosus

VI

pulmocutaneous artery

dorsal aorta (single)

FIG. 12.19 Diagram showing modification of aortic arches as found in most urodele amphibians, ventral view.

laries. They lie lateral to the aortic arches, the latter being located at the bases of the gills, each serving as a *gill bypass.* Blood may go directly through the aortic arches or through the adjacent loops. In urodeles at the time of metamorphosis the gills degenerate, and the vascular loops atrophy. The main aortic arch, which has maintained its integrity from the beginning, persists.

The fifth aortic arch persists, and the pulmonary artery arises from the fifth arch rather

than the sixth, the ventral portion of which is lacking (Fig. 12.21). Blood going to the lungs in *Necturus* has already been oxygenated by passage through the gills. The lungs, therefore, under normal conditions are of little value as respiratory organs.

Reptiles Just as in amphibians, reptiles retain aortic arches III, IV, and VI. The fifth arch may also be retained in reduced form in certain lizards, and a remnant of the radix between arches III and IV may persist in some snakes. In most reptiles further modifications occur in the aortic arches. These consist mainly of splitting of the distal portion of the conus arteriosus and part of the ventral aorta into three vessels (Fig. 12.22). The fourth aortic arch on the left side establishes a separate connection with the right side of the partially divided ventricle. Together with a portion of the radix on the left side, it becomes the *left arch of the aorta.* The sixth arch on each side gives off a pulmonary artery to the lung and in most cases loses its connection with the radix. The two pulmonary arteries then arise from a common trunk, the *pulmonary aorta,* coming from the right side of the ventricle. The remaining vessel derived from the truncus arteriosus connects to the left side of the ventricle and as it courses forward gives off the fourth aortic arch on the right side, and finally divides into the two

FIG. 12.20 Diagrammatic cross section through gill region of early frog tadpole, showing relation of aortic arches to blood vessels in the external gills. (*After Maurer.*)

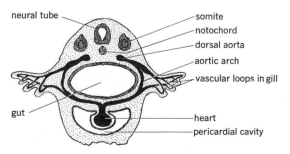

neural tube

somite

notochord

dorsal aorta

aortic arch

vascular loops in gill

gut

heart

pericardial cavity

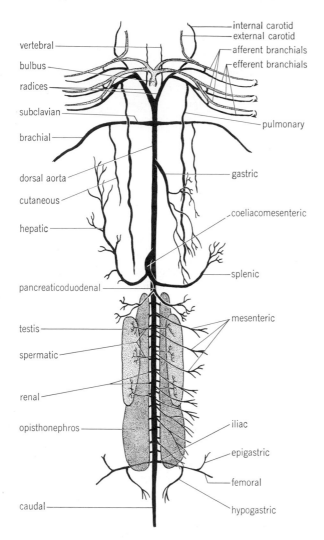

FIG. 12.21 Ventral view of arterial system of *Necturus.*

common carotid arteries with their external and internal branches. The right fourth aortic arch, together with a portion of the radix on that side, becomes the *right arch of the aorta,* which joins the posterior continuation of the left arch to form the dorsal aorta proper.

Birds The main changes taking place in

the aortic arches of birds correspond, in general, to those of reptiles. In birds, however (Fig. 12.23), the fourth arch and radix on the left side lose their connection with the dorsal aorta and finally disappear. The ventral aorta splits into two portions, a systemic aorta and a pulmonary trunk, or aorta. The systemic aorta is connected to the left ventricle, and the pulmonary aorta to the right. The fourth aortic

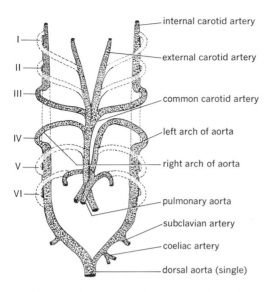

- internal carotid artery
- external carotid artery
- common carotid artery
- left arch of aorta
- right arch of aorta
- pulmonary aorta
- subclavian artery
- coeliac artery
- dorsal aorta (single)

FIG. 12.22 Diagram showing modification of the aortic arches as found in reptiles, ventral view. The ventral aorta (truncus arteriosus) has split into three vessels: right and left systemic aortae and a pulmonary trunk, or aorta.

arch on the right side leaves the systemic aorta and, by means of the radix, leads to the dorsal aorta proper. The latter supplies the entire body with oxygenated blood.

The pulmonary trunk leading from the right ventricle gives off the pulmonary arteries, which are actually outgrowths of the sixth aortic arches. Until the time of hatching there is a ductus arteriosus on the right side, representing the portion of aortic arch VI between the pulmonary artery and the right radix. This serves as a shunt from the right ventricle to the dorsal aorta at a time before the lungs are functioning. It closes at the time of hatching, and all blood from the right ventricle is then sent to the lungs for aeration. A cord of connective tissue on the right side, the *ligamentum arteriosum,* is all that remains of the former arterial shunt.

Mammals The changes in the aortic arches of mammals are rather similar to those

FIG. 12.23 Diagram showing modification of aortic arches as found in birds, ventral view.

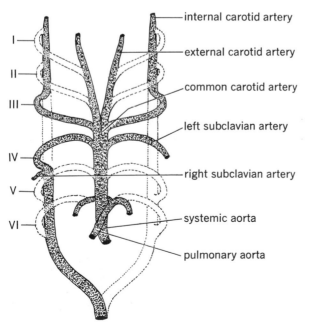

- internal carotid artery
- external carotid artery
- common carotid artery
- left subclavian artery
- right subclavian artery
- systemic aorta
- pulmonary aorta

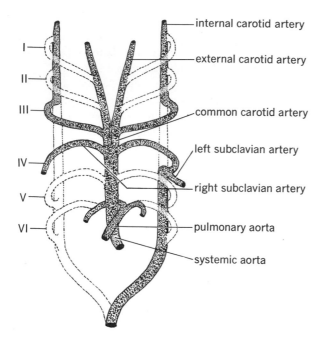

I

II

III

IV

V

VI

internal carotid artery

external carotid artery

common carotid artery

left subclavian artery

right subclavian artery

pulmonary aorta

systemic aorta

FIG. 12.24 Diagram showing modification of aortic arches as found in mammals, ventral view.

of birds except that the radix on the right side, rather than the left, loses its connection with the aorta (Fig. 12.24) The fourth aortic arch on the left side together with its radix becomes the arch of the definitive aorta. The fourth arch on the right and a portion of the right radix become the right subclavian artery. The left subclavian develops as an enlargement of one of the intersegmental arteries coming off the aorta in this region. In mammalian embryos there is at first a ductus arteriosus on each side, but the one on the right persists for only a short time. The left one, which serves as a shunt between pulmonary and systemic aortae, persists until birth, when it finally becomes occluded. A ligamentum arteriosum is finally all that remains (Fig. 12.25).

REMAINDER OF THE ARTERIAL SYSTEM The arteries supplying the head region are, for the most part, branches of the external and internal carotids. In some forms, branches of the fourth arches or of the subclavian arteries also pass to the cephalic region. Occipitovertebral and vertebral arteries are examples. The aorta continues posteriorly into the tail, where it becomes the *caudal artery*. Most of the vertebrate body is supplied with blood through branches of the aorta which may for convenience be grouped under two divisions, somatic arteries, supplying the body proper, and visceral arteries, distributed to the urogenital organs as well as to various portions of the digestive tract and associated structures.

Somatic Arteries The arteries supplying the body proper are usually paired structures which clearly show evidences of metamerism. They supply portions of the body derived from the embryonic epimere, being distrib-

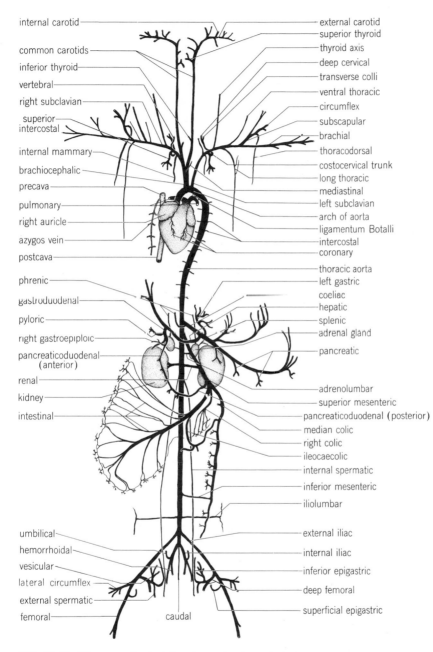

internal carotid

external carotid
superior thyroid
thyroid axis

common carotids
inferior thyroid

deep cervical
transverse colli

vertebral

ventral thoracic

right subclavian

circumflex

superior
intercostal

subscapular
brachial

internal mammary

thoracodorsal

brachiocephalic

costocervical trunk
long thoracic

precava

mediastinal
left subclavian

pulmonary

arch of aorta

right auricle

ligamentum Botalli

azygos vein

intercostal
coronary

postcava

thoracic aorta
left gastric

phrenic

coeliac

gastroduodenal

hepatic

pyloric

splenic

right gastroepiploic

adrenal gland

pancreaticoduodenal
(anterior)

pancreatic

renal

kidney

adrenolumbar
superior mesenteric

intestinal

pancreaticoduodenal (posterior)
median colic
right colic
ileocaecolic
internal spermatic
inferior mesenteric
iliolumbar

umbilical

external iliac

hemorrhoidal

internal iliac

vesicular

inferior epigastric

lateral circumflex

deep femoral

external spermatic

femoral   caudal

superficial epigastric

FIG. 12.25 Diagram of arterial system of cat, ventral view.

uted to the dorsal, or epaxial, musculature and vertebral column, where they are referred to as *parietal,* or *segmental, arteries.* Posterior branches pass through the intervertebral foramina into the neural canal to the spinal cord and its coverings. In higher forms, in which the body is divided into more or less definite regions, such terms as *intercostal, dorsolumbar,* and *sacral* are applied to these segmentally arranged vessels. Two or more segmental arteries may fuse, thus obscuring the fundamental metameric arrangement. The vessels going to the appendages, i.e., the *subclavians* to the pectoral appendages and the *iliacs* to the

pelvic, may be composed of a union of several segmental arteries, the number of vessels concerned corresponding to the number of somites involved in the formation of the limb. During development, rerouting of blood and shifting of vessels leading to the appendages occur, and the end result may show little evidence of metameric origin.

Visceral Arteries The arteries supplying the viscera are of two kinds, paired and unpaired. The paired arteries are segmentally arranged and supply portions derived from the embryonic mesomere from which the urogen-

FIG. 12.26 Diagram showing a portion of the dorsal aorta and its unpaired branches which supply the viscera of the dogfish *Squalus acanthias,* ventral view.

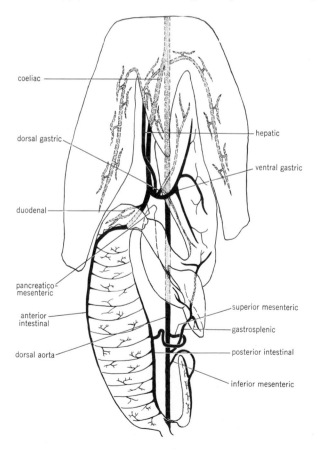

coeliac

dorsal gastric

hepatic

ventral gastric

duodenal

pancreatico- mesenteric

anterior intestinal

dorsal aorta

superior mesenteric

gastrosplenic

posterior intestinal

inferior mesenteric

ital organs and their ducts arise. Although the mesomere is, for the most part, unsegmented, the arteries supplying its derivatives show pronounced evidences of metamerism, particularly in regard to the manner in which they supply the pronephric, opisthonephric, and mesonephric kidneys. Such terms as *renal, genital, ovarian, spermatic,* and *urogenital* arteries are applied to the paired visceral arteries. Renal and genital arteries are numerous in the lower vertebrates, but the number is greatly reduced in higher forms.

The unpaired visceral arteries supplying the spleen and the digestive tract and its derivatives course through the dorsal mesentery of the gut. They branch profusely, the method of branching showing great variation even in members of the same species. There are usually three unpaired visceral arteries in vertebrates. The most anterior of these is the *coeliac artery,* supplying the anterior viscera, including stomach (*gastric*), spleen (*splenic*), pancreas (*pancreatic*), liver (*hepatic*), and duodenum (*duodenal*). The second unpaired visceral artery is the *superior mesenteric,* which supplies the entire length of the small intestine, with the exception of the pyloric end of the duodenum, which is taken care of by the coeliac artery. Branches of the superior mesenteric at its anterior end also supply a portion of the pancreas. The remainder of the vessel is distributed to the cecum and upper half of the large intestine. The third unpaired artery is the *inferior mesenteric,* supplying the posterior part of the large intestine and rectum.

Variations from the above may be accounted for by fusions or separations. For example, in frogs and other amphibians, the coeliac and superior mesenteric arteries have united into a single *coeliacomesenteric artery.* In the dogfish *Squalus acanthias* an unpaired *gastrosplenic artery* arises directly from the aorta just posterior to the origin of the superior

mesenteric, to supply the spleen and posterior part of the stomach (Fig. 12.26). It actually represents a branch of the superior mesenteric which has become secondarily connected with the aorta.

VENOUS SYSTEM

As in the case of the arterial system, a comparison of the veins in the various vertebrate groups shows that they are arranged according to the same fundamental plan and that the variations encountered form a logical sequence as the vertebrate scale is ascended. In its development the venous system of higher forms passes through certain stages common to the embryos of lower forms.

There is some difference in the formation of the earliest veins which appear in an embryo, according as a yolk sac is present or absent. In forms without a yolk sac a pair of *subintestinal veins* appears in the splanchnic mesoderm ventral to the gut. These veins soon fuse, except in the region of the anus, around which they form a loop, only to unite posteriorly, where they continue on into the tail as the *caudal vein.* In animals having a yolk sac, a pair of *vitelline veins* draining the yolk sac joins the posterior end of the heart. Indeed, their fusion is primarily responsible for heart formation. The part of the heart which the vitelline veins enter is destined to become the *sinus venosus.* The vitelline veins are concerned with obtaining nourishment used by the embryo during its development. Even though the yolk sac in mammals contains no yolk, vitelline veins (and arteries) are present. Each vitelline vein is joined posteriorly by a subintestinal vein arranged in a manner similar to that in forms lacking a yolk sac.

Before long, two additional pairs of veins, *anterior cardinals,* from the dorsal side of the head region, and *posterior cardinals,* from the posterior end of the body, establish connections with the fused vitelline veins. They join the sinus venosus on each side by a common vessel variously known as the *common cardinal* or *duct of Cuvier.* The anterior cardinals are generally referred to as jugular veins, which receive *internal* and *external* tributaries. In fishes and salamanders an *inferior jugular vein,* draining the ventrolateral portions of the

FIG. 12.27 Diagram of primitive venous complex from which the venous systems of vertebrates are derived, ventral view.

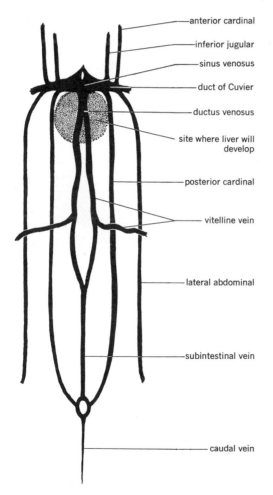

anterior cardinal

inferior jugular

sinus venosus

duct of Cuvier

ductus venosus

site where liver will develop

posterior cardinal

vitelline vein

lateral abdominal

subintestinal vein

caudal vein

head, also enters the common cardinal on each side. It apparently has no homologs in higher forms.

In lower vertebrates a pair of *ventral abdominal,* or *lateral, veins,* located in the ventral, or lateral, portions of the body wall, also enters the common cardinal veins. In amniotes these veins form secondary connections with vessels draining the allantois. Blood in the allantois or umbilical cord, as the case may be, therefore courses through the abdominal veins in the body wall and enters the sinus venosus via the common cardinal veins. The term *allantoic,* or *umbilical, veins* is then applied to these vessels in amniote embryos. Although both left and right umbilical veins are present in reptiles, only the left persists in birds and mammals. When the liver develops, the umbilical veins lose their connection to the sinus venosus and course through the liver via a large channel, the *ductus venosus,* which in turn joins the sinus venosus.

From the primitive venous complex just described (Fig. 12.27) are derived the principal veins of the body. Shifting and rerouting of the blood as it returns to the heart account for the variations observed in venous vessels, which nonetheless show a consistent arrangement within each vertebrate class. The vitelline and umbilical vessels are lost, as such, at hatching or birth, but remnants persist even in adult life.

The primitive veins so far described correspond generally to the main divisions of the arteries but are not usually so clear cut. The vitelline veins and subintestinals give rise to the *unpaired visceral veins,* which later drain the digestive tract and its associated organs and which are located in the mesentery of the gut. *Paired visceral veins* drain the urogenital organs derived from the embryonic mesomere. *Somatic veins* drain the body wall and are generally paired.

HEPATIC PORTAL CIRCULATION

All vertebrates as well as amphioxus have a hepatic portal system. It is the vitelline-subintestinal group of vessels which is concerned with its formation.

In an embryo the liver develops as an outgrowth of the gut in the region just posterior to the heart. As the liver grows out, it soon surrounds the vitelline (or subintestinal) veins. These then break up into a network of small vessels (sinusoids), which ramify through the liver. Posterior to the liver several anastomoses form between the vitelline veins. The right vessel gradually dwindles away, leaving the left vitelline vein with its subintestinal tributary as the main vessel draining the digestive tract. This then becomes the hepatic portal vein, which collects blood from all portions of the alimentary canal as well as from the spleen. It enters the liver, where it breaks up into sinusoids. The original connections of the vitelline veins to the sinus venosus become the *hepatic veins,* which carry the blood collected from the liver sinusoids to the heart.

RENAL PORTAL CIRCULATION

Many vertebrates have a renal portal vein which originates in the caudal region of the body and terminates in capillaries in the opistonephric or mesonephric kidneys. Descriptions of the renal portal vein in the various vertebrate classes are given below.

HYPOPHYSIO-PORTAL CIRCULATION

Branches of the internal carotid arteries, via the circle of Willis, supply the pituitary gland. Some go directly to the gland; others break up into capillaries in the median eminence of the hypothalamus and the hypophysial stalk. These combine to form larger hypophysio-portal venules, which empty into sinusoids in the anterior lobe, thus constituting a true portal system of veins. Hormones or neurosecretions formed in the hypothalamus or neighboring area of the brain reach the anterior lobe via the hypophysioportal pathway. These *release* and *inhibitory factors* control the hormonal output of the cells of the anterior lobe.

Venous Systems of Different Chordates

Amphioxus The veins of amphioxus (Fig. 12.28) represent the vertebrate venous system reduced to its simplest terms. The caudal vein continues forward as the subintestinal vein. It also connects to the postcardinal on each side so that blood may be returned from the tail via either channel. An anterior cardinal joins the postcardinal on each side. The two enter the posterior end of the "heart" (sinus venosus) by means of a common cardinal vein. The subintestinal vein breaks up in capillaries in the intestine. Another vein, beginning in capillaries in the intestine, leads to the hepatic cecum, where it again breaks up into capillaries. This vessel is a true hepatic portal vein. A contractile hepatic vein originating in the hepatic cecum carries blood to the "heart." A pair of parietal veins draining the body wall dorsal to the gut furnishes additional somatic vessels which enter the posterior end of the "heart."

Cyclostomes The venous system of the lamprey does not differ greatly from that of amphioxus. Two inferior jugular veins from the anteroventral region of the body unite to form a single vessel which joins the sinus venosus. Paired anterior and posterior cardinals are present. The common cardinal veins appear to have fused, or perhaps the left one has degenerated in the adult so that but a single common cardinal enters the sinus venosus on the dorsal side. The caudal vein, which divides at the cloacal region, connects

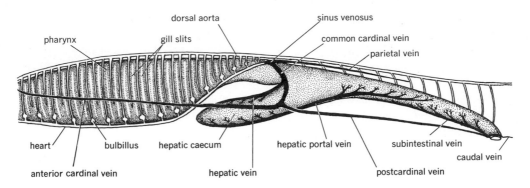

FIG. 12.28 Diagram showing the essential features of the circulatory system of amphioxus, lateral view.

to the postcardinal veins, which course on either side slightly ventrolateral to the dorsal aorta. They fuse as they approach the heart.

The anterior part of the subintestinal vein remains as a small vessel located in the typhlosole of the intestine. Its posterior end has lost all connection with the caudal vein. The subintestinal vein becomes the hepatic portal vein, which also receives a branch from the head.

A contractile *portal heart* is present in the hepatic portal vein of cyclostomes. Blood from the liver is collected by a single hepatic vein which joins the sinus venosus. No renal portal system exists in cyclostomes. Segmentally arranged veins from the kidneys and gonad enter the postcardinal veins, as do those from the body wall.

Fishes Many primitive features are retained by the venous systems of fishes, but important advances over conditions in amphioxus and cyclostomes are to be found.

The sinus venosus receives a duct of Cuvier (common cardinal) on each side (Fig. 12.29*A*). An anterior cardinal, or jugular, vein brings blood to the common cardinal from the dorsal side of the head region. In many fishes a pair of *inferior jugular veins* from the ventrolateral part of the head also enters the common cardinal veins. These are lacking in *Polypterus* and fused in *Lepisosteus*. Each common cardinal also receives a postcardinal vein from the posterior end of the body. Since fishes are the first vertebrates to possess paired appendages, subclavian and iliac veins, bringing blood from the pectoral and pelvic fins, respectively, first make their appearance in this group. In young fishes the subclavian veins may enter the anterior part of the postcardinals, but in adults a shifting has occurred so that the subclavian veins usually enter the lateral veins or the common cardinals directly. The iliac veins join the lateral abdominals, which course in the body wall to join the common cardinal.

The posterior wall of the sinus venosus usually receives two hepatic veins of similar size which return blood from the liver to the heart. The hepatic portal vein of fishes is well developed (Fig. 12.30) and, as in other forms, is derived from the left vitelline vein and its subintestinal tributary. The subintestinal vein has in most cases lost its connection with the caudal vein.

The main differences over the primitive condition to be found in the veins of fishes are concerned with changes which occur in

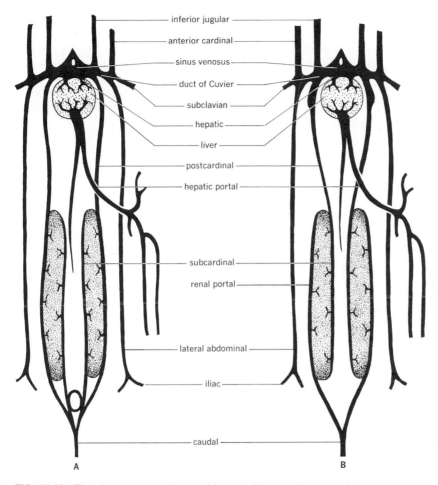

inferior jugular

anterior cardinal

sinus venosus

duct of Cuvier

subclavian

hepatic

liver

postcardinal

hepatic portal

subcardinal

renal portal

lateral abdominal

iliac

caudal

A

B

FIG. 12.29 The changes over the primitive condition which occur in the venous system of fishes, ventral view.

the posterior part of the body and which result in the development of the renal portal system.

With the development of the opisthonephric kidneys and their posterior elongation, the postcardinals grow backward, ultimately to unite with the caudal vein at the point where it divides to form a ring about the anus. A new vessel, or pair of vessels, the *subcardinal veins,* now develops between the

opisthonephric kidneys, and a connection with the caudal vein is established near the point where it is joined by the postcardinals. Small blood vessels passing through the opisthonephric kidneys connect postcardinal and subcardinal veins. They are not associated with glomeruli, which receive arterial blood. Blood in the caudal vein now has two alternative routes through which it may pass on its way to the heart. It may go into the postcardinals directly or through the subcardinals and renal veins to the postcardinals.

The next advance involves an interruption in the course of each postcardinal vein so that anterior and posterior portions are no longer continuous (Fig. 12.29B). The anterior ends of the subcardinals then join the anterior portions of the postcardinals, and the posterior

FIG. 12.30 Diagram of hepatic portal vein and its tributaries in *Squalus acanthias,* ventral view.

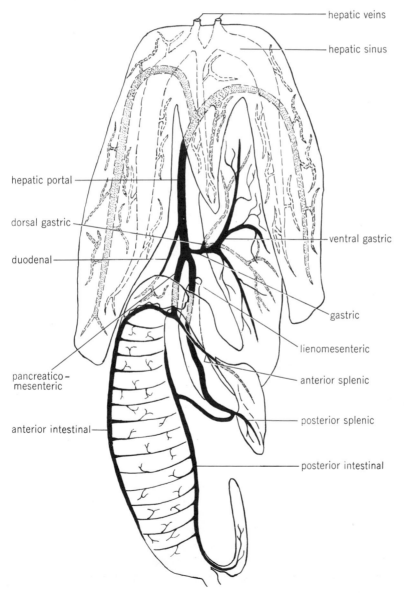

ends of the subcardinals lose their direct connection with the caudal vein. Blood from the tail now travels only by one route. It goes up the caudal vein into the original posterior portion of the postcardinals, which are now termed the *renal portal veins.* It then passes through the renal veins, which now have assumed capillary dimensions, finally to enter the subcardinal veins. The blood courses anteriorly via the postcardinals, which join the common cardinals. The postcardinals usually receive veins from the gonads.

The newly developed renal portal vein falls within the definition of a portal system

FIG. 12.31 Diagram of systemic veins and renal portal system of dogfish *Squalus acanthias,* ventral view.

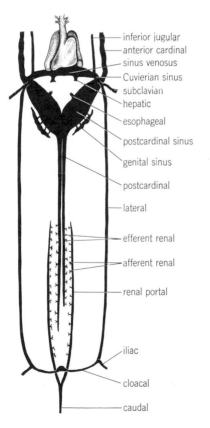

inferior jugular
anterior cardinal
sinus venosus
Cuvierian sinus
subclavian
hepatic
esophageal

postcardinal sinus

genital sinus

postcardinal

lateral

efferent renal

afferent renal

renal portal

iliac

cloacal

caudal

since it breaks up into a capillary network before the blood which it contains reaches the heart.

The changes in the venous system described above have now reached the stage found in elasmobranch fishes (Fig. 12.31). Many interesting variations are encountered in the venous systems of other fishes, only a few of which will be considered. In teleost fishes lateral abdominal veins are not present. The subclavians enter the common cardinals, and the iliacs join the postcardinals. Veins from the swim bladder usually join the hepatic portal vein. This is to be expected, since the swim bladder is a derivative of the gut. However, in certain forms the connection is with the postcardinal veins instead. In *Polypterus* they join the hepatic veins directly. In dipnoans the pulmonary veins from the swim bladder enter the newly formed left atrium of the heart, and the double type of circulatory system appears for the first time.

In the lungfishes a connecting link is found which indicates how the venous system of amphibians may have arisen in evolution from the more simple and primitive arrangement found in other fishes. The veins of *Epiceratodus* show this transition most clearly. In *Epiceratodus* a single, midventral, *anterior abdominal vein,* similar to that of amphibians, makes its appearance. The lateral abdominal veins have fused to form the anterior abdominal, which courses forward to enter the sinus venosus. In the meantime the iliac vein has formed a connection with the renal portal on each side (Fig. 12.32). The branch of the iliac which joins the anterior abdominal is called the *pelvic vein.* The right postcardinal, including its posterior portion (originally the right subcardinal), has become much larger than its counterpart on the left side. The connection with the caudal vein on both sides is retained. The larger vessel on the right side is now called the *postcaval vein.* It

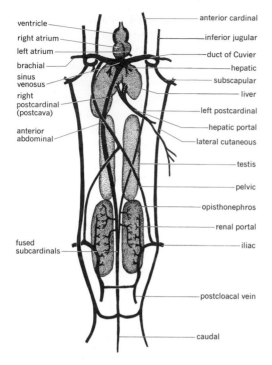

ventricle
right atrium
left atrium
brachial
sinus venosus
right postcardinal (postcava)
anterior abdominal
fused subcardinals

anterior cardinal
inferior jugular
duct of Cuvier
hepatic
subscapular
liver
left postcardinal
hepatic portal
lateral cutaneous
testis
pelvic
opisthonephros
renal portal
iliac
postcloacal vein
caudal

FIG. 12.32 Diagram of venous system of *Epiceratodus*, ventral view. (*Modified from B. Spencer.*)

passes through the liver to open into the sinus venosus. The smaller left postcardinal passes the liver and enters the left common cardinal. The presence of the postcava in *Epiceratodus* clearly foreshadows the appearance of the postcaval vein in amphibians, the anterior portion of which, however, has a different origin.

The venous system of *Protopterus*, except for the lack of an anterior abdominal vein, even more closely resembles that of amphibians than does that of *Epiceratodus*.

Amphibians The main change that has occurred in the amphibian venous system over that of *Epiceratodus* lies in the anterior connections of the anterior abdominal vein. Whereas in the lungfish this vein joins the sinus venosus directly, in amphibians it enters

the hepatic portal vein (Figs. 12.33 and 12.34). Thus the renal portal and hepatic portal systems are brought into close association, since blood from the legs must pass through one or the other.

Pulmonary veins from the lungs enter the left atrium, as in the Dipnoi. In lungless salamanders, of course, pulmonary veins are absent and the left atrium is reduced in size.

The common cardinals, which originally received the subclavian, jugular, and postcardinal veins, are further consolidated in amphibians and are now called the *precaval veins*. They enter the sinus venosus on each side. The jugular vein has external and internal tributaries. Since cutaneous respiration has developed to a high degree in amphibians,

FIG. 12.33 Diagram of venous system of a typical urodele amphibian, ventral view.

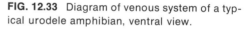
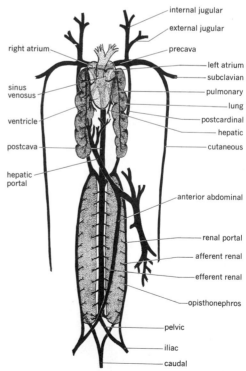

right atrium
sinus venosus
ventricle
postcava
hepatic portal

internal jugular
external jugular
precava
left atrium
subclavian
pulmonary
lung
postcardinal
hepatic
cutaneous
anterior abdominal
renal portal
afferent renal
efferent renal
opisthonephros
pelvic
iliac
caudal

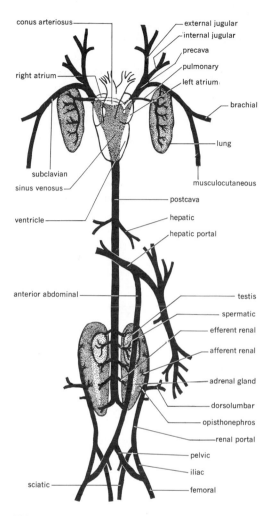

conus arteriosus
external jugular
internal jugular
precava
pulmonary
right atrium
left atrium
brachial
lung
subclavian
musculocutaneous
sinus venosus
postcava
ventricle
hepatic
hepatic portal
anterior abdominal
testis
spermatic
efferent renal
afferent renal
adrenal gland
dorsolumbar
opisthonephros
renal portal
pelvic
iliac
sciatic
femoral

FIG. 12.34 Diagram of venous system of a typical anuran amphibian, ventral view.

exceptionally large cutaneous veins are present which join the subclavians to enter the sinus venosus.

The postcava in amphibians differs from that of *Epiceratodus* in that its anterior portion is derived from the sinus venosus and portions of the original vitelline veins rather than from the right postcardinal. The posterior part, however, like that of *Epiceratodus,* comes from the old subcardinals.

Although both urodeles and anurans are similar in the above respects, they exhibit certain differences in regard to the arrangement of the postcaval-postcardinal complex. In most urodeles and a few anurans the anterior portions of the postcardinals persist in reduced form, connecting the middle portion of the postcava with the common cardinal on each side. In most adults anurans, however, the anterior portions of the postcardinals usually disappear, and the postcava furnishes the only route through which blood from the kidneys and gonads can be returned.

Reptiles Generally speaking, the venous system of reptiles shows little change over the condition in amphibians. The large systemic veins entering the heart have shifted more to the right side, following the further partitioning of the heart. Two precavae are the original common cardinals, which receive the jugular, subclavian, and postcardinal veins. The anterior portions of the postcardinals have degenerated into two small *vertebral veins.* In snakes the subclavians are lacking.

More blood from the posterior part of the body now courses through the anterior abdominal vein, which joins the hepatic portal vein anteriorly. The importance of the renal portal vein has diminished, and in some forms direct channels may even pass through the kidneys, connecting renal portal and postcaval veins. As in amphibians, blood in the hind limbs and tail reaches the heart either by the renal portal–kidney–postcaval route or by the pelvic–anterior abdominal–hepatic portal pathway.

Since cutaneous respiration does not exist in reptiles, the pulmonary circulation has assumed greater importance. The pulmonary veins from the two lungs show discrepancies in size in various forms, depending upon the degree of asymmetrical development of these respiratory organs. One pulmonary vein may even be entirely absent in certain snakes in which the left lobe of the lung is absent.

The reptilian postcava, as in amphibians, is derived partly from the subcardinals and partly from the vitelline veins. The postcardinals have practically disappeared.

Birds The reptilian character of the venous system is clearly indicated in birds. The sinus venosus has been incorporated in the wall of the right atrium, so that the two precavae and single postcava enter the right atrium directly. Each precava is formed by the confluence of subclavian and jugular veins. The original postcardinal connection is no longer in existence.

The postcava assumes an even greater importance than that of reptiles and is the chief pathway for the return of blood from the posterior part of the body. It is derived in the same manner as that previously described for amphibians and reptiles. The posterior end of the postcava receives blood directly from the limbs (Fig. 12.35) via the "renal portal" veins (old posterior cardinals). The name "renal portal" is scarcely appropriate in birds, for

FIG. 12.35 Diagram of venous system of bird, ventral view.

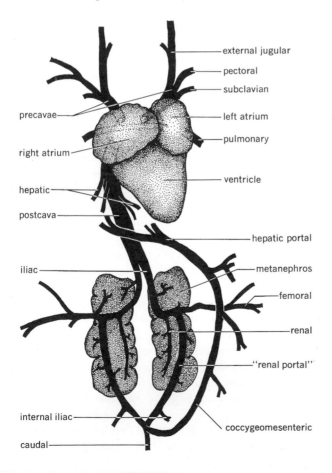

external jugular
pectoral
subclavian
precavae
left atrium
pulmonary
right atrium
ventricle
hepatic
postcava
hepatic portal
iliac
metanephros
femoral
renal
"renal portal"
internal iliac
coccygeomesenteric
caudal

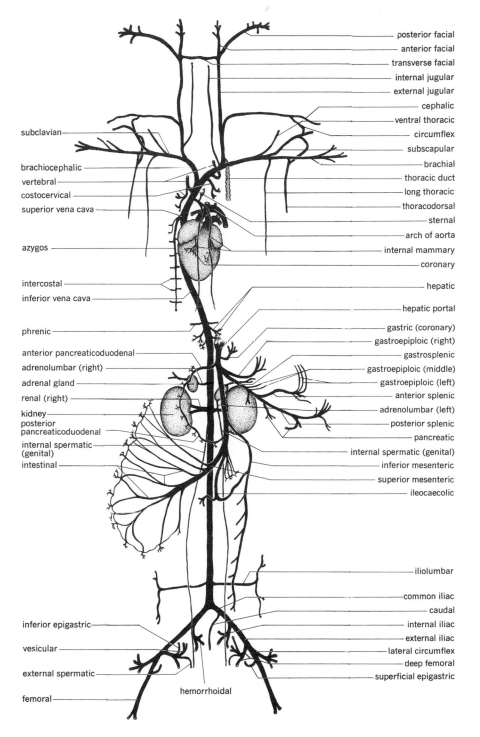

posterior facial
anterior facial
transverse facial
internal jugular
external jugular
cephalic
ventral thoracic
circumflex
subscapular
brachial
thoracic duct
long thoracic
thoracodorsal
sternal
arch of aorta
internal mammary
coronary
hepatic
hepatic portal
gastric (coronary)
gastroepiploic (right)
gastrosplenic
gastroepiploic (middle)
gastroepiploic (left)
anterior splenic
adrenolumbar (left)
posterior splenic
pancreatic
internal spermatic (genital)
inferior mesenteric
superior mesenteric
ileocaecolic
iliolumbar
common iliac
caudal
internal iliac
external iliac
lateral circumflex
deep femoral
superficial epigastric

subclavian
brachiocephalic
vertebral
costocervical
superior vena cava
azygos
intercostal
inferior vena cava
phrenic
anterior pancreaticoduodenal
adrenolumbar (right)
adrenal gland
renal (right)
kidney
posterior pancreaticoduodenal
internal spermatic (genital)
intestinal
inferior epigastric
vesicular
external spermatic
femoral
hemorrhoidal

FIG. 12.36 Diagram of venous system of cat, ventral view.

although these vessels pass through the kidneys, the passage is for the most part direct and not interrupted by a capillary network. The hepatic veins of birds join the postcava as it nears the heart.

The caudal vein in birds is greatly reduced in correlation with the decrease in size of the tail. A vein, variously known as the *inferior mesenteric, coccygeomesenteric,* and *caudal mesenteric,* connects the caudal vein with the hepatic portal vein. Since this vessel connects the two "portal" systems, it is possibly homologous with the anterior abdominal vein of amphibians and reptiles, although this homology is questioned by some. Others consider a small *epigastric vein* in birds to be the homolog of the anterior abdominal vein of lower forms. This vessel carries blood from the great omentum to one of the hepatic veins and is not located in the ventral body wall.

Mammals The shifting of the main venous channels to the right side is more clearly indicated in mammals than in other vertebrates (Fig. 12.36). As in birds, precaval and postcaval veins enter the right atrium directly, since the sinus venosus has been gathered into its walls. In some mammals two precavae are present, but in others they are joined. The vessel on each side which receives the jugular and subclavian vein is then called the *brachiocephalic vein.* The internal jugular vein of mammals is the original anterior cardinal vein. It is usually smaller than its external jugular tributary.

In mammals a portion of the anterior end of the right postcardinal persists as the *azygos vein.* This drains the intercostal muscles and enters the precava. It is homologous with the vertebral veins of reptiles. Much variation is to be found in the azygos vein in different mammals.

The greatest change in the venous system of mammals occurs in the postcava, which has become considerably simplified. No trace of a renal portal system is found in the adult, and all blood from the posterior end of the body is collected by the postcava. The embryonic development of the mammalian postcava is complicated by the appearance of new vessels called the *supracardinal veins,* which contribute to its formation. In other respects, however, its homologies with the postcava of lower forms are clear.

The anterior abdominal vein has disappeared in mammals, being found only in the monotreme *Echidna.* The allantoic, or umbilical, veins of mammalian embryos, returning blood from the placenta, are homologous with the lateral abdominal veins of elasmobranchs and with the anterior abdominal veins of amphibians and reptiles. Usually only the left umbilical vein persists, passing through the liver as the *ductus venosus* to join the postcava before it enters the heart. The umbilical vessels degenerate when they cease to function after birth.

The hepatic portal vein, usually referred to as *the* portal vein, is similar to the hepatic portal vein of lower forms.

FETAL CIRCULATION

The allantoic, or umbilical, circulation is of great importance to the embryos of amniotes. In reptiles and birds, with their large-yolked eggs, the vitelline circulation is used in obtaining nutritive substances from the yolk. Although during early development the vitelline circulation is used in respiration, it is soon superseded by the allantoic circulation, which, along with the developing allantois, has grown out to the periphery of the egg, beneath the porous shell. Here an exchange of oxygen and carbon dioxide takes place.

In mammals, the eggs contain little or no yolk. The chief role of the vitelline arteries is the establishment of the superior mesenteric artery; that of the vitelline veins lies in the formation of the hepatic portal vein. The placenta usually develops in association with the allantois. Allantoic, or umbilical, arteries carry blood from the embryo to the placenta, and similarly named veins are employed in its return. The placenta is the organ in which exchange of gases, nutrient materials, and wastes takes place between the mother and her developing young.

A pair of umbilical arteries arises from the dorsal aorta and passes out the umbilical cord to the placenta (Fig. 12.37). At first a pair of umbilical veins returns oxygenated blood to the embryo. Later, in birds, the right artery degenerates, but in mammals both persist. In both birds and mammals the right umbilical

vein degenerates, and the left one alone continues to function. From the umbilical cord it courses through the body wall for a short distance and then enters the liver through a large channel, the *ductus venosus,* to join the postcava.

The lungs do not function during embryonic life. The umbilical circulation is used to obtain oxygen and food from the placenta.

In a typical mammalian fetus, blood enters the right atrium through precaval and postcaval veins. A portion of it then goes to the right ventricle, guarded by the tricuspid valve. Instead of being pumped to the nonfunctional lungs, much of this blood is shunted to the systemic aorta via the ductus arteriosus. It will be recalled that the ductus arteriosus is a portion of the sixth aortic arch which connects

FIG. 12.37 Diagram showing arrangement of the fetal blood vessels of man, ventral view. Arrows indicate the direction of blood flow. Black represents unoxygenated blood; light stippling, oxygenated blood; intermediate stippling, mixed blood.

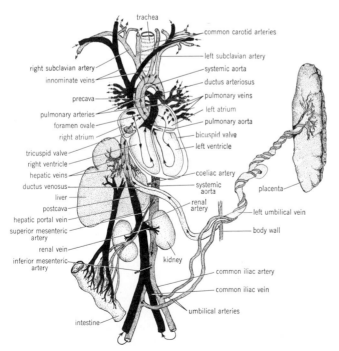

the pulmonary and systemic aortae during embryonic life. The rest of the blood in the right atrium passes through an opening in the interatrial septum (*foramen ovale*) into the left atrium. Joined by a relatively small quantity coming from the pulmonary veins, it passes over the bicuspid valve into the left ventricle, from which it is pumped into the systemic aorta. This blood, together with that coming from the ductus arteriosus, is distributed to all parts of the body, and a portion is sent out through the umbilical arteries to the placenta.

From the above description it would seem as though a mixing of unoxygenated blood from precaval and postcaval veins and oxygenated blood from the left umbilical vein, which joins the postcava, would occur in the right atrium. Although this matter has been the subject of much controversy and investigation for over a century, it now seems fairly clear that little mixing occurs. The greater part of the blood from the postcava, carrying a large proportion of oxygenated blood from the left umbilical vein, crosses the dorsal part of the right atrium to enter the left atrium through the foramen ovale. A smaller amount passes into the right ventricle. The stream of unoxygenated blood from the precavae passes, for the most part, along the ventral portion of the right atrium into the right ventricle. Although different species vary somewhat in this respect, there is actually a moderately complete separation of oxygenated and unoxygenated blood in the fetal heart.

CHANGES OCCURRING AT BIRTH
At birth the umbilical circulation ceases to function. The umbilical arteries no longer pulsate and gradually constrict. The umbilical vein ceases to drain blood from the placenta back into the fetus. The stumps of the umbilical arteries become the *hypogastric,* or *internal iliac,* arteries which supply the caudal body wall and portions of the urogenital organs.

The umbilical vein disappears, its former course being marked by a fibrous cord, the *ligamentum teres,* extending from the umbilicus (navel) to the undersurface of the liver. The remnant of the ductus venosus is found in the liver as a slender fibrous cord, the *ligamentum venosum.*

When the atrium is first partitioned, a membrane, the *septum primum,* grows down and divides it into left and right chambers. An opening, the *interatrial foramen,* appears in the septum primum, so that the cavities of the atria are in communication. Later a second septum, the *septum secundum,* semilunar in shape, grows down parallel to the septum primum on its right side. An oval aperture in the septum secundum is the *foramen ovale.*

With the cessation of the umbilical circulation, the amount of blood entering the heart through the postcava is materially decreased, since flow through the umbilical vein has stopped. At birth the lungs expand, and pulmonary respiration begins. The ductus arteriosus is gradually occluded and becomes a fibrous cord, the *ligamentum arteriosum.* Blood from the right ventricle is no longer shunted to the systemic aorta but is sent to the lungs instead. The pulmonary veins send a greatly increased amount of blood to the left atrium. The foramen ovale then closes. This is brought about by a decrease in pressure in the right atrium, which is now receiving less blood, and an increased pressure in the left atrium due to the increased flow through the pulmonary veins. The septum primum and septum secundum are pushed together, causing the foramen ovale to be obliterated. A thin area, the *fossa ovalis,* in the interatrial septum indicates the former position of the foramen ovale.

Other factors may also play a part in the changes in circulation occurring at birth. After these changes have taken place, a perfect double circulation is established.

Faulty development of the septum primum and septum secundum or other abnormalities may prevent the foramen ovale from closing properly. An open, or patent, foramen ovale is not necessarily fatal or even very harmful in man. In some cases, a mixture of oxygenated and unoxygenated blood takes place. As a result the child has a purplish color and is referred to as a "blue baby." In severe cases this defect can be corrected by heart surgery. Other anomalous conditions may result from improper closure of the ductus arteriosus or from a failure of the interventricular septum to develop properly.

LYMPHATIC SYSTEM

Lymphatic vessels develop considerably later than blood vessels and quite independently of them. Connections of lymphatic vessels with veins are secondary developments and usually take place near the heart.

The smallest units of the lymphatic system are the *lymph capillaries*. Tissue fluid passes by diffusion into the small anastomosing lymph capillaries. It is then referred to as *lymph*. Lymph capillaries end blindly at their free ends. Their bases unite to form larger vessels, which in turn combine to form vessels of still greater caliber. The largest vessels are those which in most cases open directly into the great veins near the heart. Although the lymphatic system is a closed system, it does not in itself form a complete circuit and in this respect differs from the blood-vascular system.

Colloidal material which has escaped from the blood capillaries into the tissue fluid and which cannot reenter the blood capillaries may diffuse through the walls of the small lymphatic vessels and thus be returned to the circulatory system.

Lymph capillaries are tubules of somewhat

FIG. 12.38 Lymph capillaries.

greater diameter than blood capillaries. Their caliber is not at all uniform (Fig. 12.38). No valves are present in lymph capillaries. Their walls are very thin, consisting of a single layer of flat, squamous endothelial cells.

The larger vessels formed by the union of lymph capillaries have thicker walls and contain valves. The walls are covered with a thin layer composed of elastic and collagenous fibers, as well as a few smooth-muscle cells. In the largest vessels, three layers, comparable to those of small arteries, make up the walls (page 330). These layers, which are not so distinct as those in blood vessels, bear similar names.

The valves, which prevent backflow, are more numerous than venous valves and, where present, occur at close intervals. They are similar to that of the valves in veins.

The walls of the larger lymphatic vessels are supplied by minute blood vessels similar to the vasa vasorum of the blood-vascular system. The nerve supply to the lymphatics is abundant.

Lymph nodes are masses of lymphatic tissue composed of collagenous and reticular connective-tissue fibers as well as reticuloendothelial cells. They are encapsulated and enclose large numbers of lymphocytes, plasma cells, and fixed macrophages. Lymphatic vessels ramify throughout the nodes, taking a rather circuitous course. The nodes are placed at intervals along the lymphatic vessels and arranged in such a manner that the lymph must filter through them on its way to the larger vessels. Phagocytic cells in the nodes remove and destroy various kinds of particles and bacteria which may be present. The plasma cells are important in antibody production (see page 327).

Lymphatic nodules, which may appear and disappear, should not be confused with lymph nodes. They are small, spherical masses of lymphocytes and plasma cells which are not clearly circumscribed and are not encapsulated. They may be located within a lymph node but also occur quite independently.

The small lymph capillaries of the intestine are called *lacteals.* Unlike the capillaries of the blood-vascular system, they absorb fat.

Several factors are instrumental in propelling the slowly moving lymph through lymphatic vessels and nodes. These include (1) muscular activity of various parts of the body, tending to squeeze the fluid along; (2) pulsations of neighboring arteries; (3) pressure built up in the smaller vessels by osmosis and absorption of tissue fluid; (4) the action of pulsating *lymph hearts.* The latter, when present, consist of enlargements in lymphatic vessels, with contractile walls. They are usually situated near a point where lymph enters the venous system. Valves are present which control the direction of flow. The rhythm of the beating lymph hearts bears no relation to the beat of the heart.

Not a great deal is known about the comparative anatomy of the lymphatic system. The small size and delicate structure of the vessels are barriers to detailed study. Furthermore, the great irregularity of the channels in the various classes of vertebrates precludes a study of evolutionary advance. Most studies have been made in birds and mammals. Less is known about conditions in lower classes.

Fishes Lymphatic vessels in fishes are extensively developed. Peripherally located channels extend into head, tail, and fins. Deeper channels follow the course of some of the larger veins. Several connections occur between the lymphatic and venous systems in the posterior and middle, as well as the anterior, parts of the body. Lymph hearts are usually not present, but some have been described in certain forms near the point of junction of lymphatic vessels and veins. The eel has a lymph heart in the tail. The European catfish has two caudal lymph hearts. Lymph nodes seen to be lacking in fishes.

Amphibians Two main sets of lymphatic vessels are present in urodele amphibians. Superficial vessels beneath the skin carry lymph to cutaneous and postcardinal veins. Between 14 and 20 lymph hearts have been observed along their course in various forms. Deeper channels follow the dorsal aorta on each side and enter the subclavian veins.

Anurans are characterized by the presence of large lymph sacs, or spaces, beneath the skin. Most of the lymph flows toward the heart. Two pairs of lymph hearts are usually present in adult animals. In the frog the first pair lies behind the transverse processes of the third vertebra, pumping lymph into the vertebral vein. A posterior pair, located near the end of the urostyle, pumps lymph into the transverse iliac vein.

Lymph hearts are more numerous in larval and tadpole stages. It has been reported that over 200 lymph hearts are present in caecilians, lying beneath the skin along the intersegmental veins.

Reptiles A well-developed lymphatic system is present in reptiles. A large subvertebral trunk divides anteriorly to enter the precaval veins. In snakes the lymphatic vessels and sinuses are exceptionally large and numerous. A posterior pair of lymph hearts pumps lymph into the iliac veins.

Birds The lymphatic vessels of birds ultimately enter two thoracic lymph ducts which join the precaval veins. Transitory lymph hearts, not found in adult birds, may be observed in the pelvic region during embryonic development. The bursa Fabricii, a cloacal derivative, is a lymphoid organ present in the young of most birds. It is important in the production of lymphocytes and seems to be concerned with antibody production.

Mammals Lymph hearts are altogether lacking in mammals. A main trunk, the *thoracic duct* (Fig. 12.36), drains all the lymphatic vessels of the posterior part of the body, as well as those coming from the left side of the head, neck, and thoracic regions. It lies beneath the vertebral column and courses anteriorly to open into the left subclavian vein just before the latter's junction with the internal jugular vein. The posterior end of the thoracic duct is expanded into a conspicuous enlargement, the *cisterna chyli*. A *right lymphatic duct* drains lymph from the right side of the head and neck, the right arm, and the right side of the thorax. It enters the *right* subclavian vein, near where it is joined by the internal jugular vein.

In mammals lymph nodes are numerous in superficial regions of the head and neck, axillae, and groin. Many lie within the body cavity in close association with the large blood vessels. They are unusually large and numerous in the mesentery of the intestine. In all these localities they serve to prevent the invasion of the body by bacteria. The phagocytic action of the fixed macrophages, in particular, is primarily responsible for such protection. *Peyer's patches* are nodules of lymphoid tissue in the small intestine, especially numerous in the region of the ileum.

OTHER LYMPHATIC ORGANS

Among other structures usually considered to belong to the lymphatic system are certain organs of pharyngeal origin, previously discussed in that connection. They include *tonsils, adenoids,* and the *thymus gland.* It seems that the chief function of tonsils and adenoids is to form antibodies against antigenic substances that may possibly enter the tissues in this region. They also filter tissue fluid and produce lymphocytes. The function of the thymus gland in relation to the lymphatic system and the circulatory system in general has only recently been elucidated. The best-known fact concerning the gland is that although it is a comparatively large organ in the young, it is relatively much smaller during adulthood. In man it reaches its greatest size between the eleventh and fifteenth years and then begins to decrease in size.

HEMAL NODES Certain organs of the body closely resemble lymph nodes except that they enclose blood vessels rather than lymphatic vessels. Blood, rather than lymph, filters through them. They are called *hemal nodes* and are entirely devoid of lymphatic vessels. In pigs a special kind of *hemolymphatic node* has been described which has properties of both lymphatic nodes and hemal nodes. It is believed that hemal nodes, like the spleen, play a part in the destruction of old and worn-out red corpuscles. They may even have a role in erythrocyte formation. Most hemal nodes are very small. They are abundant in the retroperitoneal tissue near the points of origin of the superior mesenteric and renal arteries. Ruminating mammals have numerous hemal nodes. Their occurrence in man is doubtful.

SPLEEN The largest lymphoid organ in the body is the spleen. It lies toward the left side of the body close to the stomach, to which it is connected by a *gastrosplenic liga-*

ment. The spleen is considered to be a hemo-lymphatic organ, since it is interposed in the bloodstream rather than in lymphatic vessels. Because of the peculiar arrangement of the blood vessels in the spleen, the blood comes in contact with phagocytes (macrophages) which engulf the fragments of disintegrating red corpuscles. Lymphocytes and plasma cells, probably originating in the thymus gland early in life, are later formed in large numbers in the spleen. The spleen also serves as a store-house for erythrocytes. Large numbers are housed by the spleen, which returns them to the blood-vascular system as requirements of the body demand. Contraction of the spleen occurs in response to the demand of the body for an increased supply of red corpuscles, which are thus forced into the bloodstream.

During embryonic development the spleen functions as a blood-forming organ. In mammals it ceases to form erythrocytes during adult life but continues to form lymphocytes. The spleen is also of importance in the production of antibodies, serving as a mechanism for defense of the body against certain diseases. Enlargement of the spleen is a feature of malaria and some other conditions in which the organ may assume relatively enormous proportions.

Among vertebrates, cyclostomes lack a discrete spleen. Some peculiar tissue in the intestinal wall is sometimes homologized with the spleen. Splenic tissue has been described in the stomach wall in *Protopterus,* which also lacks a spleen but possesses a lymphoid mass partially surrounding the stomach. The spleens of other vertebrates vary considerably in shape. They are usually compact organs. In anurans the spleen first begins to be less significant in production of red corpuscles, and in the adults of higher forms this property has probably been lost altogether.

SUMMARY

1   The circulatory systems of higher metazoans serve to carry substances of physiological utility to all parts of the body and to remove from them the waste products of metabolism. Wastes are carried to appropriate organs for elimination.

2   The vertebrate circulatory system is composed of two systems of elaborately branched tubes which ramify throughout the body. They are (*a*) the blood-vascular system, composed of heart, arteries, arterioles, capillaries, venules, and veins; (*b*) the lymphatic system, made up of lymph capillaries, vessels, sinuses, nodes, and lymphoid organs of various types.

3   The blood-vascular system is divided into arterial and venous systems, the former carrying blood away from the heart and the latter returning it. One, two, or three portal systems of veins are present. Each breaks up into a capillary network before reaching the heart. Two of these systems, known as the hepatic portal and renal portal systems, end in liver and kidneys, respectively. The third is associated with the pituitary gland and is called the hypophysio-portal system.

4   The walls of arteries, veins, and the larger lymphatic vessels are composed of three layers, the tunica intima, tunica media, and tunica adventitia. The thickness and detailed structure of these layers vary in the different types of vessels.

5   The heart is primarily a pulsating tube derived from a fusion of two vitelline veins.

It becomes divided into chambers called atria and ventricles. In addition, two accessory chambers may be present, called the sinus venosus and conus arteriosus, respectively. Cyclostomes and most fishes have two-chambered hearts with one atrium and one ventricle; Dipnoi and amphibians with three-chambered hearts have two atria and one ventricle; the three-chambered heart of most reptiles is like that of amphibians, but an incomplete partition appears in the ventricle. In crocodilians this becomes complete, forming a four-chambered heart with two atria and two ventricles. Birds and mammals have four-chambered hearts. Valves of various types regulate the direction of blood flow.

6   Animals with two-chambered hearts have the single type of circulation in which only unoxygenated blood passes through the heart, which pumps it to the gills for aeration. The double type of circulation exists in three- and four-chambered hearts, through which pass two streams of blood, one oxygenated and the other either unoxygenated or only partly oxygenated.

7   Vessels which carry blood from the heart to the lungs and back comprise the pulmonary circulation. Unoxygenated blood or partially unoxygenated blood is pumped from the right ventricle (or right side of the ventricle) to the lungs. Oxygenated blood returns from the lungs to the left atrium. The remaining vessels make up the systemic circulation, which sends oxygenated blood throughout the body and returns unoxygenated blood to the heart itself.

8   The arterial systems of vertebrates are fundamentally similar, at least during developmental stages. Usually six pairs of aortic arch vessels extend from the ventral aorta to the dorsal side of the pharyngeal region, where they fuse to form a vessel on each side. These two vessels combine posteriorly to form a single dorsal aorta. Changes in the aortic arches constitute the chief differences in the arterial systems of the separate vertebrate classes. A progressive reduction in the numbers of aortic arches occurs as the evolutionary scale is ascended. In cyclostomes and fishes, which retain the greatest number of aortic arches, each arch divides into afferent and efferent portions connected by a capillary network. The gill lamellae through which the capillaries ramify serve as respiratory organs.

9   The single dorsal aorta, as it courses posteriorly, gives off many paired somatic arteries to supply the body proper. Paired visceral vessels pass to the derivatives of the embryonic mesomere. Usually three unpaired visceral vessels are distributed to the digestive tract and spleen.

10   The primitive venous system arises from several paired veins, all of which join the sinus venosus. These include (*a*) a pair of vitelline veins, continued posteriorly as the subintestinal veins; (*b*) paired anterior and posterior cardinals, which, on each side, join to form a common cardinal vein, or duct of Cuvier, entering the sinus venosus; (*c*) lateral abdominal veins from the body wall, also entering the duct of Cuvier; (*d*) subcardinal veins, which appear between the opisthonephric or mesonephric kidneys and which are joined by the caudal vein from the tail.

11   The vitelline and subintestinal veins give rise to the hepatic portal vein, which loses its connection with the sinus venosus and, instead, breaks up in sinusoids in the liver.

12 The anterior cardinals become the internal jugulars. In fishes and urodele amphibians a pair of inferior jugular veins from the ventral head region also enters the sinus venosus. These are without homologs in higher forms.

13 The caudal vein from the tail loses its connection with the subintestinal and subcardinal veins and divides at its anterior end, each portion connecting with the posterior end of a postcardinal. The postcardinal on each side breaks up into two portions. The posterior section, joined by the caudal vein, brings blood to the kidneys and thus becomes a renal portal vein. The anterior portion on each side is joined by a subcardinal. In Dipnoi the right postcardinal is larger than the left and is called the postcava. In higher forms the postcava is derived as an outgrowth of the vitelline veins which join the subcardinals.

14 A subclavian vein from the forelimb enters the duct of Cuvier on each side, and an iliac vein from the hind limb joins the posterior part of each lateral abdominal vein.

15 Beginning with the Dipnoi the two lateral abdominal veins fuse to form an anterior abdominal vein, which then joins the hepatic portal vein near the liver. The anterior abdominal vein loses its importance in reptiles and birds and is absent in all mammals except *Tachyglossus*. The umbilical, or allantoic, veins of amniote embryos are homologous with the lateral and anterior abdominal veins of lower forms.

16 The renal portal system also becomes reduced in reptiles and birds and is absent in mammals, the postcava furnishing the main venous drainage for the posterior part of the body.

17 The ducts of Cuvier, from amphibians on, become the precaval veins. In amphibians and reptiles they enter the sinus venosus, but in birds and mammals they enter the right atrium directly, the sinus venosus having been incorporated into the wall of the right atrium. In several mammals the left precava loses its connection with the heart and a new vessel shunts blood from the left side of the anterior end of the body to the right precaval vein.

18 The lymphatic system collects tissue fluid from all parts of the body. After the fluid enters the lymphatic system, it is called lymph. Lymphatic vessels connect with the venous system at certain places, usually with the large veins near the heart. Much variation in detailed structure of the lymphatic system exists in the various groups of vertebrates. Some forms have lymph hearts which help to propel the lymph, but in others muscular movements and the pressure of newly accumulated lymph cause the fluid to flow slowly through the lymphatic vessels. Lymph nodes composed of lymphatic tissue and through which lymph filters slowly, interrupt the course of lymphatic vessels. Tonsils, adenoids, thymus glands, and Peyer's patches are lymphoid structures.

19 Many vertebrates have hemal nodes which interrupt the course of blood vessels in a manner similar to that of lymph nodes interrupting the course of lymphatic vessels.

20 The spleen is the largest lymphoid structure in the body. It functions somewhat as a hemolymphatic organ but, in addition, is able to store and release red corpuscles as the needs of the body dictate. The spleen lies in the mesentery of the gut near the stomach. It is absent in cyclostomes and *Protopterus*.

21   Blood-forming (hemopoietic) tissues include the mesoderm of the yolk sac, body mesenchyme, liver, spleen, lymphoid organs, and bone marrow. Most of these function in this capacity only during embryonic life. In adult mammals only lymphoid tissue and bone marrow ultimately have a hemopoietic function.

22   The composition of blood varies somewhat in different vertebrate groups. The main difference lies in the erythrocytes, which are flattened, oval cells in all except mammals. Only camels and llamas, among mammals, have oval erythrocytes. The red corpuscles are nucleated, except in mammals, in which erythroblasts, which form erythrocytes, extrude their nuclei. Leukocytes, formed in the bone marrow and lymphoid tissues, can wander about the body outside blood vessels. Certain kinds act as phagocytes, engulfing bacteria and particles of various kinds. Blood platelets, possibly homologous with spindle cells of lower forms, are concerned with blood clotting in mammals.

23   During fetal life in amniotes, an umbilical, or allantoic, circulation exists which performs a role in respiration. The pulmonary circulation is inconsequential before birth; at birth it first begins to function, and the umbilical circulation ceases. Umbilical arteries are branches of the dorsal aorta. The umbilical vein, formed primarily from the left umbilical vein, is homologous with the lateral and anterior abdominal veins of lower forms. An opening, the foramen ovale, in the interatrial septum, and the ductus arteriosus, connecting pulmonary and systemic aortae, make possible the peculiar routing of blood which is characteristic during fetal life. The umbilical vessels, foramen ovale, and ductus arteriosus do not exist as such after birth.

BIBLIOGRAPHY

Greep, Roy (ed.): "Histology," McGraw-Hill, New York, 1966.

Johansen, K., and D. Hanson: Functional Anatomy of the Hearts of Lungfishes and Amphibians, *Am. Zool.,* 8:191–210 (1968).

Satchell, G. H.: "Circulation in Fishes," Cambridge University Press, New York, 1971.

White, F. N.: Functional Anatomy of the Heart of Reptiles, *Am. Zool.,* 8:211–219 (1968).

13 INTEGRATING SYSTEM: NERVOUS SYSTEM

All cells are capable of responding to stimuli. The responses vary depending on the cell type: glandular cells secrete, muscles contract, nerves conduct impulses, and lymph cells ingest microorganisms. In metazoan forms some means of integration must be established to provide a meaningful and coordinated response by the total organism or from those areas which are affected. The nerve cells of the nervous system and the endocrine glands supply the two major components of this integrating system. These two components, one electrochemical and the other chemical, are referred to as the *neuroendocrine system.* The nervous system exerts its control through nerves distributed directly to the various body structures, acting at a high speed in a fraction of a second. The endocrine secretions are slow coordinators using the blood and lymphatic vessels as means of transportation and causing effects which may last for days or weeks.

The interaction of these two types of integrating system is very complex and difficult to separate. We have arranged the neuroendocrine system in three major groups and will discuss them under three separate headings: the nervous system, the endocrine organs, and the sense organs. It should be remembered that these three groupings are all part of one central coordinating system and affect the operation of each other, together providing the basic integration and response of the whole organism to stimuli.

NERVOUS SYSTEM Of all the cells in the body, those of which nervous tissue is composed have the most highly developed properties of irritability and conductivity. The greater part of the *nervous system* is composed of nervous tissue of ectodermal origin. Nervous tissue traces its beginning, for the most part, to the embryonic *neural tube* and *neural crest.* In some cases *placodes,* or thickenings of the superficial ectoderm in certain restricted areas, also give rise to nervous elements. The function of the nervous system is to receive stimuli and to send impulses from one part of the body to another. In this manner the functions of the many organs and parts of the body are coordinated and integrated. Nervous tissue is also the seat of all conscious experience.

The structural units of nervous tissue are *neurons,* namely *nerve cells,* and their processes. The part of the cell in which the nucleus lies is called the *cell body.* Masses of nerve-cell bodies located in the brain or spinal cord are referred to as *nuclei,* an unfortunate term. Such a mass of nerve-cell bodies located outside the central nervous system is called a *ganglion.*

The processes of a neuron are of two kinds: (1) *dendrites,* usually numerous, which are short processes often showing a high degree of branching close to the nerve-cell body, and (2) a single, slender *axon,* which is usually a long process with branches, the *terminal arborization,* at its end (Fig. 13.1). Collateral branches may be given off by the axon along its course, but these are frequently lacking. The number of cytoplasmic processes extending from a nerve-cell body varies and forms the basis for classifying different types of neurons (Fig. 13.1).

The axon with its enveloping sheath, or sheaths, is spoken of as a *nerve fiber.* Those nerve fibers located outside the brain and spinal cord are covered by a thin, continuous encapsulating membrane, the *neurolemma.* This is composed of *Schwann cells,* which are present in series beginning at the brain or spinal cord or at a ganglion and ending almost at the terminal arborization. They possibly represent neuroglia cells from the central nervous system (page 63) which have moved out to a peripheral position. They are not connective-tissue cells. Each sheath cell, with its cytoplasm and flattened nucleus, surrounds a section of the axon.

Nerve fibers are usually said to be either *medullated* or *nonmedullated,* according as they are or are not surrounded by a sheath of glis-tening white material called myelin, which lies between the axon and neurolemma (Fig. 13.2). Both the neurolemma and the myelin sheath are interrupted at rather regular intervals by circular constrictions, the *nodes of Ranvier* (Fig. 13.1). Sections between adjacent nodes are called *internodes.* A single sheath cell covers an internode. It is believed that the myelin sheath may serve as an insulating substance, much in the manner of the coating on an electric wire, preventing loss of energy of the nerve impulse during its passage along

FIG. 13.1 *A,* diagram of a typical medullated neuron.

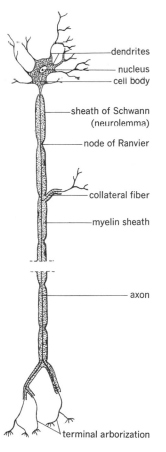

- dendrites
- nucleus
- cell body
- sheath of Schwann (neurolemma)
- node of Ranvier
- collateral fiber
- myelin sheath
- axon
- terminal arborization

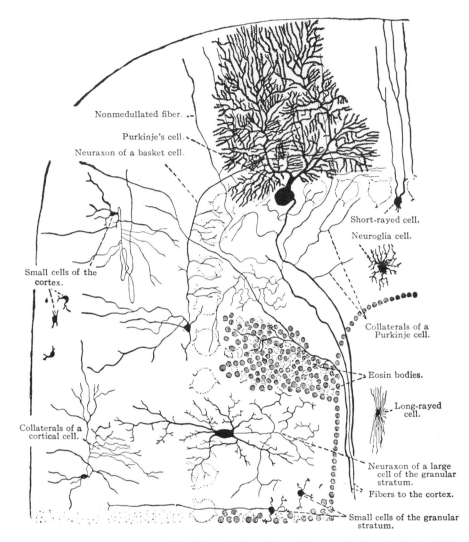

Nonmedullated fiber.

Purkinje's cell.

Neuraxon of a basket cell.

Small cells of the cortex.

Collaterals of a cortical cell.

Short-rayed cell.

Neuroglia cell.

Collaterals of a Purkinje cell.

Eosin bodies.

Long-rayed cell.

Neuraxon of a large cell of the granular stratum.

Fibers to the cortex.

Small cells of the granular stratum.

FIG. 13.1 (*Continued*) *B*, section through the cortex of the cerebellum to illustrate the differences in size, shape, and arrangement of neurons and their processes.

the fiber. It has been shown that the myelin sheath is related to the velocity of the impulse, the thickest sheathed nerve fibers conducting at the greatest speeds. Until rather recently, so-called nonmedullated fibers were believed to lack myelin sheaths, and the neurolemma was reported to surround the axon directly. Electron microscopy has demonstrated that even in such fibers a trace of myelin is present. In the following pages we shall use the term "sparsely medullated" rather than "nonmedullated" in referring to such fibers, which appear gray in contrast to the glistening white of medullated fibers.

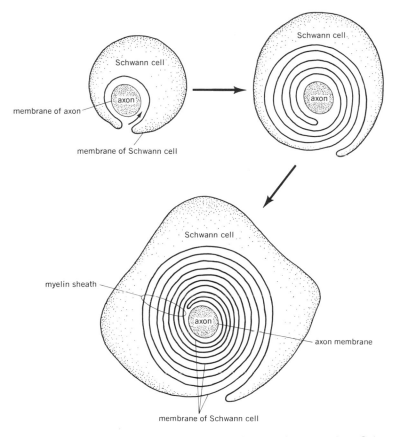

FIG. 13.2 The formation of the myelin sheath around an axon by a Schwann cell (diagrammatic).

Medullated fibers in the brain and spinal cord are not surrounded by a neurolemma. Instead, certain neuroglia cells (page 63) form very incomplete sheaths about the myelin.

A *nerve,* or *nerve trunk,* is made up of numerous nerve fibers outside the central nervous system, bound together in parallel bundles by connective tissue. Most peripheral nerves are composed of medullated fibers. Sparsely medullated fibers are found chiefly as components of nerves making up the sympathetic nervous system.

Generally speaking, the presence of medullated fibers occurs only among vertebrates,

cyclostomes being the only ones in which the nerve fibers apparently lack myelin sheaths.

GRAY MATTER AND WHITE MATTER Groups of nerve-cell bodies, their dendrites, and the proximal unmyelinated portions of their axons have a grayish appearance and form the greater part of the *gray matter* of brain and spinal cord. *White matter* is composed chiefly of bundles of glistening white medullated fibers. Such bundles, which it is possible to trace in prepared sections of brain and cord, are called *fiber tracts.* In some regions gray and white matter are intermingled, such an arrangement being known as a *reticular formation.* In general, white matter serves to conduct impulses from one part of the body to another, whereas gray matter functions in integrating impulses.

PRIMARY DIVISIONS All parts of the nervous system are structurally connected and functionally integrated. Nevertheless, there are two main divisions: (1) the *central nervous system,* composed of the brain and spinal cord; and (2) the *peripheral nervous system,* made up of nerves and ganglia connected to the brain and spinal cord. Part of the peripheral nervous system is composed of autonomic fibers which are distributed to those parts of the body under involuntary control. These in turn consist of sympathetic and parasympathetic components which often function in an opposite manner in mediating various involuntary activities of the body. The autonomic portion of the peripheral nervous system is referred to by most authorities as the *peripheral autonomic system.*

In addition to nerve cells of various types, other components called *neuroglia* are present in the central nervous system. These ectodermal cells are interspersed among the nervous elements, providing support and some degree of protection. True connective tissue is not present. Several types of neuroglia cells are recognized. Other nonnervous, supporting *ependymal cells* of an epithelial nature line the cavities of brain and spinal cord. They are often ciliated.

IMPULSE TRANSMISSION In the vertebrate body impulses always travel in one direction in nerve cells, passing from dendrites to cell body and then along the axon, but the fiber is capable of transmitting impulses in both directions. The single directiveness of the impulse is due to the nervefiber connection and results in morphological *polarity* of the neuron. The nervous system is

made up mostly of neurons arranged in chains in such a manner that impulses can be transmitted from one to another in series. The terminal arborization of one cell comes in close contact with the dendrites or cell bodies of one or more other neurons. The cells do not fuse, however. Such a point of junction is known as a *synapse.* An impulse does not travel across a synapse; instead a fresh impulse is set up on the other side of the synapse. In this manner impulses are conducted from one part of the body to another. There have been many speculations as to the nature of the synapse. It is generally accepted that a minute amount of a chemical substance, either a transmitter substance (*acetylcholine*) or an inhibitory substance, given off by the terminal fibers of the axon sets up an impulse in the dendrite of the next cell. Many of the postganglionic fibers of the sympathetic nervous system (page 413) employ a different chemical substance, *norepinephrine.*

REFLEX ARC The term *reflex action* refers to an immediate involuntary response to a sensory stimulus, the knee jerk being a familiar example. If the knees are crossed and the patellar ligament is tapped, the leg may jerk forward involuntarily because of sudden contraction of the *quadriceps femoris* muscle. In this case impulses set up in sensory fibers in the patellar ligament travel from the knee to the spinal cord, where they form synapses with motor fibers leading to the quadriceps femoris muscle. The pathway through which such impulses travel is known as a *reflex arc* (Fig. 13.3). The reflex may be modified by nerve impulses coming to the same motor nerve from other sources.

Reflex arcs make up the *functional units* of the nervous system. A reflex arc consists of two or more neurons together with a nonnervous component, called the *effector,* which in most cases is either a muscle cell or a

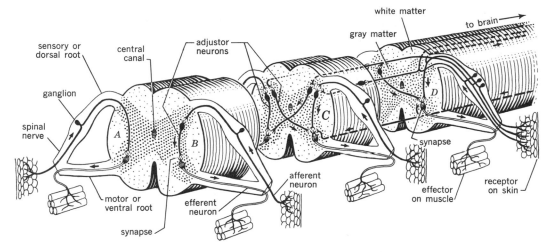

FIG. 13.3 Simplified stereogram of the vertebrate spinal cord and nerves to show relations of neurons involved in reflex arcs. Afferent neurons, solid; efferent neurons, outlined; association neurons, broken lines; receptors, as in skin; effectors, as on muscles. Arrows show path of nerve impulses. Each nerve contains many fibers. *A*, simple reflex arc; *B*, reflex arc with one adjustor neuron; *C*, reflex arc with cross connections; *D*, reflex arc with cross connections and also others to and from the brain. (*From Storer et al., "General Zoology," 5th ed., McGraw-Hill Book Company, New York, 1972, by permission.*)

glandular cell. A *receptor* element is present which receives the stimulus and transmits the impulses thus set up, via the neuron. Receptors may be sensory neurons; their dendritic ends are frequently associated with special accessory structures which facilitate the reception of stimuli. The terminal arborization of the axon of the receptor makes direct or indirect connections with the dendrites of a motor neuron within brain or spinal cord. The axons of the motor neuron terminate in the effector, the action being mediated in all probability by a neurohumor. The simplest type of reflex arc consists of a sensory neuron synapsing directly with a motor neuron which leads to an effector.

In most cases one or more *intermediate neurons* are present between receptor and effector neurons. Through these, impulses are transmitted from sensory neurons to other neurons in brain or cord. Intermediate neurons on the same side of the central nervous system are called *association neurons* (Fig. 13.3). If they cross to the opposite side, they are known as *commissural neurons*. The presence of intermediate, association, or commissural fibers determines whether a reflex will occur on the same side on which the sensory impulse travels or on the opposite side.

The interpositioning of an association neuron between the receptor and motor neurons (Fig. 13.3) provides for a greater number of choices for any given incoming stimuli. Though only one association neuron is present in a simple reflex arc, they may provide a clue to the evolution of more complicated and higher brain levels. The interpositioning of additional neurons would increase the total complexity of the system and

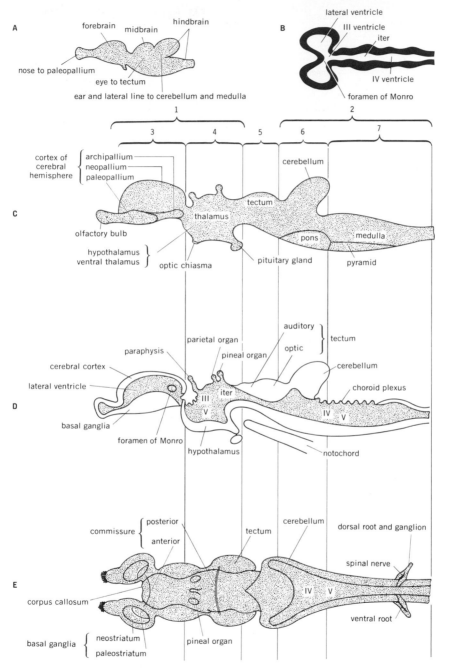

FIG. 13.4 Diagrammatic view of the major brain subdivisions and structures: *A*, the three major subdivisions of the brain and the special sense organs projecting to these structure in lower vertebrates; *B*, the ventricles of the brain; *C*, lateral view of a generalized vertebrate brain: (1) prosencephalon; (2) rhombencephalon; (3) telencephalon; (4) diencephalon; (5) mesencephalon; (6) metencephalon; (7) myelencephalon; *D*, dorsal view of a generalized brain; *E*, dorsal view of a generalized brain. (*From Waterman, "Chordate Structure and Function," The Macmillan Company, New York, 1971.*)

provide such mechanisms as feedback, modulations, and eventually centers of association which could provide a great number of responses after integrating and coordinating all incoming information.

CENTRAL NERVOUS SYSTEM

Brain and spinal cord develop from the dorsal hollow neural tube itself. The anterior end of the neural tube, destined to form the brain, enlarges from almost the very beginning. Two constrictions appear in the enlarged portion so that three primary brain vesicles are rather clearly marked off. They are called the *forebrain (prosencephalon), midbrain (mesencephalon),* and *hindbrain (rhombencephalon* (Fig. 13.4)). The remainder of the neural tube, which fails to enlarge at the same rate as the brain, becomes the *spinal cord.*

Subsequent changes in prosencephalon and rhombencephalon result in a further division of these two portions of the brain, but the mesencephalon undergoes no division. It gives rise to the *optic lobes.* Paired outgrowths from the anterior end of the prosencephalon give rise to the *telencephalon,* destined to form the *cerebral hemispheres* (in higher forms); the remainder constitutes the *diencephalon (thalamencephalon).* The anteroventral portion of the telencephalon later modifies to form the *olfactory lobes.* A dorsal projection of the *metencephalon,* or anterior end of the rhombencephalon, gives rise to the *cerebellum,* and the remainder of the hindbrain becomes the *myelencephalon,* or *medulla oblongata* (Fig. 13.5). The floor of the metencephalon in lower vertebrates does not differ from that of the myelencephalon, but in higher forms it becomes thickened by development of fiber tracts and, in mammals, forms the conspicuous *pons* (Fig. 13.13*B*). The posterior end of the brain is continuous with the spinal cord.

CAVITIES OF BRAIN AND SPINAL CORD

With the differentiation of the neural tube into brain and spinal cord, its original cavity becomes modified to form the *ventricles* of the brain and the *central canal* of the spinal cord (Fig. 13.6). With the development of the cerebral hemispheres the cavities which extend into them become the *lateral ventricles,* or *ventricles I* and *II.* The cavity of the diencephalon is then known as the *third ventricle.* Each lateral ventricle communicates with the third ventricle by means of an opening, the *interventricular foramen,* or *foramen of Monro.* In higher vertebrates a narrow canal extends posteriorly from the third ventricle through the mesencephalon. It is referred to as the *cerebral aqueduct* or *aqueduct of Sylvius.* In lower forms the aqueduct may be expanded on either side, and the term *mesocoel,* or *optic ventricles,* is then applied. Posteriorly the cerebral aqueduct communicates with the *fourth ventricle* in the rhombencephalon. In many lower forms the anterior part of the fourth ventricle becomes modified into a chamber, the *metacoel,* extending into the cerebellum. The remainder of the fourth ventricle is then known as the *myelocoel.* It is continued posteriorly as the central canal of the spinal cord. Lymphlike cerebrospinal fluid is contained within the cavities of brain and spinal cord.

THE SPINAL CORD

The spinal cord generally assumes the shape of a more or less cylindrical but slightly flattened tube which widens at the anterior end, where it is continuous with the medulla oblongata. Its posterior end usually tapers down to a nonnervous fine thread, the *filum terminale.*

In cyclostomes and fishes the spinal cord is

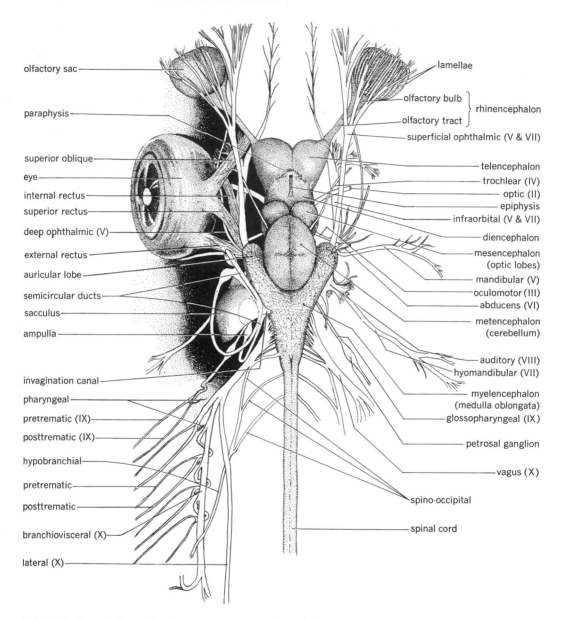

olfactory sac

paraphysis

superior oblique
eye
internal rectus
superior rectus
deep ophthalmic (V)
external rectus
auricular lobe
semicircular ducts
sacculus
ampulla

invagination canal
pharyngeal
pretrematic (IX)
posttrematic (IX)
hypobranchial
pretrematic
posttrematic
branchiovisceral (X)
lateral (X)

lamellae
olfactory bulb
olfactory tract
rhinencephalon
superficial ophthalmic (V & VII)
telencephalon
trochlear (IV)
optic (II)
epiphysis
infraorbital (V & VII)
diencephalon
mesencephalon (optic lobes)
mandibular (V)
oculomotor (III)
abducens (VI)
metencephalon (cerebellum)
auditory (VIII)
hyomandibular (VII)
myelencephalon (medulla oblongata)
glossopharyngeal (IX)
petrosal ganglion
vagus (X)
spino-occipital
spinal cord

FIG. 13.5 Dorsal view of brain, sense organs, and cranial nerves of *Squalus acanthias.*

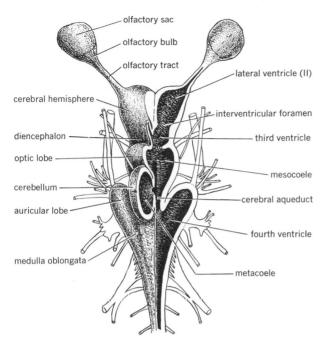

olfactory sac
olfactory bulb
olfactory tract
lateral ventricle (II)
cerebral hemisphere
interventricular foramen
diencephalon
third ventricle
optic lobe
mesocoele
cerebellum
cerebral aqueduct
auricular lobe
fourth ventricle
metacoele
medulla oblongata

FIG. 13.6 Dorsal view of brain of dogfish with part of wall removed to show cavities of the brain.

of fairly uniform diameter, but in most tetrapods two conspicuous swellings, or *enlargements,* occur where the nerves going to the limbs arise. The enlargements are actually places in which greater numbers of nerve-cell bodies are situated. The *cervical enlargement* is the more anterior of the two. It is the region where the large nerves supplying the forelimb arise. The *lumbar enlargement,* near the posterior end of the spinal cord, marks the point of origin of nerves supplying the hind limbs. In such limbless forms as snakes, neither enlargement is present.

The spinal cord of higher forms (Fig. 13.7) has a conspicuous *ventral fissure* on the ventral surface. A slight median depression, or *sulcus,* may be present on the dorsal side. From the sulcus a *dorsal septum* extends toward the interior of the cord. Dorsal septum and ventral

fissure incompletely divide the spinal cord into symmetrical halves, connected across the middle by *commissures* of nervous tissue. The central canal is very small and lies in the center of the nervous mass connecting the two halves of the cord.

LENGTH OF THE CORD During development in many vertebrates the growth of the spinal cord fails to keep pace with that of the vertebral column. As a result the spinal cord of the adult may be much shorter than the backbone. In man it averages slightly less than 18 in. in length and reaches down only to the upper border of the second lumbar vertebra. The filum terminale, however, continues posteriorly. Within the phylum there is a definite tendency toward reduction in relative length of the spinal cord.

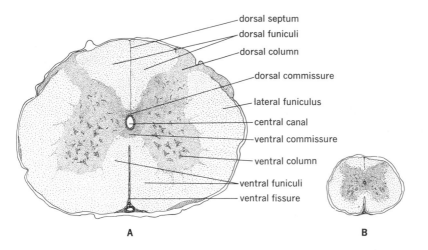

dorsal septum
dorsal funiculi
dorsal column
dorsal commissure
lateral funiculus
central canal
ventral commissure
ventral column
ventral funiculi
ventral fissure

A B

FIG. 13.7 Cross sections of spinal cords of *A*, cat, and *B*, frog, shown at same magnification. Gray matter is heavily stippled; white matter is lightly stippled.

GRAY AND WHITE MATTER OF CORD In cross section the spinal cord is seen to be composed of gray and white matter, the former being almost completely surrounded by the latter (Fig. 13.7). In amniotes the gray matter is arranged somewhat in the form of a butterfly or in the shape of the letter H. The portions corresponding to the upper bars of the H extend dorsally and are known as *dorsal columns.* The lower bars are the *ventral columns,* and the connecting bar, in which the central canal lies, forms the *dorsal* and *ventral gray commissures,* lying above and below the central canal, respectively. The small nerve-cell bodies in the dorsal columns are, for the most part, those of *association neurons.* Their dendrites form synapses with the axons of sensory, or afferent, nerve fibers which enter the spinal cord via the dorsal roots of spinal nerves. The axons of the association neurons, which are of variable length, course up or down the spinal cord, cross to the opposite side, or pass ventrally on the same side to form synapses with the dendrites of motor, or efferent, neurons, the very large

cell bodies of which are located in the ventral columns. In higher forms the axons of these anterior motor neurons typically emerge from the cord via the ventral roots of the spinal nerves (Fig. 13.16*B*), but in lower forms some may emerge through the dorsal roots as well.

The cells in the dorsal and ventral columns show a rather definite arrangement into areas each of which is associated with a certain type of function. All functions in the body may be resolved into two main categories, *somatic* and *visceral.* Somatic functions are those carried on by the skin and its derivatives, the voluntary musculature, and skeletal structures. Visceral functions are those performed by the other organ systems of the body, i.e., digestive, respiratory, etc. *Somatic sensory fibers* carry impulses from somatic tissues *to* the central nervous system. They form synapses with cells in the *upper* portion of the dorsal columns. *Visceral sensory fibers* are those carrying impulses from the visceral organs. Synapses are formed with cells in the *lower* portion of the dorsal columns. The motor neurons, the cell bodies of which are located in the ventral columns, are likewise of two types, somatic and visceral. The cell bodies of *somatic motor neurons* are located primarily in the *lower* portions of the ventral columns, whereas *visceral motor neurons* have their origin in the *upper* and *lateral* portions of the ventral columns. Thus there are four areas in the gray matter on each side of the spinal cord, arranged in a dorsoventral sequence as follows: somatic sensory, visceral sensory, visceral motor, somatic motor (Fig. 13.8*A*). The visceral areas are usually smaller than the somatic areas. A similar arrangement is present in the gray matter of the medulla oblongata (Fig. 13.8*B*).

The white matter of the cord is arranged in longitudinal columns called *funiculi.* The funiculi are composed of fiber tracts made up of medullated fibers that carry impulses up and down the cord and to and from the brain. A *lateral funiculus* is present on each side between dorsal and ventral columns. A *dorsal funiculus* lies between the dorsal septum and the dorsal column on each side, and a *ventral funiculus* is located between the ventral fissure and the ventral column of gray matter. The two ventral funiculi are in communication through the *white commissure,* which lies just below the ventral gray commissure. This serves as a bridge between the white matter of the two sides.

Fibers in the dorsal funiculi, for the most part, carry sensory nerve impulses up the cord and to the brain. Those in the ventral funiculi are primarily motor, carrying impulses down the cord and from the brain. The lateral funiculi carry both kinds of fibers.

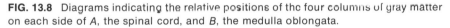

FIG. 13.8 Diagrams indicating the relative positions of the four columns of gray matter on each side of *A*, the spinal cord, and *B*, the medulla oblongata.

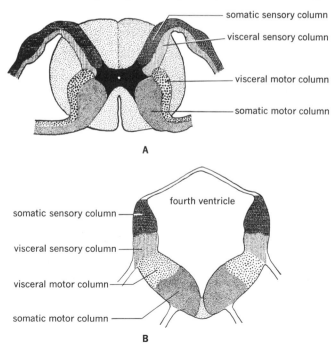

somatic sensory column

visceral sensory column

visceral motor column

somatic motor column

A

fourth ventricle

somatic sensory column

visceral sensory column

visceral motor column

somatic motor column

B

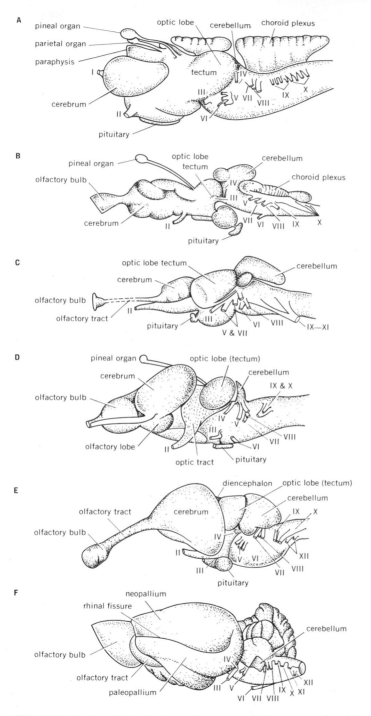

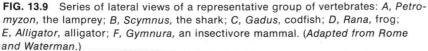

FIG. 13.9 Series of lateral views of a representative group of vertebrates: *A, Petromyzon*, the lamprey; *B, Scymnus*, the shark; *C, Gadus*, codfish; *D, Rana*, frog; *E, Alligator*, alligator; *F, Gymnura*, an insectivore mammal. (*Adapted from Rome and Waterman.*)

In general, the lower vertebrates do not show so elaborate an organization or arrangement of the columns and funiculi as the higher forms. In amphioxus there is no clear-cut distinction between gray matter and white matter, since medullated fibers have not as yet made their appearance. Gray matter and white matter are not sharply delineated in the spinal cord of cyclostomes. Here dorsal septum and ventral fissure are lacking. Shortening of the spinal cord is first observed in certain teleost fishes. It is conspicuous in anuran amphibians and most mammals but not in urodeles, reptiles, or birds.

THE BRAIN

The chordate brain appears in its simplest form in the *cerebral vesicle* of amphioxus, which is scarcely larger in diameter than the spinal cord. In higher forms it becomes larger and more complex as the evolutionary scale is ascended (Fig. 13.9).

In the primitive condition the cell bodies of the neurons constituting the central nervous system are aggregated about the central canal of the neural tube. This arrangement persists in the spinal cords of higher forms. In the brain region, however, cells have migrated to peripheral areas, so that gray matter and white matter differ in their spatial relations from the arrangement observed in the spinal cord. The degree of development of the various parts of the brain of vertebrates is correlated with their position in the evolutionary scale and with certain special requirements related to the particular environments in which they live. The brain is the center of control over most body activities aside from certain simple reflexes.

The three primary divisions of the brain—prosencephalon, mesencephalon, and rhombencephalon—make up what is often referred to as the *brainstem*. Each division

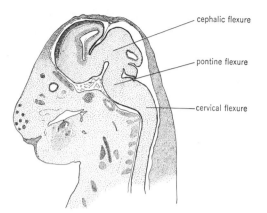

FIG. 13.10 Sagittal section through head of 18-day rat embryo, showing flexures of the brain.

may originally have developed in association with one of the three major sense organs. The sense of smell is primarily related to the prosencephalon, the sense of sight to the mesencephalon, and the sense of pressure change and equilibrium to the rhombencephalon. Further developments from the brainstem, such as cerebral hemispheres, roof of the midbrain, and cerebellum, appeared later as dorsal outgrowths into which nerve cells migrated, so that gray matter appears peripherally in these regions.

FLEXURES Only in amphioxus are "brain" and spinal cord arranged in a straight line. In vertebrates, however, because of unequal rates of growth of various parts, certain flexures occur during embryonic development which modify the original condition temporarily or permanently (Fig. 13.10). Since the brain lengthens more rapidly than other head structures, the bending is apparently influenced by space limitations. In birds and mammals the front part of the brain (cerebral hemispheres and diencephalon) comes to lie, at least partially, on top of, or over, the posterior portions.

GRAY AND WHITE MATTER OF BRAIN The posterior end of the medulla oblongata merges imperceptibly with the anterior end of the spinal cord. The region just in front of the first pair of spinal nerves is the level at which brain and spinal cord join. The structure of the medulla oblongata at its posterior end differs but little from that of the spinal cord. Farther forward in the brain, gray and some white matter lose their original relationship and are more or less intermingled. In the cerebral hemispheres, roof of the midbrain (*optic tectum*), and cerebellum of higher vertebrates, the position of these layers is reversed, so that the gray matter forms a layer over the underlying white matter. The gray matter of the brain, like that of the spinal cord, consists of nerve-cell bodies with their dendrites and proximal portions of their axons. They are usually grouped together in the form of nuclei. The white matter consists of tracts of medullated fibers connecting various parts of the brain, and of ascending and descending fibers carrying impulses to and from the spinal cord.

MYELENCEPHALON The lateral and ventral walls of the medulla oblongata thicken markedly, but the dorsal wall retains its epithelial character. The *pia mater,* a vascular membrane surrounding the brain, fuses with the thin roof, the two together being called a *tela choroidea.* This covers the large fourth ventricle of the brain. Vascular folds of the tela choroidea extend into the fourth ventricle to form the *posterior choroid plexus.*

The thickened ventral and lateral walls of the medulla contain large white fiber tracts as well as several columns of gray matter. The latter are basically arranged, like those of the spinal and cranial nerves, into seven functional types: general somatic sensory, special somatic sensory, general visceral sensory, special visceral sensory, somatic motor, visceral motor, and special visceral motor. In higher vertebrates the integrity of the columns of gray matter in the brain is lost, and a separation into various *nuclei* has occurred. Several fiber tracts in the medulla cross over (decussate) to the opposite side of the brain. The sixth through the twelfth cranial nerves are associated with the medulla oblongata.

The myelencephalon is often referred to as the oldest part of the brain since it is well developed in all vertebrates, even though other portions may be rudimentary or lacking. It contains important nerve centers which control such vital physiological processes as regulation of heartbeat, respiration, and metabolism. They represent portions of the visceral motor column.

The dorsal anterior portion of the medulla oblongata is spoken of as the *acousticolateralis area.* It is actually a development of the somatic sensory column and is continuous with the auricles of the cerebellum. This area contains nuclei associated with nerves from the lateral-line system and the inner ear. In terrestrial vertebrates it is associated with the equilibratory and auditory functions of the ear.

METENCEPHALON The dorsal part of the metencephalon becomes the elevated and thickened cerebellum. Its function is to coordinate the neuromuscular mechanism of the body. The cerebellum is highly developed in animals that are active, whether such activity occurs in water, on land, or in the air. The ventral portion of the metencephalon in lower vertebrates is composed of heavy fiber tracts which merge with those of the medulla oblongata. Its cavity, the *metacoel,* connects below with the fourth ventricle.

Prominent, irregular projections, called *auricular lobes* or *restiform bodies* (Fig. 13.5), present in certain fishes and continuous with the medulla oblongata, are actually parts of the cerebellum. They are centers of equilibra-

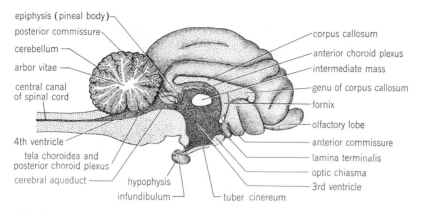

epiphysis (pineal body)
posterior commissure
cerebellum
arbor vitae
central canal of spinal cord
4th ventricle
tela choroidea and posterior choroid plexus
cerebral aqueduct
hypophysis
infundibulum
tuber cinereum

corpus callosum
anterior choroid plexus
intermediate mass
genu of corpus callosum
fornix
olfactory lobe
anterior commissure
lamina terminalis
optic chiasma
3rd ventricle

FIG. 13.11 Diagram of sagittal section of brain of cat. (*After Reighard and Jennings, "Anatomy of the Cat," Holt, Rinehart and Winston, Inc., by permission.*)

tion. The greatest advance in the cerebellum of higher forms is the appearance of a pair of *floccular lobes* near the ventral side. They correspond to the auricular lobes of fishes and first appear in crocodilians. In birds the cerebellum consists of a prominent middle portion, the *vermis,* with floccular lobes on either side. It extends posteriorly and covers the tela choroidea of the medulla oblongata. The *cerebellar cortex,* or gray matter of the cerebellum, covers the white matter, which, in birds and mammals, shows a complexly branched arrangement called the *arbor vitae.* The mammalian cerebellum (Fig. 13.11) is still more complex. The vermis is divided into anterior, middle, and posterior lobes. The middle lobe has bilateral extensions, the *cerebellar hemispheres.* The surface of the cerebellum is thrown into numerous folds, or *gyri,* separated by deep grooves, or *sulci.* The ventral side of the mammalian metencephalon, as well as that of some birds, is marked by a conspicuous *pons* composed of a prominent mass of transverse nerve fibers. Motor fibers from the cerebral cortex, which controls body movement, pass via the pons to the cerebellum. Lower mammals have a simpler form of

cerebellar organization than is found in higher forms.

MESENCEPHALON The embryonic midbrain undergoes relatively less change than other portions of the brain. The floor and side walls are thick and composed of fiber tracts, the *cerebral peduncles,* connecting forebrain and hindbrain. The roof consists of a thick layer of gray matter, the *optic tectum.* In lower vertebrates two dorsal prominences develop in the roof of the midbrain. They are the *optic lobes,* or corpora bigemina, which serve as centers for the visual sense. Each optic lobe contains a large cavity, the *optic ventricle,* the two together forming the *mesocoel* (Fig. 13.6). In higher forms, however, the lobes are almost solid structures and only the narrow cerebral aqueduct passes through the center of the mesencephalon. A transverse fissure divides the optic lobes of snakes and mammals into four prominences, the *corpora quadrigemina (tectum).* The optic lobes are primarily receptive centers for sensory nerve impulses coming from the eye. When corpora quadrigemina are present, the anterior pair, called the *superior (anterior) colliculi,* contains receptive centers for the visual sense, and the posterior pair, the *inferior (posterior) colliculi,* serves to integrate auditory impulses. The superior colliculi in mammals have become less important as visual centers with the great development of the cerebral cortex, which has taken over much of the integration and coordination of visual impulses.

Nuclei for the third (oculomotor) and fourth (trochlear) cranial nerves are located in the mesencephalon.

The optic lobes are more conspicuous features of the brains of lower vertebrates than of higher forms. Nevertheless, they are unusually well developed in birds in correlation with their highly organized visual-sensory system.

DIENCEPHALON The embryonic diencephalon undergoes considerable modification. It will be recalled that the optic vesicles, from which the sensory portions of the eye develop, arise from the diencephalon.

The anterior portion of the dorsal roof plate covering the third ventricle retains its original epithelial character and together with the pia mater forms a tela choroidea. Vascular folds of the tela choroidea extend into the third ventricle, forming the *anterior choroid plexus.* Portions may even pass through the interventricular foramina into the lateral ventricles.

Posterior to the tela choroidea the roof of the diencephalon consists of the *epiphyseal apparatus.* In several lower forms this is composed of an anterior *parapineal,* or *parietal, body* and a posterior *pineal body,* or *epiphysis.* It is probable that these structures originally consisted of a bilateral pair which in the course of evolution became shifted so that one is now situated in front of the other. Although both are present in lampreys, some fishes, frogs, *Sphenodon,* and numerous lizards, only the pineal body has persisted in most

fishes, urodeles, many reptiles, birds, and mammals. The parapineal body, when well developed, forms a small median eye. The highest degree of development of this organ occurs in *Sphenodon.* It is developed to a somewhat lesser degree in some lizards. An *interparietal foramen* in the skulls of these and certain other forms provides passage for the nerve connecting the median eye and the brain. In no case is such an eye well enough developed to be comparable to the usual vertebrate eye. The *brow spot* of the common leopard frog marks the location of a vestigial parapineal organ.

The pineal body, which lies posterior to the parapineal organ when the latter is present, is thought to represent what may originally have been another eyelike structure, the two having formed a pair. Only in lampreys are *both* parapineal and pineal organs associated with such structures. The pineal body is present in all vertebrates and generally appears to be glandular in nature. There is increasing evidence that it may be an endocrine gland. Electron-microscopic as well as experimental studies indicate that the cells of which the pineal body is composed may be responsible for certain photoreceptive processes. In lower forms, cells are present which are not unlike rod and cone cells in the retinae of normal eyes but are not organized as such. It is possible, however, that in the course of evolution there has been a change from a primitive photoreceptive type of organ, which can translate photic stimuli into physiological controls of different types, to a secretory structure which can carry out similar functions in response to stimuli affecting normal optic pathways. Thus the demonstrated effect of light on the reproductive cycles of rodents may in some manner be mediated via the pineal body.

The thickened lateral portion of the diencephalon, known as the *thalamus,* contains important relay centers and consists largely of numbers of nuclei. They are integrating centers for impulses passing to and from the cerebral hemispheres. Significant centers are the *lateral* and *medial geniculate bodies,* which relay optic and auditory impulses, respectively. In reptiles and mammals, the walls are thickened inwardly so as to meet in the center of the third ventricle. The mass of gray matter connecting the two sides is the *intermediate mass,* or *soft commissure.*

The *hypothalamus,* or ventral portion of the diencephalon, contains nerve centers which integrate the functions of the peripheral autonomic system with those of other nervous tissues. Such mechanisms as temperature regulation, genital functions, water, fat, and carbohydrate metabolism, and the rhythm of sleep are controlled by the hypothalamus. Certain release factors and release-inhibiting factors extracted from the hypothalamus affect the release of hormones by the anterior lobe of the pituitary gland (page 423) via the hypophysio-portal vein pathway. The close relationship between the nervous system and endocrine organs, referred to as the *neuroendocrine system,* should again be emphasized. Making up the hypothalamus are (1) the *optic chiasma* where the optic nerves cross; (2) the *tuber cinereum,* believed to be the parasympathetic center; (3) a pair of *mammillary bodies,* for integration of the olfactory sense; (4) the *infundibulum,* the distal portion of which contributes to the posterior lobe of the pituitary gland.

TELENCEPHALON It is in the degree of development of the telencephalon that the greatest differences in the brains of vertebrates are to be found. In the highest forms the *cerebral hemispheres,* derived from the telencephalon, cover over the greater part of the remainder of the brain. The seat of consciousness lies in the cerebral hemispheres.

Here are located the nerve centers controlling the activities which characterize the highly developed psychic life of man, such as intelligence and thought.

Early in development two lateral swellings appear at the anterior end of the prosencephalon. These grow anteriorly and dorsally. The original anterior end of the neural tube remains practically unchanged in position and becomes the *lamina terminalis,* in which the *anterior commissure,* connecting the olfactory regions of the two halves of the brain, is later located.

At the anterior end of each swelling, or cerebral hemisphere, is an outgrowth, the *olfactory lobe,* which makes contact with the posterior part of the olfactory apparatus. The lateral ventricles may or may not extend into the olfactory lobes.

The cerebral hemispheres enlarge to an increasing degree as the vertebrate scale is ascended. In all vertebrates the floor of each hemisphere early differentiates into a thickened *corpus striatum,* a reticular formation. The gray regions of the corpus striatum are often called the *basal nuclei.* The function of the corpus striatum is rather obscure. It is poorly developed in cyclostomes. The remainder of each hemisphere consists of a *pallium,* which roofs over the lateral ventricle. It is the pallium that has become so highly developed and modified in the evolution of the higher groups of vertebrates.

In cyclostomes each hemisphere is divided into an anterior olfactory bulb and a posterior olfactory lobe, the latter sometimes being called the cerebral hemisphere. It is concerned with little more than receiving impulses from the olfactory apparatus and relaying them to the diencephalon.

In fishes the pallium is thin-walled and the gray matter is present only on its inner walls adjacent to the ventricles. In teleosts a thin, nonnervous layer forms the roof of the lateral ventricles. The gray matter of the pallium has pushed laterally and downward toward the corpus striatum. In elasmobranchs and dipnoans the pallium is fairly thick, but the gray matter still lines the cavities and the white matter forms the outer surface. In fishes the telencephalon has not progressed beyond serving as an olfactory center.

As the scale is ascended there is an increasing tendency for nerve cells from the inner gray layer to migrate out into peripheral areas. In amphibians the pallium is thicker in general than in fishes, and some cells from the gray matter have moved peripherally. The pallium is divided into two general regions, a dorsal, medial *archipallium* and a more lateral *paleopallium.* Both are related to the olfactory sense. The olfactory lobes merge almost imperceptibly with the anterior ends of the cerebral hemispheres.

The first really marked change occurs in reptiles. The cerebral hemispheres have increased in size, have grown backward to cover partially the diencephalon, and are separated medially by a deep fissure. An increased

amount of gray matter has migrated to the periphery. In certain reptiles a new area, the *neopallium,* has appeared at the anterodorsal end of each hemisphere between archipallial and paleopallial areas (Fig. 13.12). It is the growth and development of the neopallium that accounts in part for the large size of the cerebral hemispheres of mammals. In crocodilians, for the first time, nerve cells migrate into the neopallium and become arranged along its outer surface, thus forming a true *cerebral cortex,* which serves as an association center. In birds, olfactory lobes are practically rudimentary. Archipallium and paleopallium are present as in reptiles, but a neopallium is lacking in most forms. There is, therefore, no cerebral cortex. The cerebral hemispheres are large because the corpus striatum of birds is of unusual size.

It is in mammals, particularly in man, that the cerebral cortex reaches the height of its development. The neopallium has increased enormously, pushing the archipallial area medially and ventrally, where it continues to serve as an olfactory center. The paleopallium

FIG. 13.12 Diagrams indicating evolutionary progress in development of the pallium.

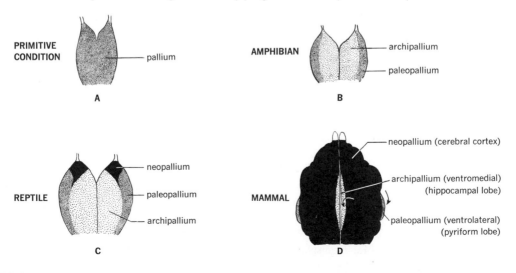

PRIMITIVE CONDITION — pallium

A

AMPHIBIAN — archipallium / paleopallium

B

REPTILE — neopallium / paleopallium / archipallium

C

MAMMAL — neopallium (cerebral cortex) / archipallium (ventromedial) (hippocampal lobe) / paleopallium (ventrolateral) (pyriform lobe)

D

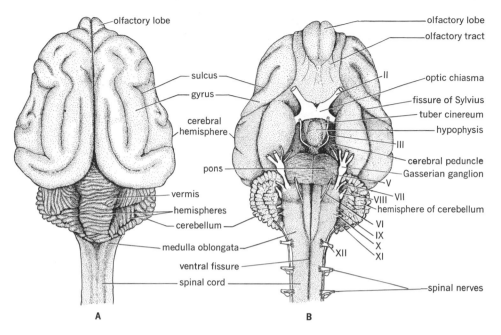

FIG. 13.13 Brain of cat from *A*, dorsal, and *B*, ventral, aspects.

is pushed ventrolaterally, where it becomes the *pyriform lobe,* a portion of the olfactory lobe. The gray nerve-cell bodies are so numerous as to form a layer of gray matter, the cerebral cortex, in the outer part of the cerebral hemispheres. This layer, however, even in man where it is best developed, is only a few millimeters thick. The mammalian corpus striatum is less conspicuous than in some lower forms.

In all vertebrates below mammals, the cerebral hemispheres are smooth. In many mammals, however, the surface becomes folded or convoluted so that ridges and depressions appear (Fig. 13.13). The ridges are called *gyri* and the depressions *sulci.* Since the cortical gray matter follows the convolutions, there is a considerable increase in surface area and in total amount of gray matter. Large mammals generally have more convolutions than smaller species. The extent to which the

cerebral cortex is convoluted is not necessarily a criterion of intelligence.

Beginning with marsupials, a broad, white mass, the *corpus callosum* appears in mammals between the two hemispheres. It is composed of a band of medullated fibers which connect nearly all parts of the neopallial cortical regions of the two sides.

The cerebral hemispheres of man are so large that they cover over all the other portions of the brain. Beneath the outer layer of gray matter is the medulla, composed of medullated fibers which either originate in cortical cells and carry outgoing impulses from the cortex or come from other parts of the brain and bring impulses to the cortex.

DECUSSATION The presence of a number of commissures in the brain and spinal cord has been frequently alluded to in the previous pages. They serve to connect similar regions of the two sides of the central nervous system and make bilateral integration possible. There are also fiber tracts in the brain, which in their course cross over, or decussate, to the opposite side. Injury to the brain on one side of the body often results in paralysis of muscles on the opposite side. Impulses set up in certain cells in the cerebral cortex travel down fibers leading to the nuclei of various cranial and spinal nerves. The fibers which course through the pyramids of the medulla oblongata cross to the opposite side. Injury anterior to the pyramidal decussation will affect structures on the opposite side of the body. Damage to fibers posterior to the decussation generally affects only the same side of the body as that on which the damage occurred.

MENINGES Both brain and spinal cord are surrounded by membranes, or *meninges* (singular, *meninx*), the complexity in arrangement of which increases according to advance in the evolutionary scale.

Cartilage and bone are covered with a tough vascular membrane known as the *perichondrium* or *periosteum,* as the case may be. The cavity of the cranium and the neural canal of the vertebral column are thus lined with perichondrium or periosteum. In either case the lining membrane is called the *endorachis.* It is not considered to be a true meninx.

In cyclostomes and fishes (Fig. 13.14A) a single membrane, the *meninx primitiva,* is present. It forms a close union with brain and spinal cord, which it covers. Between the

FIG. 13.14 Diagrams illustrating the relations of the meninges to the spinal cord: *A* and *B*, cross sections through spinal cords of dogfish and salamander, respectively; *C* and *D*, sections of spinal cord and brain of mammal, respectively.

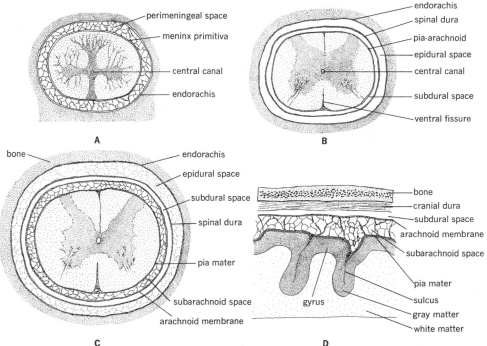

meninx primitiva and the endorachis is a *perimeningeal space,* filled with mucoid and fatty tissue. Strands of connective tissue cross the perimeningeal space.

Beginning with urodele amphibians (Fig. 13.14*B*), instead of a single meninx, two layers are present, an inner *pia-arachnoid layer* and an outer *dura mater.* The pia-arachnoid layer has its origin, at least in part, from neural-crest cells, whereas the dura mater is composed of connective tissue. The pia-arachnoid is very vascular and in intimate contact with the surface of brain and spinal cord. Between pia-arachnoid and dura mater is the *subdural space.* The cavity between the dura mater and the endorachis is the *epidural space,* filled largely with alveolar tissue, fat, and numerous veins.

In mammals (Fig. 13.14*C* and *D*) the pia arachnoid has differentiated further into two layers, an inner *pia mater* and an outer *arachnoid membrane.* A *subarachnoid space,* containing cerebrospinal fluid, lies between the two. In the brain region the *cranial dura mater* fuses with the endorachis, and the epidural space is obliterated. Over the spinal cord a fusion of *spinal dura* and endorachis does not occur, but the epidural space contains fatty and other connective tissues as well as a number of veins. The cranial dura is continuous with the spinal dura at the foramen magnum.

The cerebrospinal fluid, which is present in the ventricles of the brain, central canal of the spinal cord, and the subarachnoid space, and which is similar to tissue fluid, circulates slowly through these cavities and spaces. Elaborated for the most part by the anterior choroid plexus, the fluid passes from the lateral ventricles through the interventricular foramina to the third ventricle, and thence through the cerebral aqueduct to the fourth ventricle. Three openings in the roof of the medulla serve for passage of the fluid into the subarachnoid space. Thus the surfaces of brain and cord are bathed by cerebrospinal fluid, which is constantly being absorbed by blood vessels in the arachnoid membrane and through lymphatic vessels.

Certain modifications of the meninges are to be observed in mammals. The cranial dura sends a fold, the *falx cerebri,* down into the fissure between the two cerebral hemispheres. A similar fold, the *tentorium,* pushes down between the cerebral hemispheres and the cerebellum. In some forms, as in the cat, the tentorium becomes ossified and fused to the parietal bones. On the ventral side of the brain the dura encapsulates the pituitary gland and forms a fold, the *diaphragma sellae,* over the *sella turcica,* a depression in the dorsal face of the sphenoid bone in which the pituitary gland lies. The stalk of the infundibulum penetrates the diaphragma sellae.

PERIPHERAL NERVOUS SYSTEM

The nerves and ganglia, which form connections with the central nervous system and which are distributed to all parts of the body, constitute the peripheral nervous system. Its autonomic portion is distributed to structures under involuntary control. Connections with the central nervous system are mediated via spinal and cranial nerves. Spinal nerves, which form connections with the spinal cord, are paired metameric structures. The paired cranial nerves are connected with the brain, all but the first four being joined to the medulla oblongata.

SPINAL NERVES

Each spinal nerve connects to the spinal cord by means of two *roots, dorsal* and *ventral.* Except in amphioxus and lampreys, the two roots unite a short distance lateral to the cord to form the spinal nerve proper. The two roots have different embryonic origins.

DORSAL ROOTS The dorsal roots originate from neural crests.* On each dorsal root is a conspicuous swelling, the *dorsal root ganglion,* in which the cell bodies of the sensory neurons of the dorsal root are located. The axons of these neurons enter the spinal cord in the region of the dorsal column of gray matter. Sensory nerve impulses travel through the spinal nerves *toward* the spinal cord and in doing so pass through the dorsal roots, which, except in some of the lower forms, are strictly *sensory roots.* These sensory components are spoken of as *afferent fibers.* In certain lower vertebrates some motor fibers (visceral efferent fibers) also course through the dorsal roots.

Sensory fibers are said to be either somatic or visceral. *Somatic sensory fibers* coming from the skin and its derivatives, voluntary muscles, and skeletal structures, form synapses with cells in the somatic sensory columns of gray matter. *Visceral sensory fibers,* from visceral structures, terminate in the visceral sensory columns of gray matter.

VENTRAL ROOTS The cell bodies of neurons making up the ventral, or motor, roots lie in the gray matter of the ventral columns. Their dendrites lie within the cord, but their axons emerge at metameric intervals from the ventrolateral angles of the cord, from which they pass to peripheral voluntary muscles or to autonomic ganglia. Motor fibers are *efferent fibers. Somatic motor fibers* arise in the somatic motor columns of gray matter, whereas *visceral motor fibers* arise in the visceral motor columns. It is the latter which

* In amphibians and birds neural crests also give rise to certain cartilages in the branchial region (page 63).

pass to autonomic ganglia, where they form synapses with motor autonomic neurons.

In amphioxus and lampreys visceral motor fibers emerge from the spinal cord and pass out the *dorsal* roots, which are thus composed of both afferent and efferent fibers. In fishes and amphibians visceral efferent fibers are found in both dorsal and ventral roots. Only in amniotes is the dorsal root strictly sensory.

A spinal nerve, formed by the union of dorsal and ventral roots, is composed partly of dendritic processes of sensory cells located in the dorsal-root ganglion and partly of axons of efferent motor neurons the cell bodies of which are situated in the ventral column of gray matter in the cord. The axons of the sensory cells are short, being confined to the proximal portion of the dorsal root.

RAMI Not far from the point where dorsal and ventral roots unite three branches, or *rami* (Figs. 13.15 and 13.16), are given off: (1) a *dorsal ramus,* supplying the skin and epaxial muscles of the dorsal part of the body; (2) a *ventral ramus,* distributed to the skin and hypaxial ventral and lateral regions; and (3) a *visceral ramus,* which courses medially and in most cases forms connections with one of the chain ganglia of the peripheral autonomic system. Dorsal and ventral rami contain somatic sensory and somatic motor fibers. Some autonomic fibers going to the periphery are also included in these rami (page 413). A typical visceral ramus consists of two parts, a *white ramus* and a *gray ramus.* The white ramus contains visceral sensory and visceral motor medullated fibers. The axons of the visceral motor fibers are distributed to ganglia of the peripheral autonomic system and are referred to as *preganglionic fibers.* They form relays with sparsely medullated *postganglionic* autonomic fibers which originate in autonomic ganglia. The gray ramus, made up of such fibers, runs parallel to the white ramus.

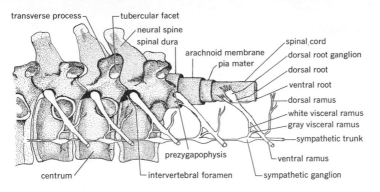

FIG. 13.15 Diagram showing relation of spinal nerves to spinal cord, meninges, vertebrae, and sympathetic trunk.

FIG. 13.16 Diagrams showing types of nerve fibers in the roots and rami of mammalian spinal nerves: *A*, distribution of somatic sensory and visceral sensory fibers; *B*, distribution of somatic motor fibers and visceral motor preganglionic and postganglionic fibers of the sympathetic nervous system; *C*, distribution of preganglionic and postganglionic fibers of parasympathetic nervous system as found in certain sacral spinal nerves.

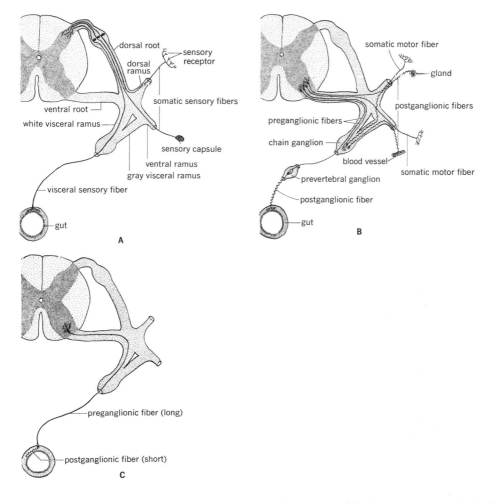

The cell bodies of these fibers are located in chain ganglia. The gray fibers join the spinal nerve and travel out the dorsal or ventral rami to supply structures under involuntary control, such as blood vessels, arrector pili muscles, and skin glands (Fig. 13.16).

PLEXUSES   In the regions of the appendages the ventral rami of certain spinal nerves are drawn out into the appendages. In such cases a more or less complicated network, or *plexus* (Fig. 13.17), may be present, formed by the connection of branches of certain nerves with those of others. In lower forms an anterior *cervicobrachial plexus* is present for each pectoral appendage and a posterior *lumbosacral plexus* for each hind limb. In higher forms a further differentiation takes place so that separate cervical, brachial, lumbar, and sacral plexuses may be present.

CAUDA EQUINA   At first, during embryonic development, each spinal nerve passes through an intervertebral foramen located at the same level as the part of the cord from which the nerve arises. Later, growth of the cord fails to keep pace with that of the vertebral column. As a result, the spinal

FIG. 13.17   Diagram of right brachial plexus of cat, ventral view. (*After Reighard and Jennings, "Anatomy of the Cat," 3d ed., Holt, Rinehart and Winston, Inc., New York, 1935, by permission.*)

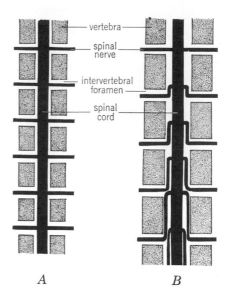

FIG. 13.18   Diagram illustrating the formation of cauda equina: *A*, the spinal nerves emerge through the vertebral column at right angles to the spinal cord; *B*, because of the disproportionate growth of vertebrae and spinal cord, the nerves emerge at some distance posterior to their point of origin.

nerves, particularly those at the posterior end, are drawn posteriorly, so that they emerge through foramina at some distance from their points of origin (Fig. 13.18). Because of the brushlike appearance of the posterior spinal nerves of higher forms and their fancied resemblance to a horse's tail, early anatomists gave to these nerves the name *cauda equina.*

Comparative Anatomy of Spinal Nerves

Amphioxus   The roots of the spinal nerves of amphioxus do not arise from the spinal cord in symmetrical pairs. Those of one side alternate with those of the other in a manner similar to that of the myotomes. Dorsal and ventral roots do not unite. The dorsal roots, which lack ganglia, carry somatic

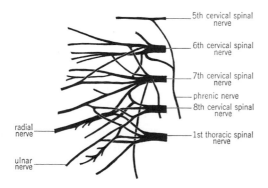

A

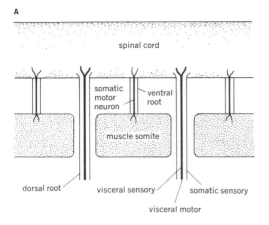

spinal cord

somatic
motor
neuron

ventral
root

muscle somite

dorsal root

visceral sensory

somatic sensory

visceral motor

FIG. 13.19 Diagram showing the condition of spinal-nerve exits in the lamprey seen from a dorsal view, right side only. Note that dorsal root exits between muscle somites. Ventral root exits directly in the middle of muscle somite. The two roots do not join each other.

sensory, visceral sensory, and visceral motor fibers. The ventral roots, which are less compact, are composed of somatic motor fibers alone.

Cyclostomes In lampreys the roots of the spinal nerves do not join each other (Fig. 13.19), but in hagfishes such a union is usual, except in the caudal region. In the gill region of lampreys fibers from adjacent dorsal roots unite, as do fibers from adjacent ventral roots, so as to form two nerves, one sensory and the other motor. These together form the *hypobranchial nerve* supplying the ventral part of the gill region.

Fishes Dorsal and ventral roots unite in fishes, but union occurs *outside* the vertebral column. In some of the anterior spinal nerves the dorsal roots have disappeared and only the ventral roots persist. They are called *spino-occipital* nerves. Fibers from these nerves contribute to the formation of a *hypobranchial* nerve (Fig. 13.5), which in this case is composed of motor fibers alone.

Amphibians The roots in amphibians unite as they pass through the intervertebral foramen, the dorsal root ganglion being located at the point of junction. In anurans *calcareous bodies* surround the spinal ganglia. In urodeles the ganglia are surrounded by spongy fatty tissue.

Between the occipital condyles and the first vertebra the ventral root of a small *suboccipital nerve* emerges on each side. This nerve, which is purely motor, may represent the *hypoglossal* cranial nerve of higher forms.

In urodeles the number of spinal nerves varies with the number of segments of the body, but in adult anurans there is a reduction in number so that in most cases only 10 or 11 remain. Cervicobrachial and lumbosacral plexuses are present. A conspicuous cauda equina is present in frogs and toads.

Reptiles The spinal nerves of reptiles show no peculiarities. As in other amniotes the dorsal roots contain only sensory fibers. Visceral motor and somatic motor fibers are confined to the ventral roots. In certain snakes and limbless lizards a distinct but poorly developed lumbosacral plexus is present indicating that these limbless forms evolved from ancestors with limbs.

Birds The arrangement of spinal nerves in birds is typical. In certain long-necked forms the nerves making up the cervicobrachial plexus arise from the cord much farther posteriorly than is usually the case. The lumbosacral plexus may be divided into separate *lumbar, sacral,* and *pudendal* plexuses. The lumbar plexus supplies the thigh. The nerves of the sacral plexus unite to form the sciatic nerve passing through the thigh to the lower leg. The pudendal plexus sends branches to the cloacal and tail regions.

Mammals The spinal nerves of mammals are named according to their relations to the vertebral column. There are, therefore, cervical, thoracic, lumbar, sacral, and caudal or coccygeal spinal nerves. Each is numbered according to the number of the vertebra which lies anterior to it, except in the case of the cervicals, the first of which emerges between the occipital bone and atlas.

The limb plexuses of mammals may be very complicated and are commonly divided into cervical, brachial, lumbar, and sacral divisions.

CRANIAL NERVES

Cranial nerves are much more specialized than spinal nerves and show little similarity to the latter in origin and distribution. Early human anatomists, not appreciating the functional characteristics of the various cranial nerves or their homologies in lower vertebrates, assigned numbers to them in an anterior-posterior sequence, a system of classification which, in the light of modern research, has been shown to be very superficial and artificial. Nevertheless, the original terminology has persisted, making it rather difficult for the student to appreciate the complexity of the component parts of various cranial nerves.

It will be recalled that the dorsal and ventral roots of the spinal nerves in amphioxus and lampreys fail to join each other. Certain cranial nerves may originally have had a similar arrangement. If the two roots of a cranial nerve unite, they do so before emerging from the medulla oblongata. In some cases dorsal roots and ganglia have apparently been lost, and in others the original ventral roots seem to have disappeared. Some purely sensory nerves are special structures not at all comparable to the sensory components of spinal nerves.

There are 10 pairs of cranial nerves in anamniotes and 12 in amniotes. Some are entirely sensory or entirely motor, but others are mixed nerves carrying both types of fibers. The cranial nerves are listed in Table 6.

In 1894 a new cranial nerve was discovered connecting to the anterior end of the cerebral hemispheres. This nerve has been found in all gnathostomes except birds and is called nerve 0 to preserve the terminology long applied to the others.

Merely stating that a nerve is sensory,

TABLE 6

Cranial nerve	Type	Function
0 Terminal		
I Olfactory	Sensory	From olfactory epithelium
II Optic	Sensory	From eyes
III Oculomotor	Motor	Innervates four of six eye muscles
IV Trochlear	Motor	To superior oblique eye muscle
V Trigeminal	Sensory	From head
	Motor	To muscles of lower jaw
VI Abducens	Motor	To lateral rectus eye muscle
VII Facial	Sensory	From taste buds and lower jaw
	Motor	To muscles of face
VIII Acoustic	Sensory	From inner ear
IX Glossopharyngeal	Sensory	From posterior region of tongue and taste buds
	Motor	To lower jaw and throat, larynx, and salivary glands
X Vagus	Sensory	From skin and taste buds
	Motor	To visceral organs of body
XI Spinal accessory	Motor	To various visceral organs; assists vagus
XII Hypoglossal	Motor	To tongue muscles

motor, or mixed has little meaning so far as functional attributes are concerned. Not all sensory nerves have similar origins, nor are all motor nerves made up of similar components. As in the case of spinal nerves, we have somatic sensory, visceral sensory, visceral motor, somatic motor, and preganglionic fibers of the autonomic system to consider. Furthermore, associated with certain cranial nerves of lower forms we have in addition sensory nerve fibers associated with the *lateral-line system,* which has no counterpart in the higher vertebrate classes. These sensory neurons arise from *placodes,* or thickenings in the superficial ectoderm, whereas most other sensory neurons are derived from neural-crest cells. Motor neurons, whether visceral or somatic, arise within the neural tube itself, growing outward. Postganglionic autonomic motor neurons, however, are derived originally from neural-crest cells.

THE ACOUSTICOLATERALIS SYS-
TEM The lateral-line system consists of certain sense organs found in cyclostomes, fishes, and aquatic urodele amphibians. The receptor organs for this system are called *neuromasts.* They have connections with sensory branches of certain cranial nerves. The cell bodies or ganglia of these nerves arise in the embryo from placodes. The inner ears also arise from placodes in the same general vicinity. Because of the similarity in origin of the inner ear to the components of the lateral-line system, it is customary to group these structures into what is called the *acousticolateralis system.* The ganglia of the lateralis system move inward and come to lie within the cranium alongside of the medulla oblongata. The nerve fibers become associated or are distributed via certain cranial nerves, namely, VII (facial), IX (glossopharyngeal), and X (vagus), of which they appear to be branches. The fibers are all sensory and form connections with the somatic

sensory columns of gray matter near the anterior end of the medulla oblongata. These nerve fibers are classified as somatic sensory fibers.

THE BRANCHIAL NERVES Four cranial nerves, spoken of as *branchial nerves,* originally supplied the gill region with sensory and motor fibers. Each is primarily associated with a visceral arch. The trigeminal nerve (V) is the nerve of the mandibular arch; the facial nerve (VII) supplies the hyoid arch; the glossopharyngeal nerve (IX) is associated with the third arch; and the vagus (X) takes care of the remaining arches (Fig. 13.20). The branchial muscles of the gill region are considered to be visceral muscles, even though they are striated and voluntary, because they are derived from the splanchnic mesoderm of the hypomere. The motor nerves supply them *directly* and are composed of visceral motor fibers which are *not* preganglionic fibers. The sensory fibers of the branchial nerves are mostly of the visceral sensory type. Some somatic sensory fibers may be present, however, coming from the skin dorsal to the gill region. The ganglia of the branchial nerves lie close to those of the lateralis system.

SPECIAL SENSORY NERVES Three cranial nerves are of a special nature. They are

FIG. 13.20 Diagram showing relationship of the brachial cranial nerves (V, VII, IX, and X) to the visceral arches and gill slits. (*Modified from Johnston.*)

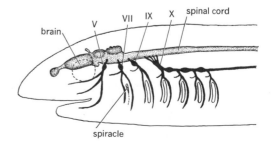

purely sensory, carrying impulses from the nose, eye, and ear, respectively. These nerves are usually classified as somatic sensory nerves, although not all anatomists are in agreement. The cell bodies of the olfactory nerve (I) are derived from a placode in the superficial ectoderm of the anterior head region. Those of the optic nerve (II) come from an outpocketing of the diencephalic region of the brain and represent an extension of the brain itself. It is not a nerve but a brain tract. The fibers of the acoustic nerve (VIII) originate from neural-crest cells.

TERMINAL NERVE (0) The terminal nerve appears to originate in the diencephalon and passes to the olfactory mucous membrane. It is a somatic sensory nerve which bears one or more ganglia. Its function is not clear. In amphibians, reptiles, and mammals it is associated with the vomeronasal, or Jacobson's organ, which seems to be an accessory olfactory structure. The nerve is best developed in elasmobranch fishes.

OLFACTORY NERVE (I) The olfactory nerve is not typical. It is composed of sparsely medullated fibers which come from neurosensory cells of the olfactory placode. The axons grow toward the brain (generally sensory neurons grow out of the central nervous system toward the target organ), forming synapses with neurons in the olfactory lobes. In cases where the olfactory lobe is drawn out into bulb and tract, the olfactory tract may be mistaken for the olfactory nerve. The nerve proper usually consists of many separate fibers which are not gathered together in a sheath. In some forms, however, the olfactory nerve is rather long and joins the olfactory lobe, which is not differentiated into bulb and tract. In mammals as many as 20 separate nerve branches on each side pass to the brain. A separate branch of the olfactory

nerve passes to Jacobson's organ in vertebrates possessing this structure (page 211).

OPTIC NERVE (II) The second cranial nerve shows certain peculiarities, for an understanding of which the reader is referred to a description of the development of the eye (page 60). It consists of a bundle of nerve fibers, the cell bodies of which are located in the retina of the eye. The fibers of each optic nerve usually cross beneath the diencephalon and pass to the lateral geniculate bodies or to the optic tectum of the mesencephalon. In mammals, impulses are relayed from the optic tectum to the cerebral hemispheres. The point at which the nerves cross is the *optic chiasma*. Crossing is complete in all vertebrates except in those mammals having binocular vision. In these, only about half the fibers in each optic nerve cross to the opposite side. Actually the optic nerve is a fiber tract of the brain rather than a cranial nerve, since it merely passes from one part of the brain (optic cup) to another (mesencephalon).

OCULOMOTOR NERVE (III) The somatic motor fibers, of which the oculomotor nerve is mostly composed, leave the ventral side of the mesencephalon and are distributed to the inferior oblique, superior rectus, inferior rectus, and internal rectus eye muscles. These are all derived from the myotome of the first prootic somite (page 194). Some fibers also pass to the muscle elevating the upper eyelid. Preganglionic autonomic motor fibers accompany the somatic motor fibers and pass to the ciliary ganglion. Postganglionic autonomic fibers then lead to the ciliary apparatus, concerned with accommodation, and to the sphincter muscles of the iris, which regulate the size of the pupil. A few somatic sensory proprioceptive fibers are borne by the oculomotor nerve.

TROCHLEAR NERVE (IV) The somatic motor trochlear nerve leaves the dorsal side of the mesencephalon near its posterior end and passes to the superior oblique eye muscle derived from the second prootic somite. The nerve bears a few somatic sensory proprioceptive fibers.

TRIGEMINAL NERVE (V) The large fifth cranial nerve arises from the lateral side of the anterior end of the medulla oblongata. It bears a large *Gasserian ganglion,* which sends fibers to the somatic sensory column of gray matter. The nerve is characteristically divided into three main branches, *ophthalmic, maxillary,* and *mandibular.* The first two of these bear somatic sensory fibers alone, but the mandibular is composed both of somatic sensory and visceral motor fibers. Visceral sensory and somatic motor fibers are lacking. The trigeminal nerve is the nerve of the mandibular arch and is usually considered to be the first of the branchial nerves. It differs somewhat from the others in that its pretrematic (maxillary) and posttrematic (mandibular) branches bear somatic sensory rather than visceral sensory fibers.

In fishes the ophthalmic branch is composed of superficial and deep portions, the latter possibly representing what may originally have been a separate branchial nerve. These branches supply the skin on the dorsal side of the head and snout with sensory fibers. In higher vertebrates the two portions of the ophthalmic branch are not distinct and only a single ophthalmic nerve is present. It is the smallest of the three branches and supplies sensory fibers to the conjunctiva, cornea, iris, ciliary body, lacrimal gland, part of the mucous membrane of the nose, and the skin of the forehead, nose, and eyelids.

The maxillary branch is the main nerve to the upper jaw, supplying the upper lip, side of the nose, lower eyelid, teeth of the upper jaw, and Jacobson's organ, if present.

The somatic sensory portion of the mandibular branch is distributed to the lower lip and teeth of the lower jaw. In mammals it also supplies the skin of the temporal region, external ear, lower part of the face, and the mucous membrane of the anterior part of the tongue. The visceral motor fibers of the mandibular branch go directly to the muscles used in chewing.

ABDUCENS NERVE (VI) The small abducens nerve arises from the ventral part of the medulla oblongata and passes to the external rectus eye muscle derived from the third prootic somite. It is a somatic motor nerve which carries a few sensory proprioceptive fibers. In the lamprey the inferior rectus eye muscle, in addition, is supplied by the abducens nerve (page 194). A branch of the abducens goes to the nictitating membrane if one is present.

FACIAL NERVE (VII) The main branch of the facial nerve, the hyomandibular nerve, descends behind the spiracle along the hyomandibular cartilage. It supplies the hyoid arc and contains visceral sensory and visceral motor components. A *geniculatee ganglion* lies at the point where the hyomandibular branch leaves the medulla oblongata. Anterior branches, the palatine ramus, innervate much of the mouth and taste-bud series.

In tetrapods the branches are modified and innervate the lower jaw and hyoid apparatus.

Visceral sensory fibers in mammals supply the taste buds of the anterior two-thirds of the tongue. These fibers form part of the *chorda tympani,* a branch of the facial nerve which passes through the middle ear.

The visceral motor components of the facial nerve pass directly to the muscles of the face, scalp, and external ear and to a few superficial neck muscles.

Preganglionic visceral motor fibers of the peripheral autonomic system accompany the chorda tympani branch and pass to the *sub-maxillary ganglion.* Here they form synapses with postganglionic autonomic fibers which supply the submaxillary and sublingual salivary glands. Other preganglionic fibers pass to the *sphenopalatine ganglion,* where they relay with postganglionic autonomic fibers which are distributed to the lacrimal gland and mucous membrane of the nose.

ACOUSTIC NERVE (VIII) The purely somatic sensory auditory nerve bears an *acoustic ganglion* which arises from neural-crest cells and is closely associated with the *geniculate ganglion* of the facial nerve in its development. In forms in which the inner ear serves both as an organ of hearing and of equilibrium, the auditory nerve is divided into two main branches, a *vestibular branch,* which carries equilibratory impulses from the vestibular portion of the inner ear, and a *cochlear branch,* for auditory impulses arising in the cochlea.

Fiber tracts pass from the eighth nerve to the cerebellum, where centers for equilibration are located. Auditory impulses pass to the inferior colliculi of the optic tectum and to the medial geniculate bodies in the roof of the diencephalon. In mammals they are then relayed to centers in the cerebral cortex.

LATERAL-LINE NERVES The nerves which are associated with the lateral-line organs of fishes are closely associated with the acoustic nerve and are of a special nature. They are derived embryonically from epithelial thickenings termed *placodes.* Typically afferent nerves are derived embryonically from neural-crest cells. The physical association of lateral-line nerves with both the facial and glossopharyngeal cranial nerves in passing

through the brain case led past investigators to assume that the lateral-line nerves were parts of these cranial nerves. However, it has been demonstrated that the association is merely one of convenience.

GLOSSOPHARYNGEAL NERVE (IX) The ninth cranial nerve is the nerve of the third visceral arch and the second gill pouch. In fishes its pretrematic branch, made up of visceral sensory fibers, passes to the anterior side of the first *typical* gill slit.* Its post-trematic branch, composed of visceral sensory and visceral motor fibers, supplies the posterior border of this gill slit. A pharyngeal branch, containing visceral sensory fibers, goes to taste buds and other receptors in the pharynx. A small somatic sensory branch to the anterior part of the lateral line is present in lower aquatic vertebrates. A prominent *petrosal ganglion* lies near the base of the nerve.

Two ganglia are present on the ninth nerve in higher forms. These are (1) a small *superior ganglion,* close to the medulla oblongata, and (2) a *petrosal ganglion,* a short distance away.

In higher vertebrates some visceral motor components pass directly to muscles of the pharynx derived from the third visceral arch region. Preganglionic visceral autonomic motor fibers pass to an *otic ganglion,* where they form synapses with postganglionic autonomic fibers supplying the parotid salivary gland. In mammals visceral sensory fibers innervate taste buds of the posterior third of the tongue as well as the mucous membrane of the pharynx and palatine tonsils.

VAGUS NERVE (X) The importance of the vagus nerve is indicated by the fact that

---

* The spiracle, found in elasmobranchs and a few other fishes, actually represents the first gill pouch. It disappears in teleosts. The first gill pouch as found in teleosts therefore really represents the second pouch of elasmobranchs.

it is largely composed of preganglionic visceral motor fibers of the parasympathetic portion of the peripheral autonomic nervous system which control such vital activities as heartbeat, respiratory movements, and peristalsis. The nerve arises from the medulla oblongata. In lower aquatic vertebrates a large somatic sensory *lateral nerve* branches off the vagus and extends the length of the body, lying underneath the lateral-line canal and located between the epaxial and hypaxial muscles of the body wall. In amphibians, with the exception of neotenic forms, the lateral nerve disappears at metamorphosis. The main part of the vagus nerve is the *branchiovisceral branch,* composed of visceral sensory and visceral motor fibers, together with a few somatic sensory fibers distributed to the skin in the ear region. Part of the branchiovisceral branch serves as a *branchial nerve* supplying the visceral arches posterior to the third, and the remaining, gill pouches. Pretrematic branches bear visceral sensory fibers alone, whereas posttrematic branches contain visceral sensory and visceral motor fibers. Pharyngeal branches supply taste buds and other sensory receptors in the pharynx. Posterior to the gills the vagus courses caudally, giving off a branch to the heart and then distributing visceral sensory and preganglionic visceral motor fibers to the coelomic viscera, except those near the posterior end. The preganglionic visceral motor fibers terminate in small autonomic ganglia in, or close to, the walls of these organs, where they form synapses with very short postganglionic autonomic fibers.

With the disappearance of gills in amniotes and after metamorphosis in amphibians, the branchial branches of the vagus are lost, for the most part. Visceral motor remnants of posttrematic branches persist to supply directly certain muscles of the pharyngeal and laryngeal regions, but the main part of the vagus is composed of preganglionic visceral motor fibers.

In the lower aquatic vertebrates the vagus nerve bears a *lateralis ganglion,* as well as a *jugular ganglion,* near its base. In higher forms the lateralis ganglion disappears, but the jugular ganglion persists. In mammals a large *nodosal ganglion* is present at the base of the vagus distal to the jugular ganglion.

SPINAL ACCESSORY NERVE (XI)
An eleventh cranial nerve is found only in amniotes and is closely related to the vagus. It is apparently composed of a cranial portion, derived from certain posterior visceral motor fibers of the vagus, and a spinal portion made up of somatic motor fibers of some of the spino-occipital nerves. Fibers of the cranial portion, which pass directly to certain muscles of the pharynx and larynx, belong to the category of branchial nerves. Fibers of the spinal portion supply the sternocleidomastoid and trapezius muscles. There is some uncertainty about the homologies of these muscles, but since they are innervated by somatic motor nerve fibers, they are most probably derived from myotomes, not from the branchial musculature.

HYPOGLOSSAL NERVE (XII) The twelfth cranial nerve, also found only in amniotes, is a purely somatic motor nerve which supplies the intrinsic muscles of the tongue, as well as several muscles in the lower jaw and neck region. The hypoglossal nerve is represented in lower forms by the hypobranchial nerve, formed by the union of two or three spino-occipital nerves. In amniotes it has acquired cranial connections. The hypoglossal nerve is best developed in mammals.

AUTONOMIC NERVES

The autonomic portion of the peripheral nervous system is composed of efferent neurons which send impulses to smooth muscles and glands in all parts of the body. It regulates the function of structures under involuntary control. The muscle of the heart, although striated, is innervated by autonomic fibers. The proper functioning of this part of the nervous system is necessary for regulating such activities as rate of heartbeat, respiratory movements, composition of body fluids, constancy of temperature, secretion of various glands, and other vital processes. Controlling centers lie in the hypothalamus. The peripheral autonomic system, being dependent upon the brain and spinal cord, cannot function independently. It is, nevertheless, involved in processes which proceed in the absence of voluntary control and outside the realm of consciousness.

The complex connections of the autonomic neurons with the central and peripheral portions of the nervous system add to the difficulties embodied in an understanding of its structure. Most investigations have been carried out in mammals, man in particular. It will best serve our purpose first to discuss the autonomic system in man and then to make comparisons with other vertebrates.

It will be recalled that somatic motor neurons pass without interruption from the somatic motor column of gray matter in the brain or spinal cord to the effector organ. A different pattern obtains in the autonomic system, in which efferent impulses must travel through *two* neurons, *preganglionic* and *postganglionic,* before they can bring about an effect (Fig. 13.16*B* and *C*).

The system is composed of these two types of neurons and a number of ganglia which serve as relay centers. The cell bodies of the preganglionic neurons are located in the visceral motor columns of gray matter in the central nervous system. These fibers make up the visceral motor components of spinal nerves and certain cranial nerves.* In spinal nerves they are constituents of the white visceral rami. Postganglionic neurons have gray, sparsely medullated axons. The cell bodies of these neurons, which are derived from neural-crest cells, lie in ganglia often located some distance from the central nervous system. It is here that the preganglionic fibers form synapses with the dendrites of postganglionic neurons. It is the axons of the postganglionic neurons which pass to smooth muscles, cardiac muscle, and glands in various parts of the body.

The autonomic portion of the peripheral nervous system is divided into sympathetic and parasympathetic systems (Figs. 13.21 and 13.22). Postganglionic fibers of both are in most cases distributed to all involuntary structures of the body. The two systems, in general, work antagonistically. Whereas one system (sympathetic) functions to strengthen an animal's defenses against adverse conditions by an expenditure of energy, the other (parasympathetic) is concerned with processes which tend to conserve and restore energy. In both systems the postganglionic fibers give off chemical substances which bring about the effects. Postganglionic sympathetic fibers, except those going to the sweat glands and uterus, produce a hormone, formerly called *sympathin,* now known to be a catecholamine named *norepinephrine.* The same amine, in different concentration, along with epinephrine, is produced by the medulla of the adrenal gland. Postganglionic parasympathetic fibers and those postganglionic sympathetic fibers supplying the sweat glands and uterus liberate

* Some visceral motor fibers of the branchial cranial nerves are exceptions, since they pass *directly* to the muscles of the gill region. These muscles are voluntary and striated, despite their origin from splanchnic mesoderm. They are not controlled by autonomic nerves.

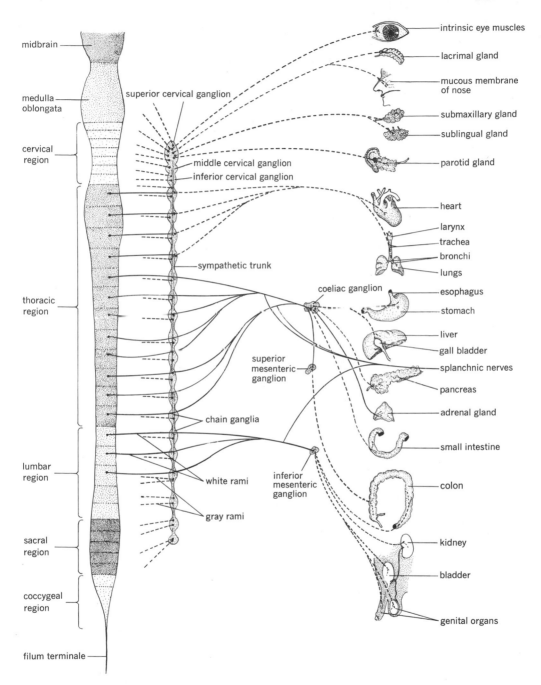

FIG. 13.21 Unilateral diagram of essential parts of sympathetic nervous system of man. Preganglionic fibers are shown in solid lines; postganglionic fibers in dotted lines.

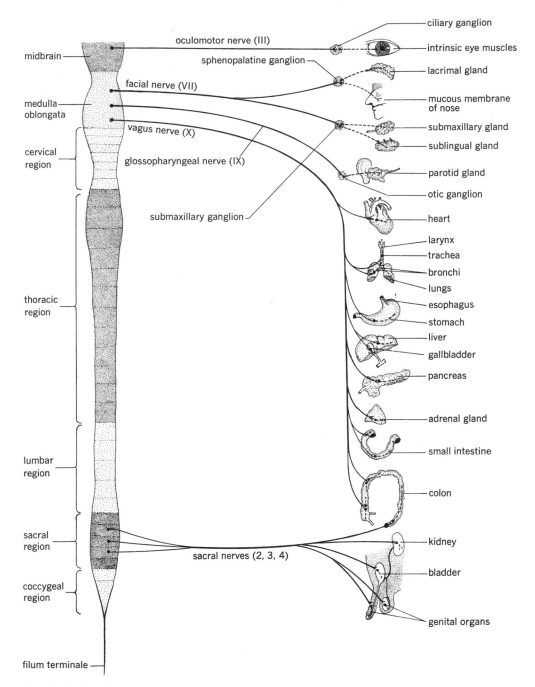

FIG. 13.22 Unilateral diagram of essential parts of parasympathetic nervous system of man. Preganglionic fibers are shown in solid lines; the very short postganglionic fibers in dotted lines.

a chemical called *acetylcholine* and are referred to as *cholinergic fibers*. Preganglionic fibers of both systems liberate acetylcholine at the synapse. Norepinephrine and acetylcholine often have opposite effects upon smooth muscles and glands.

SYMPATHETIC NERVOUS SYSTEM

(Fig. 13.21) Since the preganglionic fibers of the sympathetic system connect with the spinal cord only in the thoracic and lumbar regions, the term *thoracolumbar outflow* is applied to this portion of the autonomic system.

On either side of the ventral part of the vertebral column lies a long *sympathetic trunk*. At fairly regular intervals each trunk bears enlargements known as *chain ganglia* (Figs. 13.15 and 13.21). Groups of these ganglia at the anterior end of the sympathetic trunk have fused to form three large *cervical ganglia*.

The white visceral rami of all the thoracic and the first three lumbar spinal nerves are partly composed of preganglionic fibers which connect to the corresponding chain ganglia. They may terminate in the ganglion at the point where they enter or may send fibers up or down the sympathetic trunk for some distance. Other preganglionic fibers pass without synapses *through* ganglia of the sympathetic trunk to form connections with one or more of three large *prevertebral ganglia* located in the abdominal region in front of the lumbar vertebrae. Prevertebral ganglia, also derived from neural crests, are sympathetic ganglia of the thorax and abdomen other than those of the sympathetic trunk. The ones involved here are the *coeliac, superior mesenteric,* and *inferior mesenteric ganglia.* The nerves leading to these ganglia from the sympathetic trunk are called *splanchnic nerves.*

From the chain ganglia and prevertebral ganglia arise sparsely medullated postganglionic fibers which are distributed to various structures of the body under involuntary control.

A *gray visceral ramus* comes from each chain ganglion, whether or not a white visceral ramus is present. The postganglionic fibers making up the gray rami are distributed via the dorsal and ventral rami of spinal nerves to the skin, where they supply smooth muscles in the walls of the blood vessels, the arrector pili muscles, and skin glands (Fig. 13.16B). Other postganglionic fibers from the anterior ganglia pass to the intrinsic muscles of the eye, lacrimal glands, mucous membrane of the nose, palate, and mouth, and to the salivary glands. From the inferior cervical ganglion and first few chain ganglia arise postganglionic fibers going to heart, larynx, trachea, bronchi, and lungs. Postganglionic fibers arising in the coeliac and superior mesenteric ganglia supply esophagus, stomach, small intestine, first part of the large intestine, liver, pancreas, and blood vessels of the abdomen. The inferior mesenteric ganglion sends postganglionic fibers to the remainder of the large intestine and to the urogenital organs. Preganglionic fibers pass *directly* to the medulla of the adrenal gland. This is composed of modified sympathetic glandular cells of neural-crest origin which are homologous with postganglionic sympathetic neurons.

Function of the Sympathetic System Among the reactions brought about by stimulation of the sympathetic system are (1) constriction of cutaneous blood vessels, causing pallor; (2) contraction of arrector pili muscles, causing goose flesh and making the hair stand erect; (3) secretion of sweat glands (cold sweat); (4) dilatation of the pupil; (5) reduction in amount of saliva so that only small quantities of a thick, mucinous secretion are given off; (6) acceleration of heartbeat; (7) dilatation of the bronchi; (8) relaxation or inhibition of the smooth muscles of the digestive tract, causing a temporary cessation of peristalsis; (9) relaxation of bladder musculature; (10) lengthening of the urethral musculature (page 287); (11) increase in amount of blood sugar; (12) rise in blood pressure; (13) increase in number of red corpuscles in the bloodstream; (14) decrease in clotting time of blood.

These reactions, considered together, are usually associated with pain, fear, and anger. They serve the body beneficially in times of danger or when it is in an aggressive state.

PARASYMPATHETIC NERVOUS SYSTEM (Fig. 13.22) The term *craniosacral outflow* is used to designate the complex of preganglionic motor fibers of the parasympathetic nervous system. The term is appropriate, since only certain cranial and sacral spinal nerves are involved.

Four cranial nerves are composed at least in part of preganglionic parasympathetic fibers: the oculomotor (III), facial (VII), glossopharyngeal (IX), and vagus (X) nerves. The ganglia in which they terminate are situated close to or in the organs supplied by this system. Hence the preganglionic fibers are rather long and the postganglionic fibers are very short.

The preganglionic components of the oculomotor nerve terminate in a small *ciliary ganglion* situated in the back part of the orbit. Postganglionic autonomic fibers then pass to the eyeball, supplying the sphincter muscle of the iris as well as ciliary muscles.

The preganglionic visceral motor fibers of the facial nerve pass either to the *sphenopalatine* or to the *submaxillary ganglion* in the head. From the sphenopalatine ganglion postganglionic fibers pass to the lacrimal gland, mucous membrane of the nose, palate, and upper part of the pharynx. Some components of the chorda tympani branch of the facial nerve are preganglionic fibers which

pass to the submaxillary ganglion. Short postganglionic fibers are distributed to the submaxillary and sublingual glands.

An otic ganglion receives preganglionic visceral motor fibers of the glossopharyngeal nerve. Postganglionic fibers pass to the parotid gland and mucous membrane of the mouth.

The preganglionic visceral motor fibers of the vagus nerve terminate in very small ganglia located in the walls of the structures they supply. It is probable that those of the gut connect with the plexuses of Auerbach and Meissner. From these plexuses postganglionic fibers pass to the smooth muscles and glands of the alimentary tract. Other parasympathetic fibers of the vagus innervate the heart, larynx, trachea, bronchi, lungs, blood vessels of abdomen, liver, gallbladder, and pancreas.

The sacral outflow of the parasympathetic system is composed of efferent fibers which course through the white visceral rami of the second, third, and fourth sacral nerves, which together form the *pelvic nerve*. These preganglionic fibers terminate in ganglia in, or very near, the lower part of the large intestine, kidneys, bladder, and reproductive organs. The postganglionic fibers are relatively short.

Function of the Parasympathetic System Stimulation of the various components of the parasympathetic system brings about effects which are, in general, opposite to those secured by stimulating the sympathetic nerves supplying these same organs. Among these reactions are (1) dilatation of blood vessels (except the coronary and pulmonary vessels of the heart and lungs); (2) constriction of the pupil; (3) increase in salivary and gastric secretion; (4) constriction of bronchi; (5) contraction of the walls of the digestive tract bringing about peristalsis and other types of contractions; (6) contraction of bladder musculature; (7) shortening of the urethral musculature; (8) dilatation of the blood vessels of the external genital organs. These reactions, when considered as a group, are those associated with comfortable or pleasurable sensations in the body and conserve energy. They form the mechanism of normal functions when the body is in a receptive state.

AUTONOMIC SYSTEM OF LOWER CHORDATES The general scheme of arrangement of the autonomic system in tetrapods is similar to that of man. The lower vertebrates show a progressive complexity of the system as the evolutionary scale is ascended.

If a peripheral autonomic system is actually present in amphioxus, it is represented by the parasympathetic system alone. In cyclostomes segmental sympathetic ganglia are present, but they are not connected by sympathetic trunks. A parasympathetic system first appears in cyclostomes but is confined to the vagus nerve. The autonomic system is much better developed in elasmobranchs. Sympathetic elements are confined to the abdominal region, but the sympathetic trunks actually have a rather diffuse arrangement. Parasympathetic fibers are present in the oculomotor (III), facial (VII), glossopharyngeal (IX), and vagus (X) nerves, but those in the vagus do not seem to go to the intestine or urogenital organs. In teleosts the autonomic system shows a still greater advance and general similarity to that of tetrapods. True longitudinal sympathetic trunks are present for the first time. Parasympathetic components, however, seem to be confined to the oculomotor and vagus nerves. A sacral outflow appears for the first time in anuran amphibians.

SUMMARY

1 Almost all the nervous system is derived from the neural tube, neural crests, and ectodermal placodes, which appear very early in embryonic development.

2 The cells making up nervous tissue have the properties of irritability and conductivity more highly developed than those of other tissues in the body. They are all derived from ectoderm.

3 The central nervous system comprises gray matter and white matter. Gray matter consists of nerve-cell bodies and sparsely myelinated fibers, together with nonnervous neuroglia and ependymal cells. White matter is made up of bundles of myelinated fibers and neuroglia. In the spinal cord gray matter is centrally located and almost completely surrounded by white matter. The gray matter is arranged into a pair of dorsal sensory columns and a pair of ventral motor columns, each composed of somatic and visceral regions. Nerve fibers are, in general, of two types, medullated and very sparsely medullated. Medullated fibers are white in appearance; sparsely medullated fibers are gray. Bundles of nerve fibers are called fiber tracts or nerves, depending on their location. Aggregations of nerve-cell bodies are referred to as nuclei or ganglia, again depending upon where they are located.

4 The nervous system is divided into two main parts: (1) the central nervous system, composed of brain and spinal cord, and (2) the peripheral nervous system, made up of cranial and spinal nerves. An important part of the latter is the autonomic system supplying structures under involuntary control.

5 The anterior end of the neural tube enlarges to form the brain. It soon becomes divided to form three primary vesicles, forebrain, midbrain, and hindbrain. The first and third become secondarily divided, but the midbrain undergoes no division. The remainder of the neural tube becomes the spinal cord.

6 The original cavity of the neural tube forms the ventricles of the brain and the central canal of the spinal cord.

7 The spinal cord in tetrapods bears two enlargements at the levels where nerves going to the limbs arise. In most forms the posterior end of the spinal cord tapers down to a fine thread, the filum terminale. There is a tendency toward a shortening of the spinal cord within the phylum Chordata.

8 It is in the development of that part of the forebrain called the telencephalon that the greatest changes occur in the brains of vertebrates. A pair of swellings, the cerebral hemispheres, appears at the anterior end. In lower forms the dorsal part, or pallium, is thin-walled, the ventral portion becoming the thickened corpus striatum. An outgrowth at the anterior end of each cerebral hemisphere becomes the olfactory lobe. As the vertebrate scale is ascended there is an increasing tendency for nerve cells from the inner layer of the pallium to migrate out to the periphery. A new area, the neopallium, first appears in reptiles in the outer, anterodorsal part of each cerebral hemisphere. The growth and development of the neopallium account for the large size of the cerebral hemispheres in mammals. Its outer surface, into which nerve cells have migrated, forms a gray layer, the cerebral cortex.

9 The remainder of the forebrain, the diencephalon, has thickened ventral and lateral portions. These contain important relay centers and consist largely of nuclei of gray matter. The floor, or hypothalamus, contains centers which integrate activities of the peripheral autonomic system. A ventral evagination forms the infundibulum, the distal portion of which gives rise to the posterior lobe of the pituitary gland. The anterior part of the roof of the diencephalon remains epithelial and, together with the pia mater, forms a tela choroidea. Vascular folds of the tela choroidea become the anterior choroid plexus, where cerebrospinal fluid is liberated. The posterior portion in numerous forms gives rise to the parapineal, or parietal, and pineal outgrowths, originally associated with additional eyes. In higher forms only the pineal body persists.

10 The floor and side walls of the midbrain, or mesencephalon, are composed of fiber tracts connecting forebrain and hindbrain. The roof consists of a thick layer of gray matter, the optic tectum. A pair of optic lobes, the corpora bigemina, is present in the roof of the mesencephalon in lower forms. They serve as visual centers. In snakes and mammals, there are four prominences, the corpora quadrigemina. The anterior two serve as visual centers, and the posterior pair as auditory centers.

11 The anterior region of the hindbrain is the metencephalon. Its dorsal region becomes the cerebellum, where the neuromuscular mechanism of the body is integrated. In a few birds and in mammals a bridge of nerve fibers, the pons, appears on the ventral side of the metencephalon.

12 The remainder of the hindbrain is the myelencephalon, or medulla oblongata.

The lateral and ventral walls are thickened, but the dorsal wall retains its epithelial character. Together with the pia mater, the roof forms a tela choroidea and posterior choroid plexus. The ventral and lateral walls contain large fiber tracts as well as columns of gray matter similar to those of the spinal cord. Certain nuclei in the visceral motor column serve as centers for such vital functions as heartbeat and respiration. Cranial nerves V to X (or XII) arise from the medulla oblongata. Several fiber tracts in the posterior part of the medulla cross (decussate) to the opposite side.

13   Brain and spinal cord are surrounded by membranes called meninges. A single meninx primitiva is present in fishes. Two membranes, an inner pia-arachnoid and an outer dura mater, exist in amphibians, reptiles, and birds. In mammals the pia-arachnoid membrane splits into an inner pia mater and an outer arachnoid membrane.

14   The peripheral nervous system is made up of cranial and spinal nerves. Several kinds of nerve fibers are represented: somatic sensory, visceral sensory, somatic motor, visceral motor fibers of branchial cranial nerves, and preganglionic autonomic visceral motor fibers belonging to the autonomic system. The latter pass to outlying ganglia, where they form synapses with autonomic postganglionic neurons, the gray, sparsely medullated fibers of which go to smooth muscles and glands in the viscera and skin.

15   Spinal nerves are paired and segmentally arranged. They arise from the spinal cord by two roots, a dorsal root bearing a ganglion and a ventral root which lacks a ganglion. Most spinal nerves give off three branches, or rami. In the regions of the limbs cross connections between certain spinal nerves form plexuses.

16   The cranial nerves arising from the brain number 10 in anamniotes and 12 in amniotes. The twelfth and part of the eleventh nerve represent spinal nerves which have been taken over by the brain. Some cranial nerves are purely sensory (I, II, and VIII), and others are purely motor (III and IV). The remainder are mixed sensory and motor nerves. Another cranial sensory nerve was discovered in 1894 and called the terminal nerve (0). Its function is not clear. In lower vertebrates nerves VII, IX, and X have somatic sensory branches related to the lateral-line system. These disappear in terrestrial forms. Cranial nerves V, VII, IX, and X are branchial nerves which bear visceral sensory and visceral motor fibers in addition to others. Visceral motor fibers pass directly to the branchial muscles. Cranial nerves III, VII, IX, and X also bear preganglionic visceral motor fibers of the parasympathetic system.

17   The autonomic part of the peripheral nervous system is made up of sympathetic and parasympathetic components. Each consists of preganglionic visceral motor fibers extending from outlying ganglia to structures in the body under involuntary control. The sympathetic system is called the thoracolumbar outflow, since its preganglionic fibers arise in the thoracic and lumbar regions of the spinal cord. The parasympathetic system is referred to as the craniosacral outflow, since its preganglionic fibers are associated with certain cranial nerves (III, VII, IX, X) and sacral spinal nerves. The two parts of the autonomic system work antagonistically. The peripheral autonomic systems of lower chordates are less complicated than those of higher forms.

BIBLIOGRAPHY

Bass, A. D. (ed.): "Evolution of Nervous Control from Primitive Organisms to Man," American Association for the Advancement of Science, Washington, D.C., 1959.

Everett, N. B.: "Functional Neuroanatomy," Lea & Febiger, Philadelphia, 1971.

Herrick, C. J.: "The Brain of the Tiger Salamander, *Ambystoma tigrinum,*" The University of Chicago Press, 1948.

————: "Neurological Foundations of Animal Behavior," Hafner, New York, 1962.

Nocol, J. A. C.: Biological Review, *Camb. Phil. Soc.,* 27:1–49 (1952). A review of the autonomic system of lower chordates.

Peters, A., A. L. Palay, and H. deF. Webster: "The Fine Structure of the Nervous System," Harper & Row, New York, 1970.

Young, J. Z.: The Evolution of the Nervous System and of the Relationship of Organism and Environment, in G. R. deBeer (ed.), "Evolution," vol. I, Oxford University Press, London, 1938.

14 INTEGRATING SYSTEM: ENDOCRINE ORGANS

All the glands in the body may be classified as *exocrine* or *endocrine*. Exocrine glands have ducts which convey their secretions to epithelial surfaces of the body, where they are discharged. Endocrine glands, on the other hand, have no ducts to carry off the secretory product or products and discharge their secretions directly into the bloodstream. Both types of glands, with the exception of the interstitial cells of Leydig of the testis (see page 435), develop from epithelial surfaces as groups of cells which grow into the connective tissue beneath the epithelial surface and there proliferate and differentiate into glandular structures of one type or another. In the case of an exocrine gland the original connection between the gland and the surface is retained and differentiates into the lining of a duct. When endocrine glands develop, connection with the epithelial surface is lost. The secretions, known as *hormones*, are chemicals capable of bringing about changes in other parts of the body. They may affect the body generally or act very specifically on only a certain organ or part of an organ, such structures

being referred to as *target organs* or *target tissues.*

A vast amount of research has been carried out on the *endocrine glands*, or glands of internal secretion, and many important discoveries have been made concerning the functions of these structures in health and in disease. Of special interest to the comparative anatomist is the fact that the endocrine glands occur with some uniformity in all vertebrates. The organs making up the endocrine group are found throughout the body, vary widely in embryonic origin, and are derived for the most part from other organ systems. For these reasons they do not form a separate organ system and will not be considered as one. Though a functional interrelationship is shown by means of the hormones they produce, each gland has its particular and specific functions. Some of the glands, however, show closer interrelationships than others.

In addition to the gonads of both sexes and the placenta of pregnancy, the endocrine system includes the thyroid, parathyroid, adrenal, and pituitary glands, as well as the islet tissue of the pancreas. The pineal body and the thymus gland were formerly considered to be endocrine organs. Then, for a time, it was believed that the pineal body was nothing more than a vestigial structure. Recent findings, however (page 396), indicate that it probably belongs in the endocrine category after all. The thymus gland (page 373), on the other hand, is a lymphoid organ of primary importance in establishing mechanisms which are of great significance in combating infection and protecting the body from invasion by bacteria or foreign tissues.

Certain glands play a dual role. For example, the pancreas gives off a secretion which passes through a duct or ducts into the intestine; in addition, it gives off endocrine secretions. The gonads have an endocrine function in addition to forming ova or spermatozoa.

Numerous chemical substances formed by cells in various organs of the body are carried by the circulatory system to other parts of the body, where they bring about reactions of one sort or another. Since the cells which produce these secretions do not form discrete glandular structures and show little functional interrelationship, they are more properly discussed in connection with the various organs of which they form a part. Among them are gastrin, serotonin, secretin (page 268), pancreozymin (page 268), cholecystokinin (page 268), renin (page 286), angiotensin (page 286), erythropoietin (page 286), and others.

PITUITARY GLAND (HYPOPHYSIS)

The great importance of the pituitary gland is generally recognized, and in the past it was frequently referred to as the "master" gland. It has become increasingly evident that practically all the endocrine glands are dependent upon the action of others and that such a designation for the pituitary gland is misleading. Furthermore, the manner in which neurosecretory products from the central nervous system control the activity of the pituitary gland, by the hypothalamus, gives increasing emphasis to the concept of a neuroendocrine designation to the nervous and endocrine organs in their joint coordination of various body functions. An outstanding feature of the pituitary gland is that besides secreting agents which exert their effects on tissues other than those of an endocrine nature, it influences many other endocrine glands and is in turn influenced by them.

The pituitary gland, or *hypophysis cerebri,* lies at the base of the brain in the region of the diencephalon, connected to the brain by a *hypophysial,* or *infundibular, stalk.* It is a compound organ which in higher forms is situated in a depression in the upper face of the sphenoid bone (basisphenoid) called the *sella turcica* (Fig. 14.1).

The pituitary gland is composed of three main parts, or lobes, anterior, intermediate, and posterior (Fig. 14.2). A fourth component referred to as the *pars tuberalis* is a modified part of the anterior lobe. The pituitary body has a dual embryonic origin but is entirely ectodermal. The *neurohypophysis* originates from the *infundibulum,* a ventral evagination of the embryonic diencephalon. From this is derived the *posterior lobe, neural lobe,* or *pars nervosa,* which remains attached to the brain by the infundibular stalk. The neurohypophysis actually includes the *median eminence* of the *tuber cinereum* of the hypothalamus; a number of nuclei in the hypothalamus; the infundibular stalk, containing axons of nerve cells in these nuclei; and the posterior lobe itself. From the ectodermal epithelium of the primitive mouth cavity, or stomodaeum, arises a dorsal evagination called the *adenohypophysis* or *Rathke's*

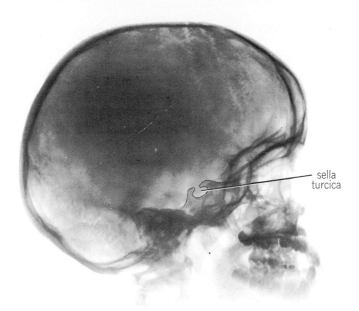

sella
turcica

FIG. 14.1 X-ray photograph of human skull, showing well-defined sella turcica in basal portion of the sphenoid bone in which the pituitary gland lies.

FIG. 14.2 Diagrammatic section through the human pituitary to show relationship of pituitary lobes.

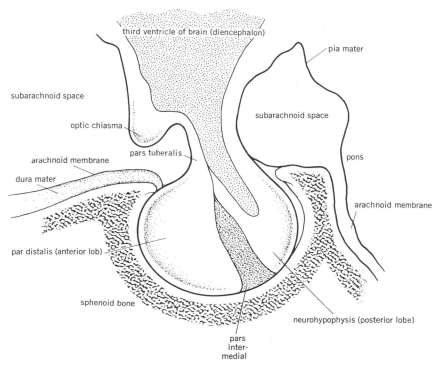

third ventricle of brain (diencephalon)

pia mater

subarachnoid space

subarachnoid space

optic chiasma

pons

pars tuberalis

arachnoid membrane

dura mater

arachnoid membrane

par distalis (anterior lob)

sphenoid bone

neurohypophysis (posterior lobe)

pars
inter-
medial

pocket. This usually constricts off the stomodaeum and becomes a closed vesicle touching the neurohypophysis. The anterior portion of Rathke's pocket enlarges and becomes modified to form the *anterior lobe,* or *pars distalis,* from which the *pars tuberalis* is derived. The posterior portion, which makes contact with the neurohypophysis, enlarges to a lesser extent to become the *intermediate lobe,* or *pars intermedia.* The original cavity of Rathke's pocket usually persists as a small space, the *hypophysial cleft,* or *residual lumen.*

The pituitary gland receives its blood supply from the *circle of Willis,* formed by branches of the internal carotid and basilar arteries. A small *hypophysio-portal system* (Fig. 14.3) of veins is associated with blood vessels draining this area.

POSTERIOR LOBE (PARS NERVOSA) This lobe does not have the histological appearance of an endocrine gland. Apparently the hormones attributed to the posterior lobe are not actually produced there but are formed by neurosecretory cells located in the supraoptic and paraventricular nuclei of the hypothalamus. Secretions, containing the hormones, pass down the axons of the nerve cells to the posterior lobe, where they are stored. Attention is again called to the increasing use of the term *neuroendocrine system* to embrace the nervous and endocrine integrating elements of the body. Six variants of the posterior lobe hormones have been recognized. The following two are typically mammalian and are the best understood at this time. *Vasopressin* promotes reabsorption

FIG. 14.3 Diagrammatic view showing the relationship of the pituitary gland to the neurosecretory cells of the hypothalamus. Secretion by the neurosecretory cells are picked up by the hypophysio-portal system in the pituitary gland.

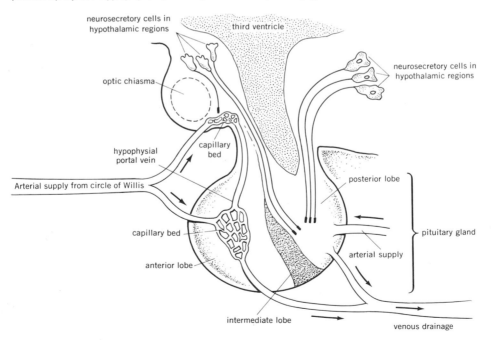

of water from the urine as it passes through the kidney tubules. It is sometimes referred to as the antidiuretic hormone. Large doses of vasopressin cause elevation of blood pressure but the dominate role is in the antidiuretic function. *Oxytocin (pitocin)* acts specifically in contracting smooth muscle of the uterus. Oxytocin also induces contraction of the branched, ectodermal, myoepithelial cells which surround the alveoli of mammary glands, thus aiding in the injection of milk.

It is of interest that the peculiar adaptability of desert rodents to paucity or even lack of water is probably related to the vasopressin-producing capacity of the neurohypophysis and its antidiuretic activity.

INTERMEDIATE LOBE (PARS INTERMEDIA)
Intermedin, the chromatophorotrophic or melanophore-stimulating hormone (MSH) of the intermediate lobe of the pituitary gland, has the effect of causing a dispersion of the pigment granules of integumentary chromatophores in most fishes, amphibians, and reptiles. Epinephrine, from the adrenal medulla, causes an aggregation of pigment granules. Variations in the activity of these hormones are responsible in large measure for changes in color and of patterns. It seems strange that intermedin should be present in the pituitary glands of mammals which do not possess chromatophores of the type mentioned here on which it can act. It apparently has no effect on melanocytes found in lower layers of the epidermis. It now appears that there are two intermedins. Both have similar properties and a chemical structure which shows certain similarities to the adrenocorticotrophic hormone (ACTH) of the anterior lobe, although their functions differ. The fact that both anterior and intermediate lobes are derived from Rathke's pocket makes it less

difficult to understand why their hormones show structural similarities. The cells which apparently secrete MSH are basophilic.

ANTERIOR LOBE (PARS DISTALIS)
Although only three kinds of secretory cells— basophils, acidophils, and chromophobes—are present in the anterior lobe, seven different hormones are known to be secreted by this portion of the pituitary gland. The existence of several other hormones has not been reported but convincingly demonstrated. In addition, intermedin may be produced by the anterior lobe in species lacking an intermediate lobe. Chromophobe cells are believed to be the precursors of both basophilic and acidophilic cells. It is probable that these may alter the character of their secretions under different conditions. Hormones believed to be secreted by the basophile cells include the *thyrotrophic,* or *thyroid-stimulating hormone* (TSH), which stimulates the thyroid gland to secrete thyroglobulin, and the *follicle-stimulating hormone* (FSH), which in the female affects the growth and development of ovarian follicles and the secretion of estrogen. In the male it may play a part in the development of mature sperm cells in the seminiferous tubules. A third hormone elaborated by the basophiles is the *luteinizing hormone* (LH), which is responsible for the preovulatory swelling of the ovarian follicles, increased secretion of estrogen, ovulation, and formation of corpora lutea. In the male it stimulates the activity of the interstitial cells of the testes. A fourth hormone secreted by the basophiles of the anterior lobe is the *adrenocorticotrophic hormone* (ACTH). The *chromatophorotrophic hormone*(MSH), if formed in the anterior lobe, is also of basophilic cell origin.

Two hormones are believed to be secreted by the acidophilic cells of the anterior lobe: (1) *somatotrophic,* or *growth, hormone* (GH), without which growth fails to occur and which may possibly stimulate the alpha cells of the pancreas to secrete glucagon; and (2) the *luteotrophic hormone* (LTH), also known as *prolactin,* which stimulates corpora lutea of certain species to secrete progesterone and also, among other things, causes properly prepared mammary tissue to secrete milk. This hormone is also responsible for the development and function of the crop glands of pigeons; it stimulates lipid synthesis in some birds, formation of brood patch, and brooding behavior. In reptiles and some amphibians prolactin promotes growth and regeneration. In fishes prolactin affects osmoregulation. Over 100 different actions are attributed to this hormone among the vertebrates. Among amphibians, the newt *Triturus viridescens* in certain areas assumes a terrestrial existence for a few years and then returns to the water for breeding. Newts in the terrestrial stage can be induced to migrate to water by administering prolactin.

The reciprocal action between the secretions of the pituitary gland and those of the organs they stimulate has already been referred to (page 420). Since the six anterior-lobe hormones seem to bear some relation to one another in respect to their sites of origin and general similarity of their chemical structure, it is quite possible that the action of such hormones as those of the thyroid and adrenal glands, as well as those from the gonads, may not act specifically or directly in inhibiting the production of one or another of the anterior-lobe hormones. They may instead affect neurosecretory cells in the hypothalamus or other nearby brain areas. These, by liberating neurosecretions into the hypophysioportal system of veins, may influence the secretions of the anterior lobe. Prolactin secretion is inhibited by hypothalamic secretions in mammals but may be stimulated by the secretions of other hormones as in birds

and mammals. Lesions of the hypothalamus have been shown to interfere with certain reciprocal actions which would occur under normal circumstances.

Certain so-called *release* factors (RF) and release inhibitory factors (RIF) have been extracted from the hypothalamus. In addition to stimulating secretion of ACTH by the anterior lobe (see below), they are effective in bringing about release of other anterior-lobe hormones (FSH, PRL, TSH, and GH). The effect upon prolactin (PRL), however, seems to be inhibitory.

A potent substance called *hypothalamic-D,* or the *ACTH-hypophysiotrophic hormone,* has been extracted from the hypothalamus (see page 395). This substance stimulates the production of ACTH by the pituitary gland and is possibly an important link in the *general-adaptation syndrome* (GAS). ACTH and the adrenocorticoids are important in protecting the body against prolonged, nonspecific stresses of various kinds. When subjected to such stresses as cold, burns, hemorrhage, starvation, etc., the body responds in a manner which serves to counteract the harmful effects induced by the stressing agent. The sum of such responses is referred to as the general-adaptation syndrome. The endocrine system is of paramount importance here, since one of the first reactions is an increase in output of ACTH, presumably in response to stimulation by a release factor from the hypothalamus. This link between the nervous and endocrine systems is of special interest. The increased secretion of ACTH causes enlargement of the adrenal cortex and an increase in output of adrenocorticoids. These, in turn, because of their various properties, increase the resistance of the body to the stressing stimulus. Overproduction or imbalance of ACTH and of the cortical steroids may lead to various diseases, collectively referred to as diseases of adaptation. Gastrointestinal ulcers, hypertension, rheumatic fever, and similar conditions are examples.

It has been suggested that a parathyrotrophic hormone from the anterior lobe of the pituitary gland may control the secretion of the parathyroid glands. There is little evidence for this.

Comparative Anatomy of the Pituitary Gland

Amphioxus Practically every structure in the head region of amphioxus has been examined with the view of homologizing it with the pituitary gland of higher forms. The most promising suggestion indicates that an ectodermal depression, the *preoral pit,* in front of the mouth of young individuals may possibly be the hypophysial homolog. In the adult the walls of the pit become ciliated and a connection with the buccal cavity is established. The structure is then known as the *organ of Müller.* Its function is to create a current of water. No possible endocrine function is suggested.

Urochordata In an adult tunicate, e.g., *Molgula manhattensis,* a nerve ganglion lies embedded in the mantle between the siphons. On the ventral side of the ganglion is the *adneural,* or *neural, gland,* which some investigators have homologized with the pituitary gland of vertebrates, the pars nervosa in particular. If such homology exists, it is far from clear.

Cyclostomes The origin of the pituitary gland from the lowest vertebrates to the highest is rather constant in its fundamental constituents and method of development. In the lamprey the single nostril on top of the head leads by a short passageway into the olfactory sac, which lies just in front of the brain. A large *nasopharyngeal pouch* extends in a posteroventral direction from the olfactory sac and terminates blindly beneath the anterior end of the notochord. Between the ventral part of the diencephalon and the nasopharyngeal pouch lies the pituitary gland, consisting of the usual components. The homologies of the structures in this region, about which there has been much confusion in the past, have been fairly well clarified. During early development a *nasohypophysial stalk* appears in close association with the developing olfactory sac at some distance from the stomodaeum. The solid stalk extends posteriorly beneath the forebrain. From the caudal tip of this stalk cells are budded off which are to become the intermediate lobe of the pituitary gland. These intermingle, with the nervous tissue forming the floor of the third ventricle. The anterior-lobe tissue is budded off somewhat later from a dorsal thickening of the remaining nasohypophysial stalk. Soon the anterior lobe becomes completely detached. The nasopharyngeal pouch is an *adult* structure formed, during metamorphosis, from the persistent remnant of the larval nasohypophysial stalk at some time after the pituitary gland has become a definite entity. It hollows out and extends caudad beneath the notochord. Although no relation of nasohypophysial stalk or nasopharyngeal pouch to the stomodaeum is evident, it is believed that at some time a separation from the stomodaeum may have occurred. The dorsal and anterior portion may then have been drawn inward with the adjacent nasal sac and have given rise to the nasopharyngeal pouch, which thus has no counterpart in other vertebrates.

In the hagfish the nasopharyngeal pouch opens into the pharynx by an aperture which appears later during larval life. The pituitary gland is represented by clusters of cells lying between the infundibulum and the nasopharyngeal pouch.

Other Vertebrates In *Polypterus* and

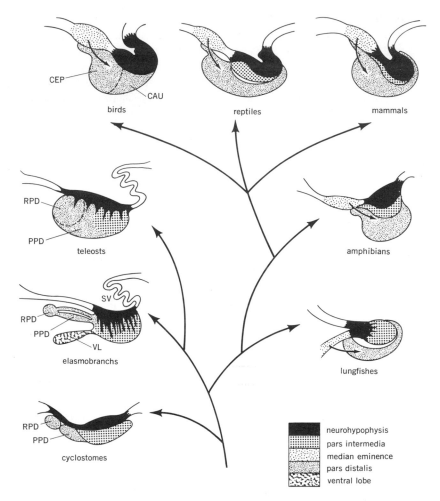

CEP

CAU

birds

reptiles

mammals

RPD

PPD

teleosts

amphibians

SV

RPD

PPD

VL

elasmobranchs

lungfishes

RPD

PPD

cyclostomes

neurohypophysis
pars intermedia
median eminence
pars distalis
ventral lobe

FIG. 14.4 Schematic diagram showing a possible evolutionary sequence for the vertebrate pituitary gland. Arrow indicates the hypophysioportal system. *CAU,* caudal division of the avian anterior lobe; *CEP,* cephalic division of avian anterior lobe; *PPD,* proximal pars distalis; *RPD,* rostral pars distalis; *VL,* ventral lobe of the elasmobranch pituitary; *SV,* saccus vasculosus. (*Adapted from C. D. Turner, "General Endocrinology," 4th ed., Saunders, Philadelphia, 1966, by permission.*)

Latimeria the pituitary gland exhibits its most primitive conditions. In *Polypterus* it connects to the mouth cavity by a persistent opening of the hypophysial cleft, or the original cavity of Rathke's pocket. The duct passes through a foramen in the parasphenoid bone. In *Latimeria,* a glandular cord about 10 cm long unites the pituitary gland with its point of origin in the roof of the mouth. Except in minor details, the pituitary glands of other vertebrates are basically similar (Fig. 14.4). Birds as well as whales, Indian elephants, and armadillos, among mammals, lack an intermediate lobe per se. Even so, their pituitary glands secrete intermedin, which seems to be derived, for the most part, from that portion of the anterior lobe adjacent to the posterior lobe.

THYROID GLAND

Perhaps the most familiar gland of the endocrine group is the thyroid. In man it is located in the lower part of the neck region, lying ventral and lateral to the trachea just posterior to the larynx. Ordinarily the thyroid gland is inconspicuous, but when enlarged, as in certain diseases, it may protrude prominently. An enlarged thyroid gland is commonly known as a goiter, of which there are several types.

The thyroid gland is almost as typical a vertebrate organ as the notochord. Its histological structure is similar in all vertebrates. The gland consists of numbers of rounded *thyroid follicles* of various sizes, separated from one another by delicate strands of connective tissue (Fig. 14.5). Each follicle is normally lined with a single layer of cuboidal epithelial cells enclosing a mass of viscid material known as *colloid*. The colloid represents the stored-up product of the secretory epithelial cells, the amount present fluctuating with changes in physiological activity of the gland. Small clumps of cells are sometimes observed lying between the cells of a follicle and its basement membrane. These *parafollicular cells* have a clear cytoplasm and are larger than the follicular cells proper. Their function, if any, is unknown.

Sufficient amounts of iodine must be available if the gland is to function normally. The colloid is composed largely of an iodized glycoprotein called *thyroglobulin*. In this form the thyroid hormones, which are amino acids, are stored. Under normal physiological conditions two of the iodinated amino acids, *thyroxine* and *triiodothyronine,* formed during hydrolysis of thyroglobulin, pass from the gland into the bloodstream. These are the hormones of the thyroid gland. Thyroxine contains approximately 65 percent iodine. Triiodothyronine has less iodine but is about

FIG. 14.5 Section through thyroid glands of *A*, salamander, and *B*, rat, both at the same magnification.

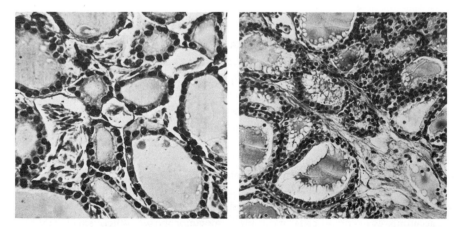

A   B

7 times more potent than thyroxine. Moreover, its action takes place much faster. The exact form of the hormone which affects the tissues of the body is uncertain. *Triiodothyropropionic acid* has been shown to be 300 times as effective as thyroxine in bringing about certain reactions. The thyroid hormones control the rate of metabolism of all tissues in the body and plays an important role in the development of the nervous system and reproductive function as well as the general growth phenomenon. Its most spectacular role is its control of metamorphosis in amphibians. Certain antithyroid or goitrogenic drugs (thiourea, thiouracil, and sulfonamides) prevent the cells of the thyroid gland from synthesizing its hormones.

PROTOCHORDATES No thyroid gland is present in the protochordates. The glandular ciliated groove in the ventral wall of the pharynx of cephalochordates and urochordates, known as the *endostyle,* has been homologized by some comparative anatomists with the thyroid gland of vertebrates. The endostyle is clearly a structure which aids in trapping food particles. Of additional significance is the fact that when cephalochordates are immersed in seawater containing radioactive iodine, bound in organic form, tends to concentrate in the endostyle. This is not true in urochordates. Hemichordates lack all traces of an endostyle.

CYCLOSTOMES A *subpharyngeal gland,* sometimes referred to as an endostyle, is present in the ammocoetes larva of the lamprey. Its function is uncertain. Five types of cells have been described in the gland. At the time of metamorphosis certain of these cells become modified, forming a typical thyroid gland. Experiments in which radioactive iodine has been used as a tracer substance show that the iodine accumulates in certain cells of the ammocoetes subpharyngeal gland even

before they become organized as thyroid follicles. These findings indicate that only a portion of the ammocoetes subpharyngeal gland, not the entire organ, may be the homolog of the thyroid gland. The fact that the thyroid first appears in the embryos of higher forms as a median, ventral diverticulum of the pharynx is also of interest in this connection. The thyroid follicles of adult lampreys show a tendency to be distributed along the ventral aorta and the arteries leading to the gills. A similar arrangement is found in the hagfish.

FISHES In other fishes a thyroid gland appears early in embryonic development. Its location in the adult shows much variation. In elasmobranchs it is a single, compact organ located just anterior to the point where the ventral aorta bifurcates into the most anterior afferent branchial arteries.

In teleosts it is usually paired and lies near the anterior visceral arches on either side. In some, e.g., the perch, it is diffuse, consisting of small masses which lie under the ventral aorta and bases of the afferent branchial arteries.

The thyroid gland of dipnoans is made up of two prominent lateral lobes connected by a constricted central portion. It lies below the tongue epithelium, just above the symphysis of the hyoid apparatus.

AMPHIBIANS In urodeles a thyroid gland appears very early in a developing embryo as an unpaired structure which soon divides into two. In the adult the glands lie on either side of the throat region along the course of the external jugular vein (Fig. 14.6). In anurans the gland is also paired and appears on each side as an oval-shaped structure located lateral to the hyoid apparatus. The presence of a functioning thyroid gland in amphibians is normally necessary for complete metamorphosis. When small tadpoles are fed with thyroid substance, they stop

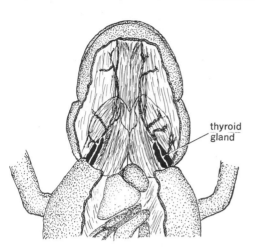

FIG. 14.6 Ventral view of lower jaw of salamander *Ambystoma texanum*. The skin has been removed to show the position of the thyroid glands.

growing and undergo a spectacular early metamorphosis, quickly being transformed into tiny frogs or toads. This reaction is one of the most sensitive tests for the presence of thyroid hormone. There is marked variation in the response of different amphibian species to the thyroid hormones. The perennibranchiate *Necturus* seems to be incapable of metamorphosing under any conditions.

REPTILES In snakes, turtles, and crocodilians the thyroid gland is unpaired, whereas in lizards it is bilobed in the young and may be paired in the adult. In lizards it lies ventral to the trachea, approximately halfway along its course. In other reptiles it lies farther posteriorly and immediately in front of the pericardium. In reptiles, as in amphibians, ecdysis is partly under control of the thyroid hormone.

BIRDS The paired thyroid glands of birds possess no noteworthy features not previously considered. They lie on either side near the region where the trachea divides into bronchi.

MAMMALS The typical mammalian thyroid consists of right and left lobes connected across the ventral side of the trachea by a narrow *isthmus* (Fig. 14.7). The lateral lobes are firmly attached to the larynx, and the isthmus is fastened to the trachea. The isthmus is the only part of the normal thyroid that is ordinarily palpable in man. If the finger is gently pressed against the trachea at the level of the suprasternal notch and the movement of swallowing is executed, the isthmus of the thyroid will be felt to move up and back again.

The thyroid gland arises in the mammalian embryo as a median ventral diverticulum of the pharynx between the first and second pharyngeal pouches. Thus for a time it is connected to the pharynx by a duct known as the *thyroglossal duct*. Under normal conditions this connection is soon lost, and the only indication of its former point of origin is a small

FIG. 14.7 Ventral view of human thyroid gland shown in relation to trachea and larynx.

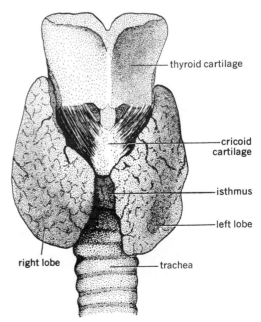

depression at the base of the tongue, the *foramen cecum* (Fig. 9.8). Occasionally small remnants of thyroid tissue remain embedded in the substance of the tongue. If these should undergo hypertrophy, a *lingual goiter* is the result.

The relationship of the thyroid gland to the other glands of the endocrine system is not entirely clear. A reciprocal action between the production of the thyroid hormones and the thyrotrophic, or thyroid-stimulating hormone (TSH), of the anterior lobe of the pituitary gland is evident. Relationships between thyroid secretion and the production of the pituitary somatotrophic and adrenocorticotrophic hormones (STH and ACTH) are also indicated.

PARATHYROID GLANDS

The parathyroid glands get their name from the fact that in mammals they lie alongside of or dorsal to the thyroid gland. Frequently they are wholly or partially em-

FIG. 14.8 Section through parathyroid and thyroid glands of rat, showing parathyroid partially embedded in thyroid tissue.

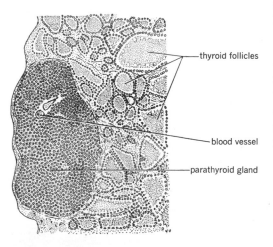

thyroid follicles

blood vessel

parathyroid gland

bedded in the thyroid or thymus glands (Fig. 14.8). In the early stages of development in all vertebrates small masses of cells proliferate from both dorsal and ventral ends of the pharyngeal pouches. The number varies in different vertebrates and vertebrate groups. It is clear that in the higher tetrapods certain of these endodermal epithelial masses give rise to the parathyroid glands.

Two types of cells have been described: *principal,* or *chief, cells* and *colloid,* or *oxyphile, cells.* It is generally believed that the chief cells are concerned with secretion of the parathyroid hormone, *parathormone* (PTH).

The glands are essential for life unless, in their absence, remedial measures are instigated. If they are removed or are badly diseased, the subject suffers from severe spasms of the muscles, a condition known as *tetany.* Death usually follows in a few days. Their function seems to be to regulate the amount of calcium and phosphorus in the blood, acting primarily on the skeleton, which contains most of the calcium and phosphorus of the body. This is done through liberation or deposition of these substances, as the case may be. The importance of the parathyroids in bone and tooth formation should be evident. It is also probable that parathormone stimulates the excretion of phosphorus by the kidneys.

CYCLOSTOMES AND OTHER FISHES Because of their small size and diffuse nature and the difficulty in extirpating these structures completely, it has not been conclusively determined whether any of the masses in cyclostomes and fishes actually represent parathyroid tissue.

AMPHIBIANS The parathyroids are first identifiable as such in amphibians. They, like the thymus glands, are usually derived from the third and fourth pairs of pharyngeal pouches. In anurans, however, the thymus glands come almost entirely from the second pouches. Whereas the thymus glands arise as dorsal buds, the parathyroids come from the ventral portions of the pouches. In urodeles they lie lateral to the aortic arches and ventral to the thymus glands. One, two, or three parathyroids may appear on each side. They are rather widely separated from the thyroids in urodele amphibians. The parathyroids of anurans are small, rounded, reddish bodies, two of which lie on either side of the posterior portion of the hyoid cartilage next to the inner ventral surface of the external jugular vein.

Apparently, the parathyroids in amphibians function in a manner similar to those of mammals, controlling the level of calcium and phosphorus salts in the blood. Seasonal changes in these glands have been described in some species of frog.

REPTILES The parathyroids of reptiles, like those of amphibians, arise from the ventral ends of pharyngeal pouches III and IV. They are frequently confused with other bodies of pharyngeal origin. There are usually two pairs present on the sides of the neck located somewhat posterior and lateral to the thyroids. In snakes the glands are situated nearer the skull than in other members of the class.

BIRDS In birds the parathyroid glands are also derived from ventral diverticula of pharyngeal pouches III and IV. In adults they appear as small, paired bodies, slightly posterior to the thyroid glands on either side. There are one or two pairs. Sometimes they lie on the dorsal surface of the thyroids.

MAMMALS In mammals the parathyroid glands arise from the *dorsal* ends of pharyngeal pouches III and IV. In man there are usually two on each side, although additional glands are sometimes found. They are roughly oval and of a lighter color than the thyroid gland, with which they are closely associated anatomically if not functionally. The names *superior* and *inferior parathyroids* are generally applied to them, the superior pair being more cranial in position. In some species (rat, mouse, pig, etc.) only a single pair is present.

ADRENAL GLANDS

The adrenal glands derive their name from the fact that in man they are situated in close proximity to the cranial borders of the kidneys. In mammals they are compact bodies made up of an inner *medulla* of ectodermal origin derived, along with the sympathetic ganglia, from neural-crest cells, and an outer *cortex* derived from mesoderm. The cortex arises from mesenchymatous tissue between the dorsal mesentery of the gut and the medial surface of the mesonephric kidney. The genital ridges, or gonad primordia, also make their appearance in this same general region. No functional relationship between cortex and medulla has been demonstrated unless both are concerned with stress reactions (page 424).

Cytoplasmic granules in the cells of the medulla stain a brownish yellow with certain salts of chromic acid. This is called the *chromaffin reaction* and is correlated with the presence of the hormone *epinephrine,* or *adrenalin.* Chromaffin tissue has also been found in such other parts of the body as the prostate gland, seminal vesicles, cervix of the uterus, and carotid body. Here the tissue is closely related to that of the sympathetic nervous system.

In the lower classes of vertebrates the homolog of medulla and cortex are usually completely divorced from each other. The parts which are homologous with the medulla of higher forms are called *chromaffin bodies.* This name is generally misleading, however, and the term *steroidogenic tissue* is now recognized as being more appropriate. Certain steroid hormones are secreted by this tissue.

Epinephrine, or *adrenalin,* from the adrenal medulla, was the first of all hormones to be discovered. It can be synthesized and is one of the most widely used drugs in the practice of medicine. Even minute quantities of epinephrine, when injected, cause marked rise in blood pressure, more forceful beat of the heart, elevation of blood-sugar level, dilatation of air passages, blanching of the skin, and other reactions.

What, in the past, have seemed to be pure extracts of the adrenal medulla are now known to contain two different, but closely related, substances, epinephrine and *norepinephrine.* They are catecholamines. The proportions of the two are relatively constant for each species. The norepinephrine content seems to be higher in species having a naturally aggressive temperament (cats, lions, tigers). In fishes, amphibians, and reptiles capable of changing color so as to blend in with the environment (metachrosis), epinephrine is believed to be responsible for the concentration of pigment granules around the nucleus of the integumentary chromatophores.

The medullary portion of the gland is innervated by preganglionic sympathetic fibers. It is generally held that the cells of the medulla are modified postganglionic cells of the sympathetic nervous system which are secretory in nature.

The proper functioning of the adrenal cortex (steroidogenic tissue) is necessary for life unless therapeutic measures are applied. Almost 50 different steroids, closely allied chemically to ovarian, testicular, and certain placental hormones, have been isolated from the adrenal cortex. They are referred to as *adrenocorticoids.* Among them are *cortisone, cortisol (compound F), corticosterone,* and others. These are known as *glucocorticoids* and are important in carbohydrate and protein metabolism. Others (11-*deoxycorticosterone* (DOC) and 11-*deoxycortisol,* or *compound S)* affect the sodium and potassium content of the blood and control electrolyte and fluid shifts in the body. One adrenocorticoid, called *aldosterone,* is more effective than others in electrolyte and water metabolism. Still another group is composed of adrenocorticoids having androgenic, estrogenic, or progesteronelike properties similar to those of hormones elaborated by the gonads. Among them is *adrenosterone.* A reciprocal action exists between certain adrenocorticoids and ACTH secreted by the anterior lobe of the pituitary gland. The action is undoubtedly mediated via the hypothalamus.

CYCLOSTOMES Adrenal elements are found for the first time in cyclostomes. In the lamprey these are two clearly distinguished series of bodies. The steroidogenic, or cortical, series consists of small, irregular structures situated along the postcardinal veins, renal arteries, and the arteries lying above the opisthonephric kidneys. The chromaffin, or medullary, series consists of small bits of tissue extending along the course of the dorsal aorta and its branches. Chromaffin tissue has been found in the hagfish, but no steroidogenic tissue has been described.

FISHES In some fishes steroidogenic bodies lie between the posterior ends of the opisthonephric kidneys and are generally paired. In others they are located within or anterior to the kidneys. Paired chromaffin bodies are located on the segmental branches of the dorsal aorta or in the walls of the postcardinal veins, in close relationship to the sympathetic ganglia.

AMPHIBIANS The two components of the adrenal glands first become closely associated in anuran amphibians. The familiar orange or yellow bands on the ventral surfaces of the opisthonephric kidneys of the frog are composed of the two types of tissue closely intermixed. In urodeles, however, the adrenal glands are not so compactly arranged. The two components are intermingled and located in strips of tissue which extend the entire length of the opisthonephros and even anterior to it.

REPTILES In most reptiles steroidogenic and chromaffin constituents are intricately intermixed. In crocodilians and chelonians the arrangement is much like that of birds.

BIRDS The adrenal glands of birds are yellowish bodies lying on either side of the postcaval vein just anterior to the kidneys and close to the gonads. The two components are closely interwoven.

MAMMALS In mammals the adrenals lie at the cranial ends of the kidneys. Only in this class is there a true cortex and a medulla. The latter occupies the central portion of the gland. Accessory "cortical" and "medullary" bodies are frequently found in the vicinity of the adrenal glands. In some species, e.g., the rat, the adrenals of the female are considerably larger than those of the male, increase in cortical tissue being responsible for the discrepancy in size. In the nutria, a rodent, the left gland is 50 percent larger than the right. In monotremes, cortex and medulla are not so distinctly separated as in other mammals. When viewed under the microscope the typical adrenal cortex is found to consist of three or more distinct layers, or zones (Fig. 14.9).

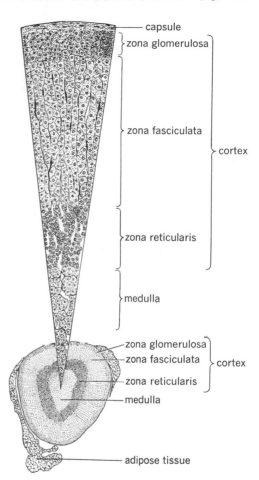

FIG. 14.9 Segment of adrenal gland of rat, showing zonation. The gland is shown below in cross section. The wedge above indicates the microscopic appearance of the medulla and cortical layers.

PANCREATIC ISLANDS

The dual role of the pancreas as both a ducted and ductless gland has been referred to. This organ is made up of two distinct kinds of tissue, which, although anatomically associated, are functionally separated. The

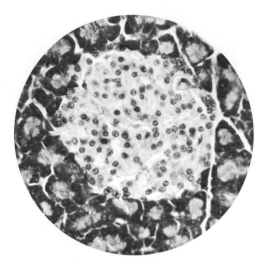

FIG. 14.10 Photograph of a section through the pancreas of a rat, showing an island of Langerhans surrounded by alveolar tissue.

comparative anatomy of the pancreas as a ducted gland was discussed in Chap. 9, Digestive System. The endocrine portion consists of isolated masses of cells, known as the islands (islets) of Langerhans (Fig. 14.10).

Two hormones are formed in the pancreas: *insulin,* secreted by the *beta cells* of the islets; and *glucagon,* believed to be elaborated by *alpha cells.* Insulin has the effect of lowering the blood-sugar level, whereas glucagon is a blood-sugar-raising principle. The discovery of insulin in 1921 has been of paramount importance in prolonging the lives of innumerable persons suffering from *diabetes mellitus,* in which the islets of Langerhans do not produce insulin. The chemical synthesis of insulin was first announced in 1963.

There is some evidence that the islets may be subject to some degree of nervous control via parasympathetic fibers located in the right vagus nerve. The existence of *neuromuscular complexes* has been demonstrated.

CYCLOSTOMES No discrete pancreas is present in the lamprey. A few small cellular

masses, buried in the substance of the liver and wall of the intestine, possibly represent endocrine pancreatic tissue. Islet tissue has been described in the hagfish in the area where the bile duct joins the intestine.

FISHES In most fishes, teleosts in particular, the pancreas is rather diffuse. Islands of Langerhans are relatively few but large. In many teleosts there is an encapsulated island of Langerhans called the *principal island,* of unusually large size and constant occurrence.

OTHER CLASSES In amphibians, reptiles, birds, and mammals the islands of Langerhans are scattered throughout the pancreas and no unusual features are to be noted. In man there are from 200,000 to 1.8 million islets, constituting between 1 and 2 percent of the total pancreatic tissue.

UROHYPOPHYSIS

Numerous fishes possess a peculiar structure, called the *urohypophysis* (urophysis), associated with the posterior end of the spinal cord. Neurosecretory cells within the spinal cord send axons posteriorly which end with enormous bulbs and give the cord a bulge. It is best developed in teleosts. It is generally agreed that the urophysis is analogous to the neurohypophysial division of the pituitary gland. Some eight neurosecretions have been identified biochemically, but the physiological effects are not clear. Data suggest roles in hydromineral balance, reproduction, osmoregulation, and smooth-muscle contraction in many organs. It may possibly bear some relation to the swim bladder in regulating buoyancy. Elasmobranchs have large neurosecretory cells in the posterior end of the spinal cord even though a urohypophysis is lacking.

TESTES

The male gonads, besides producing spermatozoa, are endocrine organs. To the comparative anatomist the importance of the endocrine secretion, or hormone, of the testes lies in the fact that the *secondary sex characters* of the male, which in many cases distinguish him markedly from the female, are dependent upon it. Furthermore, the normal development, growth, and functioning of the *accessory sex organs* (page 294) are also partly under control of the hormone *testosterone.*

The nature of the tissue which actually produces the endocrine secretion of the testes in lower vertebrates is rather obscure. It is generally agreed that *interstitial cells,* lying in the spaces between seminiferous ampullae or seminiferous tubules, are responsible, even though in many forms they are irregular in appearance.

Most vertebrates are seasonal breeders, having a definite reproductive cycle and breeding but once a year. In some, sexual dimorphism is not very clear cut, but in others marked secondary sex characters are apparent. Such features are most prominent at the breeding period, after which they usually regress. In male fishes the changes take the form of altered color patterns, modifications of fins, appearance of excrescences on the skin, etc. In amphibians they consist of such modifications as swelling of the thumb pads, enlargement of the mental gland, changes in color pattern, development of a dorsal crest, and enlargement of cloacal glands. Reptiles generally do not exhibit conspicuous sexual dimorphism, but seasonal changes in skin pigmentation and certain other features have been noted in some. In birds, changes in beak color, plumage changes, etc., denote a seasonal periodicity. In all the above-mentioned vertebrates the accessory sex organs undergo pronounced seasonal change.

MAMMALS In the testes of mammals the *interstitial cells of Leydig* are a constant feature. Masses of these cells with their abundant blood supply are present in the interstices between the seminiferous tubules (Fig.

FIG. 14.11 Photograph of a section through the testis of a rat, showing cells of Leydig lying in the interstices between adjacent seminiferous tubules.

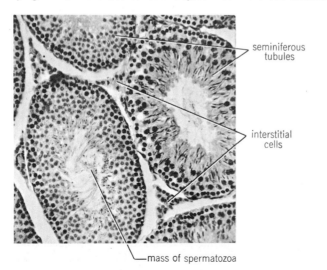

seminiferous tubules

interstitial cells

mass of spermatozoa

14.11). Interstitial tissue represents an unusual type of endocrine tissue since it is diffuse and does not arise from an epithelial surface like the others but is of mesenchymal origin. Electron-microscopic studies have shown that the cytoplasm of the interstitial cells contains a network of very fine interconnecting tubular structures as well as droplets of lipoid substance and some crystalline elements exhibiting a highly organized microstructure. Assignment of male-hormone production to the interstitial cells is well established. Even in cryptorchid individuals with undescended testes in which the seminiferous tubules are degenerate and spermatozoa do not form, the secondary sex characters and the accessory sex organs usually develop normally. In such individuals the interstitial cells appear to be normal, and are not influenced, at least to the same degree as the seminiferous tubules, by the higher body temperature. They secrete the hormone *testosterone* in a normal manner.

Secretion of testosterone by the interstitial cells is regulated by a gonadotrophic hormone from the anterior lobe of the pituitary gland (see page 423). It is known as ICSH (interstitial-cell-stimulating hormone). There is a reciprocal relationship between the production of testosterone by the testes and ICSH by the pituitary gland. Neurosecretory cells in the hypothalamus undoubtedly play an important role in this relationship.

The testes of several kinds of mammals, including man, are fairly large at birth and shrink markedly in size within a few days. The larger size is due to masses of interstitial cells which have developed in the fetus, probably in response to placental hormones. When the source of placental hormones is withdrawn at birth, the interstitial cells shrink and do not become active again until puberty.

Testosterone was first extracted from bull testes. It can now be prepared synthetically. Chemical substances having properties similar to those of testosterone are known as *androgens*.

Castration of various domestic animals has long been practiced to improve the quality of flesh or to make them more docile. A castrated bull is known as a *steer;* a castrated stallion is called a *gelding.* A *barrow* is a castrated boar; and a castrated man is a *eunuch.* If animals or human beings are castrated *before* the age of puberty, typical masculine structures and characteristics fail to develop. Castration *after* the age of puberty leads to a regression of such accessory sex organs as seminal vesicles and prostate and bulbourethral glands. Secondary sex characters such as the voice and beard in human beings are altered little, if at all. Among skeletal structures influenced by testosterone are the pelvic bones of certain species, the larynx of man, antlers of most species of deer, and the generally larger size of the male skeleton compared with that of the female. The levator ani muscle of the rat and the masticating muscles of the male guinea pig seem to be markedly affected by castration.

Another hormone, *androstenedione,* has recently been reported to be produced by the testes.

OVARIES

Like testes in the male, the ovaries of the female are both cytogenic and endocrine in nature. These two functions are mediated through tissues which have a much closer morphological relationship than the corresponding parts of the testes. Much more is known about the endocrine activity of the ovaries of mammals than about those of members of the lower vertebrate classes. The endocrine secretions of the mammalian ovary include *estrogens, progestogens, androgens,* and *relaxin.* The first three of these are steroids; the last is not.

It is still not certain which components of the mammalian ovary are responsible for the secretion of estrogen. The cells of the theca interna and possibly those of the stratum granulosum and discus proligerus seem to be the primary source of estrogen secretion. The principal estrogen elaborated by these cells is *estradiol.* All substances having the properties of estradiol are referred to as estrogens. Several chemicals which are not steroids have estrogenic properties.

The term *progestogen* (or *progestin*) is used collectively to include steroid substances which bring about progestational changes in the mammalian uterus. Whereas estrogens are effective in inducing *growth* of certain tissues, progestogens bring about their differentiation. *Progesterone,* secreted by the cells of the corpus luteum, is the chief naturally occurring progestogen. Normally it works synergistically with estrogen. A synthetic progestogen, 19-*norprogesterone,* when injected, is 5 times as effective as progesterone itself.

The source of ovarian androgens remains obscure. Some have estrogenic or progestogenic effects on the female but in massive doses may stimulate portions of the male genital tract.

The exact site of secretion of the nonsteroid hormone, *relaxin,* is not known. Its potencies have been demonstrated in a number of mammals (see page 438).

Ovarian estrogen is primarily responsible for the development of the secondary sex characters of the female. The development and maintenance of the organs making up the female reproductive system are at least partially under control of this hormone. In seasonal breeders the waxing and waning of secondary sex characters and accessory sex organs in the female parallel those of the male. The changes accompany the gradual appearance and disappearance of the female sex hormone.

Although in most fishes following ovulation the emptied follicles shrink and are resorbed, it seems that in certain oviparous and ovoviviparous elasmobranchs and teleosts, corpora lutea form from nurse cells of the ova at the sites of follicular rupture. Their presence seems to be in some way related to extended retention of eggs in the oviducts in oviparous forms and with the retention of developing embryos in the uteri of ovoviviparous species.

Amphibians as a group are noted for their oviparity. A few salamanders and an African toad, *Nectophrynoides,* give birth to living young. In the latter, corpora lutea develop in the ovaries and the young are retained in the uterus of the mother for as long as 9 months.

In the *male* toad, a peculiar structure called *Bidder's organ* lies at the anterior end of each testis (Fig. 10.10). If the testes are removed, Bidder's organs will develop into functional ovaries, thus bringing about a complete sex reversal. The otherwise rudimentary oviducts and uteri enlarge, seemingly in response to estrogen elaborated by the transformed Bidder's organs.

In certain ovoviviparous snakes, well-developed corpora lutea form in the ovaries following ovulation. All lizards possess corpora lutea in their ovaries during pregnancy. The function of the corpora lutea is not clear, but it seems to be of less importance than in mammals.

In most birds only the left ovary develops into a functional adult organ. If the left ovary is removed, the accessory sex organs regress and the otherwise rudimentary right gonad becomes modified into a testislike organ. Although spermatozoa seldom appear in such gonads, they apparently secrete androgen, which causes the appearance of masculine characters.

Progesterone is present in the blood of

laying hens, but its source is unknown. After ovulation in the domestic fowl, the collapsed follicle (calyx) is retained for a time and has an effect upon the time of the subsequent ovulation. If the calyx is removed surgically, the egg which came from it is retained in the uterus, or shell gland, for an abnormally long time. Progesterone may possibly be secreted by the follicle, even after ovulation when only the calyx remains.

MAMMALS Estrogen has been shown to be at least partially responsible for the development of the secondary sex characters and accessory sex organs of the female. Progesterone and, in certain cases, relaxin are concerned only with changes occurring in preparation for and during pregnancy.

Functional corpora lutea must be present during the early stages of pregnancy in all mammals and, in most species, through the entire gestation period. Removal of the ovaries or of the corpora lutea alone results in the resorption or abortion of the developing young. In some, however, as in the human being, mare, and guinea pig, removal of the ovaries or corpora lutea during the latter part of pregnancy has no such effect. In these animals and probably certain others, progesterone secreted by the placenta may take over, supplementing that from the ovaries. Since progesterone is necessary for bringing about certain essential changes in the lining of the uterus, for the early development of the placenta, and for complete development of the mammary glands, its importance can be appreciated. A previous sensitization of the uterus and mammary glands by estrogen is necessary before progesterone can become fully effective.

The hormone *relaxin* has been extracted from the corpora lutea of sows, from the blood serum of pregnant rabbits, dogs, cows, and other mammals, and from rabbit pla-

centas. It has not been identified chemically. It is effective in relaxing the public ligaments of female guinea pigs which have been previously sensitized with estrogen. It is also effective to a slight extent in relaxation of human pelvic ligaments during the latter part of pregnancy.

The cyclic activity of the reproductive organs in all species of mammals, including the phenomenon of menstruation in the human being and many other primates, depends upon the periodic fluctuations in the levels of estrogen and progesterone secreted by the ovaries. The periodicity itself depends upon a reciprocal action of these hormones and the gonadotrophic hormones of the anterior lobe of the pituitary gland.

At least three gonadotrophic pituitary hormones affecting the ovaries are recognized: FSH (*follicle-stimulating hormone*), LH (*luteinizing hormone*), and LTH (*luteotrophic hormone*). LH in the male is referred to as ICSH (page 436). LTH is also sometimes called *prolactin* or the *lactogenic hormone* since it has an effect upon the secretion of milk by properly prepared mammary tissue.

FSH, as the name implies, brings about growth and development of the ovarian follicles and secretion of estrogen by the follicular cells. LH causes the preovulatory swelling of the follicles and augments the secretion of estrogen. In mammals with restricted mating periods, it is at this time that the female is receptive to the male. The ripened follicles then rupture, ovulation takes place, and corpora lutea begin to develop. LTH is now released by the pituitary gland. It stimulates the corpora lutea to secrete progesterone. Unless pregnancy occurs, secretion of progesterone soon diminishes. Estrogen and progesterone both have an inhibiting effect upon the pituitary gland via the hypothalamus. When their inhibiting influence is withdrawn, FSH and LH production and release begin once more and the whole cycle is repeated.

During pregnancy all cyclic manifestations of the reproductive system cease. The placenta in certain mammalian species is the source of estrogenic and progestogenic hormones responsible for inhibition of ovarian activity via the hypothalamus and the pituitary gland.

In recent years a reliable and inexpensive method of preventing conception in women has been devised which involves only the oral administration of a pill at stipulated times. The pill consists of various chemical forms of estrogen and progesterone in varying concentrations. One pill is taken orally each day for 20 successive days, beginning with the fifth day of the menstrual cycle. After the twentieth daily dose, administration is discontinued. Menstruation generally occurs a few days later. Then on the fifth day a new regimen is begun. There may be unfortunate side effects. Recent evidence indicates that the use of the pill may cause an increasing chance of blood clotting. Other symptoms such as nausea, swelling of the breasts or other body parts, headaches, allergic rashes, depression, and irritability have been reported. Such pills are available only by a doctor's prescription and are not to be used indefinitely. The principle involved in this method of contraception is suppression of release of gonadotrophic hormones by the pituitary gland. The same material is of value in treating some forms of infertility. When treatment is stopped and the pituitary gland is no longer inhibited, the ovaries actually seem to become more active than they were formerly.

In all human females at about middle age the ovaries cease to function both as cytogenic and as endocrine organs. The time at which this occurs is variously known as the *menopause, change of life,* or *climacteric.* In human females it generally occurs around the forty-fifth year. In other mammals it is proportionately correlated with the life span. Changes in certain secondary sex characters occur at the menopause, and there is an accompanying atrophy of the genital organs and a cessation of periodic manifestations.

Extirpation of the ovaries at any time prior to the menopause will bring about a premature climacteric, in which all the symptoms of the natural phenomenon are manifested.

PLACENTA

The *placenta* is an important endocrine organ in addition to its other functions. It is a transitory structure, developing within the uterus early in pregnancy and being expelled at the time of parturition. Hormones are known to be elaborated by the placentas of various mammals, but much more systematic study must be made before all components responsible for their secretions are definitely known. Mammalian species vary greatly as to types of placentas, kinds of placental hormones, and times when they are elaborated. Placental hormones are, in general, similar to ovarian hormones and pituitary gonadotrophins. The endocrine role of the placenta, then, is highly specialized, rather than important in the general body economy.

Several types of placentas are to be found among mammals. Whether the more primitive types, such as the yolk-sac and allantoic placentas of metatherians, produce endocrine secretions, has not been established. Even in some of the higher mammals the endocrine function of the placenta remains obscure.

In a number of mammals the placenta is the source of extraovarian estrogen produced during pregnancy. Progesterone is also secreted by the placenta in a number of species. In addition, the placenta is the source of gonadotrophic hormones found in the blood and urine of certain mammals during pregnancy. Placental gonadotrophins are chemically and physiologically different from those produced by the pituitary gland. In man and other primates the term *chorionic gonadotrophin* is used to denote the chorionic hormone elaborated by the placenta. It has properties similar to those of both LH and prolactin (LTH) of the pituitary gland. The presence of excreted chorionic gonadotrophin in human urine is used as a basis for several pregnancy tests, including the Friedman, Aschheim-Zondek, and Galli-Mainini tests, as well as some others. Another gonadotrophic hormone, present in the blood serum of pregnant mares from the fortieth to the one-hundred-eightieth days of pregnancy, is referred to as *equine gonadotrophin* or *PMS* (*pregnant mare serum*). It has properties of a mixture of FSH with a small amount of LH. This hormone actually seems to be uterine, rather than placental, in origin, being formed in the endometrial cups of the pregnant uterus. PMS is probably responsible for the formation of accessory corpora lutea found in the ovaries of pregnant mares during part of the gestation period. It is noteworthy that the ovaries of the blue antelope of India and the Indian elephant also possess accessory corpora lutea during pregnancy. Gonadotrophin found in the urine of pregnancy is often used successfully in lowering the undescended testes of cryptorchid boys into their normal position in the scrotum. In some species, as in the rabbit, the placenta seems to be the source of the hormone relaxin. Certain steroids similar to adrenocorticoids and androgens apparently are elaborated by the placenta under certain conditions, but their significance in this connection is obscure.

The role played by the placenta as an endocrine organ seems to be that of an accessory ovary and pituitary gland, at least insofar as the gonadotrophic properties of the latter are concerned.

SUMMARY

1 The endocrine and nervous systems, sometimes together referred to as the neuroendocrine system, are the dominating and coordinating systems of the body. Humoral distribution of secreted chemical substances is utilized entirely by the endocrine system in bringing about its effects.

2 The endocrine organs consist of a number of ductless glands which secrete substances called hormones into the blood or lymph. They are carried by the circulating fluids to all parts of the body. The secretions are of a chemical nature and bring about certain changes which are of either a specific or a general character.

3 Each glandular component of the group has its particular and specific functions, yet there is a close interrelationship among them.

4 The endocrine glands are derived embryologically from epithelial surfaces with the exception of the interstitial cells of the testes.

5 Certain endocrine organs have a dual function and serve either as exocrine and endocrine glands (pancreas) or as cytogenic and endocrine glands (ovaries and testes).

6 It is doubtful whether the thymus gland should be included in the endocrine system. The pineal body, however, seems to belong in this category.

7   The pituitary gland plays a dominant role in the endocrine group. It is usually made up of three parts, or lobes, which either affect the body generally or stimulate other endocrine glands to activity. A reciprocal relationship exists between the pituitary and several of the other endocrine glands via the hypothalamus.

8   The thyroid gland, consisting of a number of rounded thyroid follicles, secretes two hormones, thyroxine and triiodothyronine, which control the state of metabolism of the entire body.

9   The parathyroid glands, which first make their appearance with certainty in the class Amphibia, usually lie in close proximity to the thyroids. The hormone of the parathyroids is known as parathormone. It is concerned with maintaining normal levels of calcium and phosphorus in the bloodstream.

10   The adrenal glands of mammals consist of two portions, an inner medulla of ectodermal origin and an outer, mesodermal cortex. In other forms the homologs of these regions do not show similar anatomical relationships. In the lower classes they are separate entities, called chromaffin bodies and steroidogenic bodies, respectively. In anurans, reptiles, and birds, the two types of tissues are inextricably intermingled.

The hormones of the medulla (chromaffin tissue) are called epinephrine, or adrenalin, and norepinephrine, or noradrenalin. Among other things they bring about an increase in blood pressure and amount of sugar in the blood. They dilate the air passages leading to the lungs. Hormones of the cortex or steroidogenic tissue are known as adrenocorticoids. They have to do with carbohydrate metabolism, maintenance of the volume of circulating blood, and the control of sodium balance and fluid shifts in the body. The normal functioning of this part of the adrenal complex is necessary for life.

11   The islets of Langerhans in the pancreas are believed to secrete two hormones, both affecting carbohydrate metabolism. One of these, insulin, has the effect of lowering the blood-sugar level; the other, glucagon, has the opposite effect.

12   The interstitial cells of the testes are believed to secrete a hormone called testosterone which controls the normal development and function of the accessory sex organs of the male and the masculine secondary sex characters.

13   The ovaries of all vertebrates secrete a hormone referred to as estrogen. It is responsible for the development and maintenance of the accessory sex organs and secondary sex characters of the female. In mammals and in several ovoviviparous species among the lower vertebrates, one and often two additional hormones have been recognized: progesterone and relaxin. Both hormones, at least in some species, are probably formed by the cells of the corpus luteum, although the exact origin of relaxin is still in doubt. Progesterone is concerned only with changes occurring before and during pregnancy. Relaxin, in general, brings about changes which facilitate the delivery of young. The cyclic activity of the female reproductive system is controlled by the ovarian hormones and their interaction with the anterior lobe of the pituitary gland, direct or indirect.

14   The placenta, in addition to its other functions, secretes hormones. Its role is that of an extremely specialized organ concerned only with certain phenomena associated

with pregnancy. The hormones which it elaborates are similar to certain pituitary and ovarian hormones.

15 With minor variations the several endocrine organs are similar in all classes of vertebrates. Very little progressive evolution is to be observed in the endocrine organs. The appearance of functional corpora lutea in mammals, ovoviviparous elasmobranchs, lizards, and snakes, and their relation to placental nourishment of intrauterine young, is perhaps the most significant development.

BIBLIOGRAPHY

Barrington, R. J. W.: "Hormones and Evolution," English Universities Press, London, 1964.

―――― and C. B. Jorgensen (eds.): "Perspectives in Endocrinology," Academic, New York, 1968.

Gorbman, A., and H. A. Bern: "A Textbook of Comparative Endocrinology," Wiley, New York, 1962.

Turner, C. D., and J. T. Bagnara: "General Endocrinology," 5th ed., Saunders, Philadelphia, 1972.

15 INTEGRATING SYSTEM: SENSE ORGANS

Associated with the nervous system are certain sensory receptor organs capable of responding to various stimuli by setting up impulses which are in turn transmitted by nerve fibers to the central nervous system. In the higher centers of the brain such impulses are interpreted as sensations. The diencephalon (thalamus) serves as such a center in most vertebrates, but a well-developed cerebral cortex, if present, may be the seat of further sensory integration. The receptor organs themselves do not perceive anything but merely serve as a means of access to the nervous system. With rare exceptions, the actual receptor cells, no matter how complex the sense organ, are derived from ectoderm. This is what one would expect, since the ectoderm is the primitive outer layer of the body and most likely to be affected by environmental change.

Similarity in structure of the sense organs of man and other vertebrates leads us to assume that similar functions, and even sensations, are experienced by all. Most sense organs are stimulated by environmental disturbances of various kinds. They are spoken of as *external sense organs,* and their receptors are called *exteroceptors. Internal sense organs,* on the other hand, are affected by stimuli originating within the body itself. Certain internal sensory receptors, known as *proprioceptors,* are located in tendons, joints, skeletal muscles, heart, and other areas. The structures in which they lie are not in direct contact with the environment but may nevertheless be affected by factors of environmental origin.

TERMINOLOGY The different kinds of receptors are named according to the nature of the stimulus by which they are affected. Table 7 includes the most common.

Each sense organ functions only within circumscribed limits, which vary in different species of animals. Sounds of high frequency, for example, which can be detected by some mammals, may be inaudible to man.

In some cases the application of a different kind of stimulus to an exteroceptor will affect it in a manner similar to that produced by the usual type of stimulus. Thus, mechanical or electrical stimulation of the retina of the eye,

TABLE 7

External senses

Sight	Exteroceptors
Hearing	Photoreceptors
Smell	Phonoreceptors
Taste	Olfactoreceptors
Touch (transient contact)	Gustatoreceptors
Pressure (sustained contact)	Tangoreceptors
Temperature	Thermoreceptors
Heat	Caloreceptors
Cold	Frigidoreceptors
Pain	Algesireceptors
Currents of water	Rheoreceptors

Internal senses

Muscle position	Proprioceptors

even if performed in total darkness, gives a sensation of light. When menthol is applied to the skin, it produces an impression of cold.

Sensory nerve fibers which convey impulses from receptor organs to the central nervous system may be grouped in two categories, *somatic sensory* and *visceral sensory*. Somatic sensory fibers are those associated with exteroceptors, proprioceptors, and statoreceptors. Visceral sensory fibers, from the digestive organs, are associated with interoceptors.

SENSE OF SIGHT

Eyes are complicated photoreceptors, but not all photoreceptors are eyes. The eyes of vertebrates are highly specialized structures which have no true homologs in members of other phyla. They appear for the first time in cyclostomes, their evolutionary history being obscure. The unpaired median eyelike structures of certain vertebrates have been referred to previously (see page 394).

THE VERTEBRATE EYE

As an introduction to the discussion of the vertebrate eye we shall first consider that of man, since its structure and function are better

FIG. 15.1 Diagram of sagittal section of human eyeball.

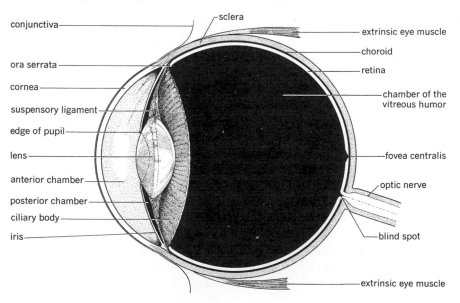

understood than are those of other vertebrates. The human eyeball (Fig. 15.1) is almost spherical in shape. It is composed of three coats, of which only the outer one is complete. The others are modified near the exposed portion of the eyeball.

The outermost coat protects and gives shape to the eyeball. It is called the *fibrous tunic.* This thick, tough layer is divided into two regions: a transparent, exposed portion, the *cornea,* and a larger opaque, posterior section, the *sclera,* the anterior part of which is commonly called the white of the eye. The muscles which move the eyeball are attached to the sclera. A thin membrane, the *conjunctiva,* continuous with the inner surface of the eyelids, covers and is fused to the outer surface of the cornea, constituting the *epithelium* of the latter. The cornea is practically circular in outline. It is somewhat thicker than the sclera and bulges out in front of it. The curved surface of the cornea is of importance in focusing light rays, supplementing the lens in this respect. Irregularities in shape of the cornea are often responsible for *astigmatism,* an optical defect causing imperfect images or indistinctness of vision.

The middle coat is composed of three regions: (1) a vascular, pigmented *choroid coat,* closely adherent to the sclera; (2) a portion which forms part of the thickened *ciliary body,* located anteriorly near the region where sclera and cornea join; (3) a portion which forms part of the *iris,* a thin, circular disc located at the anterior end of the ciliary body where the uvea turns sharply inward and away from the fibrous tunic. An opening in the center of the iris is the *pupil.*

The innermost coat of the eyeball is the *retina.* The portion in contact with the choroid coat is thickened and is sensitive to light. The remainder of the retina is thin and nonsensory. It contributes to the ciliary body and iris and terminates at the rim of the pupil. The light-sensitive part of the retina ends abruptly at the ciliary body.

A biconvex *crystalline lens* lies immediately behind the iris. The lens has a clear, glassy appearance during life, but in preserved specimens it usually becomes opaque. Its posterior side is more convex than its anterior surface. A thin *lens capsule* invests the lens closely. From the ciliary body arise numerous radially arranged fibers which are attached to the lens capsule. They form the *suspensory ligament,* which keeps the lens in place. The shape of the lens is not fixed. *Accommodation,* or the adjustment of the eye which enables it to focus on objects at various distances, is accomplished by changing the shape of the lens (page 450).

The cavity in front of the lens is partially divided by the iris into *anterior and posterior chambers.* A clear watery fluid, the *aqueous humor,* fills these cavities, which are continuous with each other at the pupil. The large cavity in back of the lens is called the *vitreal cavity,* or *chamber of the vitreous humor.* It contains a transparent, semigelatinous substance, the *vitreous body,* or *vitreous humor.*

The optic nerve leaves the eye from behind, piercing the retina, choroid, and sclera. The point at which it emerges from the retina is the *blind spot.*

When light enters the eye, it first passes through the thin conjunctival epithelium covering the cornea, then through the cornea proper, aqueous humor, pupil, lens, and vitreous humor, finally to impinge upon photoreceptors in the retina.

EARLY DEVELOPMENT The various parts of the eye arise from three rudiments: (1) the retina and optic nerve are derived from the diencephalon; (2) the lens differentiates from the superficial ectoderm of the head; (3) the other portions come from neighboring mesenchyme.

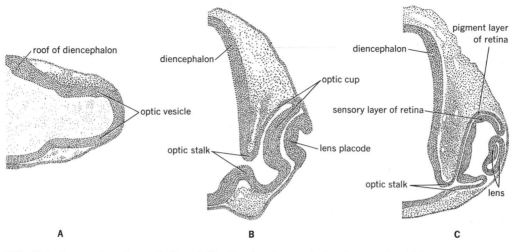

FIG. 15.2 Successive stages *A, B,* and *C* in the development of optic cups in chick embryo: *A*, 33-hour embryo; *B*, 48-hour embryo; *C*, 72-hour embryo. (*Drawn by R. Speigle.*)

The first indication of eye formation (Fig. 15.2) is the appearance of a pair of lateral expansions, the *optic vesicles,* arising from the forebrain in the region that is later to become the diencephalon. The distal end of each optic vesicle soon enlarges and comes in contact with the superficial ectoderm of the head, but the proximal portion remains relatively unchanged. It is referred to as the *optic stalk.* The distal portion of the swollen vesicle next becomes indented in such a manner as to form a temporarily imperfect (page 447) double-walled *optic cup*. Pigment granules appear in the cells forming the outer layer of the optic cup, which now is called the *pigment layer of the retina*. The inner layer gives rise to the *sensory,* or *nervous, portion of the retina.* The two layers later fuse.

A thickening of the superficial ectoderm occurs at a point where the optic cup underlies it. This is the *lens placode,* which gives rise to the crystalline lens. It invaginates at the same time that the optic cup is forming

and soon becomes a closed vesicle, which comes to lie within the optic cup (Fig. 15.3).

The indentation which forms the optic cup is continued along the ventral side of the cup, as well as along the ventral border of the optic stalk. This is the *choroid fissure,* which appears as a gap extending from the optic cup toward the brain. For a time the choroid fissure remains open, furnishing a pathway for blood vessels which are important for normal lens development (Fig. 15.4). Before long the edges of the choroid fissure fuse, thus forming a small and narrow tube within the already tubular optic stalk. The vessels within the optic stalk are retained as the *central artery* and *vein* of the retina. The axons of nerve cells which develop in the sensory layer of the retina later converge toward the optic stalk and grow through the wall of the optic stalk to the brain, thus forming the *optic nerve* (II). The central artery and vein actually course *through* the optic nerve.

The inner layer of the retina near the rim

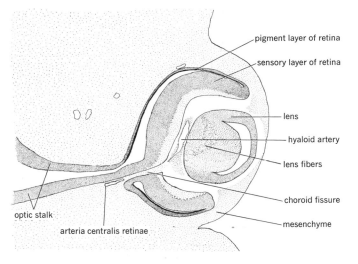

pigment layer of retina

sensory layer of retina

lens

hyaloid artery

lens fibers

choroid fissure

mesenchyme

optic stalk

arteria centralis retinae

FIG. 15.3 Sagittal section through optic cup of 12-mm pig embryo.

of the optic cup remains thin, whereas the remainder thickens as its walls become differentiated to form the nervous, or sensory, portion. The irregular line which marks the boundary of the two areas is the *ora serrata*. The thin portion of the inner layer fuses with the pigment layer overlying it, and these two, together with a part of the uveal coat, form the ciliary body and iris.

The mesenchyme surrounding the original optic cup differentiates into the uveal and fibrous coats of the eyeball. The cornea is derived from mesenchyme which grows in between the optic cup and superficial ectoderm, the latter forming the epithelial layer of the cornea.

FIBROUS TUNIC The tough, inelastic,

FIG. 15.4 Developing optic cup, showing choroid fissure through which the central artery of the retina courses. (*Drawn by R. Speigle.*)

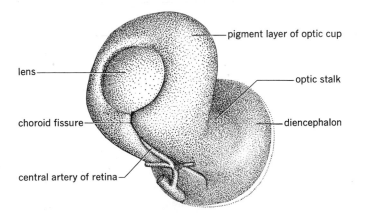

pigment layer of optic cup

lens

optic stalk

choroid fissure

diencephalon

central artery of retina

connective-tissue fibers of which the sclera is composed keep the eyeball constant in shape and resistant to external or internal pressures. In many ray-finned fishes, certain reptiles, and in birds a ring of *scleral ossicles,* or *sclerotic bones,* is embedded in the portion of the sclera near the cornea. One noteworthy feature of the cornea is the rapidity with which it is repaired after injury. Sensory receptors for pain are numerous in the corneal epithelium, but other sensory receptors seem to be lacking.

Choroid Coat The portion of the uvea adherent to the sclera consists of a thin, soft, dark brown vascular membrane. The pigment serves to keep light from entering the eyeball at random and also prevents internal reflections.

Ciliary Body At the place where the choroid portion of the uveal membrane becomes thicker, less vascular, and less heavily pigmented, numerous involuntary ciliary-muscle fibers are found. The inner face of the ciliary body is characterized by numerous radiating folds, the *ciliary processes.* They are formed partly from the uvea and partly from the nonnervous layer of the retina lying beyond the ora serrata.

Iris The iris is composed of an outer layer, which is part of the uvea, and an inner layer, derived from the nonnervous portion of the retina. In this region it is the *inner* of the two retinal layers (that nearest the lens) which is heavily pigmented, a condition just opposite to that obtaining in the remainder of the retina. The outer part of the retinal portion of the iris, as well as the uveal portion, contains relatively little pigment. The dark pigment in the inner portion of the iris is the only pigment present in individuals with blue eyes. Because of a physical effect, known as the Tyndall-blue phenomenon, this pigment, as seen through the colorless tissue covering it, gives the impression that the iris is blue. The phenomenon involves interference of light rays. In brown and black eyes additional pigment is present in the outer, or uveal, region. The pink eye of the albino is devoid of all pigment. It owes its color to blood in the numerous vessels of the iris. Although the albino condition is frequently encountered among vertebrates, true albinism rarely occurs in man.

The iris contains two groups of involuntary, *ectodermal* muscle fibers called *sphincter* and *dilator muscles,* respectively. Contraction of the sphincter elements results in a diminution in size of the pupil. The pupil is the aperture surrounded by the iris and through which light travels to reach the lens. The shape of the pupil is generally round or in the form of a slit, but in some vertebrates it takes the shape of a complex keyhole. The dilator muscles, arranged in a radial manner, cause a dilation of the pupil when contracted. The involuntary contraction of these two sets of muscle fibers regulates the amount of light that enters the eyeball under varying conditions of illumination. The dilator muscles are innervated by postganglionic fibers of the sympathetic nervous system; the sphincter muscles are supplied by postganglionic fibers of the parasympathetic system.

RETINA The pigment layer of the retina is closely applied to the uveal layer throughout its entire extent. The retina consists of three regions: (1) the *optic region,* or the thickened region, in contact with the choroid coat; (2) the *ciliary region,* which together with the ciliary portion of the uvea forms the ciliary body; (3) the *iridial portion,* contributing to the iris. It is in the optic region that the photoreceptors of the vertebrate eye are situated.

The term *optical axis* is used to refer to an imaginary line passing through the center of the cornea, pupil, and lens to the posterior pole of the eye. A small area of the optic portion of the retina in direct line with the optical axis is the *area centralis,* often called the *macula lutea,* or *yellow spot.* In the center of the macula lutea is a shallow depression, the *fovea centralis,* which is the area of most acute vision.

The structure of the optic portion of the retina is highly complex. A detailed description will not be attempted in this volume. Neurosensory photoreceptors called *rods* and *cones* are located in the *outermost* part of the sensory portion of the retina, i.e., that which is next to the choroid coat and farthest from the source of light. Rods and cones differ in shape, the rods being slender and filamentous and the cones short and thick (Fig. 15.5). In man and higher primates rods are believed to distinguish different degrees of light and darkness, as well as movement and direction, whereas cones are chiefly concerned with visual acuity and color vision. In the human

eye, rods far outnumber cones, the proportion being about 20 to 1. From these cells slender nerve fibers extend toward the vitreous humor, forming synapses with dendrites of *bipolar neurons.* The axons of the bipolar cells synapse in turn, with the dendrites of neurons forming the *ganglionic layer* of the retina. The axons of the ganglionic layer course over the inner surface of the retina next to the vitreous humor. They converge at a single point a short distance from the fovea centralis, bend sharply, and then run parallel to each other to form the optic nerve. The nerve then passes *through* the retina, choroid, and sclerotic coats on its way to the brain. At the point where the fibers converge to form the optic nerve, neurosensory cells are lacking. This area is called the *blind spot,* since it is responsible for a gap in the visual field.

A true macula lutea, or yellow spot, is found only in man and some other primates. In other vertebrates the term *area centralis* is more appropriate. Only cone cells are present in the macula lutea. They are longer, more slender, and more numerous than those in

FIG. 15.5 Simplified diagram showing relation of rod and cone cells to bipolar and ganglionic layers of retina. Note convergence of several rod cells onto a single bipolar cell. Bipolar cells may converge in a similar manner onto a single ganglionic cell.

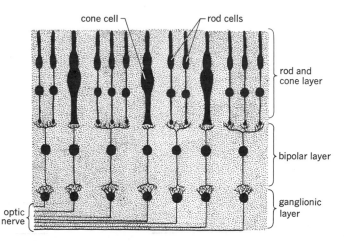

other parts of the retina. With the increase in number of cones, the number of bipolar and ganglionic cells is increased in this area, since here a single cone cell synapses with a bipolar cell and a single bipolar cell connects with a ganglionic cell. Each cone cell thus has its own line of access to the brain and is individually more important than if several such cells connected to a single bipolar cell. The latter condition is typical of rods. Because of the shape of the fovea and the manner in which light rays are refracted there, images which come to a focus on the fovea are magnified to a considerable extent and thus are perceived more clearly than those in other portions of the retina.

Rhodopsin The outer segment of each rod cell in land vertebrates and most marine fishes contains a reddish pigment called *rhodopsin* or *visual purple.* The presence of rhodopsin is necessary for rod vision in dim light. In bright light it is partially bleached so that the threshold for vision is raised. In going from bright light into darkness or semidarkness, rhodopsin is rapidly regenerated by the rod cells. This decreases the threshold for vision, since much less light is required to stimulate the rods. Several minutes must elapse before the eyes become adapted to such changed conditions. Sufficient quantities of vitamin A must be available for the body to synthesize rhodopsin. A different chemical, porphyropsin, is found in the eye of freshwater fishes, lampreys, and larval amphibians.

Cones contain three visual pigments, each sensitive to a different wavelength, which are comparable to the two types found in rods.

Inversion of the Retina In the vertebrate eye light must pass *through the various layers* of the transparent retina before it can stimulate the sensory rod and cone cells. Impulses thus set up must pass *through the cell* constituting these layers to reach the optic nerve. This peculiar state of affairs occurs only in the vertebrates, which are said to possess an *inverted eye.*

Several theories have been advanced to explain the inversion of the retina and its evolutionary origin. The most satisfactory of these states that rods and cones are actually modified ependymal cells which line the floor and walls of the cavity of the neural tube and received light through translucent openings in the skull of early chordates. The position of the sensory cells in the optic cup and optic stalk may be traced through the configurations of cup and stalk and are continuous with the ependymal layer of the neural tube proper. Ependymal cells, at least during embryonic stages, are frequently flagellated. If the flagella are photosensitive, as indeed they are in the so-called infundibular organ of amphioxus, then the receptive elements of the rod and cone cells are homologous with the flagella of ependymal cells. As these cells became outfolded to form the optic cup, they retained the original orientation, and therefore lie in what appears to be a reversed position in the fully formed vertebrate retina (Fig. 15.6).

ACCOMMODATION Stimulation of rods and cones may produce several sensations: light, various colors, form, perspective, and motion. In order to perceive objects and patterns in detail, it is essential that a picture of the object or pattern to be perceived fall upon the retina in perfect focus so that the photoreceptive rods and cones can be stimulated in the correct manner. The curved surfaces of the cornea and lens are of primary importance in this connection. The cornea is responsible for placing the image upon the retina; the lens makes minor adjustments in sharpening the focus. The eye must adapt itself in order to bring objects at various distances into focus. This is known as *accommo-*

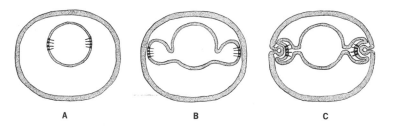

FIG. 15.6 Diagram illustrating Studnicka and Walls' theory of retinal inversion.

dation. In higher vertebrates accommodation is brought about by changing the shape of the lens. In some, as in birds and possibly in certain reptiles, changing the degree of curvature of the cornea is employed as well. In fishes the entire lens can be moved forward or backward without changing the curvature of the cornea. Altering the shape of the lens or its distance from the retina changes its focal distance so that objects at various distances from the eye can be brought to a focus on the retina. The only variable lenses in existence are those in the eyes of amniote vertebrates.

If the eye follows a moving object, the lens changes its curvature, always in such a manner as to form a sharply focused image upon the retina. As the object approaches the eye, the lens becomes more convex. Conversely, as an object moves away from the eye, the lens tends to flatten. Thus when the eyes are focused upon distant objects, the lens is flattened and said to be at rest. Focusing upon near objects requires some effort in order to increase the curvature of the lens. The ciliary body (Fig. 15.7) is the structure responsible for accommodation in mammals. It extends from the ora serrata to a point near the circumference of the lens, being attached to the lens capsule by the suspensory ligament.

The lens capsule is an elastic membrane.

FIG. 15.7 Anterior section of mammalian eye as viewed from behind after cutting the eyeball into anterior and posterior halves.

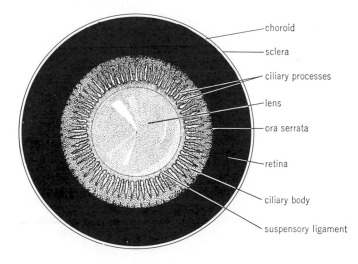

choroid

sclera

ciliary processes

lens

ora serrata

retina

ciliary body

suspensory ligament

Any pull on the capsule by the suspensory ligament would tend to flatten the lens. When the eye is at rest, i.e., focused for distant objects, the ciliary muscles are relaxed. Under such conditions the pull upon the suspensory ligament is greatest and the lens is flattened. When the ciliary muscles contract, as they do for near objects (those within the critical distance of 20 ft), they tend to pull the choroid coat forward. This relieves the tension on the suspensory ligament and lens capsule, and the lens tends to bulge and becomes more rounded. Pressure of the vitreous humor against the posterior surface of the lens prevents it from changing its shape to more than a slight degree.

ACCESSORY STRUCTURES OF THE EYE Numerous structures associated with the eye play an important role in its efficient functioning. These include the extrinsic eye muscles, eyelids, eyebrows, and the various parts of the lacrimal apparatus.

Extrinsic Muscles Six broad, strap-shaped muscles are inserted on the sclera and are so arranged that when they contract, they cause the eyeball to rotate for a variable distance. They are arranged in two groups. The first, known as the *rectus group,* is composed of four elements named the superior rectus, inferior rectus, internal (medial) rectus, and external (lateral) rectus, respectively. Each derives its name from the region of the eyeball on which it inserts, the points of insertion being rather evenly spaced around the equatorial region of the eyeball but in front. All four muscles converge to their point of origin in the posteromedial portion of the orbit. The second, or *oblique, group* of muscles is made up of two elements, superior oblique and inferior oblique. They insert on the sclera very near the insertions of the superior rectus and inferior rectus, respectively, but their respective points of origin are at the posteromedial and anteromedial wall of the orbit.

Eyelids Two transverse folds of skin, the upper and lower eyelids, lie in front of the eye, serving to protect it. Each eyelid is lined on the inside with conjunctiva continuous with that reflected over the surface of the cornea. A reinforcing band of dense connective tissue inside each eyelid is called the *tarsal plate.* The opening between the lids is the *palpebral fissure.* Both lids are movable in most vertebrates, but the upper lid in man and some other forms has a greater range of movement than the lower. A muscle, the *levator palpebrae superioris,* functions in elevating the upper lid.

In many vertebrates a third eyelid, the *nictitating membrane,* is present. It lies beneath the other two and passes from the inner angle out over the surface of the eye. It is usually transparent. In man, a small fold, the *plica semilunaris,* in the inner corner of the eye, is believed by some to be a homolog of the nictitating membrane. The nictitating membrane of amphibians is not homologous with those of other vertebrates.

In man the edges of each eyelid are provided with three or four rows of hairs, the *eyelashes.* Modified sweat glands, the *glands of Moll,* open into the follicles of the eyelashes, as do small sebaceous *glands of Zeis.* On the margins of the lids just inside the lashes are the openings of the sebaceous *Meibomian,* or *tarsal, glands.*

Eyebrows In mammals, thickened areas of skin may be situated above the upper edge of the orbit and provided with numerous thick, stiff hairs.

Lacrimal Apparatus The lacrimal apparatus of each eye consists of a *lacrimal gland,* which secretes a watery fluid, and a system of small canals by means of which the fluid is conveyed from the medial corner of the eye to the nasal passage. In mammals the lacrimal gland lies beneath the lateral portion of the upper eyelid, where several small ducts penetrate the conjunctiva. *Tears,* or *lacrimal fluid,* moisten the surface of the eyeball and provide nourishment for the nonvascular cornea. The fluid is secreted continuously, even when the eyelids are closed, and serves to keep the surface of the eyeball moist and clean by washing away dust and other foreign particles.

Miscellaneous Structures Other structures associated with the eye include: (1) the *optic pedicel,* a cartilaginous rod, found in most elasmobranchs, lying among the converging rectus muscles and giving support to the eyeball; (2) *Harder's gland,* a sebaceous gland lying behind the eyeball, secreting an oily fluid which lubricates the nictitating membrane, and found in many amphibians, reptiles, and birds, as well as in some mammals (in snakes its secretion contributes substantially to the saliva, thus facilitating the lubrication of the captured prey prior to swallowing); (3) such additional muscles as the *retractor bulbi,* and *levator bulbi,* which lower or raise the eyeball, as the case may be; (4) *scleral ossicles,* forming a circlet of thin, overlapping bones which are actually embedded in the sclera but appear to cover its exposed portion in certain fishes, birds, and most reptiles.

COMPARATIVE ANATOMY OF THE VERTEBRATE EYE

The eyes of all vertebrates are constructed on the same general plan, but many variations are to be found. In some forms the eyes are primitive; in others they are degenerate and functionless.

Cyclostomes The eye of the hagfish is degenerate, functionless, and scarcely more

than a millimeter in diameter. The optic nerve is degenerate, and other nerves as well as muscles and lens are lacking. Cornea, sclera, and choroid are not differentiated.

The eye of the lamprey, although primitive, is well developed. It offers no clue to the possible evolutionary origin of the vertebrate eye. In the ammocoetes larva stage the eyes are undeveloped and without function, but they develop markedly at the time of metamorphosis. Among important features are a flattening of the outer surface of the eyeball; a thin membranous sclera; thin cornea *not* fused to the skin; lack of suspensory ligament and ciliary apparatus; fixed size of the pupil; and a permanently spherical lens, held in place by the vitreous humor. Eyelids are lacking. Rod cells far outnumber cones.

Fishes The eyes of the cartilaginous fishes are exceptionally large, those of the deep-sea-dwelling holocephalians being the largest of all in relation to body size. Eyelids are merely folds of skin with little power of movement. An optic pedicel is to be found in most elasmobranchs. The eyeball is elliptical in shape, with a short anterior-posterior axis. Conspicuous features include cartilage in the sclera, lack of intrinsic muscles in the ciliary body, and a large, permanently spherical lens. In most elasmobranchs the surface of the choroid coat adjacent to the retina contains a layer of light-reflecting crystals of *guanin.* The layer is called the *tapetum lucidum.* Light rays striking the tapetum are reflected back, causing the eyes to "shine." Cones are absent in most elasmobranchs, and it is probable that color vision is lacking. An area centralis is present for the first time, but it lacks pigment. There is no fovea. Accommodation is accomplished by moving the lens.

In *Latimeria* rods are abundant, but cones are extremely scarce.

Features of interest in the eyes of other fishes include cartilage or even bony plates in the sclera, a tapetum lucidum, and a layer, the *argentea,* composed of silvery guanin crystals surrounding the *outer* portion of the uvea. This layer covers the dark pigment of the uvea and, by reflecting light, aids in rendering the almost transparent young fish inconspicuous. The argentea loses its significance later in life. The visual cells of the retina consist of exceptionally long rods and two kinds of cones, single and "twin." The latter are found only in teleosts. Many bony fishes have an area centralis, and color vision seems to be widespread. Some fishes even possess a fovea. Ciliary muscles and functional iris are lacking, accommodation being accomplished, by moving the spherical lens.

In flat fishes, e.g., the flounder and halibut, both eyes in an adult are on the same side of the head. Early in life the eye on the side on which the animal will later come to lie migrates over or through the head.

Amphibians Aquatic amphibians have no problem to contend with in keeping their corneas moist, but terrestrial forms have had to make adjustments to keep the surface of the eye from drying. In such forms, movable eyelids, moistened with glandular secretions, appear for the first time. There is a thick upper eyelid and a thinner, more movable lower lid. The upper border of the lower lid in anurans is transparent and moves upward when the eye is closed. It is called the nictitating membrane but is probably not homologous with the similarly named structures of higher vertebrates. Closure of the eye is not accomplished merely by moving the eyelids but is caused partly by retracting the entire eye within the orbit by means of the *retractor bulbi* muscle. Protrusion of the eyeball is effected by means of the *levator bulbi.*

The eyes of anurans are best developed among amphibians. The ciliary body is more complex than in any of the lower vertebrates, and a suspensory ligament is present, lying entirely within the vitreous body. Two small muscles of mesodermal origin, one dorsal and the other ventral, are responsible for accommodation. Each courses from the cornea to the inner part of the ciliary body. When these *protractor lentis* muscles contract, they draw the lens more closely to the cornea.

No argentea or tapetum lucidum is present in amphibians, despite the fact that the eyes of certain anurans "shine."

There are four types of photoreceptor cells in the retina: red rods, green rods, single cones, and double cones. Red rods contain the reddish pigment rhodopsin (*visual purple*); green rods contain a greenish substance, the *visual green.* The two possibly function in the same manner. It is doubtful whether amphibians have color vision. Frogs possess an area centralis, but no fovea is present. The shape of the pupil shows many variations.

The eyes of urodeles are, on the whole, smaller and less complex than those of anurans. Eyelids are absent in purely aquatic forms and are poorly developed in others. A nictitating membrane never develops beyond the rudimentary stage.

Caecilians, considered by many to be blind, actually have very small, almost microscopic eyes. The eyeball, which is only a fraction of a millimeter in diameter, lies beneath a transparent area of skin, to which it is fused. The retina is surprisingly well developed.

Reptiles The eyes of reptiles, with the exception of snakes, show a marked similarity in structure. The chief changes to be observed over amphibians are those associated with further adaptation to terrestrial life. Except in snakes and a few other reptiles, the eyelids have become increasingly movable, but the lower lid is still the larger. A true transparent nictitating membrane is present, lubricated by a well-developed Harderian gland. With the exception of *Sphenodon,* chameleons, and snakes, a lacrimal gland is present. A very important difference in the reptilian eye is the modified ciliary apparatus, which alters the shape of the lens and cornea by a peripheral squeezing action. Cones have increased in number, and most reptiles have an area centralis. Several possess a fovea. Color vision is of doubtful occurrence in crocodilians and snakes but is believed to exist in turtles and lizards. Colored oil droplets are found in the cones of birds and turtles. Their function may be to filter light for the reception of one of the basic colors, but the process is not clear.

The eyes of snakes differ in several ways from those of other reptiles and all vertebrates. The eye in general is highly developed but in a different way and lacks many of the muscles and accessory structures found in other vertebrates. One of the differences lies in the fusion of the eyelids. Here the skin over the eye forms a transparent window, or *spectacle,* responsible for the fixed stare characteristic of these animals. In certain lizards the eyelids also have fused and are transparent. In snakes true lacrimal glands are lacking. Instead of lacrimal fluid the oily secretion of Harder's gland flows in the space between cornea and spectacle. It drains through a duct into nose and mouth, where it is added to the salivary secretions. Many authorities believe that the snake eye has arisen along different evolutionary lines from those followed by other vertebrates due to the lack of conformity in the embryonic development of the eye of snakes.

Birds The eyes of birds are strikingly uniform in structure. There is little deviation in essential features from the condition found in reptiles. The eyeball of the bird is very large in proportion to body size. A medium-sized hawk may have an eyeball as large as a man's.

The eyeball is partly concave (Fig. 15.8), the concave area being caused by a ring of sclerotic bones (see page 448) which surrounds the eyeball in the region of the ciliary apparatus.

A highly developed nictitating membrane is present in birds. It is well lubricated and

FIG. 15.8 Diagrammatic sagittal section through the eye of a bird of prey.

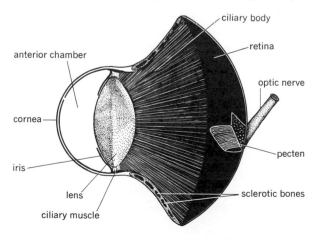

cleans the lining of the other lids as well as the corneal surface. During flight the nictitating membrane is said to be closed, offering protection to the delicate cornea.

Accommodation in the bird's eye is controlled by the ciliary apparatus, as in reptiles. It involves deformation of the cornea and increasing the curvature of the anterior surface of the lens by a squeezing action.

Rods and two kinds of cones, single and double, are present in the retina. A conspicuous fovea is present in the area centralis. Some birds have more than one fovea. Color vision is apparently widespread.

Perhaps the most interesting feature of the bird's eye is the *pecten,* a serrated, fan-shaped structure which extends into the vitreal cavity. The base of the pecten is attached to the blind spot and along the line formed by the fusion of the lips of the choroid fissure, from which it is derived. The function of the pecten has long been the subject of speculation. Among the more plausible suggestions are (1) that a shadow of the pecten falls upon the retina, thus aiding in the perception of movement; (2) that the pecten is a supplemental nutritive device which supplies the retina with necessary substances by diffusion through the vitreous humor.

In many birds, as in numerous other vertebrates, the eyes are laterally located, each covering a different visual field. The term *uniocular vision* refers to this arrangement. Even in such animals, however, the two visual fields usually overlap to some degree. Birds of prey, such as owls and, to a lesser degree, hawks and eagles, have *binocular vision.* Both eyes can be focused upon the same object, the two visual fields overlapping to a much greater degree than in animals having uniocular vision. This is of advantage in judging distance, and therefore in capturing prey. In uniocular vision a greater visual field is covered, enabling its possessor to detect danger coming from almost any direction.

Mammals Although the eye of man, as previously described, is more or less typical of mammals, there are, among mammals, certain deviations worthy of mention. The upper eyelid is larger and more movable than the lower. Eyelashes and Meibomian glands are usually present. Lacrimal glands are located under the upper lid at the outer angle. Harderian glands are lacking in most forms but are present in whales and certain semiaquatic forms, as well as in mice and shrews. A well-developed nictitating membrane is present in the platypus, aardvark, horse, scaly anteater, caribou, and panda. There are many variations in shape of the mammalian pupil, but the round opening is most common. Capacity for color vision among mammals seems to be limited to the higher primates. Many mammals possess binocular vision, but this is best developed in man and other primates.

SENSES OF EQUILIBRIUM AND HEARING

The vertebrate ear usually functions in a dual capacity, serving, at least in higher forms, both as an organ of hearing and of equilibrium. Its equilibratory function is the older and more fundamental. The structures responsible for this are to be found in all vertebrates with but little variation. The portion of the ear concerned with hearing first begins to differentiate in fishes and becomes more and more complex as the evolutionary scale is ascended.

When referring to the ear, our attention is usually directed to the special apparatus found in higher vertebrates for reception of sound waves and their transmission to sensory receptors located in the head. In mammals this apparatus is composed of three portions referred to, respectively, as the *outer, middle,* and *inner ears.* The outer and middle ears are concerned only with receiving, amplifying, and transmitting sound waves. They are lacking in cyclostomes, fishes, salamanders, and some others. It is the inner ear in which the sensory receptors are located, whether the ear serves only for equilibration or for both equilibration and hearing. The sensory receptors are composed of supporting cells and sensory cells with long hairlike processes. The latter so closely resemble the sensory neuromasts of the lateral-line system that a close relationship is highly suggestive. For this and other reasons the senses of equilibrium and hearing and of the lateral line are often referred to altogether as the *acousticolateral system.*

ACOUSTICOLATERAL ORGANS

The lateral-line system of the acousticolateral sense organs occurs only in cyclostomes, fishes, and aquatic amphibians. Among the functions ascribed to the lateral-line organs, which are sometimes referred to as *rheoreceptors,* are reception of deep vibrations in the water and of stimuli caused by currents or movements of water, including those minor local currents produced by the animal itself.

In their primitive condition the organs consist of sensory papillae called *neuromasts* (Fig. 15.9), composed of groups of sensory cells surrounded by supporting cells. The neuromasts are usually arranged in rows or lines which the nerves follow. In cyclostomes, certain fishes, aquatic amphibians, and amphibian larvae the neuromasts are located on the surface of the skin, particularly in the head region. In other fishes they sink into the skin in depressions, or grooves, which in most forms become closed over to form tubes or canals (Fig. 15.9). In holocephalians (Fig. 15.10) the grooves remain open. The closed tubes of most fishes open onto the surface of the skin through minute pores which, in bony fishes, penetrate through the scales (Fig. 5.19). The canals are filled with a watery fluid, but the canal openings are covered with mucus. In embryonic and larval stages there are typically three longitudinal lateral-line canals, dorsal, ventral, and lateral, extending the length of the body. The dorsal and ventral canals disappear in the trunk region, so that only the lateral canal persists. On the head, however, the original arrangement remains in the form of three main branches of the lateral line proper. The specific arrangement of the canals varies considerably in different fishes.

Of interest is the fact that in the newt *Tri-*

FIG. 15.9 Diagram showing relation of neuromasts to lateral-line canal and lateral branch of vagus nerve (X), as found in most fishes.

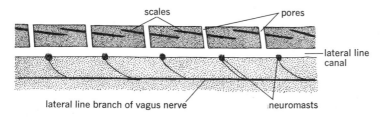

scales — pores
lateral line canal
lateral line branch of vagus nerve — neuromasts

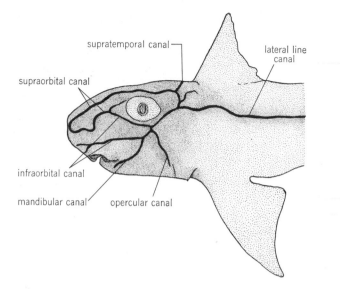

supratemporal canal

lateral line canal

supraorbital canal

infraorbital canal

mandibular canal

opercular canal

FIG. 15.10 Semidiagrammatic lateral view of head of *Chimaera monstrosa,* showing arrangement of grooves of lateral-line system in the head region.

turus viridescens the lateral-line organs disappear when the animal takes to terrestrial existence, only to reappear when it later returns to water for breeding. Aquatic reptiles and mammals have no trace of a lateral-line system.

The lateral-line system is a somatic sensory system supplied by branches of three cranial nerves, the seventh, ninth, and tenth. The seventh nerve is associated with canals in the head region, the ninth with a restricted area at the base of the supratemporal canal, and the tenth with the lateral-line canal proper. All three nerves bear ganglia derived in part from thickened placodes in the skin. Because the inner ears arise from similar placodes and because the sensory cells of the inner ear are rather similar in structure to neuromasts of the lateral-line system, the term *acousticolateral system* is used to embrace both systems. The two are believed to have a common evolutionary history.

Other peculiar sense organs found only in fishes are closely related to the lateral-line system. *Pit organs,* consisting of individual neuromasts which have sunken into small pits in the skin, are present in cyclostomes. *Ampullae of Lorenzini* (Fig. 15.11), found in elasmobranchs, certain other fishes, and even in some amphibians, are present in large numbers on the surface of the head. Each consists of a lobulated bulblike enlargement containing sensory cells supplied by the facial nerve, lying at the base of an elongated, fluid-filled tube, or canal, which connects to the surface by a small pore. It has been demonstrated that the ampullae of Lorenzini act as electroreceptors. The reception of electrical stimuli enables animals to detect enemies and prey and to determine direction. *Vesicles of Savi,* found only on the ventral surface of the electric ray, *Torpedo,* are closed sacs separated from the rest of the epidermis. They are receptors of electromagnetic stimuli acting in

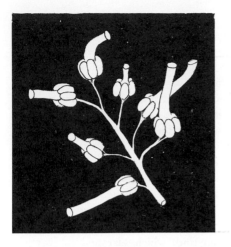

FIG. 15.11 A group of sensory ampullae of Lorenzini.

response to local changes or displacement of electric fields.

The importance of the lateral-line organs in connection with the aquatic mode of life is indicated by their highly developed supply of branches of cranial nerves and by the fact that they have entirely disappeared in terrestrial forms.

THE EAR

The epithelium lining the internal ear is of ectodermal origin. It first arises as a thickened placode in the superficial ectoderm of the head. Each of the paired placodes invaginates in the form of an *auditory pit,* which soon closes over to form a hollow *auditory vesicle.* It loses its connection with the superficial ectoderm (Fig. 15.12), lies close to the brain, and is entirely surrounded by mesenchyme. Extensive changes occur as the otocyst develops further. Its various divisions together with their supporting elements of fibrous connective tissue make up a delicate and complex structure, the *membranous labyrinth.* This consists of a closed series of tubes and sacs lined with epithelium of ectodermal origin.

The primary importance of the internal ear seems at first to have been in equilibrium, the

FIG. 15.12 Successive stages *A, B,* and *C* in the development of auditory vesicles in chick embryo: *A,* 33-hour embryo; *B,* 48-hour embryo; *C,* 72-hour embryo. (*Drawn by R. Speigle.*)

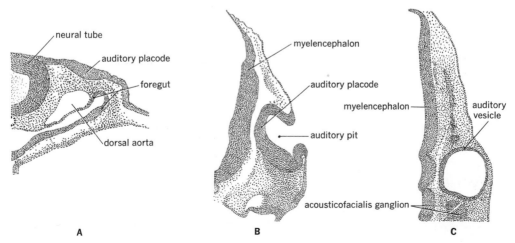

A   B   C

reception of sound appearing later in vertebrate evolution.

EQUILIBRIUM The typical membranous labyrinth consists of two chamberlike enlargements, an upper *utriculus* and a lower *sacculus,* connected by a constricted area, the *sacculoutricular duct.* A dorsomedial evagination of the otocyst, the *endolymphatic duct,* joins either the sacculus or the sacculoutricular duct. Three narrow tubes, the *semicircular ducts,* connect at both ends with the utriculus (Fig. 15.13). They are arranged in planes approximately at right angles to one another and are called the *external, anterior,* and *posterior semicircular ducts.* The external duct lies in a horizontal plane, the other two being vertical and at right angles to each other. At its lower end each duct bears an enlargement, the *ampulla.* The ampulla of the horizontal duct is situated anteriorly, close to that of the anterior duct. In lower forms a slight projection of the ventral wall of the sacculus may be present. This is the *lagena,* the forerunner of the auditory portion of the ear of higher forms.

The membranous labyrinth is filled with *endolymph,* a fluid the viscosity of which is 2

to 3 times that of water. Almost completely surrounding the membranous labyrinth is the *perilymphatic space,* filled with fluid *perilymph,* which is actually cerebrospinal fluid, the perilymphatic space being in communication with the subarachnoid space. Surrounding the perilymphatic space is cartilage or bone, depending upon the species. In higher forms a bony labyrinth, situated in the temporal bone, encloses the membranous labyrinth and follows all its configurations. The membranous labyrinth is attached to the bony labyrinth in certain places (page 463). The *semicircular canals* are those portions of the bony labyrinth which surround the semicircular ducts. They are filled with perilymph.

The actual receptors for the sense of equilibrium are elevated patches of sensory cells, the *cristae ampullares (acousticae)* and the *maculae acousticae.* The former are located in the ampullae of the semicircular ducts. They are made up of *supporting cells* and neuromast cells provided with long processes. The maculae, of which there are typically two, are also composed of supporting cells and neuromast cells, the latter bearing short, hairlike processes. One, the *macula utriculi,* lies in the wall of the utriculus; the other, the *macula sacculi,* is located in the wall of the sacculus. In those vertebrates having a lagena, a third macula, the *macula lagenae,* lies at its base. Both cristae and maculae are innervated by fibers of the vestibular branch of the acoustic nerve (VIII). A thickened mass forms in connection with each macula and covers its surface. This *otolithic membrane* is composed of gelatinous material into which groups of "hairs" from the neuromast cells penetrate. Small crystalline bodies called *otoconia* are deposited in the outer part of the otolithic membrane. They are composed of a mixture of calcium carbonate and protein. These deposits may be rather extensive, each forming a compact mass, the *otolith.*

FIG. 15.13 Right membranous labyrinth of rabbit. (*Modified from Retzius.*)

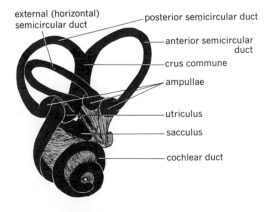

external (horizontal) semicircular duct

posterior semicircular duct

anterior semicircular duct

crus commune

ampullae

utriculus

sacculus

cochlear duct

Movements of the endolymph affect the cristae, stimulating the sensory hair cells, thus setting up impulses transmitted to the brain by branches of the acoustic nerve. The cristae give an awareness of movement (kinetic sense), being affected by rotational movements.

Changes in position of the otoliths associated with the maculae, in response to gravitational forces, affect the hair cells and give information regarding the position of the body when at rest (static sense) or of changes in velocity.

Comparative Anatomy of the Inner Ear

Cyclostomes The membranous labyrinth in cyclostomes (Fig. 15.14) is considered to be primitive. In the hagfish only one vertical semicircular duct is present, bearing an ampulla at either end. The lamprey has two vertical semicircular ducts, each having an ampulla.

In all forms above cyclostomes the three semicircular ducts are present without any basic modification.

Fishes In elasmobranchs the membranous labyrinth shows certain peculiarities (Fig. 15.15), there being two utriculi. A small projection from the ventral portion of the

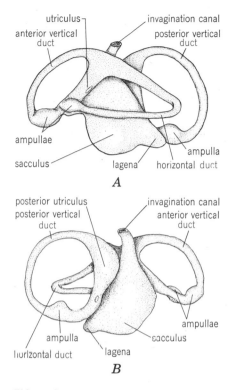

FIG. 15.15 Membranous labyrinth of dogfish shark: *A*, lateral view of left ear; *B*, medial view of left ear.

sacculus forms the lagena. From the dorsal side of the sacculus in many elasmobranchs a small tube extends upward, opening on the surface of the head. It is called the *invagination*

FIG. 15.14 *A*, membranous labyrinth of lamprey (*modified from Krause*); *B*, membranous labyrinth of hagfish (*modified from Retzius*).

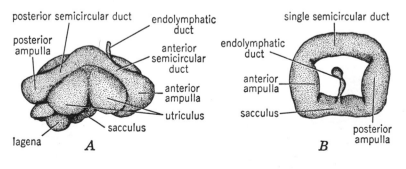

canal and is *not* homologous with the endolymphatic duct of higher forms.

In other fishes the original invagination canal closes over and an endolymphatic duct forms as an outgrowth from the saccular region. Both utricular and saccular otoliths are present, the latter being more prominent and in some cases so large that they almost fill the saccular chamber. The utricular otolith is called the *lapillus;* that of the sacculus is the *sagitta.* It is possible that the saccular macula may be stimulated by low-frequency vibrations and thus assumes the beginning of an auditory function. In teleosts an otolith, the *asteriscus,* is also present, associated with the macula lagenae.

Amphibians In some frogs the sacculus is divided into upper and lower portions. A small lagena (Fig. 15.16) projects from the posterior portion of the sacculus. This is destined to give rise to the *cochlear duct* of higher forms. A *macula (papilla) lagenae* is located in the lagena. An additional sensory patch, the *papilla basilaris,* apparently derived from the papilla lagenae, is present in many amphibians. Others lack this but possess a small *papilla amphibiorum* in the wall of the sacculus. Urodele amphibians may possess both types of papillae. The papilla amphibiorum functions as a hearing device in amphibians but is not the forerunner of the auditory organs of amniotes. It is the papilla basilaris which seems to be of special significance in the phylogenetic development of the more elaborate hearing organs of amniotes.

Amniotes The significant changes in the membranous labyrinth in reptiles, birds, and mammals involve further development and lengthening of the lagena and an elaboration of the basilar papilla. These structures are concerned with the sense of hearing. The equilibratory portion of the inner ear remains relatively unchanged.

HEARING The development of the inner ear, discussed above, indicates that it may have originated phylogenetically as a specialized section of the anterior part of the lateral-line system which sank down into the skull. There is some evidence that in fishes parts of the internal ear became associated with sound.

Teleosts of the order Cyprinoformes have a chain of small bones, the *Weberian ossicles* (page 227) interposed between the swim bladder and the perilymphatic space, thus relating the function of the swim bladder with the auditory senses. Weberian ossicles are derived from the four most anterior vertebrae. In other fishes direct connections may exist between the swim bladder and the inner

FIG. 15.16 Medial view of membranous labyrinth of frog *Rana esculenta.* (*Modified from Gaupp.*)

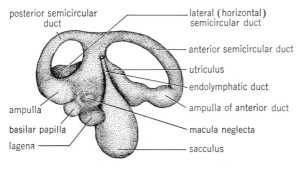

posterior semicircular duct — lateral (horizontal) semicircular duct

— anterior semicircular duct

— utriculus

— endolymphatic duct

ampulla — ampulla of anterior duct

basilar papilla —

lagena — — macula neglecta

— sacculus

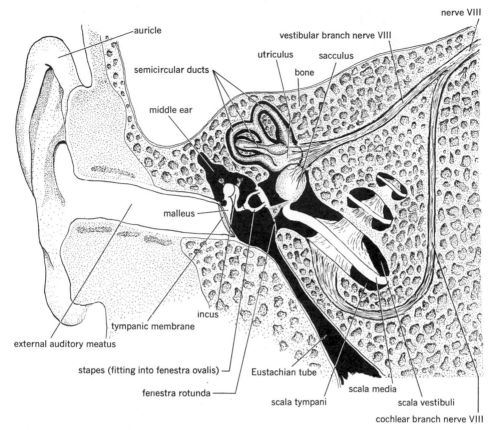

FIG. 15.17 Diagram of the ear region of man.

ears via a long anterior extension of the air bladder which lies along the membrane system of the ear.

In tetrapods the lagena is drawn out, or lengthened, to form the *cochlear duct,* and the sensory basilar papilla elaborated into the *organ of Corti,* which is the actual receptor organ for the sense of hearing. The cochlear duct begins to assume a helical form in mammals (Fig. 15.17). The degree of coiling varies, reaching its epitome in the alpaca, in which it makes five distinct turns. In man, only 2¾ turns are present. The cochlear duct is fastened to the bony labyrinth on either side but is free from it above and below. The cochlea proper is divided into three longitudinal spaces, or chambers, called the scalae (Fig. 15.18). The upper one is the *scala vestibuli* and the lower the *scala tympani.* These are filled with fluid perilymph. Between the two is the *scala media,* or *cochlear duct.* Like other parts of the membranous labyrinth, it is filled with endolymph. The three scalae together make up the *cochlea.* The floor of the scala media is called the *basilar membrane.* This separates the endolymph in the scala media from the perilymph in the scala tympani. The thin, sloping roof of the scala media is referred to as the *vestibular,* or *Reissner's, membrane.* It separates the endolymph in the scala

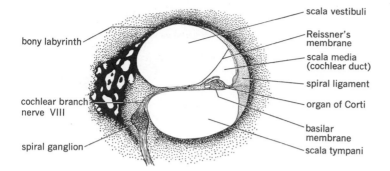

FIG. 15.18 Semidiagrammatic section through mammalian cochlea, showing the three scalae surrounded by the bony labyrinth.

media from the perilymph in the scala vestibuli. At the apex of the cochlea the scala vestibuli is continuous with the scala tympani. The point of junction is the *helicotrema.* The scala media comes to a point and ends blindly at the helicotrema.

The basilar membrane supports the organ of Corti. The latter is composed of numerous neuromast cells which have connections at their bases with dendrites of the cochlear branch of the auditory nerve. The organ of Corti in birds, although structurally different from that of mammals, appears to serve the same function.

MIDDLE EAR

Urodele amphibians and a few anurans lack a middle ear. This is a specialized condition and not primitive. Beginning with the majority of anurans, however, a special mechanism is present in most vertebrates by means of which sound waves are transmitted to the auditory portion of the membranous labyrinth. The *middle ear,* or *tympanic cavity,* as this portion of the ear is called, is concerned only with the auditory function of the ear. The middle ear is a modification of the first

pharyngeal pouch. It traces its origin to the spiracular (hyomandibular) cleft which opens onto the surface of the head in elasmobranchs and a few other fishes. It should be thought of as a pharyngeal pouch which has failed to break through to the outside. The membrane between the pouch and the external surface becomes the *tympanic membrane,* or *eardrum,* upon which sound waves impinge. The outer portion of this pharyngeal pouch becomes expanded to form the tympanic chamber. The portion nearest the pharynx becomes the narrow *Eustachian tube,* which opens into the pharynx by a small aperture. The Eustachian tube provides a means by which atmospheric pressure on the two sides of the tympanic membrane can be equalized, thus permitting the membrane to vibrate freely.

STRUCTURE In most tetrapods two openings are present between the tympanic cavity and the cochlea. The upper one, or *fenestra ovalis (vestibular window),* connects with the scala vestibuli. A small bone, the stapes, generally serves as a plug which fits into the fenestra ovalis. The lower opening, or *fenestra rotunda (cochlear window),* lies between the tympanic cavity and the scala tympani. A delicate membrane over the opening separates the two cavities. Its function is to relieve pressure in the perilymph fluid.

In the majority of anurans a small rod-shaped bone, the *columella,* extends across the cavity of the middle ear from the center of the tympanic membrane directly to the fenestra ovalis. The rounded base of the columella, referred to as the *stapedial plate,* fits into the fenestra ovalis. In reptiles and birds the columella is usually divided into two parts. The basal portion, or *stapes* proper (sometimes called the *plectrum*), fills the fenestra ovalis. The remainder is the extra-columella, which is attached to the tympanic membrane.

It should be recalled that the stapes is derived from the hyomandibular cartilage along with the shift from the hyostylic to the autostylic method of jaw suspension. In mammals (Fig. 15.17) two additional bones, the *malleus* and *incus,* are present between the tympanic membrane and the stapes. Malleus, incus, and stapes are called the *auditory ossicles*. The malleus, derived from the *articular* bone of lower forms, connects at one end with the tympanic membrane and with the incus on the other. The incus, which is homologous with the *quadrate* bone of lower vertebrates, lies between the malleus and stapes. The auditory ossicles serve to transmit and amplify sound vibrations from tympanic membrane to perilymph in the scala vestibuli.

Comparative Anatomy of the Middle Ear

Amphibians Caecilians, urodeles, and a few anurans lack a middle ear. Nevertheless, a vestigial columella is usually present. In most amphibians at the time of metamorphosis an additional bony element, the *operculum,* appears, fitting into the fenestra ovalis along with the columella. It is not represented in other vertebrates. In strictly aquatic forms vibrations are transmitted to the ear via the jaws, suspensory apparatus, columella, and fenestra ovalis to the perilymph. In terrestrial species that lack a middle ear the pathway is via the forelimbs, shoulder girdle, operculum, and fenestra ovalis to perilymph.

Most anurans have a middle ear although both tympanic membrane and middle ear are lacking in certain burrowing toads. In the tongueless toads the two Eustachian tubes open by a single median aperture into the pharynx.

Reptiles Tympanic membrane, middle ear, and Eustachian tube are altogether lacking in snakes, yet their sense of hearing is acute. The extracolumella is attached at its outer end to the quadrate bone. Hence the path of sound vibrations is via the jaws, extracolumella, plectrum, fenestra ovalis, to perilymph. The tympanic membrane in aquatic turtles is thin and delicate, but in terrestrial species it is thick and covered with skin. In crocodilians the two Eustachian tubes open into the pharynx by a single median aperture.

In most reptiles the tympanic membrane is flush with the surface of the head, but in some lizards and in crocodilians it lies at the base of a depressed area.

Birds A well-developed middle ear is present in birds. The tympanic membrane lies at some distance from the surface of the head at the bottom of a narrow passage. The Eustachian tubes have a single, median pharyngeal outlet.

Mammals The tympanic membrane in mammals is situated at the end of a fairly long passageway, the *external auditory meatus*. In many, the middle ear is surrounded by a bony *tympanic bulla* which is part of the temporal bone. The Eustachian tubes open separately. In whales, air sacs, which are diverticula of the Eustachian tubes, regulate pressures on the tympanic membrane and are associated with hearing.

EXTERNAL EAR

A true external ear which catches and directs sound waves into the auditory canal is found only in mammals. It is called the _pinna_ or _auricle._ Tufts of feathers, present in some birds near the external ear opening, may function in a similar manner but do not constitute a pinna. There seems to be a tendency to eliminate the external ear in aquatic mammals. Valves are present which prevent the entrance of water into the ear passage. The pinna is also degenerate in most burrowing mammals.

The deep depression within the pinna is the _concha._ In some forms the pinna is capable of little if any movement, but in others it has a wide range of movement.

The external auditory meatus is lined with skin continuous with that covering the concha and pinna. _Ceruminous,_ or _wax, glands,_ as well as protective hairs, are integumentary derivatives found in the auditory canal.

Mechanism of Auditory Function

Sound waves are caught by the pinna, concentrated by the concha, and directed into the external auditory meatus in mammals. In all terrestrial form the sound impinges on the tympanic membrane and causes it to vibrate. Vibrations of the tympanic membrane are transmitted and amplified by the middle-ear bones (Fig. 15.19) to the fenestra ovalis. The

FIG. 15.19 Diagrammatic section through ear of higher generalized vertebrate to show pathway of auditory stimuli to inner ear.

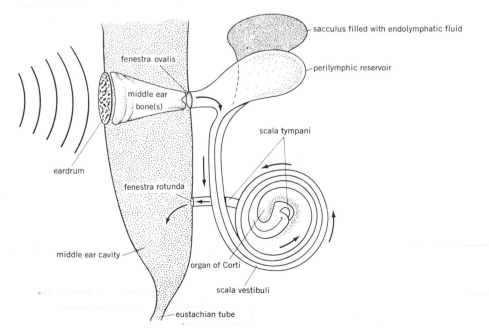

sacculus filled with endolymphatic fluid

fenestra ovalis

perilymphic reservoir

middle ear bone(s)

scala tympani

eardrum

fenestra rotunda

middle ear cavity

organ of Corti

scala vestibuli

eustachian tube

movement of the membrane causes movement in the perilymph of the scala vestibuli. Movement of the fluid travels up to the helicotrema at the apex of the cochlea and down the perilymph in the scala tympani, ultimately to be released into the middle-ear cavity by the movement of the membrane covering the fenestra rotunda. Vibrations in the perilymph are transmitted to the endolymph in the scala media and through the basilar membrane to the perilymph of the scala tympani. Whether the sensory receptors of the organ of Corti are stimulated by endolymph movement or by the movement of the basilar membrane is not clear. In either case stimulation of the neuromast cells of the organ of Corti sets up impulses in the auditory nerve, which then travel to the brain, where they are interpreted as sound.

OLFACTORY SENSE

The usual close association between the olfactory and respiratory organs of air-breathing vertebrates has been discussed previously. In most lower aquatic vertebrates with the gill type of respiration the olfactory and respiratory mechanisms are not related.

The *olfactoreceptors* consist of individual bipolar neurons derived from placodes in the superficial ectoderm. They are located in the olfactory epithelium, which, in addition to the olfactory neurons themselves, is composed of *basal cells* and *supporting cells*. The relation of the olfactory nerve to the brain is discussed in Chap. 13, Nervous System.

The dendritic ends of the olfactoreceptors are brushlike in appearance, each cell bearing from 5 to 12 delicate olfactory hairs at its tip. They are covered with the secretion produced by *Bowman's glands,* the ducts of which pass through the olfactory epithelium. The secretion, in addition to being protective, serves as a moist medium in which volatile substances are dissolved before stimulating the olfactoreceptors. In both chemical senses, smell and taste, it is essential that substances be in solution before they can be detected.

Comparative Anatomy of Olfactory Organs

Amphioxus A small *flagellated pit* near the pigment spot at the anterior end of amphioxus has been suspected of possessing an olfactory function, but no nervous connections between it and the brain vesicle have been discovered and its function remains in doubt. Hair cells in the velum and buccal cirri may possibly be involved in olfaction.

Cyclostomes In cyclostomes a single, median nasal aperture opens onto the surface of the head. It leads by means of a short passageway to the blind *olfactory sac* lined with numerous folds covered with *olfactory epithelium.* Paired olfactory nerves, indicating a bilateral origin of the olfactory apparatus, pass through a fibrocartilaginous *olfactory capsule* which surrounds the olfactory sac and which lies just anterior to the brain. A long tube, the *nasopharyngeal pouch,* projects ventrally from the anterior edge of the olfactory sac. This ends blindly in the lamprey but is connected with the pharynx in the hagfish. In the lamprey water is alternately forced in and out of the nasopharyngeal pouch and olfactory apparatus, the intake and outflow coinciding with respiratory movements in the gill region. This enables the animal to detect chemical substances in solution.

Fishes The olfactory organs of fishes are paired. They may be nothing more than blind pits lined with olfactory epithelium but are constructed to permit the passage of water in and out. In many elasmobranchs a prominent *oronasal groove* anticipates the condition first found in the Sarcopterygii, in which a true connection between the nostrils and mouth is to be found. In the latter the olfactory epithelium, consisting of a few large folds, is located in the dorsal part of the nasal passage which connects the external and internal nares.

In most fishes the olfactory epithelium is commonly thrown into numerous folds which greatly increase the epithelial surface.

Amphibians The olfactory passage in amphibians is short. In aquatic forms the passage is lined with folds. Olfactory sense cells lie in the depressions between the folds, and ciliated epithelium covers the ridges.

The olfactory epithelium in terrestrial species is located in the upper medial portion of the nasal passages. It is not folded to any extent. In some forms a shelflike fold from the lateral wall foreshadows the appearance of the *conchae,* or *turbinal folds,* which become highly developed in more advanced vertebrates. Glandular areas in the nasal passages serve to keep the olfactory epithelium moist.

A new structure, the *vomeronasal,* or *Jacobson's organ,* first appears in amphibians. It arises as a ventromedial or ventrolateral evagination of the nasal passage, with which it connects by means of a short duct. This organ is believed to be used in testing food substances held in the mouth. It is supplied with branches of the terminal (0), olfactory (I), and trigeminal (V) cranial nerves.

Reptiles With the development of palatal folds in most reptiles and of a complete secondary palate in crocodilians, the nasal passages are elongated and the choanae open in a more posterior portion of the roof of the mouth. A single concha is present in the lateral wall of each nasal passage. It consists of a projection of the maxillary bone and serves to increase the olfactory surface.

Jacobson's organ is insignificant in turtles and crocodilians but is highly developed in *Sphenodon,* snakes, and lizards, in which it has lost its connection with the nasal passage and opens directly into the mouth cavity. The tongue is cleft into two prongs and is used as an accessory organ. The tongue darts out of the mouth gathering olfactory clues and returns. The tips of the tongue are inserted into the organ pockets. Presumably olfactory data are received by the sensory epithelium lining these pockets.

Birds The main advance to be observed in the nasal passages of birds is in the greater elaboration of the conchae. Three conchae are present in the lateral wall of each nasal passage, but only one, the *superior concha,* is covered with olfactory epithelium. The olfactory sense in birds, except in marine species, is rather poorly developed. Jacobson's organ exists only as a transitory embryonic rudiment.

Mammals It is among mammals that the olfactory sense reaches the epitome of development, although in some it is poorly developed or lost altogether.

In most mammals the elongated nasal passage contains a complicated mass of folds derived from the three conchae. These greatly increase the surface of epithelium exposed in each nasal passage. The labyrinth is supported by cartilages or by *turbinate bones,* called the *ethmoturbinate, maxilloturbinate,* and *nasoturbinate,* according to the bone from which they arise. The nasoturbinates and particularly the ethmoturbinates are the major support of the olfactory epithelium. The more complicated they are, the keener the sense of smell.

Large air spaces called *sinuses* are located within certain bones of the skull in eutherian mammals. The sinuses communicate with the nasal passages but are not concerned with the olfactory sense. Their function, if indeed they have any, is obscure. The presence of sinuses makes the skull lighter. The principal sinuses are those in the frontal, ethmoid, sphenoid, and maxillary bones. The maxillary sinus in man is the largest and is frequently the site of infection.

SENSE OF TASTE

The receptors for the sense of taste are spoken of as *gustatoreceptors* or taste buds. They are composed of two kinds of cells, *supporting cells* and *neuroepithelial,* or *taste, cells* (Fig. 15.20). The long axes of the cells extend from the base to the surface of the taste bud. The number of taste cells in a taste bud varies. Estimates in man range from 4 to 20. Each taste cell bears a minute *taste hair* at its tip. The taste hairs project into a small, pitlike depression, the *taste pore.* Taste pores are found only in mammals.

It has been established that only four fundamental taste sensations exist bitter, sweet, salty, and sour, or acid. Sometimes two additional tastes, alkaline and metallic, are added to the list. Variations from the above may be due to combinations of two or more of the fundamental tastes and to complications brought about by olfactory stimuli.

DISTRIBUTION OF TASTE BUDS

In certain fishes the skin covering the entire

FIG. 15.20 A taste bud.

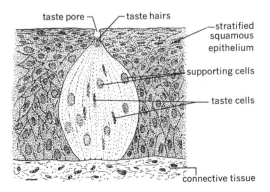

body is abundantly supplied with taste buds. However, in most vertebrates the gustatory receptors are confined to the oral region, being particularly numerous on the tongue. Here they are usually associated with small elevations of the surface epithelium, the *lingual papillae.*

Comparative Anatomy of Gustatoreceptors

Amphioxus On the velum and buccal cirri of amphioxus are groups of *hair cells* which may serve as gustatoreceptors. It is of interest that this primitive animal shows an aggregation of sensory receptors at the anterior end of the body.

Cyclostomes In adult cyclostomes taste buds are present on the surface of the head, in addition to those present in the pharyngeal region. In larval forms they are confined to the pharyngeal lining.

Fishes In elasmobranchs taste buds are associated with papillae in the epithelium lining the mouth and pharynx. Higher fishes show a greater diversity in taste-bud distribution. In some they are present on the surface of the head as well as mouth and pharynx. In such scavenger fishes as carp, suckers, and catfish, taste buds are present over the entire body surface, including fins and tail. Beginning with the Dipnoi in the evolutionary scale, they are confined to the lining of the mouth, pharynx, and tongue.

Amphibians Taste buds of amphibians are present on the roof of the mouth, tongue, and lining of the jaws. In frogs they lie on the free surfaces of fungiform papillae. The more numerous filiform papillae do not have taste buds associated with them.

Birds The horny tongues of most birds lack taste buds, which are situated mainly in the lining of mouth and pharynx. The large, rather fleshy tongue of the parrot bears numerous taste buds.

Mammals Although taste buds in mammals are generally present in various parts of the lining of the mouth and pharynx, most are associated with lingual papillae. In man taste buds are present on the soft palate, posterior surface of the epiglottis, and even on the vocal cords. They lie in the mucous membrane covering these structures. Four types of lingual papillae are recognized, *filiform, fungiform, foliate,* and *circumvallate* (Fig. 9.8). Filiform papillae almost never have taste buds associated with them, but the other types of papillae have an abundant supply.

NERVE SUPPLY Taste buds are innervated by visceral sensory branches of several cranial nerves, but there are no separate nerves for the sense of taste. The nerves involved are the facial (VII), glossopharyngeal (IX), and vagus (X). In such forms as the catfish and buffalofish, which have a general cutaneous distribution of gustatoreceptors, the visceral sensory area on either side of the medulla oblongata is expanded markedly, forming the paired *vagal lobes.* The pharyngeal taste buds of elasmobranchs are reported to be of endodermal origin.

COMMON CHEMICAL SENSE

In lower aquatic vertebrates the entire surface of the body is sensitive to mildly irritating chemicals. The moist integument of amphibians also comes under this category. In terrestrial vertebrates only those surfaces normally kept moist are sensitive to chemical irritants. Free nerve endings of spinal nerves and certain cranial nerves are the receptors concerned with receiving information regarding

chemical irritants. The sense has no relation to the chemical senses of taste and smell.

OTHER SENSORY RECEPTORS

CUTANEOUS RECEPTORS

In addition to the sense organs already referred to, other external sensory receptors are abundant in the skin covering the entire surface of the body. The sensations resulting from the stimulation of these receptors are referred to as *cutaneous sensations.* They include touch, heat, cold, and pain. Some cutaneous receptors consist merely of free endings of nerve fibers. Others are encapsulated nerve endings, and still others, in mammals at any rate, terminate in hair follicles. Until rather recently biologists believed that each type of cutaneous receptor subserved a particular function and was specifically concerned with one sensation only.

SENSE OF TOUCH Touch is the sense by which contact with objects gives evidence of certain of their qualities. The sensory endings for the sense of touch are called *tangoreceptors.* Branches of the dendritic processes of sensory nerve cells, penetrating among the epithelial cells of the skin, cornea, and mucous membranes (Fig. 15.21*A*), are considered to be tangoreceptors. The nerve endings for the sense of pain (*algesireceptors*) also terminate freely in the skin. It is difficult to distinguish between these two kinds of receptors. In the skin of the pig's snout, nerve fibers which penetrate into the epidermis terminate in small expanded discs which are in contact with special epithelial cells (Fig. 15.21*B*). They are tangoreceptors known as *Merkel's tactile discs* or *corpuscles.*

Several types of tangoreceptors consist of modified or encapsulated nerve endings. They

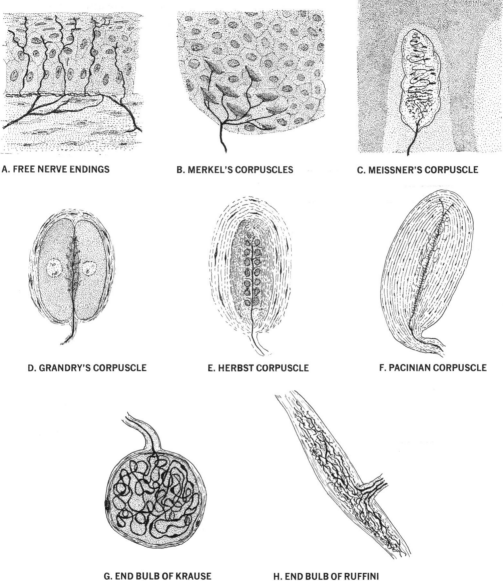

A. FREE NERVE ENDINGS

B. MERKEL'S CORPUSCLES

C. MEISSNER'S CORPUSCLE

D. GRANDRY'S CORPUSCLE

E. HERBST CORPUSCLE

F. PACINIAN CORPUSCLE

G. END BULB OF KRAUSE

H. END BULB OF RUFFINI

FIG. 15.21 Various types of cutaneous receptors found in vertebrates. Those depicted here are *not* drawn to the same scale.

lie in the dermis of the skin for the most part. Among them are *Meissner's corpuscles* (Fig. 15.21C) found in the skin of primates, particularly in areas which normally undergo friction. *Genital corpuscles* are special nerve endings in the skin of the external genitalia and of the nipples. *Grandry's corpuscles* (Fig. 15.21D) and *Herbst's corpuscles* (Fig. 15.21E) are found on the beaks and mouth parts of birds, respectively. Herbst's corpuscles are also present in areas of the skin where feathers are few. Rather large *Pacinian* (*Vater-Pacinian*) *corpuscles* (Fig. 15.21F) are to be found in the deep layers of the dermis, as well as in tendons and periosteum. It is believed that these encapsulated nerve endings are stimulated by pressure. The sense of pressure is interpreted as being evoked by a stimulus greater in degree than that which causes the sensation of touch. It is not a true cutaneous sensation. *Corpuscles of Golgi* and *Mazzoni* are present in the subcutaneous tissues of the fingers.

Tactile nerve endings terminating in hair follicles are, of course, found only in mammals. The nerve endings are confined to the root of the hair and base of the follicle. Noteworthy among sensory hairs are the specialized *vibrissae* so well developed in mammals of nocturnal habit.

Touch or pressure causes a deformation of the skin with a consequent alteration in shape of the sensory endings. This may be the important factor involved in stimulating tangoreceptors.

TEMPERATURE SENSES Thermoreceptors are confined for the most part to the skin. They are practically lacking in the exposed surface of the eyeball. There is some disagreement concerning the specific end organs for temperature sensations. Free nerve endings may possibly serve as thermoreceptors, but it is more likely that specific encapsulated nerve endings are responsible for receiving thermal stimuli.

Cold The *end bulbs of Krause* (Fig. 15.21G) are usually considered to be cold receptors.

Heat *End organs of Ruffini* (Fig. 15.21H) are generally believed to serve as caloreceptors for the sensation of warmth or heat. The endings for this sense lie somewhat deeper in the dermis than those for cold.

The pit organs of pit vipers are heat-sensitive and aid in increasing the directional sensitivity to the stimuli. The pit is lined with a highly innervated membrane supplied by the trigeminal nerve (V). These organs are most sensitive to infrared wavelength of 15,000 to 40,000 Å. The field of each pit overlaps the other, providing the animal with a 180° field in front of him.

SENSE OF PAIN Algesireceptors which mediate pain sensations consist of naked nerve endings lying in the deeper layers of the epidermis. They are also present in the cornea, where tangoreceptors are lacking.

INTERNAL RECEPTORS

PROPRIOCEPTIVE SENSE Receptors located in tendons, joints, and skeletal muscle give information regarding the position and movements of various parts of the body. These *proprioceptors* may consist of free nerve endings or of spindles and corpuscles of various kinds.

DEEP PAIN SENSE Free nerve endings may be found lying between individual muscle fibers. Deep pain sensations are believed to arise from their stimulation. Pacinian corpuscles located in mesenteries, omenta, and visceral peritoneum probably give information concerning stretching of mesenteries, distention of the colon, and the like.

SENSE OF EQUILIBRIUM Although a portion of the inner ear is chiefly concerned with the sense of equilibrium, other sensory organs may also be involved. Proprioceptors and tangoreceptors, particularly those on the soles of the feet in tetrapods, also give information regarding the position of the body.

SUMMARY

1   Receptor organs are structures capable of responding to certain types of stimuli by setting up impulses which are transmitted to the central nervous system. In the higher centers of the brain they are interpreted as sensations. The actual receptor cells with few exceptions are of ectodermal origin.

2   Receptor organs are classified as external or internal sense organs. External sense organs, or exteroceptors, are stimulated by environmental disturbances of various kinds. They include those for the senses of sight, hearing, smell, taste, touch, pressure, temperature, and pain. The lateral-line organs of lower aquatic vertebrates are exteroceptors. Internal sense organs, affected by stimuli arising within the body, include statoreceptors for the sense of equilibrium, proprioceptors for the sense of joint position, and interoceptors for such internal sensations as hunger, thirst, and nausea. Internal receptors for deep pain sensations are also recognized.

3   The vertebrate eye has no true homolog among members of other phyla and appears for the first time in cyclostomes. Its evolutionary history is obscure. The retina, or sensory portion, develops as a cuplike outgrowth of the diencephalic region of the forebrain. The lens is derived from superficial ectoderm. Mesenchyme surrounding the optic cup differentiates into the uveal and fibrous coats of the eyeball.

In the vertebrate eye, light must pass through the various layers of the retina before it can stimulate the sensory receptors, or rod and cone cells. Impulses thus set up must pass back through the cells comprising these layers to reach the optic nerve. The retina is thus said to be inverted.

Accommodation is accomplished in different ways by various vertebrates. Lower forms change the position of the lens; higher forms change the shape of the lens.

Eye movements are brought about by six strap-shaped muscles supplied by the somatic motor, third, fourth, and sixth cranial nerves. Other accessory eye structures in various vertebrates serve the animal in adapting itself to its own environmental needs.

4   Lateral-line organs of fishes and aquatic amphibians are related to the aquatic mode of life. They are believed to receive deep vibrations in the water and also to receive information concerning currents or movements of water. The receptors consist of neuromasts located on the surface of the skin or inside grooves or canals in the skin. The lateral line proper usually extends the length of the body but terminates on the head in several branches. The system is innervated by the seventh, ninth, and tenth cranial nerves. All three nerves bear ganglia derived in parts from placodes in

the skin. Because the inner ears arise from similar placodes, and since the sensory cells of the inner ear are similiar in structure to neuromasts, the term acoustico-lateralis system is used to include both systems. Other structures belonging to the lateral-line system are pit organs, ampullae of Lorenzini, and vesicles of Savi.

5 The vertebrate ear, at least in higher forms, functions in a dual capacity as an organ of equilibration and hearing. The equilibratory function is more fundamental and primitive. The portion of the ear in which the sensory receptors are located is the inner ear, or membranous labyrinth. The part concerned with equilibration shows but little variation in different forms. It consists typically of utriculus, sacculus, endolymphatic duct, and three semicirular ducts arranged at right angles to one another. Each semicircular duct bears an ampulla at its lower end. The sensory receptors for the sense of equilibrium are cristae ampullares and maculae acousticae, located in the ampullae and in the utriculus and sacculus, respectively. An otolithic membrane containing crystalline bodies covers each macula. Movements of the endolymph affect the cristae, whereas changes in position of the otolithic membrane in response to the pull of gravity stimulate the maculae. In cyclostomes the membranous labyrinth is degenerated and only one or two semicircular ducts are present.

The lagena, a slight projection of the ventral wall of the sacculus, is usually present in lower forms. It evolves into the auditory portion of the ear in higher vertebrates. A macula is generally present in the wall of the lagena. A basilar papilla, derived from the macula lagenae, first appears in amphibians and is the forerunner of the organ of Corti, the receptor for the sense of hearing in higher forms. In crocodilians and birds the lagena begins to form an elongated spiral, the cochlear duct, filled with endolymph. In mammals the cochlear duct (scala media) becomes more complicated. It contains the organ of Corti, with its hairlike cells, which are stimulated by vibrations transmitted from the environment.

Beginning with anuran amphibians a special mechanism is present in most vertebrates by means of which sound waves can be transmitted to the auditory portion of the membranous labyrinth. The middle ear, as this portion is called, is derived from the first pharyngeal pouch. A tympanic membrane separates the middle ear from the outside. Small bones transmit vibrations from the tympanic membrane to fluid perilymph surrounding the membranous labyrinth. The perilymph in turn affects the endolymph within the membranous labyrinth. A Eustachian tube connects the cavity of the middle ear with the pharynx. It serves to equalize pressure on the two sides of the tympanic membrane. An external ear is present only in mammals. It catches and directs sound waves, which then impinge upon the tympanic membrane.

6 Receptors for the sense of smell consist of bipolar neurons derived from placodes in the superficial ectoderm. In aquatic vertebrates they are confined to the olfactory epithelium lining the olfactory sacs, which are blind invaginations of the outer epithelium. In crossopterygian fishes a connection between the nares and mouth first appears. The olfactory epithelium is located in the dorsal part of this nasal passage, which is primarily a respiratory organ. Further advances are to be found in a separation of the respiratory and olfactory regions from the mouth cavity by the development of a sec-

ondary palate. Outgrowths from the lateral walls in the form of folds and scrolls (turbinates or conchae) increase the respiratory and olfactory surfaces. Chemical substances must be in solution before they can stimulate the moist, mucus-covered olfactory epithelium.

7   Taste receptors consist of taste buds, small structures composed of neuroepithelial cells and supporting cells. Substances to be tasted must be in solution. In some fishes gustatory receptors are distributed throughout the skin covering the entire body, but in most vertebrates they are confined to the oral and pharyngeal regions, being particularly numerous on the tongue. Here they are associated with lingual papillae.

8   The entire surface of the body of lower aquatic vertebrates is sensitive to mildly irritating chemicals. In terrestrial forms only those surfaces normally kept moist are sensitive to such stimuli. The receptors for this common chemical sense, which is not related to smell or taste, consist of free nerve endings.

9   Cutaneous receptors for the senses of touch, pressure, temperature, and pain consist of free nerve endings in some cases and encapsulated nerve endings of various types in others. No specific nerves are concerned with the cutaneous senses.

10   Internal receptors for the proprioceptive sense consist of free nerve endings or spindles and corpuscles of various kinds. They are located in muscles, tendons, joints, and periosteum. Free nerve endings between muscle fibers may mediate deep pain sensations. Pacinian corpuscles in mesenteries, omenta, and visceral peritoneum probably give information concerning stretching of mesenteries and distention of the alimentary tract.

BIBLIOGRAPHY

Cahn, P. H. (ed.): "Lateral Line Detectors," Indiana University Press, Bloomington, 1967.

Case, J.: "Sensory Mechanisms," Macmillan, New York, 1966.

Griffin, D. R.: "Listening in the Dark," Yale University Press, New Haven, Conn., 1958.

Matthews, L. H., and M. Knight: "The Senses of Animals," Museum Press, London, 1963.

Polyak, S.: "The Vertebrate Visual System," University of Chicago Press, Chicago, 1958.

Prince, J. H.: "Comparative Anatomy of the Eye," Thomas, Springfield, Ill., 1956.

van Bergeijk, W. A.: The Evolution of Vertebrate Hearing, in W. Neff (ed.), "Contributions to Sensory Physiology," vol. 2, Academic, New York, 1967.

Walls, G. L.: "The Vertebrate Eye and Its Adaptive Radiation," Cranbrook Institute of Science, Bloomfield Hills, Mich., 1942.

16 SUMMARY: CHARACTERISTICS AND ADVANCES

The evolution of the chordates spans some 500 million years and represents, both in the past and present, a tremendous variety of forms. This is evident from the fact that the chordate type of organization has been able to withstand and adapt to a wide variety of circumstances.

The earliest chordates show a rather low level of organization and efficiency in all systems. Sequential evolution has produced forms which have reached higher and higher levels of biological efficiency and organization, ending, at least for now, with the mammals. This increased complexity is reflected at all levels of morphology and physiology.

In dealing with chordate anatomy we have restricted ourselves to discussion of living forms, with only passing reference to the vast numbers of extinct chordates. This is not to imply that these forms are not important, for indeed they play a very important role in providing the transitional or, in some cases, the final stages of chordate evolution.

Nevertheless in our discussion of living forms there is a suggestion that evolutionary changes in the vertebrates have followed certain trends and patterns. Aquatic vertebrates tend to have streamlined body shapes, paddle-like limbs, and flat-surfaced vertebrae. Running forms show elongation of limbs and reduction in the number of bones in their limbs. Flying vertebrates tend to lighten the body with light or hollow bones, and burrowing forms lose their limbs. These are obvious trends, but certainly such principles could probably be cited for all organ systems.

A summary of the characters discussed within the various chordate groups follows. Starting with the protochordates and finishing with the mammalian groups, we present the major changes in most organ systems. Advancement and specialization are included to reflect the changes that have occurred in the evolution of the chordates.

PHYLUM CHORDATA

UNIQUE FEATURES A notochord; a hollow dorsal nerve tube; gill slits connecting to the pharynx, or traces of them, present sometime during life.

OTHER FEATURES Bilateral symmetry; cephalization; metamerism; true body cavity, or coelom; ventral-anterior direction of blood flow from heart. These features are also characteristic of numerous invertebrate animals.

SUBPHYLUM I. HEMICHORDATA

Marine, burrowing forms, body wormlike and unsegmented with proboscis, collar, and trunk; numerous gill slits connecting pharynx with outside; a stiffened, forward extension of the gut into the proboscis probably *not* homologous with notochord; dorsal and ventral nerve strands, the dorsal one being larger and having a tubular structure only in the collar region; presence of giant nerve cells in anterior region suggests the beginning of a brain; dioecious; tornaria larva of certain forms closely resembling larval echinoderms. Example: *Dolichoglossus kowalevskii*.

SUBPHYLUM II. CEPHALOCHORDATA

Marine; free-swimming; notochord well developed and extending entire length of body; no well-defined head; nerve cord slightly enlarged at anterior end, forming a cerebral vesicle; pharyngeal gill slits connecting indirectly with outside through an atrium and atriopore; metamerism very clearly indicated; dioecious; gonads arranged metamerically; hepatic cecum off ventral side of digestive tract homologous with the liver of higher forms; heart a ventral pulsating tube, sometimes said to be one-chambered; hepatic portal vein; metapleural folds suggest possible origin of paired appendages of vertebrates (doubtful); epidermis one cell layer thick; integumentary glands all of unicellular type; spinal nerves metameric, but those of one side alternate with those of the other; dorsal and ventral nerve roots not united; endostyle lying along midventral line of pharynx; no reproductive ducts. Example: Amphioxus (*Branchiostoma*).

SUBPHYLUM III. UROCHORDATA (TUNICATA)

Marine, some free-swimming forms, but others becoming sessile after a free-swimming larval period; excurrent and incurrent siphons; pharynx large with numerous gill slits; notochord only in larva and then confined to tail region; nervous system of adult reduced to a small ganglion; hermaphroditic; endostyle in midventral part of pharynx perhaps forerunner of thyroid gland of higher forms; adneural gland around nerve ganglion possibly homologous with pituitary gland of higher forms; covering of body, or tunic, composed of tunicin (cellulose); heart showing reversal of beat; mantle cavity present. Examples: Ascidians (*Molgula manhattensis, Botryllus, Doliolum, Salpa*).

SUBPHYLUM IV. VERTEBRATA (CRANIATA)

Anterior end of dorsal nerve cord enlarged to form a brain which is three-lobed during early development; protective and supporting endoskeleton; a cranium surrounding the brain, and a segmented spinal column composed of vertebrae enclosing the spinal cord; paired sense organs in head—olfactory, optic, and otic—are typical.

Superclass I. Pisces

Aquatic; in most cases respire throughout life by means of gills associated with pharyngeal gill slits.

Class I. Agnatha

Absence of jaws; absence of paired appendages even in embryo; a single median nostril representing a fusion of two; poorly developed cranium. In ostracoderms, known only through fossil remains, there were heavy dermal plates in the skin; some had an endoskeleton of a sort; some possessed lateral body lobes posterior to the head region, from which pectoral appendages may have evolved.

ORDER I. CYCLOSTOMATA Living forms; body rounded, but tail laterally compressed; round suctorial mouth; median fins supported by cartilaginous fin rays; no scales in skin; skin soft with numerous unicellular mucous glands present; 6 to 14 pairs of gill pouches; poikilothermous; poorly developed brain not completely surrounded by cranium; vertebrae also poorly developed; protocercal tail; tongue specialized for rasping; epidermal horny teeth; persistent notochord; skeleton entirely cartilaginous; two-chambered heart; lateral-line system, but no true lateral line; no genital ducts; primitive in many respects, highly specialized in others. Examples: Lampreys and hagfishes.

SUBORDER I. PETROMYZONTIA Inhabiting rivers, lakes, and the sea; dioecious; ventral suctorial funnel at anterior end beset with horny teeth; seven pairs of gill pouches, each opening separately to the outside; ammocoetes larvae undergo metamorphosis; pharynx a blind pouch guarded by a velum; pancreas essentially lacking; opisthonephric kidneys; dorsal and ventral roots of spinal nerves not united; eyes primitive but well developed; only two semicircular ducts (the vertical ones) present in inner ear; nasal sac gives off nasopharyngeal pouch which ends blindly. Example: Lampreys.

SUBORDER II. MYXINOIDEA All marine; hermaphroditic; mouth nearly terminal and surrounded by four tentacles; no buccal funnel; few teeth; parasitic habit; 6 pairs of gill pouches in *Myxine* with a single opening to the exterior on each side; an asymmetrical esophageo-cutaneous duct on left side; 6 to 14 pairs of gill pouches in *Bdellostoma;* archinephros in embryo, pronephros and opisthonephros in adult; eyes degenerate and functionless; one semicircular duct in inner ear with an ampulla at either end; nasopharyngeal pouch connects with nasal sac in front and with pharynx behind; no metamorphosis. Example: Hagfishes.

The remaining vertebrates typically have paired pectoral and pelvic appendages; true upper and lower jaws; paired nostrils; well-developed endoskeleton; closed cranium.

In those members of the superclass Pisces, higher in the evolutionary scale than members of the class Agnatha, paired appendages are present in the form of fins; median fins are also present; dermal scales of various types are in the skin; unicellular and multicellular mucous skin glands are abundant; they are dioecious; poikilothermous; have opisthonephric kidneys (persistent pronephros in a few); a two-chambered heart [a few (lungfishes) have a three-chambered heart]; 10 pairs of cranial nerves in addition to the terminal nerve (0); genital ducts.

Class II. Placodermi

Fossil forms only; most primitive jawed fishes; primitive jaws and paired fins; bony armor of dermal plates in skin; some with lunglike pharyngeal diverticula. Example: *Bothriolepis.*

Class III. Chondrichthyes

Marine with few exceptions; cartilaginous skeleton; placoid scales; ventral, subterminal mouth; heterocercal tail; pelvic fins of male modified to form claspers used in copulation; no swim bladder; spiral valve in small intestine; lower jaw skeleton composed of Meckel's cartilage, upper jaw of palatoquadrate bar; oviparous, ovoviviparous, and viviparous forms; eggs large with abundant yolk; meroblastic cleavage; variable number of aortic arches.

Subclass I. Elasmobranchii

Five to seven pairs of gill slits opening separately to the outside; a pair of spiracles, representing the first pair of gill slits, usually opening on top of head; hyostylic method of jaw attachment; persistent notochord partially replaced by cartilaginous vertebrae; cloaca.

ORDER I. SELACHII Pectoral fins distinctly marked off from cylindrical body; gill slits lateral in position; tail used in locomotion. Examples: Sharks and dogfish.

ORDER II. BATOIDEA Dorsoventrally flattened bodies; pectoral fins not sharply marked off from body; demarcation between body and tail distinct; gill slits ventral in position; spiracle large and well developed; undulatory movements of pectoral fins used in locomotion. Examples: Skates (oviparous) and rays (ovoviviparous).

Subclass II. Holocephali

Persistent notochord; poorly developed vertebrae; operculum; open lateral-line canals; no spiracles; no cloaca. Example: Chimaeras.

Class IV. Osteichthyes

Skeleton bony, at least to some degree; scales usually present of ctenoid, cycloid, or ganoid types; terminal mouth; operculum; swim bladder usually present; tail usually homocercal but sometimes heterocercal or diphycercal; Meckel's cartilage invested by bone; dermatocranium roofs over most of chondrocranium; four pairs of aortic arches; oviparous, ovoviviparous, and viviparous forms; no cloaca. Example: Bony fishes.

Subclass I. Actinopterygii

All fins, paired and unpaired, supported by dermal skeletal fin rays. Example: The ray-finned fishes.

SUPERORDER I. CHONDROSTEI Fin rays of dorsal and anal fins exceeding in number the supporting skeletal elements; skeleton mainly cartilaginous; tail diphycercal or heterocercal; clavicles present; nostrils not connecting with mouth cavity; spiracles; spiral valve in small intestine. Examples: So-called "ancient" fishes, or primitive ray-finned fishes. (*Polypterus*).

SUPERORDER II. HOLOSTEI Dermal rays of dorsal and anal fins equal in number to internal radial skeletal elements; marine forms extinct; living species are freshwater forms; internal skeleton partly cartilaginous; abbreviated heterocercal or homocercal tail; pelvic fins usually posteriorly located; no spiracles; no clavicles. Examples: The intermediate ray-finned fishes; *Amia,* the freshwater dogfish; *Lepisosteus,* the gar.

SUPERORDER III. TELEOSTEI Dominant fishes in the world today; homocercal tail; internal skeleton completely ossified; no spiracles; paired fins small; pectoral fins well up on sides of body; pelvic fins far forward in many; scales bony, thin, flexible, and rounded; well-developed jaws. Examples: Most familiar fishes; perch sunfish, herring, etc.

Subclass II. Sarcopterygii

Clavicles present. Examples: Lobe-finned fishes and lungfishes.

ORDER I. CROSSOPTERYGII Fins
borne on fleshy, lobelike scaly stalks; skeleton of pectoral and pelvic fins with single point of attachment to girdles; premaxillary and maxillary bones present. Example: Lobe-finned fishes.

SUBORDER I. RHIPIDISTIA Fossil
forms only; of importance in evolution of tetrapod limb. Example: Eusthenopteron.

SUBORDER II. COELACANTHINI
Head short and deep; hollow spinal column extending far forward into head; ancient fishes, until 1938 known only by fossil remains; first dorsal fin in living forms is fan-shaped, rest are lobate; hypophysial duct connects pituitary gland with mouth cavity; brain small and of simple construction; internal nares lacking in living species. Example: *Latimeria chalumae.*

ORDER II. DIPNOI Swim bladder ba-
sically bilobed and physostomous with ventral connection to esophagus and used as lung; internal as well as external nares in all; pulmonary circulation; three-chambered heart with double type of circulation; larval forms with external gills; sound production; diphycercal tail; cranium mostly cartilaginous; autostylic method of jaw attachment; spiral valve in intestine; premaxillary and maxillary bones lacking. Examples: The true lungfishes; *Protopterus, Lepidosiren,* and *Epiceratodus.*

Superclass II. Tetrapoda

Paired appendages are limbs which are typically pentadactyl; cornified outer layer of epidermis; lungs; skeleton bony for the most part; sternum usually present; number of skull bones reduced; reduction in visceral skeleton.

Class I. Amphibia

Transition from aquatic to terrestrial life clearly indicated in this group; eggs laid in water or in moist situations and covered with gelatinous envelope; metamorphosis; larvae with integumentary gills; respiration in adults (with few exceptions) by means of lungs and smooth, moist, vascular integument; buccopharyngeal respiration; poikilothermous; multicellular mucous and poison glands in skin; five pairs of pharyngeal pouches in embryo; lungs simple in structure, appearing at time of metamorphosis; opisthonephric kidneys; cloaca; reduction in number of bones in skull; skull flattened; two occipital condyles; three-chambered heart; pentadactyl limbs usually modified; no nails or claws; 10 pairs of cranial nerves in addition to terminal nerve (0). Examples: Caecilians, salamanders, newts, frogs, and toads.

ORDER I. ANURA Absence of tail in
adult; head and trunk fused, large, wide mouth; two pairs of well-developed limbs, the hind limbs being fitted for leaping and swimming; toes webbed; tympanic membrane flush with head; middle ear present for first time; true sound-producing organs; amplexus and external fertilization, therefore oviparous; larvae do not resemble parents; larvae with horny jaws in lieu of teeth; nine vertebrae plus urostyle; movable eyelids; lacrimal glands; well-developed sternum and limb girdles. Examples: Frogs and toads.

ORDER II. URODELA (CAUDATA)
Body divided into head, trunk, and persistent tail regions; two pairs of weak limbs in most species; larvae closely resembling parents; internal fertilization (except *Cryptobranchus*); oviparous (except black salamander); larvae with true teeth in both upper and lower jaws;

no middle ear; parietal muscles distinctly seg-mented; sternum poorly developed and primi-tive; lateral line lost at metamorphosis except in perennibranchiates. Examples: Newts and salamanders.

ORDER III. APODA (GYMNOPHI-ONA) Snakelike bodies; no limbs or limb girdles; very short tail; anus almost at posterior end of body and ventral; in some, fishlike dermal scales embedded in skin; no gills or gill slits in adult; extremely small eyes em-bedded in skin; no eyelids; internal fertil-ization; male with protrusible copulatory organ; intestine not differentiated into large and small regions; compact skull. Example: Caecilians.

Class II. Reptilia

Many fossil forms. Living species with fol-lowing characteristics. internal fertilization; oviparous and ovoviviparous forms; large-yolked eggs laid on land; three embryonic membranes, amnion, chorion, and allantois, appearing for the first time; poikilothermous; skin dry with dead corneal layer well devel-oped; epidermal scales, some with bony plates underlying them; skin almost devoid of glands; limbs typically pentadactyl, ter-minating in claws (not in snakes or limbless lizards); respiration entirely by means of lungs except for cloacal respiration in aquatic che-lonians; no metamorphosis; at least a partial separation of nasal and oral cavities; five pairs of pharyngeal pouches in embryo; vertebral column divided into cervical, thoracic, lumbar, sacral, and caudal regions (except snakes and limbless lizards); metanephric kidneys; one occipital condyle; three-chambered heart (ex-cept in crocodiles and alligators in which it is four-chambered); no lateral-line system; 12

pairs of cranial nerves in addition to terminal nerve (0).

ORDER I. CHELONIA Living forms represent a very old group which has per-sisted with little change for 175 million years. Short wide bodies; no teeth, each jaw being covered by a horny scale; shell composed of plastron and carapace covering body and made of bone covered with large epidermal scales; poikilobaric respiration; single penis in male; oviparous; lumbar region of vertebral column lacking; thoracic vertebrae and ribs fused to carapace; solidity of skull bones well marked; limbs pentadactyl except in marine forms; terrestrial, freshwater, and marine species; anus a longitudinal slit. Examples: Turtles, tortoises, and terrapins.

ORDER II. RHYNCHOCEPHALIA Well-developed parapineal eye; no copulatory organs in male; anal opening a transverse slit; diapsid condition in regard to temporal fossae; amphicoelous vertebrae; lizardlike body form with scaly skin; gastralia; only one living species. Example: *Sphenodon punctatum.*

ORDER III. SQUAMATA Skin cov-ered with horny epidermal scales; anal opening a transverse slit; paired eversible hemi-penes in male as copulatory organs; large palatal vacuities in roof of mouth; quadrate bone forms a movable union with squamosal; vertebrae usually procoelous; gastralia lacking. Examples: Lizards and snakes.

SUBORDER I. SAURIA (LACTER-TILIA) Visible external earpits; tympanic membrane not at surface of head; movable eyelids and nictitating membrane; urinary bladder usually present; tail readily detachable; metachrosis well exemplified in most; usu-ally two pairs of pentadactyl limbs terminat-ing in claws; two halves of lower jaw firmly united and size of mouth opening restricted; well-developed protrusible tongue; air sacs in some; some with a well-developed parapineal eye; oviparous and ovoviviparous forms. Example: Lizards.

SUBORDER II. SERPENTES (OPHIDIA) No limbs, but a few have vestiges of limb girdles; no sternum; no external ear openings or tympanic membranes; middle ear lacking; eyelids immovably fused and transparent; loose ligamentous attachment of jaws; ventral scales used in locomotion; left lung smaller than right and often absent altogether; hemipenes as copulatory organs; vertebral column divided into precaudal and caudal portions; tongue long, forked, protractile; no urinary bladder. Example: Snakes.

ORDER IV. CROCODILIA (LORICATA) Tail laterally compressed; two pairs of short legs; five toes on forefeet and four on hind feet; toes webbed; tympanic membrane exposed but protected by a fold of skin; teeth set in sockets in jawbones; oviparous; anal opening a longitudinal slit; single penis in male; four-chambered heart; first appearance of true cerebral cortex; adapted for amphibious life; diapsid condition in regard to temporal fossae; gastralia; dorsal and ventral scales reinforced with bony plates; tongue not protrusible; no urinary bladder. Examples: Alligators, crocodiles, caymans, gavials.

Class III. Aves

Feathers; legs and feet covered by horny skin or scales of reptilian type; light, hollow bones; loss of right ovary and oviduct except in certain birds of prey; very well developed eyes; nictitating membrane; forelimbs modified to wings; homoiothermous; sexual dimorphism often highly developed; oviparous;

large size of eggs; meroblastic cleavage; skin almost devoid of glands; one occipital condyle; four-chambered heart; claws at ends of digits of hind limbs; at least partial separation of nasal and oral cavities; four pairs of pharyngeal pouches in embryo; syrinx for sound production; air sacs from lungs; metanephric kidneys; fertilization internal; copulation by cloacal apposition except in a few forms with a single penis; rigidity of spinal column with posterior thoracic, lumbar, sacral, and proximal caudal regions firmly united; right aortic arch 4 persisting, left one disappearing; toes reduced to four or less; intratarsal joints in legs; flexible neck; no urinary bladder; 12 pairs of cranial nerves; absence of terminal nerve (0). Example: Birds.

Subclass I. Archaeornithes

Fossil forms only; long, jointed tail with 18 to 20 separate caudal vertebrae; gastralia; three clawed digits on wings; thecodont method of tooth attachment; poorly developed sternum. Example: *Archaeopteryx.*

Subclass II. Neornithes

Thirteen or fewer compressed caudal vertebrae; metacarpal and distal carpal bones fused; no free, clawed digits on wings (except young hoatzin); well-developed sternum; uncinate processes on ribs.

SUPERORDER I. ODONTOGNATHAE Fossil forms only; teeth present in jaws and set in grooves or shallow sockets. Examples: *Hesperornis,* a flightless, swimming bird; *Ichthyornis,* a flying, toothed form.

SUPERORDER II. PALEOGNATHAE (RATITAE) Flightless; wings poorly developed or rudimentary; no keel on sternum except in the tinamous; distal caudal vertebrae free; toothless; horny beaks, or bills. Examples: Walking or running birds; ostrich, rhea, kiwi, cassowary.

SUPERORDER III. NEOGNATHAE Keeled (carinate) sternum; wings well developed; five or six free caudal vertebrae terminating in a pygostyle; no teeth; horny beaks or bills. Examples: Most modern birds; penguins use wings for swimming, not for flight.

Class IV. Mammalia

Hair present but sometimes scanty; homoiothermous; mammary glands; sweat glands and oil glands in skin with few exceptions; pentadactyl limbs often modified; claws, nails, or hoofs at ends of digits; movable lips and tongues in most forms; highly developed cerebral cortex; four-chambered heart; palate partially separates nasal and oral cavities; four pairs of pharyngeal pouches in embryo; metanephric kidneys; urinary bladder; single penis for copulation; seven cervical vertebrae (four exceptions); two occipital condyles; increased cranial capacity of skull; reduction in number of skull bones; long movable tail usually present; 12 pairs of cranial nerves in addition to terminal nerve (0); pinna usually present; lungs for respiration; diaphragm; nonnucleated red corpuscles except in embryo; left aortic arch 4 persisting, right one disappearing; blood platelets instead of spindle cells; testes located in scrotum in most cases; three auditory ossicles; dentition thecodont and usually heterodont.

Subclass I. Prototheria

ORDER I. MONOTREMATA Oviparous: eggs incubated outside body; cloaca present; no nipples or teats; oviducts distinct, undifferentiated, and opening separately into cloaca; temperature-regulating mechanism not highly integrated; rather poorly developed brain; separate coracoid and precoracoid bones; epipubic bones; no pinna; males with intra-abdominal testes; penis used only for sperm transport; monophyodont. Examples: Egg-laying mammals; *Ornithorhynchus anatinus, Tachyglossus.*

Subclass II. Theria

Mammary glands with nipples or teats; no cloaca (except pika); viviparous; eggs practically microscopic in size; testes usually in scrotum; penis for passage of both urine and spermatozoa; oviducts differentiated into Fallopian tubes, uterus, and vagina; pinna usually present; coracoid represented by a process on scapula. Examples: Marsupials and placental mammals.

Infraclass I. Metatheria

ORDER I. MARSUPIALIA Young born in immature condition, further development usually in marsupium; usually no placental attachment; epipubic bones generally present; paired uteri and vaginae; a third vaginal canal often present; mammary glands provided with nipples; no cloaca; testes in scrotum; brain having corpus callosum for first time. Most species confined to Australia and surrounding islands. Examples: Marsupials; opossum, kangaroo.

Infraclass II. Eutheria (Placentalia)

Developing young nourished by means of an allantoic placenta; no marsupial pouch or epipubic bones; single vagina. Examples: The placental mammals.

Living eutherian, or placental, mammals are grouped in the following orders:

ORDER I. INSECTIVORA Small size;

elongated snouts; prism-shaped molar teeth; pentadactyl; insectivorous diet usually. Examples: Moles, shrews, hedgehogs.

ORDER II. DERMOPTERA Patagium; webbed feet; nocturnal; arboreal; herbivorous. Example: *Galeopithecus,* the flying lemur.

ORDER III. CHIROPTERA Pentadactyl forelimbs modified to form wings used in flight; keeled sternum; nocturnal; diet primarily insects, nectar, and fruit. Example: Bats.

ORDER IV. PRIMATES Well-developed nervous system, particulary the cerebral portion; long pentadactyl limbs; eyes directed forward and completely encircled by bony orbits; thumb and big toe usually opposable; plantigrade foot posture; usually only one young at a birth. Examples: Lemurs, tarsiers, monkeys, apes, and man.

SUBORDER I. LEMUROIDEA Long tails not prehensile; second digit of hind foot bearing claw, other digits bearing nails; thumb and hallux well developed; arboreal; crepuscular or nocturnal. Example: Lemurs.

SUBORDER II. TARSIOIDEA Elongated heel bone; small size; large protruding eyes and ears; second and third digits of hind foot bearing claws, nails on others; tails not prehensile; arboreal; nocturnal. Example: Tarsiers.

SUBORDER III. ANTHROPOIDEA Flattened or slightly rounded nails usually on all digits; arboreal or terrestrial; diurnal; tendency to upright posture; specialization of nervous system with highly developed, convoluted cerebral hemispheres. Examples: Monkeys, apes, and man.

ORDER V. EDENTATA Adults lacking teeth or having only poorly developed molar teeth. Examples: South American anteater, sloths, armadillos.

ORDER VI. PHOLIDOTA Body covered with large, overlapping, horny scales with a few hairs interspersed; teeth lacking. Example: Scaly anteater, the pangolin.

ORDER VII. RODENTIA Two incisor teeth on both upper and lower jaws, used for gnawing, grow throughout life; canine teeth lacking; diastema between incisors and premolars; foot posture plantigrade or approximately so; claws. Examples: Rats, mice, squirrels, beavers, gophers, capybara.

ORDER VIII. LAGOMORPHA Four incisor teeth in upper jaw, one pair behind the other, grow continuously; canine teeth lacking; tails short and stubby. Pikas are the only members of the subclass Theria to have a cloaca in the adult. Examples: Rabbits, hares, pikas.

ORDER IX. CARNIVORA Three pairs of small incisor teeth in both jaws; canine teeth well developed; flesh-eating mammals; poorly developed clavicles, sometimes absent; mobility of limbs; bipartite uterus; zonary type of placenta.

SUBORDER I. FISSIPEDIA Toes separated. Examples: Cats, dogs, weasels, martens.

SUBORDER II. PINNIPEDIA Aquatic carnivores; toes webbed; appendages in form of flippers; males larger than females; short tail. Examples: Seals, walrus, sea lions.

ORDER X. CETACEA Pectoral appendages webbed to form flippers; no claws; pelvic appendages lacking; tail (flukes) flattened dorsoventrally and notched; males with intra-abdominal testes; nasal openings on top of head; homodont dentition in toothed forms. Examples: Whales, dolphins, porpoises.

ORDER XI. TUBULIDENTATA Thick-set body; large, pointed ears; long snout; thick skin; scanty hair covering; few permanent teeth; those present lacking enamel; no incisors or canines; very long claws; plantigrade foot posture. Example: Aardvark.

ORDER XII. PROBOSCIDEA Great size; massive legs; nose and upper lip form proboscis, or trunk with nostrils at tip; upper incisor teeth are tusks; thick skin with scant hairy coat; males with intra-abdominal testes. Example: Elephants.

ORDER XIII. HYRACOIDEA Somewhat resemble guinea pigs in size and shape but more closely related to ungulates; small ears; short tail; persistent growth of upper incisor teeth; four toes on forefeet; three toes on hind feet; second toe on hind feet bearing claw, the rest being hooflike nails; males with intra-abdominal testes; canine teeth lacking. Example: Conies.

ORDER XIV. SIRENIA Herbivorous, aquatic forms; heavy dense bones; forelimbs in the form of flippers; hind limbs lacking; tail bearing horizontally arranged flukes which are not notched; no external ears; skin sparsely haired; males with intra-abdominal testes. Examples: Manatees and dugongs.

ORDER XV. PERISSODACTYLA Odd-toed, hoofed forms; unguligrade foot posture; axis of leg passes through middle toe; herbivorous; gallbladder lacking. Examples: Horses, donkeys, zebras, tapirs.

ORDER XVI. ARTIODACTYLA Even-toed, hoofed forms; unguligrade foot posture; axis of leg passing between third and fourth toes; herbivorous; true horns and antlers only in this group; incisors and canines of upper jaw usually absent (except in pigs). Examples: Cattle, pigs, hippopotamuses, camels, deer.

Glossary

Many anatomical terms are either common Latin (L.) or Greek (G.) words. The following list identifies the root sources of the principal terms used in this book. The alert student will discover that many roots are employed again and again in different combinations. Only brief definitions are provided here. Each term may also be defined or discussed in the body of the text. This glossary does not obviate the need for a good medical or unabridged dictionary.

Abdomen (L. *abdere,* to hide) portion of the body cavity containing the digestive system.

Abducens (L. *ab,* away, + *ducere,* to lead) name of the sixth cranial nerve.

Abductor (L. *abducere,* to draw away) something that draws a part away from the axis of the body.

Acetabulum (L. *acetabulum,* vinegar cup) cup-shaped depression on the innominate bone, in which the head of the femur fits.

Acoustic (G. *akoustikos,* of hearing) relating to hearing or perception of sound.

Adductor (L. *adducere,* to bring toward) something that draws a part toward the median line.

Adrenal (L. *ad,* to, + *renes,* kidneys) endocrine gland near or upon the kidney.

Alisphenoid (L. *ala,* wing, + G. *sphen,* wedge, + *oeidos,* form) one of the sphenoid bones; in the human skull, the greater wing of the composite sphenoid bone.

Allantois (G. *allas,* sausage, + *oeidos,* form) an extraembryonic saclike extension of the hindgut of amniotes, serving excretion and respiration.

Alveolus (L. *alveolus,* little hollow) any small lobular or pitlike chamber.

Amnion (G. *amnion,* fetal membrane) inner membrane surrounding the embryo.

Ampulla (L. *ampulla,* flask) saccular dilatation of a canal.

Analogy (G. *analogos,* proportionate) similarity in function between two organs or parts in different species of animals.

Anastomosis (G. *anastomosis,* opening) natural communication between two blood vessels.

Anus (L. *anus,* ring) rear opening of the digestive tract.

Archenteron (G. *arche,* beginning, + *enteron,* intestine) embryonic digestive tube.

Arcual (L. *arcua,* a bow) like an arch.

Auditory (L. *audire,* to hear) relating to the perception of sound.

Auricle (L. *auricle,* little ear) the external ear; receiving chamber of the heart.

Autonomic (G. *autos,* self, + *nomos,* law) self-controlling, independent of outside influences.

Autostylic (G. *autos,* self, + *stylos,* pillar) type of jaw suspension in which the jaws articulate directly with the cranium.

Axial (L. *axis,* axle) relating to the central part of the body as distinguished from the appendages.

Azygous (G. *a,* not, + *zygos,* yoke) unpaired.

Basal (L. *basis,* footing or base) relating to a base.

Basi- (L. *basis,* base) pertaining to the base.

Basibranchial (L. *basis,* base, + *branchia,* gill) median-ventral component of a gill arch.

Basihyoid (L. *basis,* base, + G. *hyoeides,* Y-shaped) median-ventral component of the hyoid arch.

Blastema (G. *blastema,* offshoot) primitive cell aggregate from which an organ develops.

Blastocoel (G. *blastos,* shoot, + *koilos,* hollow) cavity in the blastula.

Blastocyst (G. *blastos,* shoot, + *kystis,* bladder) hollow sphere of early mammalian development.

Blastoderm (G. *blastos,* shoot, + *derma,* skin) primitive cellular plate at the beginning of embryogeny.

Blastodisc (G. *blastos,* shoot, + L. *discus,* plate) plate of cytoplasm at the animal pole of the ovum.

Blastomere (G. *blastos,* shoot, + *meros,* part) one of the cells into which the egg divides during the cleavage phase of development.

Blastopore (G. *blastos,* shoot, + *pores,* opening) opening into the gastrocoel.

Blastula (G. *blastos,* shoot) the cells of an early embryo which are commonly arranged in a hollow sphere.

Brachial (G. *brachialis,* arm) relating to the arm.

Branchial (G. *branchia,* gill) relating to gills.

Branchiomerism (G. *branchia,* gill, + *meros,* part) segmentation corresponding to the visceral arches.

Bronchus (G. *bronchos,* windpipe) one of the two branches off the trachea leading to a lung.

Buccal (L. *bucca,* cheek) relating to the mouth; the surface toward a cheek.

Cancellous (L. *cancellus,* latticework) spongy, as of bone.

Capillary (L. *capillaris,* relating to hair) microscopic blood vessel, intermediate between arteries and veins.

Capitulum (L. *caput,* head) small head or rounded extremity of a bone.

Cardiac (G. *kardia,* heart) relating to the heart.

Cardinal (L. *cardinalis,* important) of special importance.

Carpus (G. *karpos,* wrist) collection of bones constituting the wrist.

Caudal (L. *cauda,* tail) relating to the tail or rear.

Cavernous (L. *caverna,* cave) relating to a cavity.

Cecum (L. *caecus,* blind) structure ending in a blind sac.

Cephalic (G. *kephale,* head) relating to the head.

Cerebellum (L. *cerebrum,* brain) brain mass deriving from the roof of the metencephalon.

Cerebrum (L. *cerebrum,* brain) brain mass deriving from the roof of the telencephalon.

Cervical (L. *cervix,* neck) relating to the neck.

Chiasma (G. *chiasma,* letter) the crossing of fibers.

Chondro- (G. *chondros,* cartilage) cartilaginous.

Chondroblast (G. *chondros,* cartilage, + *blastos,* shoot) embryonic cartilage-forming cell.

Chondroclast (G. *chondros,* cartilage + *klastos,* broken) cell concerned in the destruction of cartilage.

Chorion (G. *chorion,* skin) outer membrane surrounding the embryo.

Choroid (G. *chorion*, skin, + *oeidos*, form) relating to a coat or membrane.

Chromatophore (G. *chroma*, color, + *phorus*, bearer) pigment-containing cell.

Chromosome (G. *chroma*, color, + *soma*, body) deeply staining rod-shaped or threadlike body in the cell nucleus; bearer of genes.

Cloaca (L. *cloaca*, sewer) combined urogenital and rectal receptacle.

Coeliac (G. *koilia*, belly) relating to the abdomen.

Coelum (G. *koiloma*, a hollow) cavity bounded by mesodermal epithelium.

Commissure (L. *commissura*, connection) bond of tissue connecting corresponding halves of brain or spinal cord.

Condyle (G. *kondylos*, knuckle) rounded articular surface at the extremity of a bone.

Constrictor (L. *constringere*, to draw together) anything which binds or squeezes a part.

Copula (L. *copula*, yoke) a part connecting two structures.

Coracoid (G. *korax*, raven, + *oeidos*, form) shaped like a crow's beak.

Corneum (L. *corneus*, horny) a horny layer.

Coronary (L. *coronarius*, a crown) denoting various encircling anatomical parts.

Coronoid (G. *korone*, a crow, + *oeidos*, form) process shaped like a crow's beak.

Corpus (L. *corpus*, the body) any body or mass.

Cortex (L. *cortex*, bark) outer portion of an organ.

Costal (L. *costa*, rib) relating to a rib.

Cranium (G. *kranion*, skull) bones of the head collectively.

Ctenoid (G. *kteis*, a comb, + *oeidos*, form) type of scale featuring toothlike projections.

Cuticle (L. *cutis*, skin) outer horny layer of the skin.

Cycloid (G. *kyklos*, circle, + *oeidos*, form) type of scale, featuring circular growth rings.

Cystic (G. *kystis*, bladder) relating to a bladder or cyst.

Deferens (L. *de*, away, + *ferens*, carrying) carrying away.

Depressor (L. *depressus*, to press down) something serving to lower or pull down.

Dermatome (G. *derma*, skin, + *tome*, cut) one of the fetal skin segments.

Diaphragm (G. *diaphragma*, partition) the partition between the abdominal and thoracic cavities.

Diaphysis (G. *diaphysis*, a growing through) shaft of a long bone.

Diapophysis (G. *dia*, through, + *apophysis*, offshoot) upper articular surface of the transverse process of a vertebra.

Diapsid (G. *di*, double, + *apsis*, arch) reptile with two temporal openings on each side.

Digit (L. *digitus*, finger) finger or toe.

Distal (L. *distalis*, distant) farthest from the center or median line.

Dorsal (L. *dorsalis*, back) relating to the top or back.

Ductus (L. *ductus*, leading) tubular structure giving exit to or conducting any fluid.

Duodenum (L. *duodeni*, twelve) first division of the small intestine.

Ectoderm (G. *ektos*, outside, + *derma*, skin) outermost of the three primary germ layers.

Effector (L. *efficere*, to bring to pass) organ which reacts by movement, secretion, or electrical discharge.

Efferent (L. *ex*, out, + *ferre*, to bear) conducting outward or centrifugally.

Endocardium (G. *endon*, within, + *kardia*, heart) lining of the cavities of the heart.

Endochondral (G. *endon*, within, + *chondros*, cartilage) within a cartilage or cartilaginous tissue.

Endocrine (G. *endon*, within, + *krinein*, separate) denoting internal secretion.

Endoderm (G. *endon*, within, + *derma*, skin) innermost of the three primary germ layers.

Endolymph (G. *endon*, within, + L. *lympha*, water) fluid contained in the membranous labyrinth of the inner ear.

Endometrium (G. *endon*, within, + *metra*, uterus) mucous membrane lining the uterus.

Endoskeleton (G. *endon*, within, + *skeletos*, dried up) internal framework of the body.

Enterocoel (G. *enteron*, intestine, + *koilos*,

hollow) coelom originally in communication with the lumen of the gut.

Epicardium (G. *epi*, upon, + *kardia*, heart) visceral peritoneum enveloping the heart.

Epidermis (G. *epi*, upon, + *derma*, skin) outer epithelial portion of the skin.

Epididymis (G. *epi*, upon, + *didymos*, twin) first, convoluted, portion of the excretory duct of the testis, passing from above downward along the posterior border of this gland.

Epiglottis (G. *epi*, upon, + *glotta*, tongue) structure which folds back over the aperture (glottis) of the larynx.

Epimere (G. *epi*, upon, + *meris*, part) dorsal region of the mesoderm on each side of the neural tube.

Epiphysis (G. *epi*, upon, + *physis*, growth) separately developed terminal ossification of a long bone; the pineal body.

Epithalamus (G. *epi*, upon, + *thalamos*, room) dorsal portion of the thalamus of the brain.

Epithelium (G. *epi*, upon, + *thele*, nipple) cellular layer covering a free surface.

Erythrocyte (G. *erythros*, red, + *kytos*, cell) red blood corpuscle.

Esophagus (G. *oisein*, to carry, + *phagein*, to eat) portion of the gut between the pharynx and stomach.

Ethmoid (G. *ethmos*, sieve, + *oeidos*, form) resembling a sieve; name of the ethmoid bone.

Extensor (L. *extendere*, to stretch out) something serving to straighten a part.

Extrinsic (L. *extrinsecus*, on the outside) originating outside of the part upon which it acts.

Falciform (L. *faliz*, sickle, + form) having a curved or sickle shape.

Fascia (L. *fascia*, a band or fillet) fibrous tissue investing an organ or area.

Follicle (L. *folliculus*, small bag) vesicular body in the ovary, containing the ovum; any crypt or circumscribed space.

Foramen (L. *foramen*, aperture) perforation through a bone or membrane.

Fossa (L. *fossa*, trench) depression below the level of the surface of a part.

Gamete (G. *gametes*, husband; *gamete*, wife) germ cell; ovum or spermatozoon.

Ganglion (G. *ganglion*, swelling under the skin) aggregation of nerve cells along the course of a nerve.

Gastrocoel (G. *gaster*, stomach, + *koilos*, hollow) cavity within the gastrula.

Gastrula (L. *gastrula*, little belly) embryo in the stage of development following the blastula, consisting of a sac with a double wall.

Genital (L. *genitalis*, pertaining to birth) related to reproduction or generation.

Glans (L. *glans*, acorn) a mass capping the body of a part.

Glomerulus (L. *glomus*, skein) tuft formed of capillary loops at the beginning of each uriniferous tubule in the kidney.

Glottis (G. *glottis*, aperture of the larynx) opening to the larynx and trachea.

Gonad (G. *gone*, seed) reproductive gland; ovary or testis.

Granulosa (L. *granular*, full of grains) layer or region having a granular appearance.

Gyrus (G. *gyros*, circle) rounded elevation on the surface of the brain.

Haploid (G. *haplous*, simple, + *oeidos*, form) having a reduced number of chromosomes in the gamete.

Hemal (G. *haima*, blood) relating to the blood or blood vessels.

Hemi- (G. *hemi*, *half*) one-half.

Hemichordata (G. *hemi*, half, + *chorde*, string) phylum of animals closely related to echinoderms and chordates.

Hemisphere (G. *hemi*, half, + *sphaira*, ball) lateral half of the cerebrum or cerebellum.

Hepatic (G. *hepatikos*, relating to the liver) relating to the liver.

Heterodont (G. *heteros*, different, + *odont*, tooth) having teeth of varying shapes.

Hippocampus (G. *hippocampus*, monster) elevation on the floor of the lateral ventricle of the brain.

Holo- (G. *holos,* entire) entire or total.

Holoblastic (G. *holos,* whole, + *blastos,* shoot) denoting the involvement of the entire egg in cleavage.

Holocephali (G. *holos,* entire, + *kephale,* head) small group of fishes included in the Chondrichthyes.

Homodont (G. *homos,* same, + *odont,* tooth) having teeth all alike in form.

Homology (G. *homos,* same, + *logos,* ratio) similarity in structure and origin of two organs or parts in different species of animals.

Homotypy (G. *homos,* same, + *typos,* type) correspondence in structure between two parts or organs in one individual.

Hormone (G. *hormon,* stirring up) chemical substance formed in one organ or part and carried in the blood to another organ or part which it stimulates to functional activity.

Humor (L. *humor,* fluid) any fluid or semifluid substance.

Hyalin (G. *hyalos,* glass) of a glassy, translucent appearance.

Hyoid (G. *hyoeides,* Y-shaped, second visceral arch; tongue bone.

Hyomandibula (G. *hyoeides,* Y-shaped, + L. *mandibula,* jaw) uppermost segment of the hyoid arch.

Hyostylic (G. *hyoeides,* Y-shaped, + *stylos,* pillar) jaw suspension wherein the hyomandibula is inserted between the jaws and cranium.

Hyper- (G. *hyper,* above) excessive; above normal.

Hypo- (G. *hypo,* under) beneath; below normal.

Hypoglossal (G. *hypo,* beneath, + *glossa,* tongue) beneath the tongue, usually denoting twelfth cranial nerve.

Hypomere (G. *hypo,* under, + *meris,* part) most ventral subdivision of mesoderm.

Hypophysis (G. *hypophysis,* an undergrowth) endocrine gland lying at the base of the brain.

Hypothalamus (G. *hypo,* under, + *thalamos,* chamber) ventral portion of the thalamus of the brain.

Infra- (L. *infra,* below) denoting a position below a designated part.

Infundibulum (L. *infundibulum,* a funnel) funnel-shaped structure or passage.

Integument (L. *integumentum,* covering) skin covering the body.

Inter- (L. *inter,* between) between or among.

Interstitial (L. *inter,* between, + *sistere,* to stand) relating to spaces or structures within parts.

Intestine (L. *intestinum*) digestive tube passing from the stomach to the anus.

Intrinsic (L. *intrinsecus,* on the inside) belonging entirely to a part.

Involution (L. *involvere,* to roll up) turning of a part around a margin or edge.

Iso- (G. *isos,* equal) equal.

Isolecithal (G. *isos,* equal, + *lekithos,* yolk) having a uniform distribution of yolk within an egg.

Jugular (L. *jugulum,* throat) relating to the throat or neck.

Keratin (G. *keras,* horn) hard, relatively insoluble protein or albuminoid, present largely in cutaneous structures.

Labium (L. *labium,* a lip) any lip-shaped structure.

Lacuna (L. *lacus,* a hollow) small depression, gap, or space.

Lamina (L. *lamina,* thin plate) thin plate or flat layer.

Larynx (G. *larynx,* upper part of windpipe) box composed of several cartilages, at the upper end of the trachea.

Ligament (L. *ligamentum,* a band or bandage) band or sheet of fibrous tissue connecting two or more skeletal parts.

Lumbar (L. *lumbus,* loin) relating to the loins, or part of the back and sides between the ribs and pelvis.

Luteum (L. *luteus,* yellow) yellow.

Macrolecithal (G. *makros,* large, + *lekithos,* yolk of egg) having a large amount of yolk stored in the egg.

Meatus (L. *meatus,* passage) canal or channel.

Medulla (L. *medulla,* marrow, pith) inner, or central, portion of an organ or part.

Melanin (G. *melas, melanos,* black) dark pigment in the skin.

Meninx, pl. meninges (G. *meninx,* membrane) membranous envelope of the brain and spinal cord.

Meroblastic (G. *meros,* part, + *blastos,* shoot) denoting the involvement of a restricted cytoplasmic area at the animal pole of the egg in cleavage.

Mes-, meso- (G. *meso,* middle) middle.

Mesenchyme (G. *meso,* middle, + *enchyma,* infusion) embryonic connective tissue.

Mesentery (L. *mesenterium,* a membranous support) double layer of peritoneum enclosing an organ.

Mesocardium (G. *mesos,* middle, + *kardia,* heart) mesentery supporting the heart.

Mesoderm (G. *mesos,* middle, + *derma,* skin) middle layer of the three primary germ layers.

Mesolecithal (G. *mesos,* middle, + *lekithos,* yolk of egg) having a moderate amount of yolk stored in the egg.

Mesonephron (G. *mesos,* middle, + *nephros,* kidney) tubular unit of a mesonephros.

Mesonephros (G. *mesos,* middle, + *nephros,* kidney) the second stage in the development of the amniote kidney.

Meta- (G. *meta,* after) behind or after something else in a series.

Metamerism (G. *meta,* after, + *meros,* part) segmentation resulting in a series of homologous parts.

Metanephron (G. *meta,* after, + *nephros,* kidney) tubular unit of a metanephros.

Microlecithal (G. *mikros,* small, + *lekithos,* yolk of egg) having a small amount of yolk stored in the egg.

Mucosa (L. *mucosus,* mucus) membrane containing or secreting mucus.

Myelin (G. *myelos,* marrow) fatty sheath of a nerve fiber.

Myocomma (G. *myos,* muscle, + *komma,* separation) partition of connective tissue separating muscle segments.

Myotome (G. *myos,* muscle, + *tomos,* cutting) prospective or actual muscle segment.

Naris (L. *naris,* nostril) nostril.

Neopallium (G. *neos,* new, + L. *pallium,* cloak) the pallium of the brain cortex of recent origin.

Nephron (G. *nephros,* kidney) tubular unit of the kidney.

Nephrostome (G. *nephros,* kidney, + *stoma,* mouth) funnel-shaped opening by which a kidney tubule communicates with the nephrocoel.

Nephrotome (G. *nephros,* kidney, + *tomos,* a slice) segment of mesomere, forerunner of a kidney tubule.

Neural (G. *neuron,* nerve) pertaining to the nervous system.

Neurilemma (G. *neuron,* nerve, + *lemma,* husk) delicate sheath surrounding the myelin substance of a nerve fiber or the axis of a nonmyelinated fiber.

Neuroglia (G. *neuron,* nerve, + *glia,* glue) supporting tissues of the brain and spinal cord.

Neurohypophysis (G. *neuron,* nerve, + *hypophysis,* an undergrowth) posterior lobe of the hypophysis (pituitary).

Neuron (G. *neuron,* nerve) cellular unit of the nervous system.

Oculomotor (L. *oculus,* eye, + *motus,* motion) movements of the eyeball; name of the third cranial nerve.

Odontoblast (G. *odont,* tooth, + *blastos,* shoot) one of a layer of cells lining the pulp cavity of a tooth which form dentine.

Olfactory (L. *olfactum,* smelled) pertaining to the sense of smell.

Omentum (L. *omentum,* membrane enclosing the bowels) mesentery passing from the stomach to another abdominal organ.

Ontogeny (G. *on,* being, + *genesis,* origin) development of the individual; distinguished from phylogeny, the evolutionary development of the species.

Oocyte (G. *oon,* egg, + *kytos,* cell) primitive ovum in the ovary.

Operculum (L. *operculum,* cover or lid) any part resembling a lid or cover.

Ophthalmic (G. *ophthalmos,* eye) relating to the eye.

Opistho- (G. *opisthe,* behind) behind or to the rear.

Opisthonephros (G. *opisthe,* behind, + *nephros,* kidney) adult kidney in the anamniota.

Optic (G. *opsis,* sight) relating to the eye or vision.

Organogenesis (G. *organon,* organ, + *genesis,* production) formation of organs.

Otic (G. *otikos,* belonging to the ear) relating to the ear.

Otolith (G. *otos,* ear, + *lithos,* stone) calcified body in the sacculus of the inner ear.

Ovary (L. *ovarium,* egg receptacle) one of the reproductive glands in the female, containing the ova or germ cells.

Ovoviviparous (L. *ovum,* egg, + *vivus,* alive, + *parere,* to bear) retaining eggs within the body, where they are developed without placental attachment.

Paleopallium (G. *palaios,* ancient, + L. *pallium,* cloak) primitive cerebral cortex.

Pallium (L. *pallium,* cloak) cerebral cortex with the subjacent white substance.

Pancreas (G. *pan,* all, + *kreas,* flesh) abdominal digestive and endocrine gland.

Panniculus (L. *pannus,* cloth) sheet of tissue.

Papilla (L. *papilla,* pimple) elevation or nipplelike process.

Para- (G. *para,* alongside) alongside, near.

Paraphysis (G. *para,* beside, + *physis,* growth) outgrowth from the roofplate of the telencephalon.

Pectoral (L. *pectoralis,* pertaining to breastbone) pertaining to the breastbone or chest.

Pellucid (L. *per,* through, + *lucere,* to shine) translucent.

Pelvis (L. *pelvis,* basin) any basinlike or cup-shaped part, e.g., the pelvis of the kidney or the pelvic girdle.

Peri- (G. *peri,* around) around or surrounding.

Pericardial (G. *peri,* around, + *kardia,* heart) surrounding the heart.

Perichondrium (G. *peri,* around, + *chondros,* cartilage) fibrous membrane covering cartilage.

Perimysium (G. *peri,* around, + *myos,* muscle) fibrous sheath enveloping each of the primary bundles of muscle fibers.

Periosteum (G. *peri,* around, + *osteon,* bone) fibrous membrane covering bone.

Peritoneum (G. *peritonaios,* stretched across) lining of the body cavity and covering of organs.

Phallus (G. *phallos,* penis) shaft of the penis.

Pharynx (G. *pharynx,* throat) segment of gut between the mouth and esophagus.

Phylogeny (G. *phyle,* a tribe, + *genesis,* origin) evolutionary development of any plant or animal species.

Pineal (L. *pinea,* pinecone) relating to an outgrowth from the roofplate of the diencephalon.

Pituitary (L. *pituita,* slime) an endocrine gland at the base of the diencephalon.

Placenta (L. *placenta,* a flat cake) organ of physiological communication between mother and fetus.

Placode (G. *plax,* plate, + *oeidos,* form) any thickened platelike area.

Placoid (G. *plax,* plate, + *oeidos,* form) relating to a plate.

Pleura (G. *pleura,* side) the membrane enveloping the lungs.

Pleuro- (G. *pleura,* side) on the side.

Plexus (L. *plexus,* a braid) network of parts, e.g., blood vessels or nerves.

Portal (L. *porta,* gate) relating to any opening.

Pre- (L. *prae,* before) anterior or before, in space or time.

Pro- (L. *pro,* before) before or forward.

Proctodeum (G. *proktos,* anus, + *hodaios,* a way) terminal portion of rectum formed in the embryo by an ectodermal invagination.

Pronephron (L. *pro,* before, + *nephros,* kidney) tubular unit of a pronephros.

Prostate (L. *pro,* in front, + *stare,* to stand) glandular body surrounding the beginning of the urethra in the male.

Protoplasm (G. *protos,* first, + *plasma,* thing formed) living matter, the substance of which animal and plant tissues are composed.

Proximal (L. *proximus,* next) nearest the trunk or point of origin.

Quadrigeminal (L. *quadrigeminis,* fourfold) divided into four parts.

Radius (L. *radius,* spoke) line from the center to periphery of a circle; bone of forearm.

Receptor (L. *recipere,* to receive) sensory nerve ending in skin or a sense organ.

Rectus (L. *rectus,* straight) straight.

Renal (L. *renes,* kidneys) relating to a kidney.

Rete (L. *rete,* net) network of nerve fibers or small vessels.

Retina (L. *rete,* net) light-sensitive layer of the eye.

Retractor (L. *re,* back, + *trahere,* to draw) something serving to withdraw a part.

Sacculus (L. *sacculus,* little bag) small pouch; the smaller of the two sacs in the inner ear.

Sacrum (L. *sacer,* sacred) bone formed by welding together sacral vertebrae.

Sarcolemma (G. *sarx,* flesh, + *lemma,* husk) sheath enclosing a muscle fiber.

Schizocoel (G. *schizo,* split, + *koilos,* hollow) coelom formed by the splitting of an originally solid layer of mesoderm.

Sclera (G. *skleros,* hard) tough outer coat of the eye.

Sclerotome (G. *skleros,* hard, + *tomos,* cutting) segment of the skeleton-producing cells derived from a mesodermal somite.

Scrotum (L. *scrotum,* scrotum) musculocutaneous sac containing the testes.

Sebaceous (L. *sebum,* tallow) oily.

Seminiferous (L. *semen,* seed, + *ferre,* to carry) conducting the semen.

Septum (L. *septum,* partition) thin wall dividing two cavities or masses of softer tissue.

Serosa (L. *serosus,* serous) peritoneal coat of a visceral organ.

Sinus (L. *sinus,* bay) space in organ or tissue.

Somatic (G. *somatikos,* bodily) the wall of the body.

Spermatid (G. *sperma,* seed) rudimentary spermatozoon derived from the division of the spermatocyte.

Spermatogenesis (G. *sperma,* seed, + *genesis,* origin) formation of spermatozoa.

Spermatozoon (G. *sperma,* seed, + *zoon,* animal) male sexual cell.

Sphincter (G. *sphinkter,* band) muscle which serves to close an opening.

Splanchnic (G. *splanchnon,* entrail) referring to the viscera.

Splanchnopleure (G. *splanchnon,* entrail, + *pleura,* side) embryonic layer formed by the union of the visceral layer of mesoderm with the endoderm.

Stomodeum (G. *stoma,* mouth, + *hodaios,* on the way) ectodermal invagination forming the mouth cavity.

Stratum (L. *stratus,* layer) layer of differentiated tissue.

Sub- (L. *sub,* under) beneath; less than.

Sulcus (L. *sulcus,* furrow) groove or furrow.

Supinator (L. *supinus,* lying on the back) something serving to turn a part upward or forward.

Supra- (L. *supra,* above) above.

Symphysis (G. *syn,* together, + *physis,* growth) union or meeting point of any two structures.

Synapse (G. *syn,* together, + *hastein,* to fasten) close approximation of the processes of different neurons.

Tarsus (G. *tarsus,* wickerwork) ankle or instep.

Telolecithal (G. *telos,* end, + *lekithos,* yolk) having the yolk accumulated at one pole.

Testis (L. *testis,* testicle) one of the male reproductive glands.

Thalamus (G. *thalamos,* room) side walls of the diencephalon.

Theca (G. *theke,* box) sheath.

Thoracic (G. *thorax,* breastplate) relating to the thorax or chest.

Thrombocyte (G. *thrombos,* clot, + *kytos,* cell) small platelike blood cell involved in the clotting mechanism.

Thymus (G. *thymos,* sweetbread) ductless gland in the neck.

Thyroid (G. *thyreos,* oblong shield, + *oeidos,* form) name of a gland or cartilage.

Trabecula (L. *trabs,* beam) supporting bar or band.

Trachea (G. *tracheia,* rough artery) air tube extending from the larynx to the bronchi.

Trigeminal (L. *trigeminis,* triplet) name of the fifth cranial nerve.

Trophoblast (G. *trophe,* nourishment, + *blastos,* shoot) outer wall of the blastocyst, concerned with nutrition of embryo within.

Tuberculum (L. *tuber,* nodule) protuberance.

Tympanic (L. *tympanum,* drum) relating to the eardrum.

Ultimo- (L. *ultimus,* last) last of a series.

Umbilical (L. *umbilicus,* navel) relating to the umbilicus.

Uriniferous (L. *urina,* urine, + *ferre,* to carry) conveying urine.

Uropygial (G. *orros,* end of sacrum, + *pyge,* rump) gland at the base of the tail in a bird.

Uterus (L. *uterus,* womb) the womb; hollow, muscular organ in which the impregnated ovum develops into the fetus.

Utriculus (L. *utriculus,* small leather bottle) larger of the two sacs in the inner ear.

Vas (L. *vas,* vessel) a tube or vessel.

Vegetal pole (L. *vegetare,* to animate) the end of a telolecithal egg containing the yolk.

Ventral (L. *venter,* belly) relating to the belly.

Vesicle (L. *vesicula,* small bladder) any small sac.

Villus (L. *villus,* tuft of hair) minute projection from the surface of a membrane.

Viscus, pl viscera (L. *viscus,* an internal organ) internal organ, especially one of the large abdominal organs.

Vitelline (L. *vitellus,* yolk) relating to the yolk of an egg.

Vitreous (L. *vitreus,* glassy) having a glassy appearance.

Viviparous (L. *vivus,* alive, + *parere,* to bear) giving birth to living young.

Zygapophysis (G. *zygon,* yoke, + *apophysis,* offshoot) articular process of a vertebra.

Zygomatic (G. *zygoma,* cheekbone) relating to the cheekbone.

Index